FUNCTIONAL FOODS

FUNCTIONAL FOODS

Designer Foods, Pharmafoods, Nutraceuticals

Edited by

Israel Goldberg

A Chapman & Hall Food Science Book

An Aspen Publication®
Aspen Publishers, Inc.
Gaithersburg, Maryland
1999

The author has made every effort to ensure the accuracy of the information herein. However, appropriate information sources should be consulted, especially for new or unfamiliar procedures. It is the responsibility of every practitioner to evaluate the appropriateness of a particular opinion in in the context of actual clinical situations and with due considerations to new developments. The author, editors, and the publisher cannot be held responsible for any typographical or other errors found in this book.

Aspen Publishers, Inc., is not affiliated with the American Society of Parenteral and Enteral Nutrition

Library of Congress Cataloging-in-Publication Data

Functional foods: designer foods, pharmafoods, nutraceuticals /
p. cm.
Originally published : New York : Chapman & Hall, 1994.
Includes bibliographical references and index.

1. Nutrition. 2. Natural foods. 3. Diet therapy. I. Goldberg, Israel, 1943–
RA784.F85 1994
613.2—dc20
93-40742
CIP

For information, address Aspen Publishers, Inc., Permissions Department, 200 Orchard Ridge Drive, Suite 200, Gaithersburg, Maryland 20878.

Orders: (800) 638–8437
Customer Service: (800) 234–1660

About Aspen Publishers • For more than 35 years, Aspen has been a leading professional publisher in a variety of disciplines. Aspen's vast information resources are available in both print and electronic formats. We are committed to providing the highest quality information available in the most appropriate format for our customers. Visit Aspen's Internet site for more information resources, directories, articles, and a searchable version of Aspen's full catalog, including the most recent publications: **http://www.aspenpublishers.com**

Aspen Publishers, Inc. • The hallmark of quality in publishing
Member of the worldwide Wolters Kluwer group

Editorial Services: Ruth Bloom

ISBN 978-1-4419-5195-3

Printed in the United States of America
4 5

Dedication

This book is dedicated to Frida for her courageous struggle in fighting the unknown.

Contents

Part III. Health Functionality of Food Components

Part IV. Market and Competition

FOREWORD

"Accuse not Nature! She has done her part;
Do Thou but Thine!"
Milton, Paradise Lost 1667

The concept that nature imparted to foods a health-giving and curative function is not new. Herbal teas and remedies have been used for centuries and continue in use in many parts of the world today. In modern society, we have turned to drugs to treat, mitigate, or prevent diseases. However, since the discovery of nutrients and our increasing analytical capabilities at the molecular level, we are beginning to become more knowledgeable of the biochemical structure-function relationship of the myriad of chemicals that occur naturally in foods and their effect on the human body.

The holistic approach to medicine and diet that began in the 1970s has now seen a renewal as we realize that certain foods, because of the presence of specific biochemicals, can have a positive impact on an individual's health, physical well-being, and mental state. In fact, because of the negative image of drugs, and the grey area of supplements, the use of foods that are **"functional"** is becoming a growth area for the food industry. In Japan this concept has led to one of the largest growing markets, where they have defined **"functional foods"** as regular foods derived only from naturally occurring ingredients. The Japanese further require that the functional foods be consumed as part of the diet and not in supplement form (i.e., not as tablets or capsules). A visit to a Japanese grocery store will quickly reveal many products with advertisements extolling **"functional benefits,"** e.g., a rice wine whose properties are beneficial in preventing coronary heart disease by a specific biochemical pathway. This concept is not new. In the past these products were considered as **"health foods,"** and their users as **"health nuts."** In many cases the scientific foundation for the beneficial function was lacking, so the traditional food industry view was to keep out lest it be considered as promoting health quackery. The U.S. Food and Drug Administration's past actions on seizures of health foods, including a classic case where a purveyor was jailed and died in jail, enhanced the image of quackery and raised a flag of warning to the conservative food industry. However as we have learned more, our approach to these "functional foods" is evolving.

Dr. Israel Goldberg of the Hebrew University in Jerusalem, Israel, conceived the idea of putting together in one text, a broad sweep of the "functional food" arena. It has been noted that the food industry for a number of years looked for a way to "reposition" itself by de-emphasizing the negatives (no cholesterol, no additives, no preservatives) and by moving to extol the positive and natural health benefits of a food. Certainly in the United States this has been further supported by the U.S.D.A. Eating Right Pyramid, which recommends five daily servings of fruits and vegetables to achieve a healthy diet. However in the United States, the Japanese laissez-faire approach to product claims is not allowed. The U.S. Food and Drug Administration, in fact, prohibits claims on food packages that relate to the food affecting the structure or function of the body; such claims make the food an illegal drug. This will change in part as the FDA makes an allowance for specific dietary health claims based on the Nutritional Labeling and Education Act of 1990.

One goal of this book was to offer a comprehensive review of the developments in the functional food area as well as of specific nutrient/body interactions. This has been adequately done by an amazing collection of well-known scientists and should prove to be of value to the food and pharmaceutical industry in understanding the boundary conditions of what is possible from food. Secondly, this treatise examines the current functional food markets throughout the world including the technologies that can be employed. This will certainly benefit those food scientists and marketers who are trying to design new acceptable products. Finally, this book helps us understand how consumer's views and legal concerns will impact the kinds of products that can be made. As a consultant to the food industry, several times I have been asked to explain the promises and pitfalls of functional foods to food company executives who are business people, not scientists or engineers. I have always had that feeling of being on the borderline of health quackery. This text will help to displace that attitude and will be useful ammunition in taking a proper stand on functional foods. Dr. Goldberg has certainly brought into one place all the scientific and legal issues, and a reading of this book will convince anyone that nature indeed imparted a function to foods and it is time to do our part, as Milton so dutifully suggested over 300 years ago.

Theodore P. Labuza

PREFACE

It is becoming increasingly clear that there is a strong relationship between the food we eat and our health. Scientific knowledge of the beneficial role of various food ingredients (nutrients) for the prevention and treatment of specific diseases is rapidly accumulating. At the same time, novel technologies—including biotechnology and specifically genetic engineering—have created an era where scientific discoveries, product innovations, and mass production will be possible as never before. These developments have resulted in an increasing number of potential nutritional products with medical and health benefits, so called "functional foods." The basis of this book is a detailed discussion of the medical and clinical studies that emphasize the possible benefits of a growing range of food ingredients on various diseases. Also, a brief description of the possible contribution of biotechnology in the production of these functional food ingredients is given. (For a more detailed discussion see *Biotechnology and Food Ingredients* by Goldberg and Williams, 1991).

Functional foods, designer foods, pharmafoods, and *nutraceuticals* are synonyms for foods that can prevent and treat diseases.

Generally, a functional food can be defined as any food that has a positive impact on an individuals's health, physical performance or state of mind *in addition to its nutritive values.*[1] The Japanese have highlighted three conditions that define a functional food.

1. It is a food (not a capsule, tablet, or powder) derived from naturally occurring ingredients.
2. It can and should be consumed as part of the daily diet.
3. It has a particular function when ingested, serving to regulate a particular body process, such as:
   - Enhancement of the biological defense mechanisms
   - Prevention of a specific disease
   - Recovery from a specific disease
   - Control of physical and mental conditions
   - Slowing the aging process.[2]

In addition to functional foods there should be a recognition of "functional diets" meaning that the overall composition and choice of foods in the diet has a functional health effect. For example, the movement to a diet higher in vegetable and fruit products and lower in animal products will have a functional effect. This may be the most practical way for deliberate and increased consumption of the functional ingredients found in foods and for substantive improvement of health.

The concept of foods that combine nutritional and medical benefits is old, but the Japanese are stimulating this emerging concept in a new way that will have an immense effect on the food industry. Although functional foods were originated and are most popular in Japan, they are gaining acceptance internationally, especially in the wealthier nations of the world.

During the past seventy years dramatic changes have been observed in the types of food we eat, reflecting the application of scientific findings and technological innovations in the food industry. During the last few years the diet-health message has evolved and initiated the development of functional foods. The recently developing area of nutrition and health will most certainly revolutionize the food industry.

It is my hope that this book will serve not only to review the science and market base available for the development of functional foods, but will also be a resource to stimulate more research and development committed to this emerging field.

---

[1] J. Rabe, personal communication.

[2] PA Consulting Group, *Functional Foods: a new global added value market?* (London: PA Consulting Group 1990).

However, the information in this book is not only for those involved in food processing and food product development. It is expected that through the sharing of this level of information, the necessity of special considerations for many of these new and sophisticated food processes, products, and ingredients will also be noted by those in charge of regulatory issues. For at this moment, products can not be labeled or designated as "functional foods," nor can their specific medical benefits be claimed even though these benefits have and will continue to be documented in the years ahead.

I would like to acknowledge my indebtedness to all the authors, each distinguished in the field he or she has reported on for this book. I am grateful to Chapman and Hall for publishing *FUNCTIONAL FOODS—Designer Foods, Pharmafoods, and Nutraceuticals* with their customary excellence. And I owe special thanks to Eleanor Riemer.

Israel Goldberg

Contributors

Jonathan C. Allen
Associate Professor
North Carolina State University
Department of Food Science
Raleigh, NC 27695-7624

John J.B. Anderson
Dept. of Nutrition
School of Public Health
The University of North Carolina
CB #7400, McGavran-Greenberg
Hall, Chapel Hill
NC 27599-7400

Jeffrey S. Bland
HealthComm, Inc.
5800 Soundview Drive
Gig Harbor, WA 98335

Jeffrey B. Blumberg
Antioxidants Research Lab.
United States Dept. of Agriculture
Human Nutrition Research Center on Aging at Tufts Univ.
711 Washington Street
Boston, MA 02111

Katrina M. Brown
Rowett Research institute
Greenburn Road, Bucksdurn
Aberdeen, AB2 9SB
Scotland, U.K.

Adolph S. Clausi
c/o William H. Willis, Inc.
164 Mason Street
Greenwich, CT 06830

Garry G. Duthie
Rowett Research Institute
Greenburn Road, Bucksburn
Aberdeen, AB2 9SB
Scotland, U.K.

Jeffrey C. Gardner
Technology Catalysts
International Company
605 Park Avenue
Falls Church, VA 22046

Israel Goldberg
Dept. of Applied Microbiology
Faculty of Medicine
The Hebrew University
P.O. Box 12272
Jerusalem 91120
Israel

Tomio Ichikawa
School of Home Economics
Mukogawa Women's University
663 Nishnomiya City
Ikebiraki-cho 6-46
Japan

Sooja K. Kim
Nutrition Study Section
Division of Research Grants
National Institutes of Health
National Institute of Aging
348 Westwood Building
Bethesda, MD 20892

Theodore P. Labuza
Dept. of Food Science &
Nutrition
University of Minnesota
1334 Eckles Avenue
St. Paul, MN 55108

Harris R. Lieberman
U.S. Army Research Institute
of Environmental Medicine
Natick, MA 01760-5007

Robert I. Lin
Nutrition International
Company
6 Silverfern Drive
Irvine, CA 92715

Zecharia Madar
Dept. of Biochemistry &
Human Nutrition
Faculty of Agriculture
The Hebrew Univ.
P.O. Box 12
Rehovot 76100
Israel

Kauko K. Makinen
Dept. of Biologic and Material
Sciences
School of Dentistry
The University of Michigan
Ann Arbor, MI 48109-1708

Wayne E. Marshall
Environmental Technology
Research
United States Dept. of
Agriculture
Agricultural Research Service
Mid South Area Southern
Regional
Research Center
1100 Robert E. Lee Boulevard
P.O. Box 19687
New Orleans, LA 70179

Kristen McNutt
Consumer Choices Inc.
2272 Woodview Road, #401
Ypsilanti, MI 48198

Darrell G. Medcalf
HealthComm, Inc.
5800 Soundview Drive
Gig Harbor, WA 98335

Herbert L. Meiselman
U.S. Army Natick Research, Development and Engineering Center
Natick, MA 01760-5020

John A. Milner
Dept. of Nutrition
The Pennsylvania State Univ.
126 Henderson Building South
University Park, PA 16803

Tsuneyuki Oku
Dept. of Nutrition
Faculty of Medicine
University of Tokyo
7-3-1, Hongo, Bunkyo, Tokyo
Japan 113

Harish Padh
Northwestern University
Center of Biochemistry
2153 Sheridan Road
Evanston, IL 60208

Mary E. Sanders
Dairy & Food Culture Technologies
7119 S. Glencoe Ct.
Littleton, CO 80122

Mary K. Schmidl
Clinical Products Division
Sandoz Nutrition Corporation
Technical Center
1541 Vernon Avenue South
Minneapolis, MN 55440

Artemis P. Simopoulos
The Center for Genetics, Nutrition and Health
2001 S Street, N.W. Suite 530
Washington, D.C. 20009

Aliza Stark
Dept. of Biochemistry & Human Nutrition
Faculty of Agriculture
The Hebrew Univ.
P.O. Box 12
Rehovot 76100
Israel

Mark L. Wahlqvist
Dept. of Medicine
Monash Univ.
Monash Medical Center
246 Clayton Road
Clayton, Melbourne, Victoria
3168 Australia

Huber R. Warner
Biology of Aging Program
National Institutes of Health
National Institute of Aging
Gateway Building, Room 2C231
Bethesda, MD 20892

Margaret P. Woods
Dept. of Applied Consumer Studies
Queen Margaret College
Clerwood Terrace
Edinburgh EH12 8TS
Scotland, U.K.

Kathie L. Wrick
Arthur D. Little, Inc.
Acron Park
Cambridge, MA 02140-2390

Part I

Introduction

Chapter 1

Introduction

Israel Goldberg

Functional foods, in addition to their basic nutritive value and natural being, will contain the proper balance of ingredients which will help us to function better and more effectively in many aspects of our lives, including helping us directly in the prevention and treatment of illness and disease.

The idea of health-filled foods is, of course, not new. The modern message probably took root in the soil of the nutrition evaluations of the 1950s and the "back to nature" revolution of the 1960s. The message was to produce packaged food which retained its nutrition and appeared and tasted as close to nature's intentions as possible. In the sixties, medicine and food producers entered their first unlikely partnership. During this period, less became more, as the medical community and food producers worked together to isolate the cause and effect of overabundance of certain ingredients on the human bodily functions.

As diet centered on eliminating or reducing certain ingredients, rather than including them in the menu, issues such as fats and cholesterol levels, sugar and salt contents and food calories, were monitored and reduced with steadily increasing alertness by demanding consumer and a responding producer.

It was not much later that food additives were entered into the equation. If food could be made better and more healthful by removing ingredients, then what about assuring the added or substituted presence of certain other ingredients? Surely this could have as much of a positive impact on an individual's health, physical performance or state of mind, as those ingredients which were restricted.

Today, these concepts of nutrition, natural being and reduction and inclusion of ingredients merge into a modern field of new scientific recognition. That is, not only do food ingredients in proper combinations actually offer potential or actual benefits in preventing or treating special conditions or diseases, but also these food ingredients can occur naturally or can be produced by biological technologies such as genetic engineering. This is the targeted level of direction which this volume will discuss.

THE MARKET OF FUNCTIONAL FOODS

Japan is currently the world leader in the development of functional foods. In Japan, functional foods (currently called foods for specified health use) are considered as a major new product opportunity—a new dimension of a wide range of food and drink products (Potter 1990).

The value of the Japanese market for functional foods, even without manufacturers being able to make explicit health claims (see below), is of the order of $3 billion, but could well reach $7.5 billion, or approximately 5% of the processed food market in Japan (PA Consulting Group 1990). Another estimation predicts that the Japanese market for functional foods will grow at 8.5% per year, reaching a level of $4.5 billion by 1995 (Anon 1991; Weitz 1991).

Over 100 companies are actively selling or developing products that could in due course fit into the category of functional foods. Over 70% of the products is estimated to be drinks; the remainder are foods. The distribution of ingredients in various potential functional foods is estimated to be: dietary fiber (40%), calcium (20%), oligosaccharides (20%), lactic acid bacteria (10%), and other (10%) (PA Consulting Group 1990).

The total size of the U.S. health food market in 1987 was $18.98 billion (PA Consulting Group 1990). From an estimated U.S. sales level of about $2.5 billion in 1988, the market for functional foods is expected to reach a level of $7.5 to $9 billion by 1995, which is equivalent to an average annual growth rate of 17 to 20 percent (Anon 1991; Weitz 1991).

Less information is available for the European market. The total size of the health food market in Europe (i.e., West Germany $4.82 billion, United Kingdom $2.28 billion, and France $1.42 billion) was $8.52 billion in 1987, which is about 44 percent of the U.S. market (PA Consulting Group 1990).

CONSUMER'S INTEREST

Because of the recently accumulating scientific data demonstrating the relationship between food consumption and disease incidence, consumers are beginning to accept that, to a significant part, "health is a controllable gift."

Consumer interest in health-promoting foods will almost assuredly increase, perhaps dramatically, because of the following converging trends (Rabe, J., personal communication).

- *Clinical evidence*—The functionality of many familiar and new food components is being established by clinical and epidemiological research. Much of this research is publically funded; hence, there are political needs for these funded organizations to publicize their findings. For example, the National Cancer Institute's (NCI) Diet and Cancer Branch has recently embarked on a major initiative—$20 million for five years—to study, assess, and develop experimental processed foods that are supplemented with food ingredients naturally rich in cancer-preventing substances, i.e., aged garlic extract, licorice extract, citrus juice, a mixed-vegetable beverage, and soybean meal (*Los Angeles Times*, Sunday, 17 May 1992; Caragay 1992).
- *Age wave*—Older consumers are more interested than the younger generation in functional foods to improve health and extend life expectancy.
- *Health care cost containment*—Government and industry will continue to search for ways to reduce health care costs. Education of nutrition will be pursued as a means of improving worker's health and reducing costs.
- *Media*—More "user friendly" information on health, diet, and nutrition is being delivered by the media (print, radio, and television). The media plays a major role in communicating this in-

formation to a public that readily responds to "magic bullets" (Goldberg 1992).

- *Nutritional labeling*—In the United States the Nutrition, Labeling and Education Act of 1990 will provide consumers with "user friendly" nutrition information on almost all food products. As consumers become more health conscious, the promotion of certain nutritional attributes can be very effective for selling certain products.
- *Mission of medicine*—Medical schools are beginning to place more emphasis on the prevention of chronic diseases. Young physicians and the progressive health care clinics will make consumer education on a healthy diet a more important part of their services.
- *Food technology*—The development of NutraSweet and Simplesse and the introduction of such products as Healthy Choice, Lovin' Lite, and Haagen-Dazs yogurt are convincing consumers that they can have tasty foods that are also healthy.
- *Brand differentiation*—Most food products represent mature categories where brand differentiation is a difficult challenge. Aggressive firms with strong technical development capability will use the interest in health and nutrition as an opportunistic way to differentiate and add value to their brand of product.

FUNCTIONALITY OF INGREDIENTS

Diet is believed to play an important role in the four major diseases of our society—cardiovascular (heart and artery) diseases, cancer, hypertension, and obesity. Heart disease and cancer account for 70% of all deaths. The degree to which diet is important in the prevention of these diseases is not known, but a commonly accepted estimate among experts is that at least 1/3 of the cancer cases can be attributed to diet and that perhaps 1/2 of the cases of heart and artery diseases and hypertension are related to diet (see Chapters 2 and 3). The major food component associated with cardiovascular diseases and cancer are too much fats, particularly saturates; and the under consumption of fiber, vegetables, and fruits.

In the early 1980s, the potential medical benefits of different food ingredients such as calcium, fiber, and fish oil were substantiated. Today various ingredients are being studied for their role in preventing or treating various chronic diseases and for their far-reaching benefits in slowing the aging process and affecting mood and performance (see Part II of this book). Recent examples include the use of fish oil supplements and antioxidants to reduce damages caused by atherosclerosis, i.e., heart attack, strokes, and occlusive peripheral vascular disease (Leaf 1991); beta (β)-carotene to prevent and decrease the risk of human cancer (Lippman et al. 1987); calcium to decrease osteoporosis (Arnaud and Sanchez 1990) and help women deal with the stress and anxiety symptoms of premenstrual syndrome (*Chicago Tribune*, 4 September 1991); fiber to reduce coronary heart disease and cancer (Kromhout et al. 1982); specified fats to reduce mortality caused by coronary heart disease (Kromhout 1992); choline to lower blood pressure (Zeisel 1981); niacin to treat hyperlipidemia (DiPalma and Thayer 1991); zinc to lower susceptibility to a variety of infectious diseases and to enhance the immune response (Keen and Gershwin 1990); magnesium to reduce both acute and chronic isochemic heart disease (Seelig 1980), and many others.

The Japanese Ministry of Health and Welfare has identified 12 very broad classes of ingredients, which they consider to be health enhancing:

- Dietary fiber
- Oligosaccharides
- Sugar alcohols
- Aminoacids, peptides, and proteins
- Glycosides
- Alcohols
- Isoprenoids and vitamins
- Cholines
- Lactic acid bacteria
- Minerals
- Polyunsaturated fatty acids
- Others (e.g., phytochemicals and antioxidants).

Fairly extensive medical research has been performed on these ingredients; however, it is controversial. Different studies contain contradicting results (these studies are described in detail in Part III of this book). Several of these ingredients are already being manufactured or researched within the food industry. These ingredients appear in a full menu of food products—from breakfast cereals to salad dressings.

THE ROLE OF BIOTECHNOLOGY

The whole issue of functional foods essentially relates to ingredients, which provide both nutritional and health benefits of the foods. Because there is an understanding that these ingredients must be obtained or produced from natural sources (''natural ingredients''), there is a substantial level of research activity oriented towards identifying, manufacturing, and demonstrating the efficacy of the new generation of functional ingredients. This research activity can employ biological technologies (biotechnology) to produce modified or new ingredients comprising the groups listed above (except for minerals) (Goldberg and Williams 1991).

The ''new'' or ''modern'' biotechnology includes genetic engineering, which is the ability to transfer individual genes from one living organism to another. Genetic manipulation provides the means to cross traditional genetic boundaries between organisms and allows the construction of genetic combinations that, previously, were only theoretical. Biotechnology also includes enzyme catalysis and protein engineering, plant and animal cell and tissue culture technology, and fermentation technology. These techniques can be used to provide novel, less expensive, safer, natural, and higher quality functional ingredients. Biotechnology allows the customization of ingredients as is discussed below for dietary fibers, fats, and oils.

A diet that is rich in foods containing dietary fiber is protective against a range of diseases such as cancer (colon), atherosclerosis and associated hypercholesteremia, diabetes, diverticulosis, hypertension, and obesity (Lo et al. 1991, Chapter 9). The two sources of dietary fiber most extensively evaluated in clinical studies to lower cholesterol and glucose concentrations in blood have been oat bran and Fibrim soy fiber. Psyllium and guar gum, sources of water soluble fibers, have also been positive in lowering the levels of these blood constituents (Lo et al. 1991).

All dietary fiber ingredients have physical and chemical properties that contribute to their functionality in providing positive health benefits. Knowing the chemical (nature of the monomer, ratio between monomers, chemical bonds between monomers, degree of polymerization) and physical (water solubility, water binding, oil binding) properties of currently available dietary fibers and how these properties affect physiological functions will allow the isolation and development of newer forms of dietary fiber (Lo et al. 1991). It is expected that dietary fibers will be developed from changes in grain/

plant genetics and processing, chemical modification of existing dietary fibers, fermentation, and selective use of enzymes.

The contribution of biotechnology to the development of dietary fibers is evident from the microbial production of new biological forms of dietary fiber such as xanthan gum, produced by *Xanthomonas campestris*; gellan gum, produced by *Pseudomonas elada*; pullulan, produced by *Aureobasidium pullulans*; curdlan, produced by *Alcaligenes faecalis* var. *myxogenes*; and dextran, produced from sucrose by the use of the enzyme dextransucrase obtained from *Leuconostoc mesenteroides* (Lo et al. 1991). An example of such a novel product is Fibercel, a yeast-derived fiber, developed by researchers at Alpha-Beta Technology, Inc., Worcester, MA. This is a potential cholesterol-lowering food ingredient that may reduce the risk of cardiovascular disease. The potential of dietary fibers will continue to have a tremendous impact on food product development and human nutrition and health.

The current contribution of biotechnology to the development of fats and oils is more recognizable than its contribution to dietary fiber. There exists substantial scientific information on the relation between the physico-chemical properties of fats and oils and their health-enhancing function. It is widely recognized that replacement of certain saturated fats in the diet with either monounsaturated or polyunsaturated fats plays an important role in reducing incidence of coronary heart disease (see Chapters 2 and 16).

Biotechnology-based approaches to the development of improved fats and oils are beginning to take on a commercial significance. Three basic approaches are currently utilized to produce modified oilseed crops or modified lipids and derivatives (Laning 1991):

1. *Genetic manipulations (classic genetics or genetic engineering) using plants or tissue cultures*—The biotechnological research now in progress is targeted to achieve higher oil yield and/or desirable fatty acid compositions in oilseed plants. (For example, most sunflower varieties produce oils containing 60% to 70% linoleic acid. Hybrid seeds with up to 95% oleic acid and 0.5% linoleic acid have been developed (Ram et al. 1988; Laning 1991).

2. *Microorganisms producing commercially significant quantities of selected specialty fats and oils*—The vast majority of fat- and oil-producing microorganisms are yeasts and molds (Ratledge 1984). Glatz and colleagues (1984) have reported on the production of unique fats and oils by the yeast *Candida lipolytica* when grown on common vegetable oils such as corn and coconut oils. Some

molds may produce relatively large quantities of γ-linolenic acid (Wassef 1977), whereas *Mortiella alpina*, a fungal soil isolate, has been shown to produce a relatively large amount of arachidonic acid (Yamada et al. 1988; Laning 1991).

3. *Enzymes (e.g., lipases) used to modify existing fats and oils*—The ability of enzymes such as lipases to modify lipid structures has created much interest in recent years. Extensive research has been applied to this subject, which will not be described here. One example is the exchange of linoleic acid with a mixture of medium-chain triglycerides, which is catalyzed by lipases in an acidolysis process. This process can be used to produce specific triglycerides of nutritional interest (Jandacek 1989; Laning 1991).

EXAMPLES OF FUNCTIONAL FOODS

Following is a brief look at several categories of popular new potential functional foods.

Milk Products

Since 1908, when Elie Metchnikoff (Nobel prize laureate) suggested that consumption of milk fermented with lactobacilli would prolong one's life (Metchnikoff 1908), there has been an interest in the potential health benefits associated with these organisms (Sanders et al. 1991). The principal organisms used are species of *Bifidobacterium* and/or *Lactobacillus* (see Chapter 14). Potential health benefits that have been linked to these organisms include improved lactose digestion, control of intestinal pathogens, reduction of serum cholesterol, tumor-inhibiting effects, immune system stimulation, prevention of constipation, generation of group B vitamins, production of bacteriocins, and inactivation of some toxic compounds (Potter 1990; Sanders et al. 1991). Besides the traditional kefir and koumiss, the new products include among others: Natural Light Yogurt (Arla Mejerierna), Ophilus (Yoplait), Whole Grain Bio (Ehrman AG), Bio Seven (Kansai Luna), Bio Pruneaux (Danone), OH BA Yogurt (Chanbourcy), Yakult (Yokult Honsha), Bifidus (Kyodo), BA Live (Loseley Dairy), and a nonfat BA yogurt from Stoneyfield farm (Cole 1991; Potter 1990).

Beverages

Most of the functional foods developed are beverages. Examples are FibeMini (maybe Japan's best-selling soft drink) from Otsuka Pharmaceuticals, which contains dietary fiber supplement, minerals, and vitamins; PF21 from Asahi Beer Co., which is a protein-rich sports drink that contains collagen; Aqualibra from Callitheke, which is a Swiss formula sports drink based on alkaline-forming ingredients; Pocari Sweet Stevia from Ootsuka, an isotonic sports drink containing the sweetener stevioside (a glycoside from the leaves of *Stevia* sp.); Fibi from Coca-Cola, which is a fiber-rich soft drink aimed at general digestive health; CaD from Asahi Breweries, Ltd., a calcium-fortified drink (the "D" signifies "dairy"); Oligo CC from Calpis Food Co., which is a carbonated beverage consisting of oligo sugar extracted from soybean; Spofit from Rawlands, which is an energy-restoring drink containing sugars and maltodextrin; Tekkotsu Inryou from Suntory Co., which is an iron- and calcium-rich beverage for children and young adults through their mid-20s; and Toshu Cha from Hitachi Zosen Corp., which is tea fortified with calcium.

Foods

Current products include ready meals, breakfast cereals, biscuits, confectionery, baby foods, ice cream, and salad dressings (PA Consulting Group 1990). Examples are Oatbrand from Oatbrand Co., a bread containing soluble fiber; Common Sense Oat Bran Flakes from Kellog Company, a bran-based cereal; Oatbran from Sunwheel Natural Foods, which is a wholemeal oatbran biscuit; Fibermax 391 from Asahi Chemical Industry, a biscuit (a wafer cookie) enriched with dietary fiber; Hemace from Asahi Chemical Industry, an iron-supplemented candy; Ladys Can Fen from Meiji Seika, a functional candy that contains iron, vitamins, and fructo-oligosaccharide; Yugao Bijin from Tokyo Tanabe Co., a pasta rich in natural dietary fiber; Caluche from Nissin Food Products, a snack rich in fiber; and products such as Oligo Harmony (pork sausages with added oligosaccharides), Fibre Harmony (pork sausages with wheat fibers), and Calcium Harmony (pork sausages with added egg shell calcium) (Cole 1991).

REGULATORY ISSUES

As the distinction between food and drug becomes less precise, the regulatory issues become more complex. Key legislative issues focus on the origin, safety, use, and effect of health-enhancing ingredients. In addition, there is still a need for a precise definition of what constitutes a "health claim." Such health claims are likely to be based on the beneficial effects certain ingredients may have on the consumer's health in the context of a "sensible" balanced diet (Cathro 1990).

In Japan there are many food products on the market with advertisements praising "functional benefits" that are positioned as "more beneficial." These products can not be officially labeled or designated as "functional foods," nor can specific medical benefits be claimed until legislation is passed. Thus, the Japanese manufacturers are currently reluctant to describe any of their products as functional foods, to avoid criticism by the authorities (Kisaburo Shiko, President of V-Call, learned this the hard way when he was sentenced to jail for attributing specific medical benefits to his health drinks. These products contained vitamins B, C, and E, plus other nutrients that were said to be effective against high blood pressure and diabetes—Lake 1990).

Japan has been a pioneer in the area of functional foods and has already established a system for approving functional foods, granting such products individual health claims for specific health needs based on approved and sufficient research data (see Chapter 18). Until their approval these products should only be considered as potential functional foods. (The first functional food with a health claim was introduced by Shiseido (Tokyo)—a rice from which the protein globin was removed, to the market in 1993.)

Contrary to Japan, the functional food market has not been warmly received by the U.S. regulatory and legislative bodies. The U.S. Food and Drug Administration (FDA) does not, at this time, allow for health claims to be made in connection to food products as outlined in the Food, Drug and Cosmetic Act of 1938 (McGrath and Petzel 1991).

General Mills Inc., marketed Benefit in April 1989 and Kellogg Co. responded with Heartwise in September of the same year. Both are oat-based brands containing psyllium. This ingredient, a natural Asian grain with more than eight times the soluble fiber of oat bran, has been claimed to reduce cholesterol and to be beneficial against heart disease. The FDA has kept a close watch on developments, and although General Mills Inc. denies that pressures from the FDA led to Benefit's withdrawal, there were doubts over psyllium's safety that prompted the decision (Cathro 1990). In 1991, Kellogg Co. "voluntarily" renamed its controversial cereal, calling it Fiberwise instead of Heartwise (Lake 1990; Goldberg 1992).

New knowledge of the functional properties of foods and the commercial interest of certain food companies bring the prohibition of health claims on foods under question and attack. As a result, the negative attitude of the FDA towards functional foods has changed partially with the implementation of the 1990 Nutrition, Labeling and Education Act (NLEA). The FDA allows specific dietary health claims based on the NLEA, recognizing that recent scientific evidence shows potential links between specific dietary ingredients and chronic diseases. Four ingredients have been highlighted by the FDA (Cathro 1990; PA Consulting Group 1990): dietary fiber—cancer of colon, heart disease; lipids—cancer, heart disease; calcium—osteoporosis, and sodium—hypertension. (see Chapters 2, 3, 9 and 15)

The FDA hopes to word its own health claims for these ingredients, and the manufacturers would then be limited to this use. This is in contrast to the current situation whereby manufacturers can propose claims and submit them for the FDA's approval. In the opinion of many individuals and several institutes the FDA should act further by expanding the economic and regulatory provisions of the NLEA, and including a more appropriate mechanism for establishing new and true health claims based on innovative research (Pszczola 1992).

In Europe, current legislation is a collection of inadequate regulations dating back to the 1940s, although some were revised in 1987 (Cole 1991). Major European countries recognize this new product category and they have systems that grant exclusivity for innovative medical and health claims for foods based on proprietary research. Procedures to grant exclusivity throughout Europe have been developed by the European Economic Community (EEC) (Pszczola 1992).

FUTURE PROSPECTS

Living longer and better is on the mind of most people, especially in wealthy countries, and diet is one of the most important facets of a healthy life. Therefore it seems very likely that the concept of "functional foods," which enables the consumer to exercise a level of "self-health maintenance," will significantly influence food and drink manufacturers (Potter 1990). Functional foods reflect the recent consumer preoccupation with convenient food and healthy eating. It is expected that if the idea of functional foods catches on with the public (and there is no reason why this will not happen considering the increase in consumer demand), there will be a huge market for such foods even if currently they cannot be labeled as such.

The brand scale acceptance and use of functional foods will depend mainly on two factors. The first is the education of the consumer, which will be a long term process, to establish its credibility of the claims made. The second factor is the development of no compromise products, meaning that taste and convenience are not compromised.

A lot of research is carried out to demonstrate the relationship between dietary intake and disease incidence, and to produce functional ingredients through biotechnology. I view functional foods as an emerging science. As with most scientific disciplines, they usually have no clear beginning and no clear end. Rather, a given subject at some point achieves a level of importance to warrant being thought of as a separate and distinct discipline. I strongly believe that this is what will happen, and I look for either the medical schools or food science departments, or perhaps a hybrid of the two, to eventually establish this field as a focused discipline of study. It's that important.

The chapters in this book reflect the latest scientific findings in this area. I hope that these will catalyze the development of functional foods by stimulating government authorities to change legislation and approval procedures, encouraging investments in research, and by gaining consumer credibility, which will foster a more accepting commercial atmosphere for the development and introduction of functional foods in the marketplace.

References

Anon. 1991. *PRNewswire*, Falls Church, VA, June 24.

Arnaud, C.D., and Sanchez, S.D. 1990. The role of calcium in osteoporosis. *Ann. Rev. Nut.* 10:397–414.

Biotech Forum Europe (BFE). 1992. 9:105–106.

Caragay, A.B. 1992. Cancer-preventing foods and ingredients. *Food Technol.* April:65–68.

Cathro, J.S. 1990. Functional foods and healthy eating. *Special Report, Leatherhead Food R.A.*

Cole, M. 1991. When food meets medicine. *Food Rev.* 6:15–17.

DiPalma, J.R., and Thayer, W.S. 1991. Use of niacin as a drug. *Ann. Rev. Nut.* 11:169–187.

Glatz, B.A.; Hammond, E.G.; Hsu, K.H.; Bachman, L.; Batj, N.; Bednarski, W.; Brown, D.; and Floctenmeyer, M. 1984. Production and modification of fats and oils by yeast fermentation. In *Biotechnology for the Oils and Fats Industry*, C. Ratledge, P. Dawson, and J. Rattray, eds., pp. 163–176. Champaign, IL: American Oil Chemist's Society.

Goldberg, I., and Williams, R., eds. 1991. *Biotechnology and Food Ingredients.* pp. 1–577. New York: Van Nostrand Reinhold.

Goldberg, J.P. 1992. Nutrition and health communication: the message of the media over half a century. *Nut. Rev.* 50:71–77.

Jandacek, R.J. 1989. Lipids of nutritional interest. Paper read at American Oil Chemist's Society Short Course. Specialty Fats—Production and Application, 30 April–2 May 1989, at Kings Island, OH.

Keen, C.L., and Gershwin, M.E. 1990. Zinc deficiency and immune function. *Ann. Rev. Nut.* 10:415–431.

Kromhout, D. 1992. Dietary fats: long-term implications for health. *Nut. Rev.* 50:49–53.

Kromhout, D.; Bosschietes, E.B.; and DeLezenne Coulander, C. 1982. Dietary fiber and 10 year mortality from coronary heart disease, cancer and all causes. The Zutphen Study. *Lancet* 2:518–521.

Lake, C. 1990. Up in the air over functional foods. *Food FIPP*. June.

Laning, S.J. 1991. Fats, oils, fatty acids, and oilseed crops. In Biotechnology and Food Ingredients, I. Goldberg and R. Williams, eds., pp. 265–285, New York: Van Nostrand Reinhold.

Leaf, A. 1991. Fish oil supplements and antioxidants. *Omega 3 News* 5:1–4.

Lippman, S.M.; Kessler, J.F.; and Meyskens, F.L., Jr. 1987. Retinoids as preventive and therapeutic anticancer agents. *Cancer Treat. Rep.* 71:391–405, 493–515.

Lo, G.S.; Moore, W.R.; and Gordon, D.T. 1991. Physiological effects and functional properties of dietary fiber sources. In *Biotechnology and Food Ingredients*, I. Goldberg and R. Williams, eds., pp. 153–191. New York: Van Nostrand Reinhold.

McGrath and Petzel, Inc. 1991. *Nutrition 2000*. A business report. Edina, MN.

Metchnikoff, E. 1908. The Prolongation of Life. New York: G.P. Putnam's Sons.

PA Consulting Group. 1990. *Functional foods: a new global added value market?* London.

Potter, D. 1990. Functional foods—a major opportunity for the dairy industry? *Dairy Ind. Int.* 55:32–33.

Pszczola, D.E. 1992. The nutraceutical initiative: a proposal for economic and regulatory reform. *Food Technol.* April:77–79.

Ram, R.T.; Andreasen, T.J.; Miller, A.; and D.R. McGee. 1988. Biotechnology for *Brassica* and *Helianthus* improvement. In *Proceedings World Conference on Biotechnology for the Fats and Oil Industry*, T.H. Applewhile, ed., pp. 65–71, Champaign, IL: American Oil Chemist's Society.

Ratledge, C. 1984. Microbial oils and fats—an overview. In *Biotechnology for the Fats and Oils Industry*, C. Ratledge, P. Dawson, and J. Rattray, eds., pp. 119–127. Champaign, IL: American Oil Chemist's Society.

Sanders, M.E.; Kondo, J.K.; and Willrett, D.L. 1991. Applications of lactic bacteria. In *Biotechnology and Food Ingredients,* I. Goldberg and R. Williams, eds., pp. 433–459. New York: Van Nostrand Reinhold.

Seelig, M.S. 1980. *Magnesium Deficiency in the Pathogenesis of Disease.* pp. 1–26. New York: Plenum Publishing.

Wassef, M.K. 1977. Fungal lipids. *Adv. Lipid Res.* 15:159–223.

Weitz, P. 1991. *Nutriceutical Products and Functional Food Additives.* Falls Church, VA: Technology Catalysts International.

Yamada, H.; Shimizu, S.; Shinmen, Y.; Kawashima, H.; and Akimoto, K. 1988. Production of arachidonic acid and eicosapentaenoic acid by microorganisms. In *Proceedings World Conference on Biotechnology for the Fats and Oils Industry,* T.H. Applewhile, ed., pp. 173–177. Champaign, IL: American Oil Chemist's Society.

Ziesel, S.H. 1981. Dietary choline: biochemistry, physiology, and pharmacology. *Ann. Rev. Nut.* 1:95–121.

Part II

Health Attributes of Functional Foods

Chapter 2

Reducing the Risk of Cardiovascular Disease

Garry G. Duthie and
Katrina M. Brown

INTRODUCTION

Despite a decline in mortality from cardiovascular disease (CVD) in many countries between 1970 and 1985 (Rosenman 1992), the affliction remains responsible for 51% of human deaths in the world (Kakkar 1989). Each year in the United Kingdom, 170,000 people die from myocardial infarction, and CVD-related disorders cost the National Health Service over £500 million a year (Coronary Prevention Group 1991). These unacceptable statistics remain despite numerous intervention programs and health promotion campaigns and may be due, in part, to our incomplete understanding of the biochemical processes responsible for the pathogenesis of the disease. For example, even if it is assumed that the "classical" CVD risk factors of smoking, blood cholesterol, and hypertension are causal rather than merely adventitious, they explain only 50% of the variance in the incidence of the disease (Gey 1986). Thus, it is perhaps not surprising that medical and nutritional advice to moderate the well-recognized CVD risk factors has had little effect in reducing morbidity. However, results from recent biochemical, epidemiological, and cell culture studies suggest that some of the unexplained variation in CVD incidence may be due to inadequate intakes of micronutrients with antioxidant activity such as vitamins E, C, and beta (β)-carotene. The purpose of this chapter is to consider whether these and other potential nutritional antioxidants confer a functional role upon certain foods to reduce the risk of CVD.

PROBLEMS WITH THE "LIPID HYPOTHESIS"

Elevated blood cholesterol concentrations are regarded as a major indicator of CVD risk. This belief arose primarily from the "Seven-Countries" cross-cultural study (Keys 1980), which found a strong relationship between the consumption of saturated fatty acids and mortality from CVD. This finding subsequently evolved into the "lipid hypothesis," which suggests that the consumption by a population of high levels of saturated fat increases blood cholesterol concentrations, and that this, in turn, is related to the number of deaths from CVD. Unfortunately, the "Seven-Countries" study is the only major study to date to find a strong link between saturated fatty acid intake and CVD and may be flawed by selection bias (Rosenman 1992). In addition, few of the intervention trials aimed at reducing blood cholesterol by dietary and pharmacological means have reduced the incidence of CVD or total mortality in the intervention groups (McCormick and Skrabanek 1988). There are also many anomalies in the "lipid hypothesis." For example, in many areas of France, a high intake of saturated fat by the population is accompanied by a low incidence of CVD. In the United Kingdom, the upper social classes have the highest fat intake and highest cholesterol levels, but the lowest incidence of CVD; and Asian immigrants have a higher rate of cardiovascular disease than Caucasians, despite the fact that, on average, Asians smoke less, and have lower blood pressure and plasma cholesterol levels (James, Duthie, and Wahle 1989). Moreover, in 194 consecutive autopsies at the Providence Veterans Hospital in the United States, only 10% of cases of fatal atherosclerosis had elevated serum cholesterol; most cases had developed without evidence of elevated serum cholesterol, diabetes, or hypertension. In a recent review, Rosenman (1992) lists many other anomalies that contradict the belief that saturated fat intake is related to CVD incidence. For instance, CVD incidence was unrelated to dietary fats, saturated fatty acids, or percent calories from saturated fatty acids in a five-year follow up of 10,000 male Israelis, and the long term incidence of CVD in a major U.K. study was not associated with either saturated or polyunsaturated fatty acid intakes.

Such inconsistencies question the veracity of the lipid hypothesis of CVD. However, many of the anomalies disappear if oxidized cholesterol, rather than cholesterol *per se*, is identified as a major risk factor in the development of CVD.

THE "ANTIOXIDANT HYPOTHESIS" OF CVD

The "antioxidant hypothesis" (Gey 1986) suggests that free radical-mediated oxidation of cholesterol is a key step in atherogenesis. One attraction of this hypothesis is that it combines the epidemiologist's approach to identifying risk factors with the biochemist's mechanistic approach; this may lead to a better understanding of the cellular events responsible for the pathogenesis of CVD.

Free Radicals and Cholesterol Oxidation

Free radicals are reactive molecules with an unpaired electron; they have the potential to damage a wide range of biomolecules (Slater 1984). Production of free radicals, which can initiate damage to biological material, occurs during normal aerobic metabolism. Activated oxygen species are formed during the stepwise reduction of oxygen to water and by secondary reactions with protons and transition metals such as copper and iron. For example, the superoxide anion (O_2^-) is produced in many cell redox systems, such as those involving xanthine oxidase, aldehyde oxidase, membrane-associated NADPH oxidases, and the cytochrome P_{450} system. About 1–4% of the total oxygen taken up by mitochondria may be used for O_2^- production and about 20% of this may be ejected into the cell. Stimulated macrophages and monocytes also release large amounts of O_2^-. As this radical is not particularly reactive, it can diffuse relatively large distances through the cell until it can be converted in a metal-catalyzed reaction into the more reactive hydroxyl radical ($OH^{\cdot}$) (Halliwell and Gutteridge 1985). Potentially injurious free radicals are also present in pollutants and halogenated anaesthetics. Moreover, each puff of a cigarette contains approximately 10^{14}

free radicals in the tar phase and approximately 10^{15} in the gas phase (Church and Pryor 1985). Long-lived quinone-semiquinone radicals ($Q^{\cdot}$) associated with the particulate phase are generated by the oxidation of polycyclic hydrocarbons. In aqueous medium, $Q^{\cdot}$ can reduce oxygen to O_2^- and hydrogen peroxide (H_2O_2) and catalyze the conversion of H_2O_2 to $OH^{\cdot}$. In the smoke phase reactive carbon and oxygen centered radicals ($ROO^{\cdot}$) are continuously generated by a reaction between nitrogen dioxide (NO_2) and aldehydes and olefins. Moreover, NO_2 may react with H_2O_2 to produce $OH^{\cdot}$. Pulmonary macrophages activated by nicotine may provide a source of H_2O_2 that may then be converted to $OH^{\cdot}$ in a Fenton reaction catalyzed by free copper or iron ions (Halliwell 1989). Therefore smokers, a high CVD group in northern latitudes, are under an elevated and sustained free radical load.

Polyunsaturated fatty acids (PUFA:H) are particularly susceptible to free radical-mediated oxidation because of their methylene interrupted double bond structure. A process of lipid peroxidation can be initiated by a free radical such as $OH^{\cdot}$ extracting a hydrogen from PUFA:H with the formation of a PUFA radical ($PUFA^{\cdot}$). This is followed by rearrangement of the double bond to form a conjugated diene, which then combines with oxygen to produce a peroxyl radical ($PUFA{:}OO^{\cdot}$), which in turn reacts with more PUFA:H to form a hydroperoxide (PUFA:OOH) and another $PUFA^{\cdot}$. The reaction is now self-propagating:

$$PUFA{:}H \rightarrow PUFA^{\cdot}$$

$$PUFA^{\cdot} + O_2 \rightarrow PUFA{:}OO^{\cdot}$$

$$PUFA{:}OO^{\cdot} + PUFA{:}H \rightarrow PUFA{:}OOH + PUFA^{\cdot}$$

Moreover, in the presence of iron or copper, PUFa:OOH can undergo a fission of double bonds and one electron reduction to form more free radicals, including the highly reactive $OH^{\cdot}$.

$$PUFA{:}OOH \rightarrow PUFA{:}O^{\cdot} + OH^{\cdot}$$

Low density lipoprotein (LDL) is a cholesterol fraction that contains a high proportion of PUFA in the phospholipid layer; thus these are susceptible to free radical-mediated peroxidation. Oxidation of LDL causes a loss of PUFA and an increase in toxic products such as PUFA:OOH, aldehydes, and lysolecithin. Experiments with cell cultures suggest that such oxidation modifies the chemical and physical properties of the LDL to render it more atherogenic. The consequences of these oxidative changes include (see Wahle and Duthie 1991):

(a) Recognition and preferential uptake of modified LDL by scavenger receptors of macrophages.
(b) Enhancement of chemotactic responses with respect to other monocytes/macrophages, leading to accumulation of macrophages at specific sites.
(c) Increased tendency for platelet aggregation.
(d) Increased release of adhesion molecules by endothelial cells.

Thus, in the process of atherogenesis (Fig. 2-1), PUFA:OOH may cause the initial lesion in the artery wall that will ultimately be the site of formation of the atheromatous plaque (Yagi 1987). Uptake of oxidized LDL by monocytes which are attracted to this site of injury promotes chemotactic attraction of more monocytes. On transformation of monocytes to macrophages, the oxidized LDL appears to limit further macrophage mobility and decreases their ability to migrate away from the artery wall. The enhanced rate of uptake of oxidized LDL may then be involved in the conversion of macrophages into foam cells, which are the precursors of the plaque that ultimately occludes the artery. Because the macrophage can oxidatively modify native LDL via the respiratory burst, autocatalytic progression may lead to the continuous growth of the atheroma (Quinn et al. 1987). Elevated concentrations of PUFA:OOH also appear to promote platelet aggregation, thus eventually contributing to thrombus formation. High intakes of those saturated fatty acids that increase LDL would provide increased substrate for the above reactions.

The mechanisms by which LDL is oxidized *in vivo* have yet to be clearly characterized. Oxidation is unlikely to occur in plasma where metal ions are safely bound to proteins such as ferritin, transferrin, and ceruloplasmin, and LDL-modification may be restricted to regions of the vascular intima. Although LDL can be modified by lipoxygenases and phospholipase A_2, oxygen radical reactions with copper are likely to play a more dominant role. Copper ions effectively promote oxidation of LDL *in vitro*, indicating the possibility that release of copper from ceruloplasmin within the atherosclerotic lesion may lead to peroxidation of preformed hydroperoxides. These hydroperoxides may themselves be formed within the lesion by the action of $OH^{\cdot}$ originating from macrophage-derived O_2^- and H_2O_2. The evidence for LDL oxidation *in vivo* is still sparse, but LDL ex-

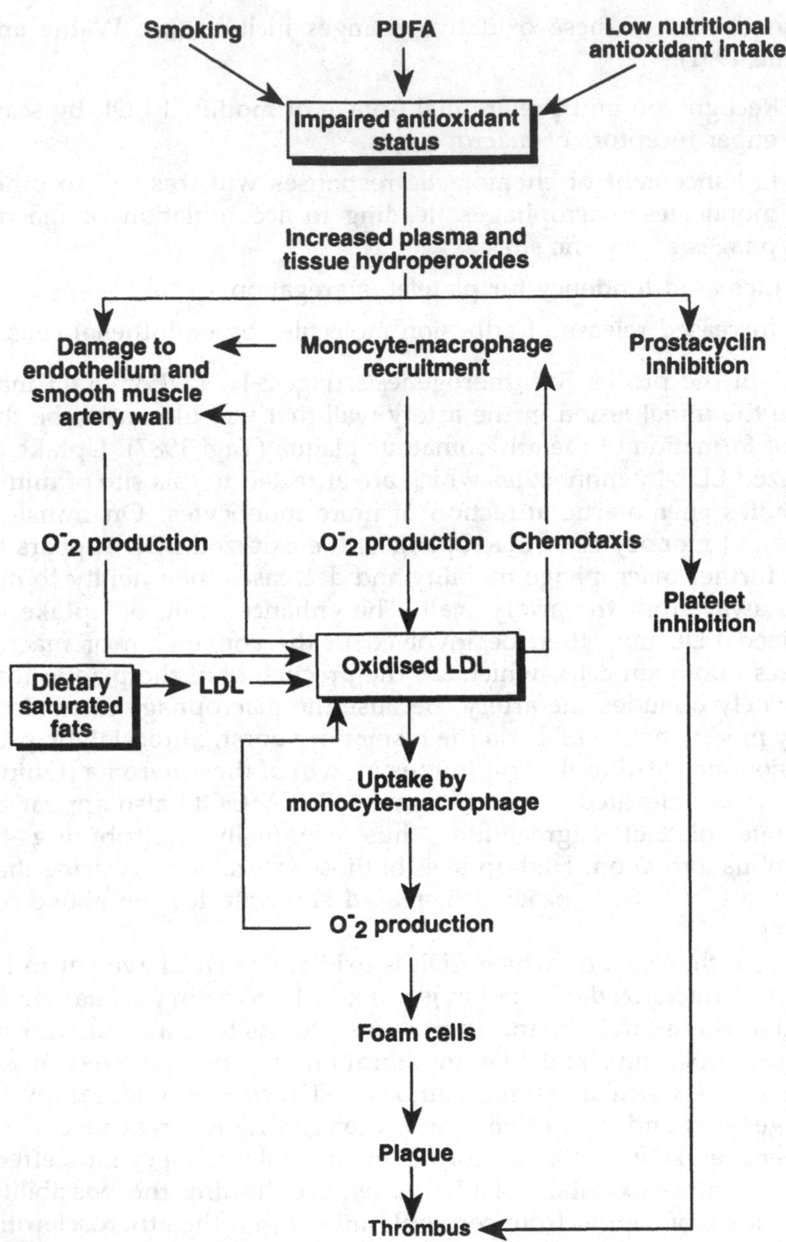

FIGURE 2-1. Flow diagram of possible interactions that contribute to the oxidative modification of low density lipoprotein (LDL) with subsequent development of the atheromatous plaque and thrombus which occlude the artery. (PUFA—Poly unsaturated fatty acids.)

tracted from aortic lesions does react with antibodies specific for oxidatively modified LDL, and gruel samples from advanced atherosclerotic lesions of human cadavers have marked pro-oxidant properties (Smith et al. 1992).

Antioxidants and Cholesterol

A complex antioxidant defense system normally protects mammalian cells and cholesterol from the injurious effects of free radicals. Antioxidants are substances that, when present at much lower concentrations than an oxidizable substrate, significantly delay or prevent its oxidation. Certain essential antioxidants are provided by the diet. These include (a) vitamin E, a major lipid-soluble antioxidant, that breaks the chain of free radical-mediated lipid peroxidation of polyunsaturated fatty acids, (b) β-carotene and other carotenoids, which may have a similar function, particularly in tissues with a low partial pressure of oxygen, and (c) vitamin C, which scavenges free radicals in the water soluble compartment of the cell and may also regenerate vitamin E. In addition several antioxidant enzymes such as glutathione peroxidase, catalase, and superoxide dismutase metabolize the toxic intermediates produced during oxidation of biological material. These enzymes often require micronutrient cofactors such as selenium, iron, copper, zinc, and manganese for catalytic activity.

Esterbauer et al. (1992) showed that the antioxidant content of LDL influences its susceptibility to oxidation *in vitro*. When LDL was exposed to a copper-mediated oxidation system, vitamin E in the outer phospholipid layer of the LDL was depleted first, followed by carotenoids such as lycopene and β-carotene. LDL oxidation was delayed if vitamin C was present in the external medium, presumably because vitamin C regenerates vitamin E. Increasing the vitamin E content of the LDL by dietary supplementation of volunteers also inhibited oxidation of LDL to the atherogenic form.

Cross-cultural investigations (Gey 1986; Gey 1992) have found that mortality due to CVD is inversely related with a cumulative antioxidant index, defined as

$$\frac{[\text{vitamin E}] \times [\text{vitamin C}] \times [\beta\text{-carotene}] \times [\text{selenium}]}{[\text{cholesterol}]}$$

where all concentrations are those in plasma. Plasma vitamin E levels alone also appear to be strongly correlated with incidence of CVD, particularly in populations with similar blood cholesterol concentrations (Gey 1990), and a recent case-control study (Riemersma et al. 1989) suggests that low plasma concentrations of vitamins E and C and β-carotene are associated with high risk of angina in middle-aged Scottish men.

By implication, if the antioxidant status of an individual is compromised because of inadequate dietary intake of antioxidant nutrients and/or increased oxidant load, such as may be caused by smoking, then free radical-mediated modification of the LDL may occur with increased risk of the onset of coronary heart disease (Fig. 2-1). It is therefore worthwhile discussing these important antioxidant nutrients in greater detail.

Vitamin E

The term "vitamin E" is often used to denote any mixture of biologically active tocopherols (see Chapter 13). Tocopherols are potent inhibitors of lipid peroxidation. The structure of the molecule (see Fig. 13-2) means that it is a highly effective antioxidant that readily donates the hydrogen atom from the hydroxyl group (OH) on the ring structure to free radicals, which then become unreactive. On donating the hydrogen, vitamin E itself becomes a free radical that is however relatively unreactive because the unpaired electron on the oxygen atom is delocalized into the aromatic ring structure, thus increasing its stability (Bisby 1990).

The eight naturally occurring vitamin E compounds are derivatives of 6-chromanol and differ in the number and position of the methyl groups on the ring structure. The four tocopherol homologues (dα-, dβ-, dγ-, dδ-) have a saturated 16-carbon phytol side chain, whereas the tocotrienols (dα-, dβ-, dγ-, dδ-) have three double bonds on the side chain. Although dα-tocopherol has the greatest biological activity (see Chapter 13), some of the tocotrienols may be particularly effective antioxidants *in vitro* because the double bonds on the phytol chain of the tocotrienols allow them to achieve a more uniforn distribution in the phospholipid membranes than the tocopherols. The tocotrienols may also have a membrane disrupting effect that stearically enhances the probability of collisions between the hydroxy group of the chromanol head and lipid radicals in the hydrophobic core of the membrane (Serbinova et al. 1991). The nutritional significance of the tocotrienols has not yet been established.

The major dietary sources of vitamin E (Table 2-1) are vegetable oils and soft margarines prepared from vegetable oils. Comparatively little vitamin E is present in butter, hard margarines, and marine oils. Wholegrain and fortified cereals are also important sources of vitamin E. The importance of foods contributing to vitamin E intake of course depends on the quantities of foods consumed, and in some countries eggs, nuts, fruits, and vegetables substantially contribute to intake. In the context of giving dietary advice, it is important to consider that the richest sources of α-tocopherol, i.e., sunflower and safflower oils, are also rich in PUFAs which may decrease the absorption of α-tocopherol (Gallo-Torres, Weber, and Wiss 1971). PUFAs may also enhance the oxidation of α-tocopherol *in vivo*, thereby inhibiting its availability and limiting any clinical benefit (Drevon 1991). In this respect, oils such as olive and rapeseed, which are rich in vitamin E and low in PUFA but high in monounsaturates, may be functionally more beneficial. Similarly, avocado and peanuts are particularly rich in both vitamin E and monounsaturates. The current U.S. recommended daily allowance for vitamin E is 8–10mg/day, but the epidemiological and biochemical studies described above indicate that protection of high risk groups from CVD could require an intake of 36–100mg/day (Esterbauer et al. 1990; Diplock 1992).

Carotenoids

Carotenoids are primarily symmetrical, C-40, polyisoprenoid structures with an extensive conjugated double bond system. They are synthesized by photosynthetic microorganisms and plants, but not by animals. Of the 600 or so carotenoids that have been characterized, about 50 serve as precursors for vitamin A (Olsen 1992). Although many carotenoids may have antioxidant function, most attention to date has been directed at β-carotene, which is particularly effective at scavenging peroxyl radicals under physiological conditions and is also a potent scavenger of singlet oxygen (Burton 1989). Recent studies indicate a possible synergistic action of β-carotene with vitamin E (Palozza and Krinsky 1992). There are, however, many polyenes and carotenoids in foods that are also found in animal tissues and whose nutritional and biochemical significance is poorly understood. These include phytoene, phytofluene, lycopene, α-carotene, β-cryptoxanthin, zeaxanthin, lutein, canthaxanthin, violaxanthin, neoxanthin, and astaxanthin (Bendich and Olsen 1989).

Table 2-1. Major Food Sources of Vitamin E as Tocopherol Homologues (mg/100 g Food Portion)

Cereal	α	γ	δ	αT3	Oils	α	γ	δ	αT3[a]	Nuts	α	δ	Fruits/ Vegetables	α	γ	δ
Breakfast/ fortified	18–105	—	—	—	Cod liver	20	—	—	—	Almonds	20	3	Avocado	3.5	—	—
Wheatgerm	22	—	—	—	Cotton	38.9	—	—	—	Brazil	6.5	11	Blackberries	3.5	4.7	4.5
Muesli	4	—	—	—	Corn	17	38.7	—	—	Hazelnuts	21	—	Cabbage	7.0	—	—
All-bran	2	1.6	—	—	Olive	12	0.7	—	—	Peanuts	8.1	8.8	Tomato	1.5	0.2	—
Puffed wheat	1.7	2.5	—	—	Palm	25.6	31.6	7.0	14.3	Walnuts	0.8	18	Spinach	2.0	—	—
					Peanut	13.0	21.4	2.1	—							
					Rapeseed	18.4	38.0	1.2	—							
					Safflower	38.7	17.4	26.4	—							
					Soya	10.1	59.3	26.4	—							
					Sunflower	48.7	5.1	0.8	—							
					Wheatgerm	13.3	26.0	27.1	2.6							

[a]T3 denotes tocotrienol.

Table 2-2. Major Food Sources of Carotene

Vegetables	μg/100 g	Fruits	μg/100 g
[a]Broccoli	2,500	Apricots	1,500
Carrots	12,000	Mangoes	1,200
[a]Spinach	6,000	Melon/cantaloupe	2,000
[a]Spring greens	4,000	Peaches	1,000
Yellow sweet potatoes	12,000		
[a]Turnip tops	6,000		

[a]Indicates cooked foods.
Note: fruits analyzed are ripe and not green.

Carotenoids such as β-carotene and lycopene inhibit the oxidation of LDL to its atherogenic form. In addition, plasma carotenoid concentrations are lower in smokers, a high CHD risk group, than in nonsmokers (Chow et al. 1986). The average daily consumption of β-carotene in the United States is about 1.5 mg (Lachance 1988), whereas the diet recommended by the U.S. National Cancer Institute (NCI) and the U.S. Department of Agriculture (USDA) provides about 6 mg of β-carotene/day. Diplock (1992) suggests an intake of β-carotene of 15–25 mg/day for high risk groups, whereas Esterbauer et al. (1990) estimate a requirement of 15 mg of total carotenoids/day.

The proportion of β-carotene in the total carotenoids in food is about 15–30% (Sies, Stahl, and Sundquist 1992). Main sources are carrots, spinach, kale, broccoli, melon, apricots, peaches, and watermelons (Table 2-2 and Seelert 1992). Lycopene, a carotenoid with particularly effective antioxidant activity, is present mainly in tomatoes.

Vitamin C

L-Ascorbic acid (vitamin C), which is synthesized by most animals but not by man, is one of the most important water soluble antioxidants (see Fig. 13-2). It efficiently scavenges O_2^-, $OH^{\cdot}$, peroxyl radicals, and singlet oxygen and may also be involved in the regeneration of vitamin E. By efficiently trapping peroxyl radicals in the aqueous phase of the plasma or cytosol, vitamin C can protect biomembranes and LDL from peroxidative damage.

Table 2-3. Major Food Sources of Vitamin C (mg/100 g)

Fresh Fruit	mg/100 g	Vegetables	mg/100 g
Avocado pears	17	Ackee	30
Banana	10	Asparagus	20
Blackberries	25	*Broccoli	45
Blackcurrants	200	Cabbage	55
Grapefruit	40	*Cauliflower	20
Guavas	180	Okra	25
Lychees	40	Peppers, green	100
Orange	50	*Potatoes, new	18
Pineapple	30	*Spinach	45
Strawberries	65	Tomatoes	20
		Turnip tops	50

*Indicates cooked foods.
Note: fruit analyzed without skin.

The incidence of CVD is inversely related to plasma vitamin C concentrations (Gey et al. 1987). Plasma vitamin C concentrations in smokers are 50% of those in nonsmokers (Chow et al. 1986; Duthie et al. 1993). In addition, the concentrations of ascorbate in atherosclerotic lesions in the human arterial wall are considerably lower than in the areas without lesions (Dubick et al. 1987), although this may be a consequence of increased pro-oxidant activity within the lesion rather than a cause of lesion formation. Ascorbate deficiency also increases plasma concentrations of lipoprotein (a), which is an important atherogenic marker. Conversely, adequate plasma and tissue ascorbate levels may inhibit the development of atherosclerosis by lowering plasma concentrations of lipoprotein (a) (Rath and Pauling 1990).

Major sources of vitamin C (Table 2-3) are fresh fruits and vegetables, particularly citrus fruit, blackcurrants, strawberries, green peppers, and new potatoes. Fruit juices and carbonated drinks are commonly supplemented with ascorbic acid. Pharmacologic doses of ascorbic acid are unlikely to be of additional benefit because tissue levels cannot be increased appreciably by ingesting doses in excess of 180 mg (Rivers 1987). The current recommended daily allowances for vitamin C in the United Kingdom and United States are 30 and 60 mg/day, respectively, although 100–150 mg/day has been suggested as an optimum intake (Esterbauer et al. 1990).

Trace Elements

Certain trace elements are required as cofactors for enzymes with antioxidant-related function; for example, glutathione peroxidase (selenium), superoxide dismutase (copper, zinc, and manganese), and catalase (iron). Inadequate dietary intakes of these trace elements may compromise the effectiveness of antioxidant defense mechanisms. Although dietary deficiency of these elements is rare in the developed world, intakes from food sources are reported to fall to inadequate levels with increasing age, and some studies report associations between low selenium status and CVD mortality (reviewed by Duthie, Wahle, and James 1989). The main food sources for essential trace elements are meat, shellfish, milk, eggs, cereals, and nuts.

"NOVEL" ANTIOXIDANTS IN FOODS

Although there is much interest in whether increased intakes of vitamin E, vitamin C, and the carotenoids reduce susceptibility to CVD, it should not be forgotten that foods contain a variety of other compounds with antioxidant activities that also may be of nutritional significance in the prevention of CVD. Some of these are briefly reviewed below.

Ubiquinone (Coenzyme Q_{10})

Ubiquinone is a lipophilic quinone (Fig. 2-2) similar in structure to vitamin E that functions as an electron carrier in the mitochondrial electron transport chain of the cell. However, it also protects membrane phospholipids and those in LDL from peroxidation (Cabrini et al. 1986). It may also protect and/or regenerate vitamin E (Mohr, Bowry, and Stocker 1992). Ubiquinone is synthesized in the body from precursors of cholesterol synthesis. For this reason it is not classed as a vitamin. However, the ability to synthesize ubiquinone decreases with age (Beyer, Nordenbrand, and Ernster 1987) and there may be an increasing dependence on food to supply the nutrient. The most abundant sources are fresh unprocessed foods, particularly meats, fish, nuts, and seed oils (Kamei et al. 1986). The average daily intake of ubiquinone is approximately 2 mg (Folkers, Vadhanavikit, and Mortensen 1985) but currently there are no data on the

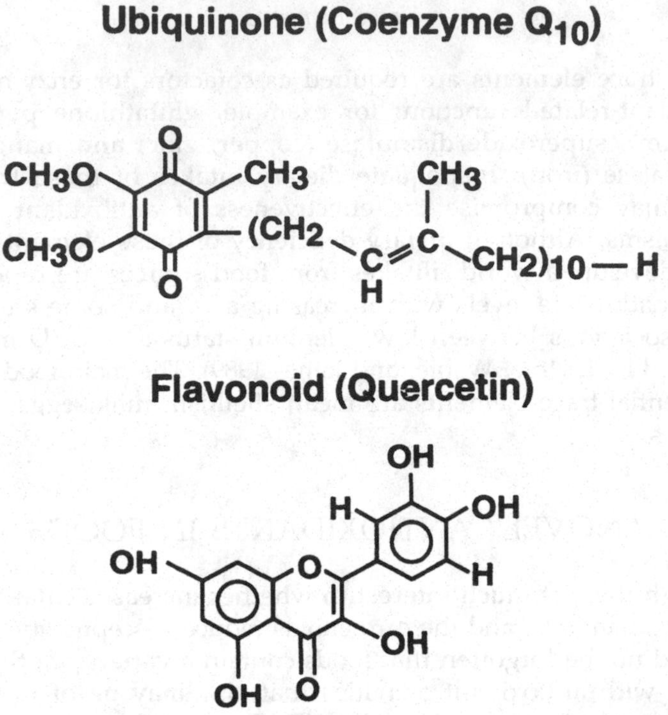

Ubiquinone (Coenzyme Q_{10})

Flavonoid (Quercetin)

FIGURE 2-2. The molecular structures of ubiquinone and of quercetin, which is an example of a flavonoid.

amount required for optimum health. Capsules containing 10 mg of ubiquinone are available from some retail outlets and remarkable claims are being made for its health-promoting effects (Bliznakow and Hunt 1989). For the present such claims should be treated with caution. However, dietary supplementation with ubiquinone increases the resistance of LDL to oxidation by free radicals (Mohr, Bowry, and Stocker 1992).

Flavonoids

Food polyphenols, such as the flavonoids, are widespread dietary components. Diets in the developed countries contain between 50 mg and 1 g/day of these benzo-γ-pyrone derivatives (Fig. 2-2), which are present in significant amounts in foods containing plant tissues such as vegetables, fruit, cereals, tea, coffee, wine, beer, and nuts. For example, a cup of brewed black Indian tea may contain over 40 mg of assorted flavonoids (Stavric and Matula 1992).

Originally regarded as being nutritionally inert, there is now increasing interest in the apparent anticarcinogenic properties of certain flavonoids, although their mechanism of action is unclear (Stavric and Matula 1992). However certain flavonoids, which have a phenolic structure similar to that of vitamin E, act as antioxidants in lipid systems, reacting with O_2^- and lipid peroxyl radicals and forming iron complexes that prevent the formation of active oxygen radicals. They also preserve vitamin C, particularly in the presence of metal ions, which can normally rapidly oxidize ascorbate. For example, quercetin, morin, myricetin, kaempferol, tannic acid, and ellagic acid have pronounced antioxidant activity (Das and Ramanathan 1992).

Flavonoids are undoubtedly important in protecting food from oxidative degeneration. However, little is known about their intestinal absorption, metabolic fate, and whether they can protect LDL from oxidative modification to an atherogenic form. Studies with cell cultures indicate that flavonoids inhibit the oxidative modification of LDL by macrophages and therefore may be natural anti-atherosclerotic components of the diet (De Whalley et al. 1990). Moreover, pharmacological doses of quercetin and rutin may possibly lower the risk of CVD by decreasing serum triglyceride concentrations and by inhibiting platelet aggregation. Other flavonoids have been reported as being effective in the treatment of atherosclerosis, hyperlipidemia, and myocardial infarction (see Stavric and Matula 1992). It has even been suggested that the protective effects of red wine against CVD (St. Leger, Cochrane, and Moore 1979) can be ascribed in part to the presence of flavonoids! Clinical and epidemiological studies have consistently demonstrated that moderate alcohol intake reduces the risk of developing CVD (Moore and Pearson 1986). Consumption of wine and beer has been associated with greater decreases in CVD than consumption of spirits (Friedman and Kimball 1986) possibly because wine and beer contain ellagic and tannic acid. Despite the reported health benefits attributed to alcohol, it must be emphasized that these relate to **moderate** daily intakes of 20–30 g.

CONCLUSIONS

Biochemical and epidemiological evidence suggests that a low antioxidant status arising from inadequate intake of foods containing antioxidant nutrients is a major CVD risk factor. An obvious protective measure is to increase fruit and vegetable intakes as these functional foods contain many of the nutrients that protect LDL from oxidative modification to an atherogenic form. Such an approach has recently been adopted by the German Society for Nutrition (DGE), which recommends a daily consumption of at least 200 g of vegetables, 75 g of salad, and 200–250 g of fruit (Seelert 1992). This is in stark contrast to a recent dietary survey of a Scottish population (Duthie et al. 1993) which showed that the total fruit and vegetable consumption of this high CVD risk population was only 348 g/day compared with the 500 g recommended above.

Another approach to achieving an increased antioxidant intake is to attempt to define the intakes of the antioxidant nutrients required for optimum health rather than to estimate recommended daily allowances, which tend to be based on the amount of the nutrient required to avoid overt deficiency. Such unofficial estimates for optimum intakes for vitamin E, vitamin C, and carotenoids invariably exceed the more widely accepted recommended daily allowances. Nevertheless, even on the basis of currently acceptable criteria, the study of Duthie et al. (1993) revealed that 49% of the population in northeast Scotland had plasma vitamin E concentrations that were indicative of a biochemical deficiency. Moreover, the dietary intake of vitamin E in over 95% of the population was less than the U.S. recommended daily allowance. On the basis of their plasma vitamin C concentrations, 40% of the population could be classed as marginally deficient in vitamin C and 50% of the subjects consumed less than the U.S. recommended daily allowance. Only 3% of the population in the study had a carotenoid intake in excess of the U.S. National Cancer Institute/U.S. Department of Agriculture recommendation. However, before health authorities will be willing to recommend increases in antioxidant intakes, it will be necessary to establish whether habitually low dietary antioxidant intakes in high CVD risk populations are causal or merely adventitious. In the first instance, this requires major intervention trials with vitamin E, vitamin C, and β-carotene. Depending on the results of these studies, dietary guidelines may then be drawn up that emphasize the functional antioxidant properties of foods. It may also transpire that the fortification of foods with antioxidant nutrients may be needed in areas with habitually low antioxidant intakes.

ACKNOWLEDGEMENTS

We thank Dr. Ian Bremner and Dr. John Arthur for their helpful suggestions on the manuscript. The authors are grateful for financial support from the Scottish Office Agriculture and Food Department (SOAFD), BASF, Germany, and the Tobacco Products Research Trust, United Kingdom.

References

Bendich, A., and Olson, J.A. 1989. Biological action of carotenoids. *FASEB Journal* 3:1927–1932.

Beyer, R.E.; Nordenbrand, K.; and Ernster, L. 1987. The function of coenzyme Q in free radical production and as an antioxidant: A review. *Chemica Scripta* 27:10–18.

Bisby, R.H. 1990. Interactions of vitamin E with free radicals and membranes. *Free Radical Research Communications* 8:299–306.

Bliznakow, E.G., and Hunt, G.L. 1989. *The Miracle Nutrient Coenzyme Q_{10}*. New York: Bantam Books.

Burton, G.W. 1989. Antioxidant action of carotenoids. *Journal of Nutrition* 119:109–111.

Cabrini, L.; Pasquali, P.; Tadolini, B.; Sechi, A.M.; and Landi, L. 1986. Antioxidant behaviour of ubiquinone and β-carotene incorporated in model membranes. *Free Radical Research Communications* 2:85–92.

Chow, C.K.; Thacker, R.R.; Changchit, C.; Bridges, R.B. Rehm, S.R.; Humble, J.; and Turbek, J. 1986. Lower levels of vitamin C and carotenes in plasma of cigarette smokers. *Journal of the American College of Nutrition* 5:305–312.

Church, D.F., and Pryor, W.A. 1985. Free-radical chemistry smoke and its toxicological implications. *Environmental Health Perspectives* 64:111–126.

Coronary Prevention Group 1991. *Coronary Heart Disease Statistics*. London: The Coronary Prevention Group/British Heart Foundation.

Das, N.P., and Ramanathan, L. 1992. Studies on flavonoids and related compounds as antioxidants in food. In *Lipid-Soluble Antioxidants*, Augustine S.H. Ong and Lester Packer, eds., pp. 295–306. Basel: Birkhäuser Verlag.

De Whalley, C.V.; Rankin, S.M.; Hoult, J.R.S.; Jessup, W.; and Leake, D.S. 1990. Flavonoids inhibit the oxidative modification of low density lipoproteins by macrophages. *Biochemical Pharmacology* 39:1743–1750.

Diplock, A.T. 1992. What recommendation can be made for an optimum intake of antioxidant vitamins and carotenoids for disease prevention? Paper read at the Vitamins and Health Symposium, 18–19 May 1992, at the German Society for Applied Vitamin Research, Bonn, Germany.

Drevon, C.A. 1991. Absorption, transport and metabolism of vitamin E. *Free Radical Research Communications* 14:229–246.

Dubick, M.A.; Hunter, G.C.; Casey, S.M.; and Keen, C.L. 1987. Aortic ascorbic acid, trace elements and superoxide dismutase activity in human aneurysmal and occlusive disease. *Proceedings of the Society for Experimental Biology and Medicine* 184:138–143.

Duthie, G.G.; Wahle, K.W.J.; and James, W.P.T. 1989. Oxidants, antioxidants and cardiovascular disease. *Nutrition Research Reviews* 2:51–62.

Duthie, G.G.; Arthur, J.R.; Beattie, J.A.G.; Brown, K.M.; Morrice, P.C.; Robertson, J.D.; Shortt, C.T.; Walker, K.A.; and James, W.P.T. 1993. Cigarette smoking, antioxidants, lipid peroxidation and coronary heart disease. *Annals of the New York Academy of Sciences*. In press.

Esterbauer, H.; Gey, F.K.; Fuchs, J.; Clemens, M.R.; and Seis, H. 1990. Antioxidative vitamine und degenerative Erkrankungen. *Deutsch Arzteblatt* 87:B2620–2624.

Esterbauer, H.; Gebicki, J.; Puhl, H.; and Jurgens, G. 1992. The role of lipid peroxidation and antioxidants in the oxidative modification of LDL. *Free Radical Biology and Medicine* 13:341–390.

Folkers, K.; Vadhanavikit, S.; and Mortensen, S.A. 1985. Biochemical rationale and myocardial tissue data on the effective therapy of cardiomyopathy with coenzyme Q_{10}. *Proceedings of the National Academy of Sciences* 82:821–827.

Friedman, L.A., and Kimball, A.W. 1986. Coronary heart disease mortality and alcohol consumption in Framingham. *American Journal of Epidemiology* 24:481–489.

Gallo-Torres, H.E.; Weber, F.; and Wiss, O. 1971. The effect of different dietary lipids on the lymphatic appearance of vitamin E. *International Journal of Vitamin and Nutrition Research* 41:504–515.

Gey, K.F. 1986. On the antioxidant hypothesis with regard to arteriosclerosis. *Bibliotheca Nutritia Dietetica* 37:53–91.

———. 1990. Inverse correlation of vitamin E and ischemic heart disease. *International Journal of Vitamin and Nutrition Research* 30:224–231.

———. 1992. Epidemiological correlations between poor plasma levels of essential antioxidants and the risk of coronary heart disease and cancers. In *Lipid-Soluble Antioxidants*, Augustine S.H. Ong and Lester Packer, eds., pp. 442–456. Basel: Birkhäuser Verlag.

Gey, K.F.; Stanelin, K.B.; Puska, P.; and Evans, A. 1987. Relationship of plasma level of vitamin C to mortality from ischemic heart disease. *Annals of the New York Academy of Sciences* 498:110–123.

Halliwell, B. 1989. Free radicals, reactive oxygen species and human disease: a critical evaluation with special reference to atherosclerosis. *British Journal of Experimental Pathology* 70:737–757.

Halliwell, B., and Gutteridge, J.M.C. 1985. *Free-Radicals in Biology and Medicine*. Oxford: Clarendon Press.

James, W.P.T.; Duthie, G.G.; and Wahle, K.W. 1989. The Mediterranean diet: Protective or merely non-toxic? *European Journal of Clinical Nutrition* 43 (Suppl. 2):31–41.

Kakkar, V.V. 1989. Mechanism, prevention and treatment of thrombotic disease. *Medical Research Council News* 45:8–9.

Kamei, M.; Fujita, T.; Kanbe, T.; Sasaki, K.; Oshibi, K. ™ani, S.; Matsui-Yuasi, I.; and Morisawa, S. 1986. The distribution and content of ubiquinone in foods. *International Journal of Vitamin and Nutrition Research* 56:57–63.

Keys, A. 1980. *Seven Countries: A Multivariate Analysis of Death and Coronary Heart Disease*. Cambridge, MA: Harvard University Press.

Lachance, P. 1988. Dietary intake of carotenes and the carotene gap. *Clinical Nutrition* 7:118–122.

McCormick, J., and Skrabanek, P. 1988. Coronary heart disease is not preventable by population interventions. *Lancet* ii:839–841.

Mohr, D.; Bowry, V.W.; and Stocker, R. 1992. Dietary supplementation with coenzyme Q_{10} results in increased levels of ubiquinol-10 within circulating lipoproteins and increased resistance of human low-density lipoprotein to the initiation of lipid peroxidation. *Biochimica et Biophysica Acta* 1126:247–254.

Moore, R.D., and Pearson, T.A. 1986. Moderate alcohol consumption and coronary heart disease: A review. *Medicine* 65:242–267.

Olsen, J.A. 1992. Carotenoids and vitamin A: An overview. In *Lipid-Soluble Antioxidants*, Augustine S.H. Ong and Lester Packer, eds., pp. 178–192. Basel: Birkhäuser Verlag.

Palozza, P., and Krinsky, N.I. 1992. β-Carotene and α-tocopherol are synergistic antioxidants. *Archives of Biochemistry and Biophysics* 297:184–187.

Quinn, M.T.; Parthasarathy, S.; Fong, L.G.; and Steinberg, D. 1987. Oxidatively modified low density lipoproteins: A potential role in recruitment and retention of monocyte/macrophages during atherogenesis. *Proceedings of the National Academy of Sciences USA* 84:2995–2998.

Rath, M., and Pauling, L. 1990. Lipoprotein (a) is a surrogate for ascorbate. *Proceedings of the National Academy of Sciences USA* 87:6204–6207.

Riemersma, R.A.; Wood, D.A.; MacIntyre, C.C.A.; Elton, R.; Gey, K.F.; and Oliver, M.F. 1989. Low plasma vitamins E and C. Increased risk of angina in Scottish men. *Annals of the New York Academy of Sciences* 570:291–295.

Rivers, J. 1987. Safety of high levels of vitamin C ingestion. *Annals of the New York Academy of Sciences* 498:445–454.

Rosenman, R.H. 1992. Diet in haste; repent at leisure. *The Biochemist* Aug.–Sept.:6–10.

Serbinova, E.; Kagan, V.; Han, D.; and Packer, L. 1991. Free radical recycling and intramembrane mobility in the antioxidant properties of alpha-tocopherol and alpha-tocotrienol. *Free Radical Biology and Medicine* 10:263–275.

Seelert, K. 1992. Antioxidants in the prevention of atherosclerosis and coronary heart disease. *Internist Prax* 32:191–199.

Seis, H.; Stahl, W.; and Sundquist, A.R. 1992. Antioxidant function of vitamins. Vitamins E and C, beta-carotene, and other carotenoids. *Annals of the New York Academy of Sciences* 669:7–20.

Slater, T.F. 1984. Free radical mechanisms in tissue injury. *Biochemical Journal* 222:1–15.

Smith, C.; Mitchison, M.J.; Aruoma, O.I.; and Halliwell, B. 1992. Stimulation of lipid peroxidation and hydroxyl-radical generation by the contents of human atherosclerotic lesions. *Biochemical Journal* 286:901–905.

Stavric, B., and Matula, T.I. 1992. Flavonoids in foods: Their significance for nutrition and health. In *Lipid-Soluble Antioxidants*, Augustine S.H. Ong and Lester Packer, eds., pp. 274–294. Basel: Birkhäuser Verlag.

St. Leger, A.S.; Cochrane, A.L.; and Moore, F. 1979. Factors associated with cardiac mortality in developed countries with particular reference to the consumption of wine. *Lancet* ii:1017–1020.

Wahle, K.W.J. and Duthie, G.G. 1991. Lipoproteins, antioxidants and vascular disease. *European Journal of Clinical Nutrition* 45S:103–109.

Yagi, K. 1987. Lipid peroxides and human diseases. *Chemistry and Physics of Lipids* 45:337–351.

Chapter 3

Reducing the Risk of Cancer

John A. Milner

DIET AND CANCER RISK

Approximately 90% of all cancer cases correlate with environmental factors, including one's dietary habits (Armstrong and Doll 1975; Wynder and Gori 1977; Doll 1992; Potter 1992). Manipulating dietary intakes appears to be one of relatively few realistic approaches to bring about a significant cancer risk reduction. While major limitations exist in defining the precise role of food constituents in the cancer process, their likelihood of significance is emphasized in both *The Surgeon General's Report on Nutrition and Health* (1989) and the *National Academy of Sciences report on Diet and Health* (1989). Although published data suggest that about 60% of cancers in women and more than 40% in men relate to food habits (Doll and Peto 1981), the actual percentage probably depends on a number of factors, including the type of tumor examined and the relative intake of both essential and nonessential nutrients. More recent estimates from cross-cultural and epidemiologic studies suggest that approximately 35% of all cancer deaths may relate to diet (Eddy 1986).

Genetic predisposition continues to be viewed as a significant cancer risk factor. However, it may only account for a small percentage of overall susceptibility (Doll and Peto 1981). Changes in death rates over a relatively short time emphasize the importance of environmental factors, rather than genetic predisposition, as determinants of cancer risk. Nevertheless, genetics may be a confounding factor that partially accounts for the observed variation in worldwide cancer risk. Because genetics determines every *in vivo* reaction, it logically determines the ability of cells to withstand chemical or viral insults. Some of the strongest evidence for an intrinsic linkage of genetics and diet comes from migration studies. These studies often show that when people relocate they rapidly acquire the risk of cancer for the new area of residence (Kolonel et al. 1981a,b). Thus, the availability of some nutrients may prevent or allow the expression and proliferation of aberrant cells, including tumors. Genetic variations make the likelihood that a simple dietary change can equally effect reduction in cancer risk in all populations almost inconceivable. However, while gathering additional information on the impact of genetics on nutrient metabolism and needs, one should allow for future recommendations to reduce cancer risk that are tailored to specific subpopulations.

CARCINOGENIC EXPOSURE AND METABOLIC ACTIVATION

Historically, some cancers have been attributed to occupational exposure to specific chemicals. Humans are inevitably exposed to compounds that are not essential for life or "natural" from the standpoint of evolution. Some of these compounds may be acutely toxic, potentially toxic following activation, or may exhibit long-term effects such as cancer promotion. The contribution of these environmental exposures to the frequency of cancer occurrence worldwide remains extremely controversial (Caddock 1990). Although elimination of environmental carcinogenic and procarcinogenic substances is a laudable goal, it would probably represent an unrealistic public health approach to reducing cancer risk. The complete elimination of these compounds is likely unachievable for a number of reasons, including the natural occurrence of some carcinogenic/mutagenic agents within the food supply. While carcinogens within the food supply may contribute to overall cancer risk, their presence does not appear to account for the majority of the observed relationships existing between diet and cancer (National Academy of Sciences 1982). Even if the presence of carcinogens or procarcinogens is a factor, the intake of specific dietary constituents may modify the way in which these compounds are metabolized and eliminated and how they ultimately influence normal, nonneoplastic cells.

Although the cause of human cancer remains unclear, exposure to N-nitroso compounds, a class of potential carcinogens, may be an important factor. N-nitroso compounds and their precursors have been detected in various parts of the environment, including some consumed foods and beverages (Caddock 1990). Considerable evidence points to the ability of humans to synthesize N-nitroso compounds from precursors occurring in the diet (Schulte 1987). Endogenous N-nitroso compound synthesis arising from the metabolism of ingested precursors is generally recognized as a major factor of human exposure to these potentially carcinogenic compounds. The ability of the human liver to metabolize nitrosamines to compounds that are known to produce DNA damage in animals supports N-nitroso compounds as possible carcinogens for humans (Montesano and Magee 1970, 1974). The *in vivo* formation of N-nitroso compounds is recognized to be accelerated in the presence of bacteria found within the human gastrointestinal and urinary tracts (Tannenbaum 1987). Diet can modify the total number and types of microorganisms that occur in humans. Vitamin C is one of relatively few dietary factors that appears to be effective in modifying microbial-mediated formation of N-nitroso compounds (Mirvish et al. 1972; Schmaehl and Habs 1980; Helser, Hotchkis, and Roe 1991). Sulfur compounds found in garlic and associated plants also appear to be effective inhibitors of nitrosamine formation (Mei et al. 1989).

While some foods may depress the nitrosamine formation (Ranieri and Weisburger 1975), others may actually promote cancers developing from these compounds. Some phenols, like those present in vegetables, or thiocyanate and iodide may promote cancer development by stimulating nitrosation reactions leading to the enhanced formation of carcinogenic nitrosoamines (Mirvish et al. 1975). Thus, it is evident that a change in the dietary habits of an individual must be considered in the context of the complete lifestyle of the individual and his/her genetic background.

Table 3-1 reveals some foods that may modify the cancer process. Foods may inhibit cancer by changing the bioactivation/detoxification of foreign compounds, altering growth regulators such as intracellular cAMP (cyclic AMP) concentrations, and/or serving as antihormones. Compounds such as selenium and vitamin C that occur in a number of foods may inhibit the conversion of procarcinogens to carcinogens (Liu et al. 1991). Compounds found in cruciferous

Table 3-1. Examples of Potential Anticarcinogenic Foods and Possible Mechanisms of Action

Mechanism	Food
I. Drug detoxification	Cucumbers, squash, parsley, carrots, lemon oil, peaches, apples, cranberry, garlic, onions, leeks, strawberries, beets, broccoli, peppers, cauliflower, kale, brussel sprouts
II. Cell proliferation	Grains, fish, carrots, squash, sweet potatoes, cucumbers, soybean, citrus fruits, apples
III. Anti-hormone function	Soybean, fennel, anise, carrots

vegetables, such as benzyl isothiocyanate and indole-3-carbinol, can induce conjugation reactions, and thus would be expected to be anti-initiatory by hastening the elimination of carcinogens. Other compounds found in foods such as vitamin A and possibly β-carotene may inhibit the cancer process by suppressing the proliferation of neoplastic tissue (Boone, Kelloff, and Malone 1990). Anti-estrogenic compounds found in soybeans may be of value in suppressing the growth of hormone-dependent cancers.

One of the encouraging and positive messages about modification of the diet is the recommendation to increase consumption of fruits and vegetables. Increased ingestion of these foods correlates with a reduction in the incidence of cancers (Block, Patterson, and Subar 1992). Although more extensively examined as a factor in gastric cancer risk, geographical correlations suggest their increased intake may depress the risk of lung, gastrointestinal tract, and mammary cancers (Block, Patterson, and Subar 1992; Kromhout 1991; Fraser, Beeson and Philips, 1991; You et al. 1989; Tuyns, Kaaks, and Haelterman 1988). While these data are insufficient to establish that there is a true protective effect, they emphasize the potential benefits of some dietary constituents, as will be discussed below. Vegetables contribute a variety of dietary constituents, e.g., fiber, vitamins, trace elements, and a host of nonessential microconstituents; thus, deciphering which constituent(s) account(s) for the reduction in cancer risk remains to be unraveled. At the present time, it is prudent to encourage increased consumption of fruits and vegetables to at least five servings daily. This simple message requires less of an educational effort than would be needed to encourage increased intake of specific nutrients.

Table 3-2. Incomplete List of Essential Dietary Components Shown to Alter Experimental Carcinogenesis

General	Amino Acids	Lipids	Vitamins	Minerals
Calories	Methionine	Linoleic	Riboflavin	Calcium
Lipids	Cysteine	ω-3	Folic acid	Zinc
Carbohydrates	Tryptophan	Stearic	B_{12}	Copper
Proteins	Arginine		Vitamins A, C	Iron
Fiber			Vitamins D, E	Selenium
Alcohol			Choline	Iodine
			Niacin	
			Pyridoxine	

ESSENTIAL AND NONESSENTIAL NUTRIENTS MAY BE INVOLVED

Specific components of foods may modify cancer development by altering the formation of specific carcinogens, altering the metabolic activation of carcinogens, or possibly by changing the occurrence of cocarcinogens or anticarcinogens. The ability of selective nutrients, such as vitamin C and selenium, to modify the metabolic activation of carcinogens is recognized and is discussed below. Compounds such as specific fatty acids and vitamin A appear to alter the proliferation of neoplastic cells as will be discussed below. While several essential nutrients are recognized as factors capable of modifying the cancer process (Table 3-2), their exact involvement remains rather ill defined. The dearth of information about dietary habits on the progression or metastatic phase of carcinogenesis deserves immediate attention.

Several nonessential nutrients also may shift the metabolism of carcinogens (Table 3-3). The metabolism of benzo(a)pyrene by nonnutrients, such as butylated hydroxytoluene, has been found to lead to a reduction in cancer in experimental animals (Wattenberg 1972; Wilson et al. 1981). Several minor constituents of commonly consumed plant foods, such as flavones, dithiothiones, thioethers, isothiocyanates, phenols, and indoles have also been reported to suppress the metabolic activation of carcinogens and thus should reduce cancer risk (Wilson et al. 1981; Wattenberg 1983, 1985). The inverse association between the risk of cancer of the gastrointestinal tract and the consumption of selected vegetables, particularly the cruciferae, suggests a possible role of some nonessential dietary constit-

Table 3-3. Some Nonessential Dietary Components Shown to Alter Experimental Carcinogenesis

Compound	Source
Benzyl isothiocyanate	Cruciferous vegetables
Diallyl sulfide	Processed garlic and related *Allium* plants
Diallyl disulfide	Processed garlic and related *Allium* plants
D-Carvone	Caraway seed oil
D-Limonene	Citrus fruit oils
Ellagic acid	Grapes and nuts
Glucobrassicin	Cruciferous vegetables
Glucotropaeolin	Cruciferous vegetables
Indole-3-carbinol	Cruciferous vegetables
Oltipraz	Cruciferous vegetables
Phenethyl isothiocyanate	Cruciferous vegetables
Protease inhibitors	Soybeans
Quercetin	Vegetables, fruits

uents in the bioactivation of carcinogens and ultimately the cancer risk of human beings.

Nutritionists have long recognized that quantity consumed does not dictate relative importance in either disease prevention or treatment. Thus, consumption of several microconstituents may be as significant in modifying cancer risk as the intake of macroconstituents such as lipids and carbohydrates. Interestingly, published articles suggest that over 500 compounds present in the food supply may influence cancer risk in animals and presumably human beings (Boone, Kelloff, and Malone 1990).

Animal models continue to be of value for evaluation of the influence of dietary change on cancer risk. While extrapolation of these data to human risk may be complicated by difficulties in equating the influence of dosages, metabolism, physiology, etc., there is overwhelming evidence that human cancers develop at the cellular level with considerable similarity to those observed in experimental animals. These similarities became evident when agents known to cause human cancers were also found to produce tumors in animals (Miller 1978; Krontiris 1983). Similarities are also evidenced by the presence of similar classes of oncogenes in tumors of rodents and human beings (Weinberg 1985).

A brief review of the impact of several major and minor dietary constituents on cancer risk is given below. Although information does exist about some other essential and nonessential nutrients, this synopsis is limited to those where sufficient data was available to extract information about the possible mechanism of action. It must be emphasized that purified constituents are not consumed and that individual nutrients must be considered as part of a complete and complex diet. Interactions among dietary constituents may at least partially explain the observed inconsistent interrelationships between specific dietary constituents and cancer risk. Although interactions among nutrients have been inadequately examined, a few examples of negative and positive interactions exist. Vitamin C has been shown to reduce the effectiveness of selenium against chemically induced colon cancer in the rat (Jacobs and Griffin 1979). Likewise, vitamin A has been shown to enhance the ability of selenium to inhibit chemically induced mammary cancer in experimental animals (Ip and Ip 1981). Greater attention to all components of the diet and their interactions should enable specific and appropriate recommendations to be made for the general population and allow for tailored recommendations for specific subgroups or individuals.

MECHANISM OF ACTION OF SELECTIVE NUTRIENTS AND NONNUTRIENTS

Caloric Intake

Several epidemiological studies report a correlation between total per capita caloric intake and cancer risk. A positive association between increased body weight or body mass index with risk of cancer at several sites has also been reported (Hirayama 1978), although others have not observed a relationship (Adami et al. 1977). In human beings the lowest overall cancer mortality is generally observed in individuals whose body weights range from approximately 10% below to 20% above the average for their age and height. While not a novel approach, it remains wise to eat a variety of foods, but in moderation.

Caloric restriction has long been recognized as a method to inhibit or significantly delay the formation and/or development of chemically or virally induced tumors (Rous 1914; Tannenbaum 1942; Birt 1987; Pollard, Luckert, and Snyder 1989; Birt et al. 1991). Skin, breast, lung, liver, colon, pancreas, muscle, lymphatic system, and endocrine system tumors have been shown to be retarded by caloric restriction. The degree of caloric restriction normally determines the magnitude of the depression in tumorigenesis. Early studies revealed the tumor latency period (appearance) was frequently delayed by rather modest caloric restriction. However, overall tumor numbers were not altered until restrictions became rather severe.

Considerable evidence points to the ability of caloric restriction to alter the promotion, rather than the initiation, phase of carcinogenesis (Birt et al. 1991). Although the mechanism by which caloric restriction inhibits tumorigenesis is unknown, a direct effect is unlikely. Changes in tumor development may result from marked physiological changes that occur as a result of inadequate caloric intake. Alterations in hormonal regulation and/or depressed tissue mitotic activity may account for the ability of caloric restriction to depress tumorigenesis (Klurfeld et al. 1989, 1991; Ruggeri et al. 1990).

The ability of *ad libitum* feeding to increase the incidence of chemically or virally induced tumors has been ascribed to the availability of excess calories. It is clear that all animals, including human beings, do not consume the same quantity of calories when food is provided free choice. Dietary intakes are recognized to modify cytochrome P_{450} enzymes and may account for some of the observed variation in tumor incidence (Yang, Brady, and Hong 1992). Studies by Clinton and colleagues (1984) demonstrated that tumor frequency in *ad libitum* fed rats, previously treated with a mammary carcinogen, was dependent upon total caloric intake, regardless of the actual composition of the diet. Furthermore, animal studies reveal that tumor development does not necessarily correlate with changes in body composition (Ruggeri et al. 1990; Klurfeld et al. 1991). Thus, considerable evidence suggests that subtle changes in energy metabolism may have a significant impact on carcinogenesis.

Dietary Fat

A significant positive association between dietary fat intake and mortality from cancer at several sites has been reported (Carroll 1975; Carroll, Gammal, and Plunkett 1968; Kolonel et al. 1981a,b; Cohen 1992). However, lipid intake does not always correlate and may be site specific, because liver and stomach cancers generally are not correlated with intakes (Carroll and Khor 1975; Welsch 1992; Willett et al. 1992). The site specific action of dietary lipids on the cancer process likely depends, at least in part, on the quantity and types of fatty acids consumed. Laboratory investigations typically reveal that increased lipid intake is associated with an increased tumor frequency, regardless of the form consumed. While caloric restriction appears to inhibit all types of tumors, the inhibitory effect of restricting dietary fat appears to be more selective. Nevertheless, polyunsaturated fats tend to enhance tumor yields to a greater degree than saturated fats (Welsch 1992). Interestingly, epidemiological data frequently reveal a strong positive relationship between cancer susceptibility and saturated fat intake, and little or no relationship with unsaturated fat intake. This discrepancy between epidemiological and laboratory investigations may relate to the minimal quantity of unsaturated fatty acids required to enhance tumorigenesis. Furthermore, total caloric intake may be more important as a risk factor than the source of the calories. Recent findings of protection provided by ω-3 fatty acids question the typical overgeneralization that all unsaturated fatty acids are equivalent (Braden and Carroll 1984; Reddy and Sugie 1988). At least some of the effects of lipids on the cancer process within some tissues may relate to the quantity of linoleic acid consumed (Fisher et al. 1992).

Studies that have examined high fat diets while controlling caloric intake have generally detected an alteration in the promotional effect of lipids on tumorigenesis (Lasekan et al. 1990; Welsch 1992). Although the initiation phase of carcinogenesis has not been examined thoroughly, limited evidence suggests this phase is altered by the quantity and type of lipid consumed (Welsch 1992). Alterations in membrane fluidity, hormonal milieu, immunocompetence, biologically active intermediates, or in the formation of by-products of fatty acid metabolism may account for lipid-induced carcinogenesis (Delicoinstantinos 1987; Lasekan et al. 1990; Welsch 1992). In view of the vast number of studies that have examined the relationship between fat intake and cancer risk, it is unfortunate that more specific dietary recommendations can not be made to the consumer. Reducing the contribution of fat to total caloric intake from the present day 38% to 30% or less appears to be prudent until additional information becomes available. A balanced intake of saturated, monounsaturated, and polyunsaturated fatty acids also appears prudent until more precise information can be gained that assesses the long-term benefits/risks of altering the proportion of dietary fatty acids.

Carbohydrates

The association between dietary fiber and cancer has received widespread attention in recent years. Several approaches have been used to address this relationship. Mixed results about the association between cereal consumption and several types of cancer have been reported, although the majority observed a protective effect of fiber-containing foods (Klurfeld and Kritchevsky 1986; Willett et al. 1992; Klurfeld 1992). Likewise, animal studies are equally inconsistent regarding the impact of dietary fiber on colon cancer risk. At least part of the variation is explained by qualitative and quantitative differences in the fiber sources, animal strains examined, and duration of the experiment (Pilch 1987). The possible action of fiber may be to reduce transit time in the bowel, thereby reducing the time the bowel is in contact with potential carcinogens; altering the intestinal microflora; binding potentially carcinogenic agents; diluting toxic compounds by virtue of its hydrophilic property; and/or changing the biological behavior of intestinal cells, including their response to gut hormones. Alterations in bile acids and short-chain fatty acids are inadequate to explain the influence of fiber on colon tumors (Klurfeld 1992).

Recommendations for the general public are to increase dietary fiber intake to 20 to 30 grams per day with an upper limit of 35 grams. Prudent use of grain products, vegetables, fruits, seeds, and nuts can be effectively used to meet current recommendations for fiber intake.

Protein

The influence of dietary protein and specific amino acids on the cancer process has not been as extensively examined as it has for several other dietary components. Epidemiological studies suggesting that the intake of high protein foods positively correlates with cancer are generally complicated by the simultaneous presence of excess fat. Thus, available data on the influence of protein and amino acids comes primarily from laboratory investigations. Increasing the dietary protein concentration from 9 to 45% had a minimal effect on cancer, as indicated by mammary, skin, or skeletal muscle tumor development (Tannenbaum and Silverstone 1949). However, Silverstone and Tannenbaum (1951) observed that a low protein diet (9%) suppressed the development of spontaneous hepatomas. The impact of dietary protein on chemically induced tumors typically reveals that tumor formation and growth are depressed when protein quality or quantity are limited. Changes in drug metabolizing enzymes are proposed to account for the observed inhibitory effect of inadequate protein or amino acid intake on tumor induction (Clinton, Truex, and Visek 1979; Singletary and Milner 1987; Yang, Brady, and Hong 1992).

The impact of dietary protein likely depends upon its content of individual amino acids. While a deficiency of an essential amino acid typically retards tumor growth, the specificity of this response is questionable. Some amino acids have been shown to modify the incidence of chemically induced tumors or modify the growth of transplantable tumors. Arginine and methionine appear to be two amino acids that when supplemented modify experimental tumorigenesis (Barbul 1986; Hawrylewics, Haung, and Blair 1991; Visek 1992). Although the mechanism by which arginine inhibits and methionine enhances tumor growth remains unknown, it may relate to changes in cellular metabolism such as polyamine biosynthesis, responsiveness to hormones, or changes in immunocompetence during carcinogenesis (Barbul 1986; Hawrylewics, Haung, and Blair 1991).

Vitamin A

Interest in retinol has stemmed from extensive epidemiologic investigations that have observed an inverse relationship between vegetable and fruit intake and the cancer risk. Vitamin A deficiency, recognized for years to increase the susceptibility of experimental animals to some chemically induced tumors, has been shown more recently to enhance the incidence of virally induced tumors (Olson 1986; Tee 1992). Because treatment with some carcinogens reduces liver stores of vitamin A, a marginal status may facilitate cellular transformation following carcinogen exposure. An inverse relationship between hepatic and plasma vitamin A concentrations with vitamin C levels has been reported. A possible explanation for this relationship is that it reflects compensatory mechanisms needed to resist changes in metabolism.

Several retinoids, some without vitamin A activity, are effective in inhibiting chemical carcinogenesis in the skin, mammary gland, esophagus, respiratory tract, pancreas, and urinary bladder of experimental animals (Olson 1986). Retinoids appear to be most effective when provided continuously in the diet (Thompson, Herbst, and Meeker 1986). Although retinoids are generally more effective when administered shortly after the carcinogen treatment, delaying the treatment frequently results in at least partial protection

(McCormick and Moon 1982). However, a point can be reached when retinoids are no longer effective. These data suggest retinoids inhibit the early phase of the promotion stage of carcinogenesis (McCormick and Moon 1982). Several vitamin A compounds and analogs are known to inhibit the *in vitro* microsomal mixed function oxidases that metabolize carcinogenic compounds (Hill and Shih 1974). Nevertheless, the greatest effect appears to involve the promotional phase of tumor growth.

The mechanism(s) by which retinoids inhibit experimental carcinogenesis remain(s) largely unknown. While vitamin A deficiency enhances carcinogenesis and vitamin A excess inhibits this process, there is little evidence that a continuum exists between the deficiency and excess states. Thus, the increased risks of animals to cancer during a deficiency state are likely mechanistically and physiologically different from the prophylactic and therapeutic effects of near toxic dosages of retinoids. Regulation of cellular differentiation is frequently cited as the mechanism of action of vitamin A and the retinoids in reducing cancer risk. The importance of this action is emphasized by the fact that most primary cancers in humans arise from epithelial tissues that depend upon vitamin A for differentiation. However, no single mechanism appears to totally explain the action of the compounds. The theories receiving the greatest attention are that retinoids control gene expression by interacting with cAMP-dependent and -independent protein kinase, and/or that retinoids control gene expression by regulating specific intracellular binding proteins in a manner similar to that of steroid hormones (Roberts and Sporn 1984).

Vitamin A may also influence immunocompetence by altering both humoral and cell-mediated immunity (Dennert 1984). In cell-mediated immunity, retinoids not only influence the total T cell population, but also stimulate cellular lysis and growth inhibition. During the early processing of the antigen by macrophages during humoral immunity, retinoids may directly interact with the B cell or activate the B cell by way of an increased number of T-helper cells (Dennert 1984).

Dietary β-carotene may also be as important in cancer prevention as preformed retinol (Matthews-Roth 1982; Santamaria et al. 1983; Ziegler et al. 1992). A reduced risk of lung cancer is consistently observed with increased dietary intake of vegetables and fruits and carotenoids in prospective and retrospective studies (Ziegler et al. 1992). The impact of carotenoids on other types of cancers has not been as extensively examined and thus remains uncertain. It must be remembered that, although β-carotene is one of the most abundant carotenoids found in biological material, it is one of more than 600 that are known to exist (Ziegler et al. 1992). The mechanism by which one or more of these carotenoids may suppress cancer remains unknown, although there is some speculation that it may be associated with antioxidant properties of the molecule. At present it is unclear whether it is vitamin A activity or the presence of constituent groups similar to that found in retinol, such as unsaturated double bonds or the aromatic characteristic, that is the most important determinant of the efficacy. At least one retinoid without vitamin A activity is capable of reducing UV radiation-induced skin cancer (Matthews-Roth 1982).

Significant progress has been made in the understanding of the role of vitamin A/retinoids in the cancer process (Tee 1992). It is hoped that future research will identify under what circumstances and by which mechanism(s) these anticarcinogenic agents function. At least some retinoids may enhance carcinogenesis (Birt et al. 1983, 1988; Heshmat et al. 1985). Thus, recommendations about vitamin A and retinoid supplements must await additional findings.

Vitamin C

Vitamin C's possible protection against cancer has been claimed for over 50 years. Its implication in human cancer prevention comes primarily from epidemiologic studies showing that consumption of foods containing high concentrations of this vitamin correlates with a reduction in cancer incidence, especially cancer of the stomach and esophagus (Glatthaar, Hornig, and Moser 1986; Block and Menkes 1988).

Experimental data from animals and cell culture systems suggest several mechanisms by which vitamin C might function in cancer prevention. An important mechanism may be inhibiting N-nitrosamine formation (Mirvish et al. 1975; Caddock 1990; Block 1991). Reed et al. (1983) demonstrated that vitamin C treatment reduced N-nitroso compound formation in a selected group of human subjects at risk for gastric cancer. Supplementation of the diet with ascorbic acid and α-tocopherol was found to significantly reduce the mutagenic compounds excreted in human feces, suggesting that antioxidants in the diet may lower the body's exposure to endogenously formed mutagens (Dion, Bright-See, and Smith 1982).

Glatthaar, Hornig, and Moser (1986) reviewed the mechanisms by which vitamin C might alter cancer incidence. Bioactivation of carcinogens by microsomal hydroxylation and demethylation systems has been observed to be decreased during ascorbic acid depletion in animals (Cameron, Pauling, and Leibovitz 1979). Overall the protection against nitroso-compound formation may be of major significance in reducing the cancer risk of humans.

The ability of vitamin C to modify tumor proliferation has been documented through studies examining synergism with various drugs, inhibition of the action of cytotoxic drugs, and interference with tumor cell metabolism (Schmidt and Moser 1985). Ascorbate may react with free copper ions, thereby leading to enhanced peroxide formation. While this metabolic effect may account for the inhibition of some tumor cells, the effectiveness of vitamin C may also depend on cellular catalase and peroxidase activities (Glatthaar, Hornig, and Moser 1986). Under some conditions ascorbic acid supplementation may reverse transformed cells to their normal-appearing morphological phenotypes (Benedict, Wheatley, and Jones 1982). An additional benefit of vitamin C may be its ability to enhance the immune system (Prasad 1980).

Vitamin D and Calcium

Recently, 1,25-dihydroxyvitamin D_3 was shown to inhibit the proliferation of some neoplastic cells including the differentiation of murine and human myeloid leukemia cells *in vitro* (Dokoh, Donaldson, and Haussler 1984). This inhibition appears to relate to membrane receptors and points to a previous unsuspected involvement of vitamin D in cell proliferation and differentiation. The mechanism of action of vitamin D in the induction of differentiation of the neoplastic cells remains largely unknown, although Wood et al. (1983) suggest the effect can not be explained by changes in ornithine decarboxylase activity.

The function of vitamin D in differentiation may relate to its action on intracellular calcium metabolism. Several studies provide indirect evidence of a possible involvement of calcium in carcinogenesis based on the effects of this mineral on the activity and pharmokinetics of carcinogens and promoters, and on the ability of carcinogen or tumor proliferation to induce disturbances in calcium homeostasis (Jaffe 1982). In animals, increased dietary calcium suppresses the hyperproliferation of the colonic epithelium cells caused by bile salts and fatty acids (Bird et al. 1986) and reduces colon-induced tumors (Appleton et al. 1987). The association of calcium-activated oxygen release and the expression of tumor promoters is of considerable interest (Green, Wu, and Wirtz 1983). Whatever the mechanism, calcium seems to have an active role in the promotion, rather than the initiation, of carcinogenesis. The ability of dietary calcium supplements to reduce colonic cell hyperplasia and increase the number of maturing crypt cells in individuals with familial colon cancer suggests that intervention trials are justified (Boone, Kelloff, and Malone 1990; Newmark and Lipkin 1992).

Vitamin E

Vitamin E (α-tocopherol) is recognized as a major radical trap in lipid membranes. Although few studies have adequately evaluated the role of the lipid phase antioxidants in carcinogenesis, their possible impact is supported by recognition that cellular damage produced by active oxygen contributes to the promotional phase of carcinogenesis. Consumption of a diet devoid of vitamin E for an extended period has been shown to alter the mixed function oxidase system that is required for carcinogen activation (Gairola and Chen 1982). Additionally, α-tocopherol has been shown to inhibit endogenous nitrosation reactions, implying that it has an anticarcinogenic effect. Providing vitamin E and ascorbic acid supplements to volunteers consuming a Western diet has been demonstrated to dramatically reduce fecal mutagenicity, again implying an anticarcinogen property of these vitamins (Dion, Bright-See, and Smith 1982).

The anticarcinogenic effects of vitamin E in laboratory investigations have largely been observed only when extremely high and nonphysiological concentrations are provided (Newberne and Suphakarn 1988). The observed effects may relate to reduced lipid peroxidation; however, other studies suggest that vitamin E may also be involved in cellular mitosis and DNA production. The merits of supplementation with both vitamin E and selenium may enhance the inhibition of carcinogenesis (Horvath and Ip 1983; Takada et al. 1992).

B Vitamins

Several B vitamin deficiencies are known to reduce the growth rate of tumor cells and interfere with the normal functioning of the organism. Because these vitamins are essential components of any adequate diet and are necessary for the continued maintenance of cellular integrity and metabolic function, these results are not surprising and probably do not reflect a specific response.

Choline deficiency will produce pathologic lesions in virtually every organ and enhances the potency of several carcinogens (Poirier 1986; Bhave, Wilson, and Poirier 1988). Choline deficiency may directly impact tumor formation because prolonged feeding of a deficient diet increased hepatic tumors, possibly by allowing the expression of existing preneoplastic lesions.

Methylation of DNA is recognized to control genetic expression. Choline deficiency enhances liver cell proliferation, reduces the supply of methyl groups, and causes hypomethylation of DNA. Riggs and Jones (1983) proposed that hypomethylation resulting from choline deficiency may activate oncogenes. The ability of choline deficiency to alter DNA methylation and gene expression has been established most recently by Wainfan and Poirier (1992). Continued research with this nutrient is likely to supply valuable information on the mechanism of carcinogenesis.

The ability of a deficiency of either vitamin B_{12} or folic acid to enhance the carcinogenicity of several chemicals suggests that these vitamins may be anticarcinogenic (Eto and Krumdieck 1986). Likewise, an improvement in cervical dysplasia in oral contraceptive users following folic acid supplementation suggests an involvement of these vitamins in the cancer process (Butterworth et al. 1982). A deficiency of either B_{12} or folic acid may cause hypomethylation, as during choline deficiency, and thus activate oncogenes. Failure of the immune system may also account for the increased tumor development occurring during a deficiency of these vitamins (Nauss and Newberne 1981).

Selenium

The ability of various forms of selenium to inhibit experimentally induced tumors is well documented (Liu et al. 1991; Milner 1985, 1986; Medina, Mukhopadhyay, and Bansal 1992; Thompson, Meeker, and Kokoska 1984; Ip, 1986 a,b; Ip and Thompson 1989). Selenium supplementation at concentrations beyond those necessary to optimize glutathione peroxidase activity is frequently effective in inhibiting the formation of chemically induced tumors in the gastrointestinal tract, liver, breast, skin, and pancreas. Although selenomethionine and several other organic selenium compounds present in foods have not been extensively examined, they appear less effective than selenite (Thompson, Meeker, and Kokoska 1984). The documented anticarcinogenic effects of selenium suggest a generalized mechanism rather than a tissue or cell specific reaction (Milner 1986). Similar to vitamin A, the continuous intake of selenium appears necessary for maximum protection (Ip 1986a,b). Also like the retinoids, during some experimental conditions, selenium supplements may actually enhance tumor development (Birt et al. 1988). However, overwhelming evidence indicates selenium is a naturally occurring anticarcinogenic trace element.

The mechanisms of selenium inhibition of carcinogenesis remain to be determined. It is unclear if one mechanism can explain its ability to alter both the initiation and promotion phases of carcinogenesis. Studies with 7,12-dimethylbenz(a)anthracene suggest part of the protection relates to an inhibition of an enzyme(s) responsible for the formation of biological active intermediates that ultimately bind to DNA (Liu et al. 1991). The intracellular form of selenium resulting in cancer inhibition remains unknown, but may relate to a metabolite formed during detoxification. Selenodiglutathione, a compound formed during selenium detoxification, has been shown to be far more effective in inhibiting carcinogen binding to DNA and tumor cell growth that equivalent quantities of selenium supplied as sodium selenite (Milner 1986; Milner, Dipple, and Pigott 1985).

Several studies reveal the interactive nature of selenium with other dietary constituents. Two of the most promising candidates for chemoprevention, vitamin A and selenium, have been shown to act additively in the inhibition of chemically induced mammary cancer (Thompson, Meeker, and Becci 1981). Likewise, vitamin E provides a more favorable environment by protecting against oxidative stress, thereby potentiating the action of selenium (Ip 1986a,b). Other nutrients may have the opposite influence on selenium, such as mentioned earlier for vitamin C.

Other Agents

1. Antioxidants

Considerable evidence indicates the liberation or generation of activated oxygen species (hydrogen peroxide, superoxide anions, hydroxyl radicals, etc.) within the cell is highly damaging and may directly or indirectly contribute to carcinogenesis (see Chapter 17) (Weisburger 1991; Statland 1992). Several metal ions, including nickel, cadmium, and cobalt, may be carcinogenic in animals under selected conditions (Sunderman 1984). Inflammatory responses caused by these metals may involve the formation of active oxygen species.

As indicated above, several dietary compounds frequently referred to as antioxidants including β-carotene, ascorbic acid, selenium, and α-tocopherol can reduce cancer incidence in experimental animals. Other agents such as synthetic antioxidants, including (butylated hydroxyanisole) BHA and BHT (butylated hydroxytoluene), also have been reported to significantly inhibit tumor formation. Although it is difficult to draw firm conclusions because numerous inconsistencies exist in both laboratory and epidemiological findings, there is a rationale for their potential protective effects as antioxidants. Nevertheless, some investigators suggest these compounds may be inhibitory by altering the metabolic activation of carcinogens unrelated to an antioxidant property.

Sulfides Found in Allium Plants

Recent scientific investigations provide evidence that garlic (*Allium sativum*) may modify cancer risk (Block 1992; Dausch and Nixon 1990; Dorant et al. 1993; Liu, Lin, and Milner 1992a,b). One of more of several sulfur compounds present in processed garlic may account for its anticarcinogenic effects. An extensive overview of the chemistry of organosulfur compounds in both intact and crushed garlic has been summarized (Block 1992).

An inverse relationship between garlic intake and human gastric cancer mortality has been reported by Mei and colleagues (1982). Residents in Cangshan county, Shandong Province, China, who habitually consume garlic (20 g) have a 10 times lower incidence of gastric cancer mortality and a lower nitrite concentration in the gastric juices than those who consume less garlic (2 g) (Mei, Wang, and Han 1985; You et al. 1989). While these are massive intakes, the influence of lower intakes remains to be determined. Likewise, data on the importance of garlic as a modifier of other types of cancers remains to be adequately evaluated.

Interestingly, garlic not only inhibits bacteria and fungi-mediated synthesis of nitrite and nitrosamines, but also directly inhibits the spontaneous synthesis of nitrosamines (Mei et al. 1989). Providing 5 g fresh crushed garlic to human subjects markedly inhibited nitrosation reactions as indicated by suppression of N-nitrosoproline formation resulting from supplemental nitrate and proline (Mei et al. 1989). Thus, at least part of the reduction in human gastric cancers correlated with garlic intake results from a reduced capacity to form nitrosamines. Binding constants between sulphydryl compounds and nitrite are greater than between amines and nitrite (Tu, Byler and Cavanaugh, 1984). A strong relationship between the sulfhydryl content of foods and their ability to bind nitrite to prevent nitrosamine formation suggests that garlic and related foods, such as onions and leeks, may have similar inhibitory effects against nitrosamine formation and induced tumors.

Recent investigations demonstrate that dietary garlic effectively inhibits liver DNA adducts resulting from the *in vivo* formation of nitrosamines and from the treatment of animals with preformed nitrosamines (Liu, Lin, and Milner 1992a,b; Lin, Liu, and Milner 1992). Thus, garlic's ability to inhibit tumors in animals, and possibly human beings, not only results from a reduction in nitrosamine formation, but also from changes in the metabolic activation/detoxification of these carcinogenic compounds.

Several sulfur compounds found in processed garlic are effective inhibitors of other types of chemically induced cancers. Diallyl sulfide, a flavor component of garlic completely inhibited N-nitrosomethylbenzylamine-induced esophageal tumors (Wargovich et al. 1988, 1990), and reduced the incidence of colon cancer in mice treated with dimethylhydrazine (Sumiyoshi and Wargovich 1990). Topical application of garlic oil substantially reduced skin cancers in dimethylbenz(a)anthracene (DMBA)-treated mice (Belman 1983). More recently, supplying processed garlic powder in the diet of rats markedly reduced the incidence of DMBA-induced mammary tumors (Liu, Lin, and Milner 1992a,b).

Some garlic compounds modify enzymatic activities involved in carcinogen activation and deactivation. Treatment with diallyl sulfide decreased the activity of hepatic cytochrome P450IIE1 (Brady et al. 1991). Glutathione (GSH) and glutathione-S-transferases (a group of isozymes involved in phase II detoxification) increased following the feeding of processed garlic powder to rats (Liu, Lin, and Milner 1992a,b; Sparnins, Barany, and Wattenberg 1988). Oral administrations of allyl methyl trisulfide, a constituent of garlic oil, promoted glutathione-S-transferase activity in the forestomach, small bowel mucosa, liver, and lung of mice. Similarly, S-allyl cysteine, a water soluble sulfur compound found in garlic, has been found to increase glutathione and glutathione S-transferase activity in liver and mammary tissue (Liu, Lin, and Milner 1992a,b). The induction of these enzymes and the increase in glutathione are consistent with an increased detoxification of carcinogenic compounds and may account for the ability of processed garlic to reduce experimentally induced cancer of the liver and mammary tissue (Amagase and Milner 1993; Liu, Lin, and Milner 1992a,b).

S-allyl cysteine also has been shown to be an effective inhibitor for *in vivo* formation of DMBA-DNA adducts (Amagase and Milner 1993). Because garlic preparations can vary in the amount of this organosulfur compound, not all garlic forms can be assumed to be equally effective inhibitors of chemically induced tumors in experimental animals, and possibly in humans. More attention needs to be given to the impact of processing on the content of the individual organosulfur compounds present in garlic and related foods. Studies from our laboratory have shown that crushed and dehydrated garlic, commercially processed high sulfur garlic, and commercially processed deodorized garlic are all effective in reducing the binding of DMBA to mammary cell DNA and therefore presumably in reducing the incidence of chemically induced mammary tumors (Amagase and Milner 1993).

Overall, several laboratory investigations demonstrate that the ability of garlic and/or its components to inhibit cancer is not limited to a specific species, to a particular tissue, or to a specific carcinogen. Thus, the intake of sulfur compounds found in processed garlic and related foods may have widespread benefits in reducing cancer risk in humans. Only after more extensive investigations can one predict how beneficial these compounds might be in reducing human cancer risk.

Ellagic Acid and other Phenols

Ellagic acid is a naturally occurring plant phenol that is related to the coumarins, a subclass of lactones found in a variety of fruits and vegetables. It occurs in particularly high concentrations in grapes and nuts. Several studies indicate ellagic acid's potential effectiveness as a chemopreventive agent (Shugar and Kao 1984; Chang et al. 1985). Although the mechanism by which ellagic acid inhibits the cancer process remains unclear, it impacts both the activation and detoxification of potentially carcinogenic agents. Its ability to alter epoxide hydrolase and increase glutathione S-transferase activity (Chang et al. 1985) may explain its ability to inhibit DNA adducts caused by carcinogenic nitrosamines and benz(a)pyrene (Mandal and Stoner 1990).

Flavonoids are not the only phenolic compounds found in fruits and vegetables that may alter the cancer process. Hydroxycinnamic acids also possess potential chemopreventive properties as indicated by Shugar and Kao (1984). However, caffeic acid and related compounds are only minimally effective inhibitors of experimentally induced tumors.

Indoles

A variety of indole derivatives occur naturally in cruciferous vegetables (Loub, Wattenberg, and Davis 1975). Several of these indoles have been examined for their ability to alter experimentally induced carcinogenesis. These studies suggest the effect of feeding rather large quantities of indole (1 to 3%) depends on the carcinogen used and may increase or decrease tumors. However, the tumors generally develop later and are more benign (Oyasu et al. 1963). Studies of indole-3-carbinol, indole-3-acetonitrile, and 3,3′-diindolylmethane are recognized for their ability to induce mixed function oxidases (Loub, Wattenberg, and Davis 1975). Although inconsistent results have been reported, indole-3-carbinol has been generally found to inhibit 7,12-dimethylbenz(a)anthracene-induced mammary tumors, reduce the metabolism of N-nitrosodimethylamine as indicated by a reduction in DNA binding, and enhance pathways for carcinogen removal as indicated by an increase in glutathione S-transferase activity (Shertzer 1983).

Thiocyanates and Isothiocyanates

Cruciferous vegetables are particularly noteworthy for their high content of isothiocyanates. One of these compound, phenethyl isothiocyanate, is present in rather high quantities in cabbage, brussels sprouts, cauliflower, kale, and turnips. Feeding of cabbage and broccoli were effective in inhibiting DMBA-induced mammary tumors in rats (Wattenberg 1992).

Early studies revealed isothiocyanic esters inhibited the growth of Ehrlich ascites carcinoma cells inoculated into mice (Daehnfeldt 1968). Since that time, extensive studies by Wattenberg and colleagues have found that isothiocyanates and related compounds are effective inhibitors of experimental induced tumors of the mammary gland, forestomach, and lungs (Wattenberg 1983). A number of thiocyanates or isothiocyanates also appear to be effective inhibitors of carcinogen metabolism as reflected by a reduction in carcinogenic binding to DNA in the target tissue (Sousa and Marletta 1985).

Flavonoids

The flavonoids are a group of organic molecules ubiquitously distributed in vascular plants. Approximately 2,000 individual members of flavonoids have been described. As typical phenolic compounds, they can act as potent antioxidants and metal chelators. Although two of the most common flavonols, quercetin and kaempferol, do exhibit some mutagenicity in the Ames assay (Bjeldanes and Chang 1977), they also appear to be effective modifiers of cancer risk. Overall, several of these flavonoids appear to be effective antipromoters (Kato et al. 1983). In addition, some appear to be effective in inhibiting the metabolism of some carcinogens, such as DMBA (Verma et al. 1988).

Terpenes

D-limonene is probably the terpene most extensively examined for its ability to inhibit cancer. This compound, found as a natural constituent of citrus oils, has been shown to inhibit skin carcinomas induced by benzo(a)pyrene (Wattenberg 1992). Feeding with D-limonene depressed the incidence mammary tumors in rats treated with DMBA (Benko, Tiboldi and Bardos 1963; Elegbede et al. 1986). Interestingly, this compound increases the regression of chemically induced mammary tumors (Elegbede et al. 1986).

Another monoterpene, D-carvone, is a major constituent of caraway seed oil. Both D-carvone and caraway seed oil inhibited the activation of dimethylnitrosamine and decreased the induction of forestomach tumors (Wattenberg, Sparnins, and Barany 1989).

SUMMARY AND CONCLUSIONS

Epidemiological and laboratory findings provide rather convincing evidence that dietary habits can significantly modify the cancer risk.

Interestingly, a rather large number of chemical compounds found in foods appear to offer protection against the cancer process. These essential and nonessential nutrients are apparently modifying the carcinogenic process at specific sites including: carcinogen formation and metabolism; initiation, promotion, and tumor progression; cellular and host defenses; cellular differentiation; and tumor growth. Unquestionably, dietary habits are not the sole determinant of cancer, but they do represent a significant point for which intervention is possible. Adjustment of dietary practices to conform to generalized dietary goals may not be necessary, or even appropriate, for all segments of the population. Sophisticated techniques and procedures are desperately needed to adequately assess the potential merit of nutritional intervention for each individual in relationship to his/her cancer risk. A thorough appreciation or understanding of how dietary components contribute to or modify the cancer process will require the continuation of carefully controlled and probing investigations. It is hoped that future research will lead to the recognition of the critical sites where nutrition intervention would be appropriate and allow for sound and realistic recommendations for dietary practices. Until this information is available, it remains prudent to eat a variety of foods and to avoid overindulging.

References

Adami, H.O.; Rimsten, A.; Stenkvist, B.; and Vegelius, J. 1977. Influence of height, weight and obesity on risk of breast cancer in an unselected Swedish population. *Brit. J. Cancer* 36:787–792.

Amagase, H., and Milner, J.A. 1993. Impact of various sources of garlic and their constituents on 7,12-dimethylbenz(a)anthracene binding to mammary cell DNA. *Carcinogenesis* (accepted).

Appleton, G.V.N.; Davies, P.W.; Bristol, J.B.; and Williamson, R.C.N. 1987. Inhibition of intestinal carcinogenesis by dietary supplementation with calcium. *Brit. J. Surg.* 74:523–525.

Armstrong, B., and Doll, R. 1975. Environmental factors and cancer incidence and mortality in different countries with special reference to dietary practices. *Int. J. Cancer* 15:617–631.

Barbul, A. 1986. Arginine: biochemistry, physiology, and therapeutic implications. *J. Parenteral Enteral Nutr.* 10:227–238.

Belman, S. 1983. Onion and garlic oils inhibit tumor promotion. *Carcinogenesis* 4:1063–1067.

Benedict, W.F.; Wheatley, W.L.; and Jones, P.A. 1982. Differences in anchorage-dependent growth and tumorigenicities between transformed C3H/10T1/2 cells with morphologies that are or are not reverted to a normal phenotype by ascorbic acid. *Cancer Res.* 42:1041–1045.

Benko, A.; Tiboldi, T.; and Bardos, J. 1963. The effect of painting with cyclical terpenes on the skin of white mice and on the skin carcinoma developed by benzopyrene painting. *Acta Un. Int. Cancer* 19:786–788.

Bhave, M.R.; Wilson, M.J.; and Poirier, L.A. 1988. Cis-II-ras and c-K-ras gene hypomethylation in the livers and hepatomas of rats fed methyl-deficient, amino acid-defined diets. *Carcinogenesis* 9:343–348.

Bird, R.P.; Schneider, R.; Stamp, D.; and Bruce, W.R. 1986. Effect of dietary calcium and cholic acid on the proliferative indices of murine colonic epithelium. *Carcinogenesis* 7:1657–1661.

Birt, D.F. 1987. Fat and calorie effects on carcinogenesis at sites other than the mammary gland. *Am. J. Clin. Nutr.* 45:203–209.

Birt, D.; Davies, M.H.; Pour, P.M.; and Salmasi. 1983. Lack of inhibitions by retinoids of bis(2-oxypropyl)nitrosamine-induced carcinogenesis in Syrian hamsters. *Carcinogenesis* 4:1215–1220.

Birt, D.F.; Julius, A.D.; Runice, C.E.; White, L.T.; Lawson, T.; and Pour, P.M. 1988. Enhancement of BOP-induced pancreatic carcinogenesis in selenium-fed Syrian golden hamsters under specific dietary conditions. *Nutr. Cancer* 11:21–33.

Birt, D.F.; Pelling, J.C.; White, L.T.; Dimitroff, K.; and Barnett, T. 1991. Influence of diet and calorie restriction on the initiation and promotion of skin carcinogenesis in the Sencar mouse model. *Cancer Res.* 51:1851–1854.

Bjeldanes, L.G., and Chang, G.W. 1977. Mutagenic activity of quercetin and related compounds. *Science* 197:577–578.

Block, E. 1992. The organosulfur chemistry of the genus *Allium*—Implications for the organic chemistry of sulfur. *Angew. Chem. Ed. Engl.* 31:1135–1178.

Block, G. 1991. Vitramin C and cancer prevention: The epidemiologic evidence. *Am. J. Clin. Nutr.* 53:270s–282s.

Block, G., and Menkes, M. 1988. Ascorbic acid in cancer prevention. In *Nutrition and Cancer Prevention: the Role of Micronutrients,* T. Moon and M. Micozzi, eds. New York: Marcel Dekker.

Block, G.; Patterson, B.; and Subar, A. 1992. Fruit, vegetables, and cancer prevention: A review of the epidemiological evidence. *Nutr. Cancer* 18:1–29.

Boone, C.W.; Kelloff, G.J.; and Malone, W.E. 1990. Identification of candidate cancer chemopreventive agents and their evaluation in animal models and human clinical trials: A review. *Cancer Res.* 50:2–9.

Braden, L.M., and Carroll, K. 1984. Dietary polyunsaturated fat in relation to mammary carcinogenesis in rats. *Lipids* 21:285–288.

Brady, J.F.; Wang, M.H.; Hong, J.Y.; Xiao, F.; Li, Y.; Yoo, J.H.; Ning, S.M.; Lee, M.J.; Fukuto, J.M.; Gapac, J.M.; and Yang, C.S. 1991. Modulation of rat hepatic microsomal monooxygenase enzymes and cytotoxicity by diallyl sulfide. *Toxic. Appl. Pharmacol.* 108:342–354.

Butterworth, C.E.; Hatch, K.D.; Gore, H.; Mueller, H.; and Krumdieck, C.L. 1982. Improvement in cervical dysplasia associated with folic acid therapy in users of oral contraceptives. *Am. J. Clin. Nutr.* 35:73–82.

Caddock, V.M. 1990. Meeting report. Nitrosamines, food and cancer: Assessment in Lyon. *Food Chem. Toxicol.* 28:63–65.

Cameron, E.; Pauling, E.; and Leibovitz, B. 1979. Ascorbic acid and cancer. A review. *Cancer Res.* 39:663–681.

Carroll, K.K. 1975. Experimental evidence of dietary factors and hormone-dependent cancers. *Cancer Res.* 35:3374–3383.

Carroll, K.K.; Gammal, E.B.; and Plunkett, E.R. 1968. Dietary fat and mammary cancer. *Can. Med. Assoc. J.* 98:590.

Carroll, K.K., and Khor, H.T. 1975. Dietary fat in relation to tumorigenesis. *Prog. Biochem. Pharmacol.* 10:308–353.

Chang, R.L.; Huang, M-T.; Wood, A.W.; Wong, C-Q.; Newmark, H.L.; Yagi, H.; Sayer, J.M.; Jerina, D.M.; and Conney, A.H. 1985. Effect of ellagic acid and hydroxylated flavonoids on the tumorigenicity of benzo(a)pyrene on mouse skin and in the newborn mouse. *Carcinogenesis* 6:1127–1133.

Clinton, S.K.; Imrey, P.B.; Alster, L.M.; Simon, J.; Truex, C. R.; and Visek, W.J. 1984. The combined effects of dietary protein and fat on 7,12-dimethylbenz(a)anthracene-induced breast cancer in rats. *J. Nutr.* 114:1213–1223.

Clinton, S.K.; Truex, C.R.; and Visek, W.J. 1979. Dietary protein, and hydrocarbon hydroxylase and chemical carcinogenesis in rats. *J. Nutr.* 109:55–62.

Cohen, L.A. 1992. Lipids in cancer: An introduction. *Lipids* 27:791–792.

Daehnfeldt, J.L. 1968. Cytostatic activity and metabolic effects of aromatic isothiocyanic acid esters. Fibiger Lab., Kongens Lyngby, Denmark. *Biochem. Pharm.* 17:511–518.

Dausch, J.G., and Nixon, D.W. 1990. Garlic: A review of its relationship to malignant disease. *Prev. Med.* 19:346–361.

Delicoinstantinos, G. 1987. Physiological aspects of membrane lipid fluidity in malignancy. *Anticancer Res.* 7:1011–1021.

Dennert, G. 1984. Retinoids and the immune system: Immunostimulation by vitamin A. In *Retinoids,* Vol. 2, M.B. Sporn, A.B. Roberts, and D.S. Goodman, eds., pp. 373–390. New York: Academic Press.

Dion, P.W.; Bright-See, E.B.; and Smith, C.C. 1982. The effect of dietary ascorbic acid and α-tocopherol on fecal mutagenicity. *Mutat. Res.* 102:27–37.

Dokoh, S.; Donaldson, C.A.; and Haussler, M.R. 1984. Influence of 1,25-dihydroxyvitamin D_3 on cultured osteogenic sarcoma cells: Correlation with the 1,25-dihydroxyvitamin D_3 receptor. *Cancer Res.* 44:2103–2109.

Doll, R. 1992. The lessons of life: Keynote address to the nutrition and cancer conference. *Cancer Res.* 52:2024s–2029s.

Doll, R., and Peto, R. 1981. The causes of cancer: Quantitative estimates of avoidable risks of cancer in the United States today. *J. Natl. Cancer Inst.* 66:1192–1308.

Dorant, E.; van den Brandt, P.A.; Goldbohm, R.A.; Hermus, R.J.J.; and Sturmans, F. 1993. Garlic and its significance for the prevention of cancer in humans: A critical view. *Brit. J. Cancer* 67:424–429.

Eddy, D.M. 1986. Setting priorities for cancer control programs. *J. Natl. Cancer Inst.* 76:187–199.

Elegbede, J.A.; Elson, C.E.; Tanner, M.A.; Qureshi, A.; and Gould, M.N. 1986. Regression of rat primary mammary tumors following d-limonene. *J. Natl. Cancer Inst.* 76:323–325.

Eto, I., and Krumdieck, C.L. 1986. Role of Vitamin B_{12} and folate deficiencies in carcinogenesis. *Adv. Exp. Med. and Biol.* 206:313–330.

Fisher, S.M.; Leyton, J.; Lee, M.L.; Locniskar, M.; Belury, M.A.; Maldve, R.E. Slaga, T.J.; and Bechtel, D.H. 1992. Differential effects of dietary linoleic acid on mouse skin-tumor promotion and mammary carcinogenesis. *Cancer Res.* 52:2049s–2054s.

Food and Nutrition Board, Commission on Life Sciences, National Research Council, *Diet and Health: Implications for Reducing Chronic Disease Risk.* 1989. Committee on Diet and Health. Cancer. pp. 593–613. Washington, D.C.: National Academy Press.

Fraser, G.E.; Beeson, W.L.; and Phillips, R.L. 1991. Diet and lung cancer in California Seventh-Day Adventists. *Am. J. Epidemiol.* 133:683–693.

Gairola, C., and Chen, L.H. 1982. Effect of dietary vitamin E on the aryl hydrocarbon hydroxylase activity of various tissues in rats. *Int. J. Vitam. Nutr. Res.* 52:398–401.

Glatthaar, B.E.; Hornig, D.H.; and Moser, U. 1986. The role of ascorbic acid in carcinogenesis. *Adv. Exp. Med. and Biol.* 206:357–377.

Green, T.R.; Wu, D.E.; and Wirtz, M.K. 1983. The oxygen-generating oxido-reductase of human neutrophils: Evidence of an obligatory requirement for calcium and magnesium for expression of catalytic activity. *Biochem. Biophys. Res. Commun.* 122:734–739.

Hawrylewicz, E.J.; Haung, H.H.; and Blair, W.H. 1991. Dietary soybean isolate and methionine supplementation affect mammary tumor progression in rats. *J. Nutr.* 121:1693–1698.

Helser, M.A.; Hotchkiss, J.H.; and Roe, D.A. 1991. Influence of fruit and vegetable juices on the endogenous formation of N-nitrosoproline and N-nitrosothiazolidine-4-carboxylic acid in humans on controlled diets. *Carcinogenesis* 13:2277–2280.

Heshmat, M.Y.; Kaul, L.; Kovi,J.; Jackson, M.A.; Jackson, A.G.; Jones, G.W.; Edson, M.; Enterline, J.P.; Worrell, R.G.; and Perry, S.L. 1985. Nutrition and prostate cancer: A case-control study. *The Prostate* 6:7–17.

Hill, D.L., and Shih, T.W. 1974. Vitamin A compounds and analogues as inhibitors of mixed-function oxidases that metabolize carcinogenic polycyclic hydrocarbons and other compounds. *Cancer Res.* 34:564–570.

Hirayama, T. 1978. Epidemiology of breast cancer with special reference to the role of diet. *Prev. Med.* 7:173–195.

Horvath, P., and Ip. C. 1983. Synergistic effect of vitamin E and selenium in the chemoprevention of mammary carcinogenesis in rats. *Cancer Res.* 43:5335–5341.

Ip, C. 1986a. The chemopreventive role of selenium in carcinogenesis. *J. Am. Col. Toxicol.* 5:7–20.

Ip, C. 1986b. The chemopreventive role of selenium in carcinogenesis. In *Essential Nutrients in Carcinogenesis*. L.A. Poirier, P.M. Newberne, and M.W. Pariza, eds., p. 431. New York: Plenum Press.

Ip, C., and Ip, M. 1981. Chemoprevention of mammary tumorigenesis by a combined regimen of selenium and vitamin A. *Carcinogenesis* 2:915–918.

Ip, C., and Thompson, H.J. 1989. New approaches to cancer chemoprevention with difluoromethylornithine and selenite. *J. Natl. Cancer Inst.* 81:839–843.

Jacobs, M.M., and Griffin, A.C. 1979. Effects of selenium on chemical carcinogenesis, comparative effects of antioxidants. *Biol. Trace Element Res.* 1:1–13.

Jaffe, L.F. 1982. Eggs are activated by a calcium explosion: Carcinogenesis may involve calcium adaptation and habituation. In *Ions, Cell Proliferation and Cancer.* A.L. Boynton, W.L. McKeehan, and J.F. Whitfield, eds. New York: Academic Press.

Kato, R.; Nakadate, T.; Yamamoto, S.; and Sigimura, T. 1983. Inhibition of 12-O-tetra-decanoylphorbol-13-acetate-induced tumor promotion and ornithine decarboxylase activity by quercetin: Possible involvement of lipoxygenase inhibition. *Carcinogenesis* 4:1301–1305.

Klurfeld, D.M. 1992. Dietary fiber-mediated mechanisms in carcinogenesis. *Cancer Res.* 52:2055s–2059s.

Klurfeld, D.M., and Kritchevsky, D. 1986. Dietary fiber and human cancer: Critique of the literature. *Adv. Exp. Med. and Biol.* 206:119–135.

Klurfeld, D.M.; Welch, C.B.; Davis, M.J.; and Kritchevsky, D. 1989. Determination of the degree of energy restriction necessary to reduce DMBA-induced mammary tumorigenesis in rats during the promotion phase. *J. Nutr.* 119:286–291.

Klurfeld, D.M.; Lloyd, L.M.; Welch, C.B.; Davis, M.J.; Tulp, O.L.; and Kritchevsky, D. 1991. Reduction of enhanced mammary carcinogenesis in LA/N-cp (Corpulent) rats by energy restriction. *PSEBM* 196:381–384.

Kolonel, L.N.; Hankin, J.H.; Nomura, A.M.Y.; Lee, J.; Chu, S.Y.; Normura, A.M.Y.; and Hinds, M.W. 1981a. Nutrient intakes in relation to cancer incidence in Hawaii. *Brit. J. Cancer* 44:332–339.

Kolonel, L.N.; Nomura, A.M.Y.; Hirohata, T.; Hankin, H.H.; and Hinds, M.W. 1981b. Association of diet and place of birth with stomach cancer incidence in Hawaii, Japanese and Caucasiona. *Am. J. Clin. Nutr.* 34:2478–2485.

Kromhout, D. 1991. Essential micronutrients in relation to carcinogenesis. *Am. J. Clin. Nutr.* 45:1361–1367.

Krontiris, T.G. 1983. The emerging genetics of human cancer. *N. Engl. J. Med.* 309:404–409.

Lasekan, J.B.; Clayton, M.K.; Gendron-Fitzpatrick, A.; and Ney, D. M. 1990. Dietary olive and safflower oils in promotion of 7,12-dimethylbenz(a)anthracene-induced mammary tumorigenesis in rats. *Nutr. Cancer* 13: 153–163.

Lin, X.Y.; Liu, J.Z.; and Milner, J.A. 1992. Dietary garlic powder suppresses the *in vivo* formation of DNA adducts induced by N-nitroso compounds in liver and mammary tissue. *FASEB J.* 6:A1392.

Lippman, S.M.; Kessler, J.F.; and Meyskens, F.L., Jr. 1987. Retinoids as preventive and therapeutic anti-cancer agents. Part 1. *Cancer Treat. Rep.* 71:391–405.

Liu, J.; Gilbert, K.; Parker, H.; Haschek, W.; and Milner, J.A. 1991. Inhibition of 7,12-dimethylbenz(a)anthracene induced mammary tumors and DNA adducts by dietary selenite. *Cancer Res.* 51:4613–4617.

Liu, J.Z.; Lin, R.I.; and Milner, J.A. 1992a. Inhibition of 7,12-dimethylbenz(a)-anthracene induced mammary tumors and DNA adducts by garlic powder. *Carcinogenesis* 13:1847–1851.

Liu, J.Z.; Lin, X.Y.; and Milner, J.A. 1992b. Dietary garlic powder increases glutathione content and glutathione S-transferase activity in rat liver and mammary tissues. *FASEB J.* 6:A1493.

Loub, W.D.; Wattenberg, L.W.; and Davis, D.W. 1975. Arylhydrocarbon hydroxylase induction in rat tissues by naturally occurring indoles of cruciferous plants. *J. Natl. Cancer Inst.* 54:985–988.

Mandal, S., and Stoner, G.D. 1990. Inhibition of N-nitrosobenzylmethylamine-induced esophageal tumorigenesis in rats by ellagic acid. *Carcinogenesis* 11:55–61.

Matthews-Roth, M.M. 1982. Antitumor activity of β-carotene, canthaxanthin and phytoene. *Oncology* (Basel) 39:33–37.

McCormick, D.L., and Moon, R.C. 1982. Influence of delayed administration of retinyl acetate on mammary carcinogenesis. *Cancer Res.* 42:2639–2643.

Medina, D.; Mukhopadhyay, R.; and Bansal, M.P. 1992. New approaches to study of selenium chemoprotective properties. In *Breast Cancer: Biological and Clinical Progress,* L. Dogliotti, A. Sapino, and G. Bussolati, eds., pp. 225–232. Boston: Kluwer Academic Publishers.

Mei, X.; Wang, M.L.; and Han, N. 1985. Garlic and gastric cancer II—The inhibitory effect of garlic on the growth of nitrate reducing bacteria and on the production of nitrite. *Acta Nutrimenta Sinica* 7:173–176.

Mei, X.; Wang, M.L.; Xu, H.X.; Pan, X.Y.; Gao, C.Y.; Han, N.; and Fu, M.Y. 1982. Garlic and gastric cancer I—The influence of garlic on the level of nitrate and nitrite in gastric juice. *Acta Nutrimenta Sinica* 4:53–56.

Mei, X.; Lin, X.; Liu, J.; Lin, X.; Song, P.; Hu, J.; and Liang, X. 1989. The blocking effect of garlic on the formation of N-nitrosoproline in humans. *Acta Nutrimenta Sinica* 11:141–145.

Miller, E.C. 1978. Some current perspectives on chemical carcinogenesis in humans and experimental animals: Presidential address. *Cancer Res.* 38:1479.

Milner, J.A. 1985. Effect of selenium on virally induced and transplantable tumor models. *Fed. Proc.* 44:2568–2572.

———. 1986. Inhibition of chemical carcinogenesis and tumorigenesis by selenium. *Adv. Exp. Med. and Biol.* 206:449–464.

Milner, J.A.; Dipple, A.; and Pigott, M.A. 1985. Selective effects of sodium selenite on 7,12-dimethylbenz(a)anthracene-DNA binding in fetal mouse cell cultures. *Cancer Res.* 45:6347–6354.

Mirvish, S.S.; Wallcave, L.; Eagen, M.; and P. Shubik. 1972. Ascorbate-nitrite reaction: Possible means of blocking the formation of carcinogenic N-nitroso compounds. *Science* 177:65–68.

Mirvish, S.S.; Cardesa, A.; Wallcave, L.; and Shubik, P. 1975. Induction of mouse lung adenomas by amines or ureas plus nitrite and by N-nitroso compounds: Effect of ascorbate, gallic acid, thiocyanate, and caffeine. *J. Natl. Cancer Inst.* 55:633–636.

Montesano, R., and Magee, P.N. 1970. Metabolism of dimethylnitrosamine by human liver slices *in vitro. Nature* 228:173–175.

———. 1974. Comparative metabolism *in vitro* of nitrosamines in various animal species including man. In *Chemical Carcinogenesis Essays,* R. Montesano and L. Tomatis eds. *IARC Sci. Publ.* 10:39–56.

National Academy of Sciences. Diet, Nutrition and Cancer 1982. Committee on Diet, Nutrition and Cancer. Washington, D.C.: National Academy Press.

Nauss, K.M., and Newberne, P.M. 1981. Effects of dietary folate, vitamin B_{12} and methionine/choline deficiency on immune function. *Adv. Exp. Med. and Biol.* 135:63–91.

Newberne, P.M., and Suphakarn, V. 1988. Nutrition and cancer: A review, with emphasis on the role of vitamins C and E and selenium. *Nutr. Cancer* 5:107–119.

Newmark, H.L., and Lipkin, M. 1992. Calcium, vitamin D and colon cancer. *Cancer Res.* 52:2067s–2070s.

Olson, J.A. 1986. Some thoughts on the relationship between vitamin A and cancer. *Adv. Exp. Med. and Biol.* 206:379–398.

Oyasu, R.; Miller, D.A.; Mcdonald, J.H.; and Hass, G.M. 1963. Neoplasms of rat urinary bladder tumorigenesis in hamsters by coadministration of 2-acetylaminofluorene and indole. *Cancer Res.* 32:2027–2033.

Pilch, S.M. 1987. *Physiological effects and health consequences of dietary fiber*. Bethesda, MD: Federation of American Societies for Experimental Biology.

Poirier, L.A. 1986. The role of methionine in carcinogenesis *in vivo*. *Adv. Exp. Med. and Biol.* 206:269–282.

Pollard, M.; Luckert, P.H.; and Snyder, D. 1989. Prevention of prostate cancer and liver tumors in L-W rats by moderate dietary restriction. *Cancer* 64:686–690.

Potter, J.D. 1992. Epidemiology of diet and cancer: Evidence of human maladaptation. In *Macronutrients. Investigating Their Role in Cancer*. M.S. Micozzi and T.E. Moon, eds., pp. 55–84. New York: Marcel Dekker.

Prasad, K.N. 1980. Modulation of the effects of tumor therapeutic agents by vitamin C. *Life Sci.* 27:275–280.

Ranieri, R., and Weisburger, J.H. 1975. Reduction of carcinogens with ascorbic acid. *Ann. NY Acad. Sci.* 258:181–189.

Reddy, B.S., and Sugie, S. 1988. Effect of different levels of w-3 and w-6 fatty acids on azonymethane-induced colon carcinogenesis in F344 rats. *Cancer Res.* 48:6642–6647.

Reed, P.I.; Summers, K.; Smith, P.L.R.; Walters, C.L.; Bartholomew, B.A.; Hill, M.J.; Vennit, S.; Hornig, D.; and Bonjour, J.P. 1983. Effect of ascorbic acid treatment on gastic juice nitrite and N-nitroso compound concentrations in achlorhydric subjects. *Gut* 24:492–493.

Riggs, A.D., and Jones, P.A. 1983. 5-Methylcytosine, gene regulation, and cancer. *Adv. Cancer Res.* 40:1–30.

Roberts, A.B., and Sporn, M.B. 1984. Cellular biology and biochemistry of the retinoids. In *Retinoids*. Vol 2, M.B. Sporn, A.B. Roberts, and D.S. Goodman, eds., pp. 2109–2186. New York: Academic Press.

Rous, P. 1914. The influence of diet on transplanted and spontaneous tumors. *J. Exp. Med.* 20:433–451.

Ruggeri, B.A.; Klurfeld, D.M.; Kritchevsky, D.; and Furlanetto, R.W. 1990. Caloric restriction and 7,12-dimethylbenz(a)anthracene-induced mammary tumor growth in rats: Alterations in circulating insulin, insulin-like growth factors I and II, and epidermal growth factor. *Cancer Res.* 49:4130–4134.

Santamaria, L.; Bianchi, A.; Annaboldi, A.; Anderoni, L.; and Bermond, P. 1983. Dietary carotinoids block photocarcinogenic enhancement of benzo(a)pyrene and inhibit its carcinogenesis in the dark. *Experientia* (Basel) 39:1043–1045.

Schmaehl, D., and Habs, M. 1980. Carcinogenicity of N-nitroso compounds in human nutrition. *Oncology* 37:237–242.

Schmidt, K., and Moser, U. 1985. Vitamin C—a modulator of host defense mechanism: An overview. *Int. J. Vitam. Nutr. Res.* 27:363–379.

Schulte, H.R. 1987. Biochemical events in nitrosamine-induced hepatocarcinogenesis: Relevance of animal data to human. *Carcinogenesis IARC Sci. Publ.* 84:17–19.

Shertzer, H.G. 1983. Indole-3-carbinol protects against covalent binding of benzo(a)pyrene metabolites to mouse liver DNA and protein. *Food Chem. Toxicol.* 21:31–35.

Shugar, J., and Kao, J. 1984. Effect of ellagic and caffeic acids on covalent binding of benzo(a)pyrene to epidermal DNA of mouse skin in organ culture. *Int. J. Biochem.* 16:571–573.

Silverstone, H., and Tannenbaum, A. 1951. Proportion of dietary protein and the formation of spontaneous hepatomas in the mouse. *Cancer Res.* 11:442–446.

Singletary, K.W., and Milner, J.A. 1987. Influence of prior dietary protein intake on metabolism, DNA binding and adduct formation of 7,12-dimethylbenz(a)anthracene in isolated rat mammary epithelial cells. *J. Nutr.* 117:587–592.

Sousa, R.L., and Marletta, M.A. 1985. Inhibition of cytochrome P450 activity in rat liver microsomes by the naturally-occurring flavonoid, quercetin. *Arch. Biochem. Biophys.* 240:345–347.

Sparnins, V.L.; Barany, G.; and Wattenberg, L.W. 1988. Effects of organosulfur compounds from garlic and onions on benzo(a)pyrene-induced neoplasia and glutathione S-transferase activity in the mouse. *Carcinogenesis* 9:131–134.

Statland, B.E. 1992. Nutrition and Cancer. *Clin. Chem.* 38:1587–1594.

Sumiyoshi, H., and Wargovich, M.J. 1990. Chemoprevention of 1,2-dimethylhydrazine-induced colon cancer in mice by natural occurring organosulfur compounds. *Cancer Res.* 50:5084–5087.

Sunderman, F.W., Jr. 1984. Recent advances in metal carcinogenesis. *Ann. Clin. Lab. Sci.* 14:93–122.

Takada, H.; Hirooka, T.; Hatano, T.; Hamada, Y.; and Yamamoto, M. 1992. Inhibition of 7,12-dimethylbenz(a)anthracene-induced lipid peroxidation and mammary tumor development in rats by vitamin E in conjunction with selenium. *Nutr. Cancer* 17:115–122.

Tannenbaum, A. 1942. The genesis and growth of tumors. II. Effects of caloric restriction per se. *Cancer Res.* 2:460–467.

Tannenbaum, A., and Silverstone, H. 1949. The genesis and growth of tumors. IV. Effects of varying the proportion of protein (casein) in the diet. *Cancer Res.* 9:162–173.

Tannenbaum, S.R. 1987. Endogenous formation of N-nitroso compounds: a current perspective. In *Relevance of N-nitroso Compounds to Human Cancer: Exposures and Mechanisms*. IARC Sci. Pub. No. 84, H. Bartsch, I.K. O'Neill, and R. Schulte-Hermann, eds., pp. 292–296. Lyon: IARC.

Tee, E-S. 1992. Carotenoids and retinoids in human nutrition. *Crit. Rev. Food Sci. Nutr.* 31:103–163.

Thompson, H.J., Herbst, E.J.; and Meeker, L.D. (1986). Chemoprevention of mammary carcinogenesis: A comparative review of the efficacy of a polyamine antimetabolite, retinoids, and selenium. *J. Natl. Cancer Inst.* 77:595–598.

Thompson, H.J.; Meeker, L.D.; and Becci, P. 1981. Effect of combined selenium and retinyl acetate treatment on mammary carcinogenesis. *Cancer Res.* 41:1413–1416.

Thompson, H.J.; Meeker, L.D.; and Kokoska, S. 1984. Effect of inorganic and organic form of dietary selenium on the promotional stage of mammary carcinogenesis in the rat. *Cancer Res.* 44:2803–2806.

Tu, S-I.; Byler, D.M.; and Cavanaugh, J.R. 1984. Nitrite inhibition of acyl transfer by coenzyme A via the formation of an S-nitrosothiol derivative. *J. Agric. Food Chem.* 32:1057–1060.

Tuyns, A.J.; Kaaks, R.; and Haelterman, M. 1988. Colorectal cancer and the consumption of foods: A case-control study in Belgium. *Nutr. Cancer* 11:189–204.

U.S. Dept. Health and Human Services, Public Health Service. 1989. *The Surgeon General's Report on Nutrition and Health.* DHHS (PHS) Publication No. 88-50210, pp. 177–247. *Cancer.* Washington, D.C.

Verma, A.K.; Johnson, J.A.; Gould, M.N.; and Tanner, M.A. 1988. Inhibition of 7,12-dimethylbenz(a)anthracene- and N-nitrosomethylurea-induced rat mammary cancer by dietary flavonol quercetin. *Cancer Res.* 48:5754–5758.

Visek, W.J. 1992. Nitrogen-stimulated orotic acid synthesis and nucleotide imbalance. *Cancer Res.* 52:2082s–2084s.

Wainfan, E., and Poirier, L.A. 1992. Methyl groups in carcinogenesis: Effects on DNA methylation and gene expression. *Cancer Res.* 52:2071s–2077s.

Wargovich, M.J.; Allnutt, D.; Palmer, C.; Anaya, P.; and Stephens, L.C. 1990. Inhibition of the promotional phase of azoxymethane-induced colon carcinogenesis in the F344 rat by calcium lactate: Effect of stimulating two human nutrient density levels. *Cancer Lett.* 53:17–25.

Wargovich, M.J.; Woods, C.; Eng, V.W.S.; Stephens, L.C.; and Gray, K.N. 1988. Chemoprevention of nitrosomethylbenzylamine-induced esophageal cancer in rats by the thioether, diallyl sulfide. *Cancer Res.* 48:6872–6875.

Wattenberg, L.W. 1972. Inhibition of carcinogenic and toxic effects of polycyclic hydrocarbons by phenolic antioxidants and ethoxyquin. *J. Natl. Cancer Inst.* 48:1425–1430.

———. 1983. Inhibition of neoplasia by minor dietary constituents. *Cancer Res.* 43:2448s–2453s.

———. 1985. Chemoprevention of cancer. *Cancer Res.* 45:1–8.

———. 1992. Inhibition of carcinogenesis by minor dietary constituents. *Cancer Res.* 52:2085s–2091s.

Wattenberg, L.W.; Sparnins, V.L.; and Barany, G. 1989. Inhibition of N-nitrosodiethylamine carcinogenesis by naturally occurring organosulfur compounds and monoterpenes. *Cancer Res.* 49:2689–2692.

Weinberg, R.A. 1985. The action of oncogenes in the cytoplasm and nucleus. *Science* (Washington, D.C.) 230:770–776.

Weisburger, J.H. 1991. Nutritional approach to cancer prevention with emphasis on vitamins, antioxidants and carotenoids. *Am. J. Clin. Nutr.* 53:226s–237s.

Welsch, C.W. 1992. Relationship between dietary fat and experimental mammary tumorigenesis: A review and critique. *Cancer Res.* 52:2040s–2048s.

Willett, W.; Hunter, D.J.; Stampfer, M.J.; Colditz, G.; Manson, J.E.; Spiegelman, D.; Rosner, B.; Hennekens, C.H.; and Speizer, F.E. 1992. Dietary fat and fiber in relation to risk of breast cancer. An 8-year follow-up. *JAMA* 268:2037–2044.

Wilson, A.G.E.; Kung, H.C.; Boroujerdi, M.; and Anderson, M.W. 1981. Inhibition *in vivo* of the formation of adducts between metabolites of benzo(a)pyrene and DNA by aryl hydrocarbon hydroxylase inducers. *Cancer Res.* 41:3453–3460.

Wood, A.W.; Chang, R.L.; Huang, M.-T.; Uskokovic, M.; and Conney, A.H. 1983. 1 α-25-Dihydroxyvitamin D3 inhibits phorbol ester-dependent chemical carcinogenesis in mouse skin. *Biochem. Biophys. Res. Commun.* 116:605–611.

Wynder, E.L., and Gori, G.B. 1977. Contribution of the environment to cancer incidence: An epidemiologic exercise. *J. Natl. Cancer Inst.* 58:825–832.

Yang, C.S.; Brady, J.F.; and Hong, J-Y. 1992. Dietary effects on cytochrome P450, xenobiotic metabolism, and toxicity. *FASEB J.* 6:737–744.

You, W.C.; Blot, W.J.; Chang, Y.S.; Qi, A.; Henderson, B.E.; Fraumeni, J.F.; and Wang, T.G. 1989. *Allium* vegetables and reduced risk of stomach cancer. *J. Natl. Cancer Inst.* 81:162–164.

Ziegler, R.G.; Subar, A.F.; Craft, N.E.; Ursin, G.; Patterson, B.H.; and Graubard, B.I. 1992. Does β-carotene explain why reduced cancer risk is associated with vegetable and fruit intake? *Cancer Res.* 52:2060s–2066s.

Chapter 4

Functional Foods in the Control of Obesity

Mark L. Wahlqvist

THE PROBLEM

Obesity continues to increase in the developed and developing world (World Health Organization Study Group 1990). In Australia, from 1983 to 1989, the prevalence of obesity (body mass index or BMI ≥ 30 kg/m^2) rose from 7.9 to 9.6% in men and 9.9 to 11.4% in women aged 25 to 64 years. For overweight (Body Mass Index-BMI ≥ 30 kg/m^2) prevalence rose from 39.2 to 41.2% for men and 21.2 to 22.6% for women (Risk Factor Prevalence Study Management Committee 1985, 1990).

NONTRADITIONAL FOODS

There has generally been a definable and enduring usage of food in human cultures, so that the foods could be regarded as "traditional," even "characterizing," of a culture. Such foods generally had a recognizable origin in basic food commodities derived from animals (meat, fish, eggs, milk, and dairy products) or plants (fruits, nuts, vegetables, cereals) and, occasionally, other sources, such as insects or fungi. Recipes have also been defined by their basic food commodity content. Sometimes these traditional foods were perceived as having therapeutic as well as health-giving or physiological roles—the grey area between food and medicine (Wahlqvist 1988). But with the increased understanding of nutritional physiology and biochemistry, and of food chemistry, has come the possibility of creating novel foods to attend to physiological needs, and sometimes therapeutic opportunities. We are increasingly having to think about and legislate for foods that are traditional, and nontraditional (or novel), the latter being "functional"—with a physiological purpose—or "medical"—with a therapeutic purpose (Codex Alimentarius Commission 1993; Health and Welfare Canada 1992; Wahlqvist 1992b).

FACTORS OF POTENTIAL IMPORTANCE IN MANAGING BODY FATNESS

Table 4-1 lists categories of food factors that might alter body fatness. They are either physico-chemical or chemical, with mechanisms of action that alter energy intake, energy expenditure, or the deposition of fat in various anatomical sites. Once understood, these factors may be taken advantage of by food technologists (Wahlqvist 1992b). Whether the final food product will ultimately work to affect body fat, however, needs to be established by clinical trials and population-based studies (Rolls and Shide 1992).

ALTERING ENERGY INTAKE BY FOOD

There are a number of mechanisms known to affect food intake (and clearly ones that are not known) that may be manipulated by food factors (Table 4-2). Factors that suppress appetite may operate at cortical or subcortical levels, at the level of the hypothalamus and pituitary (Castonguay and Stern 1990). Examined this way, "functional foods" may be seen to take advantage of hedonic factors (palatability, taste, texture, and odor), learning, and socio-cultural influences. Food memory appears to be located in the amygdala and

Table 4-1. A Classification of Factors in Food of Potential Importance for Energy Balance and Fat Distribution in Humans

Physico-chemical factors
Dietary fiber with effects on particle size and viscosity Extrusion technology to alter particle size and viscosity
Chemical factors

1. Macronutrient
• Carbohydrate
• Dietary fiber
• Protein
• Fat
  • Amount
  • Quality (degree of polyunsaturation, omega-3 or omega-6 fatty acid content)
2. Micronutrient
• Thiamin
• Zinc
3. Nonnutrient
• Caffeine
• Phyto estrogens
• Capsaicin

hippocampus (Morely et al. 1983). The work of Nishiso and Ono (1992) demonstrates that the composite of a number of food characteristics allows discrimination between food and nonfood, and between different types of food, through to the subtleties of different cheeses or wines.

These food memories can be recruited or extinguished by factors included in food, like salt. At the same time, there are specific appetites for food components like sodium (Welsinger et al. 1989) and probably for fat and carbohydrate. Culture can be of overwhelming importance in encouraging or discouraging food usage, but with newer ways of formulating foods, these preferences or prejudices may be transcended. In the case of the specific appetite for fat, the peptide opioids and galanin (Castonguay and Stern 1983) may play a role, and for carbohydrate, amino acids. (Rolls, Kim, and Federoff 1990; Schiffman 1987)

Various anorectic agents, or their analogs, in food like the xanthines (caffeine in tea and coffee and theobromine in cocoa), small peptides (Hansky 1988; Zioudrou, Steaty, and Klee 1979; Brantl et

Table 4-2. Potential Sites and Mechanisms of Action of Food Factors on Energy Balance and, Consequently, Body Fatness

	Endogenous System	Exogenous Analog or Modulator
Appetite	Hypothalamic-pituitary inputs (e.g., special sense; peripheral from abdominal girth and gut distension; galanin from hypothalamus on pituitary) and outputs (e.g., pituitary peptide beta-cell tropin: growth hormone)	Caffeine ? peptides Micronutrients (zinc, thiamin) Fat quality (omega-3, omega-6)
Taste, smell, look texture and sound of eating (crunch, crackle, etc)		Food factors with organoleptic properties (e.g., Maillard products) or conferring physico-chemical properties (e.g., dietary fiber, viscosity-altering, extrusion technology)
Energy utilization	Autonomic nervous system (ANS) Hormonal (catecholamine, glucocorticoids, sex hormones)	Caffeine Thermogenic factors (e.g., capsaicin) Fat quality (omega-3, omega-6)

al. 1981; Chang et al. 1983; Donnelly and McNaughton, 1992; Kanarek, Marks-Kaufman, Lipeles 1980), with opioid activity, are known as exorphins, and are derived from, for example, gluten, β-casein, or those compounds found in coffee with roasting or brewing (Boublik et al. 1983; Wynne et al. 1987; deMan 1980a,b).

In the case of xanthines they presumably act dominantly at the cortical level, but also probably on the reticular activating system to affect appetite. In the case of the small peptides, they may act both centrally and peripherally in the gut. There is growing interest in ways in which specific appetites like those for fat may be influenced (Foltin et al. 1990, 1992). It is generally agreed that fat intake is a significant factor in allowing energy imbalance (Foltin et al. 1990, 1992; Kendall et al. 1991; Rolls and Shide 1992; Rolls et al. 1992), and that reduction in fat intake together with total energy intake can favorably influence body fat (Richman, Steinbeck, and Caterson and 1992; Sheppard, Kristal, Kushi 1991; Wahlqvist 1992a; Wahlqvist, Welborn, and Newgreen 1992; Welborne and Wahlqvist 1989).

However, not all fat may have the same effect on energy balance, although it is generally ascribed the same Atwater factor (9 calories or 37 kilojoules per gram) (Carey and Read 1988). In reality, polyunsaturated fats of the omega (ω)-6 and ω-3 kind, entering pathways in the formation of prostaglandins, and becoming involved in membrane structure and function, including receptor function, are likely to have a disproportionately large effect on energy balance. There is growing evidence that the quality of fat intake may be exploited to favorably affect energy balance (Borkman et al. 1989; Natarajan, Fong, and Angel 1990; Parish, Pathy, and Angel 1990; Parrish et al. 1991; Stewart et al. 1990; Storlien and Bruce 1989). The opportunities to influence the organoleptic properties of food (Table 4-2), taste, smell, look, and texture and, indeed, even the sound of eating (Schiffman 1987), and consequently, energy intake are major. Here ways of cooking are also important; for example, in generations of Maillard products, the amino acid and glucose derived compounds not only allow browning of food, but also flavor and aroma. Of growing currency in food technology is the alteration of the physicochemical properties of food by dietary fiber, modified starches, and by extrusion technology to alter particle size, viscosity, and ultimate shape and presentation (Hansky 1988).

Several studies show that it is easier to reduce energy intake and, therefore improve energy balance, by reducing fat intake than by changing other macronutrients (Kendall et al. 1991; Lissner et al. 1987; Prewitt et al. 1991; Rolls and Shide 1992; Rolls et al. 1992; Shide and Rolls 1993; Storlien and Bruce 1989). However, there is also evidence that a diet accompanied by low fat can also allow reduced energy intake and more favorable energy balance (Hannah, Dubey, and Hansen 1990). It is often difficult to sort out the preferred arrangement amongst the macronutrients, fat, carbohydrate, protein, and dietary fiber. It should also be remembered that dietary fiber that is fermented in the large intestine, principally soluble dietary fiber, itself has an energy value of about 3 calories (13 kilojoules) per gram (Wahlqvist et al. 1981). The reasons that fat seems such a problem, when ingested in larger amounts, for energy balance, is probably predicated on its high energy density (9 calories or 37 kilojoules per gram) and on how quickly after it has been placed in the mouth it can be in the gut without feedback loops operating to suppress appetites. It also delays gastric emptying (Ganong 1989). This is unlike the lower energy density food items containing carbohydrate (4 calories or 16 kilojoules) and dietary fiber, where the energy density is as low as 0.8 to 1 calorie or about 4 kilojoules per gram) of food (e.g., for rice or boiled potatoes). Thus the ability of a given energy intake to cause abdominal distension is much greater for unrefined carbohydrates than for fat, and this may be a key consideration in achieving feedback signals to reduce appetite. There are currently many efforts to produce fat substitutes that will satisfy the organoleptic and textual interest in food, but at a lower energy density. Those now proved by food regulatory authorities are dominantly the microparticulat proteins (Simplesse) and there is ongoing research on sucrose polyesters (Olestra). Evidence points in the direction of a favorable outcome for energy balance with these substitutes, but there is a measure of controversy on this matter (Rolls and Shide 1992; Rolls et al. 1992; Blundell 1991; Drewnowski 1990).

Until recently, it has been assumed that the Atwater factor for fat applied equally to all kinds of fat, whether saturated, monounsaturated, or polyunsaturated in its various forms, and irrespective of chain length (Carey and Read 1988). But there are various ways in which the quality of fat, its chain length and saturation, may alter energy balance, including by way of appetite. In the case of chain length, shorter chain fatty acids go directly up the portal circulation, rather than as chylomicrons, via the thoracic duct to the central venous circulation, and, as a consequence, undergo rapid metabolism in the liver, altering their overall energetics, and involvement in lipid metabolism (Gollaher et al. 1992; Mascioli et al. 1991; Pscheidl et al. 1992; Swenson et al. 1991). ω-3 and ω-6 fatty acids can enter pathways for formation of the metabolic regulatory prostaglandins, contribute to membrane formation (with its receptor and secondary messenger generating properties and affect lipoprotein transport, which is itself an energy-delivering system.

An increase in the polyunsaturated (principally ω-6) saturated fat ratio in serum phospholipids in overweight men has been found to predict a decrease in body mass index (BMI in kilogram per meter^2) over six months (Jones and Schoeller 1988; Stewart et al. 1990), and ω-3 fatty acids are less prone than other fat to increase fatness in rodents. (Moreel et al. 1992; Natarajan, Fong, and Angel 1990; Parrish, Pathy, and Angel 1990; Parrish et al. 1991; Storlien and Bruce 1989) Thus, an alteration in both quality and quantity of food is likely to be of benefit to humans in achieving energy balance.

ALTERING ENERGY EXPENDITURE BY FOOD

One component of 24-hour energy expenditure can be attributed directly to food. The thermic effect of feeding (TEF) measures increased energy expenditure after food ingestion (Wahlqvist and Marks 1991). There has been growing interest in factors that are thermogenic from both a nutritional and pharmocological point of view. There is increasing evidence, for example, that beta (β)-3 receptor agonists are more specifically directed towards increasing thermogenicity than are other β agonists, although so far a suitable agent for pharmocological use in humans has not been forthcoming (Cawthorne 1990).

Particular interest has been shown in capsaicin (Henry and Emery 1986), found in chili peppers, which may increase thermogenicity, principally by way of effects on the vasculature (Cameron-Smith et al. 1990). It may well be a prototype for such agents. Its effects seem to be to increase the thermic response to food. Differences in the thermic responses to food may also be in part, a basis for preference in macronutrient choice, with fat being least and protein most wasteful of energy in this respect.

It may well be as important or more important to increase energy expenditure than to decrease energy intake to achieve energy balance. At higher energy throughputs, it is possible to eat more and to obtain more of essential nutrients and other factors of biological factors from food. Studies that have prospectively examined the predictive power of energy intake for cardiovascular mortality and total mortality favor a higher plane of energy nutrition for optimal health and longevity (Kromhout, Bosschieter, and de Lezenne Coulander 1984, 1985; Kushi et al. 1985; Lapidus and Bengtsson 1986; Morris, Marr, and Clayton 1977).

ALTERING FAT DISTRIBUTION BY FOOD

Not only is the total amount of body fat important for human health—so too is the distribution of body fat. Abdominal obesity, and prob-

Table 4-3. Factors that are Predictive of Body Fatness in Food Intake Models, by Gender[a]

	Food intake	
	Positive	Negative
BMI (body mass index)		
Men	Citrus/apples pears/ bananas, light snacks, tropical fruit, poultry	Mushrooms
Women	Wine, nuts potatoes	Rice
WHR (waist-hip ratio)		
Men	Seaweeds	Citrus/apples/pears/ bananas wine, pastry, biscuits
Women	Processed seafoods, melon, carrots.	

[a]Taking into account the residual effect of WHR in the total fatness (BMI) model and the residual effect of BMI in the abdominal fatness (WHR) model.

ably intra-abdominal at that, seems most adverse for the likelihood of greater total mortality, coronary mortality, stroke, diabetes, gallbladder disease, and breast cancer (Bjorntorp 1985; Vague 1991). Thus, food factors might behave like endogenous sex hormones, to alter the distribution of fat. The most likely candidates are estrogenic compounds which come ultimately from plant foods, but which may be eaten by grazing animals and appear in their milk and carcasses. Phytoestrogens, from soya flour and its products, clover and beanshoots, and various seeds and nuts seem likely candidates for altering human biology at least in estrogen-deficient women, if not men (Aldercreutz 1990; Kaldas and Hughes 1989; Rose 1992; Wilcox et al. 1990).

Preliminary evidence in Chinese populations has pointed to different foods and food patterns within this ethnic group, accounting for distribution of body fatness in ways different to total body fatness (see Table 4-3).

SELECTIVITY IN FUNCTIONAL FOOD USE FOR BODY FATNESS

It may well be that not all subgroups in the population would be equally advantaged by functional food use for body fatness. Factors such as age, gender, genetic predisposition, and physical activity should be considered.

Age

With age often comes a decrease in lean mass and an increase in fat mass. In this situation, the most likely candidate, other than physical activity, to prevent this undesirable change, is growth hormone or the regulation of its secretion. Genetically engineered human growth hormone has now been shown to have a favorable effect on the relationship between fat and lean mass (Arvat et al. 1992; Bodkin et al. 1991; Dubey et al. 1988; Guistina et al. 1992a,b; Rudman et al. 1990; Seamark 1991; Vance 1990), and there is increasing evidence that the hypothalamic hormone galanin, which regulates pituitary function and growth hormone secretion, may do the same (Arvat et al. 1992; Evans and Shine 1991; Guistina 1992b; Ulman et al. 1992). It may well be possible to identify peptides or other analogs of these compounds in foods to advantage.

Gender

Note can be taken of the fact that women tend to have a more favorable body fat distribution, with less central and more peripheral fat than men. In some cultures, like the Oriental, 50% of energy intake may be accompanied by phytoestrogens; thus, there is the potential for men as well as women to be advantaged by these factors in food as far as body fat distribution is concerned. Risk of other related endpoints such as bone density, proneness to breast and prostatic cancer, and cardiovascular disease, let alone the expression of the menopause in women, may also be improved (Wilcox et al. 1990). The common experience of an increase in body fatness, especially body fatness in postmenopausal women, as a feature of estrogen deficiency gives further credence to this view (Risk Factor Prevalence Study Management Committee 1990).

Genetic Predisposition to Obesity or Abdominal Fatness

Although genetic factors are unquestionably strong in predisposing to obesity and abdominal fatness (Bouchard 1993; Depres 1993; Hills and Whalqvist 1993), not all individuals who are genetically predisposed express the problem. Therefore, environmental factors, such as food and what it contains may also be of importance for these people.

Nonfood Environmentally Protective or Enhancing Factors (e.g., Physical Activity)

It is sobering to consider that the need for manipulation of food intake may well be predicated simply on the advent of adverse lifestyles in the other direction, such as sedentariness, and cigarette smoking, the latter of which predisposes to abdominal fatness, while tending to decrease total body weight (Williamson et al. 1991). Whatever the ultimate value of food manipulation, it is likely that these other lifestyle considerations will remain paramount, because their adversities often extend beyond body fatness, its distribution, and its outcomes.

Genetic Predisposition to Obesity Outcomes

Work like that of Fruchart in relation to obesity and cardiovascular disease, and that of Zimmet in relation to obesity and diabetes indicates that there are other factors to consider in how consequential obesity is for individuals. With Fruchart's work, the kind of apoprotein (relatively more apoB-48 of the "hypervariable region" kind) has been shown to increase the risk of premature coronary mortality in the obese (Hagen et al. 1991; Moreel et al. 1992). For transitional island populations, Zimmet has shown that those obese who are more physically active are less likely to express diabetes for all its genetic determination, than those who are not obese (Tuomilehto, Tuomilehto-Wolf, and Zimmit 1992; Zimmet, Dowse, and Sergeanston 1990; Zimmet 1988).

CONCLUSIONS

There are unquestionable opportunities for identifying factors found in traditional foods for the development of novel or functional foods, such that the growing problem of obesity in society may be minimized. Some groups in the population may be more advantaged by this than others. It will remain important not to presume that favorable outcomes will apply on theoretical grounds, but to test products in clinical trials and in the marketplace. Again, it will need to be acknowledged that each time a novel food is created and sold, it will be a significant human experiment and that all due monitoring will need to be in place. Such monitoring will require, for its effectiveness and value, a great deal of innovative planning and design in its own right.

References

Aldercreutz, H. 1990. Western diet and Western diseases: some hormonal and biochemical mechanisms and associations. *Scand. J. Clin. Lab. Invest. Suppl.* 201:3–23.

Arvat, E.; Ghigo, E.; Nicolosi, M.; Boffano, G.M.; Bellone, J.; Yin-Zhang, W.; Mazza, E.; and Camanni, F. 1992. Galanin reinstates the growth hormone response to repeated growth hormone-releasing hormone administration in man. *Clin. Endocrinol.* 36(4):347–350.

Bjorntorp, P. 1985. Obesity and the risk of cardiovascular disease. *Ann. Clin. Res.* 17:3–9.

Blundell, J. 1991. Pharmacological approaches to appetite suppression. *Trends Pharmacol. Sci.* 12:147–57.

Bodkin, N.L.; Sportsman, R.; DiMarchi, R.D.; and Hansen, B.C. 1991. Insulin-like growth factor-I in non-insulin-dependent diabetic monkey: basal plasma concentrations and metabolic effects of exogenously administered biosynthetic hormone. *Metabolism* 40(11):1131–1137.

Borkman, M.; Chisholm, D.J.; Furler, S.M.; Starlien, L.H.; Kraegen, E.W.; Simons, L.A.; and Chesterman, C.N. 1989. Effects of fish oil supplementation on glucose and lipid metabolism in NIDDM. *Diabetes* 38(10):1314–1319.

Boublik, J.H.; Quinn, M.J.; Clements, J.A.; Herrington, A.C.; Wynne, K.N.; and Funder, J.W. 1983. Coffee containing potent opiate receptor binding activity. *Nature* 301:246–248.

Bouchard, C. 1993. Genetic variation in exercise and obesity. In *Fitness and Fatness* A.P. Hills and M.L. Whalqvist, eds. (in press).

Brantl, V.; Teschemacher, H.; Blasig, J.; Henschen, A.; and Lottspeich, F. 1981. Opioid activities of β-casomorphins. *Life Sciences* 28:1903–1909.

Cameron-Smith, D.; Colquhoun, E.Q.; Ye, J.M.; Hettiarachchi, M.; and Clarke, M.G. 1990. Capsaicin and dihydrocapsaicin stimulate oxygen consumption in the perfused rat hindlimb. *Int. J. Obesity* 14:259–270.

Carey, J.B., and Read R.S.D. 1988. Energy. In Wahlqvist ML (editor), *Food and Nutrition in Australia.* 3d edition. pp. 190–214. Melbourne: Thomas Nelson.

Castonguay, T.W., and Stern, J.S. 1983. The effect of adrenalectomy on dietary component selection by the genetically obese Zucker rat. *Nutrition Reports International* 28:725–731.

———. 1990. Hunger and appetite. In *Present Knowledge in Nutrition*. 6th edition, M.L. Brown, ed., pp. 13–22. Washington D.C.: International Life Sciences Institute, Nutrition Foundation.

Cawthorne, M.A. 1990. Thermogenic drugs: The potential of selective β-adrenoceptor agonists as anti-obesity drugs. In *Progress in Obesity Research 1990*. Y. Oomura, S. Tarui, S. Inoue, and T. Shimazu, eds., pp. 567–74. Proceedings of the 6th International Congress of Obesity. London: John Libbey and Co.

Chang, K.J.; Su, Y.F.; Brent, D.A.; and Chang, J.K. 1983. Isolation of a specific μ-opiate receptor peptide, morphiceptin, from an enzymatic digest of milk proteins. *J. Biol. Chem.* 260:9706–712.

Codex Alimentarius Commission. 1993. Report of 18th session of the Codex Committee on nutrition and foods for special dietary uses. Bonn-Bard Godesberg, Germany, 28 September–2 October 1992. Published jointly by Food and Agriculture Organization of the United Nations and World Health Organization. Alinorm 93/26, pp. 4–23.

deMan, J.M. 1980a. Colour. In *Principles of Food Chemistry*. J.M. deMan, ed., pp. 186–226. Westport, CT: AVI.

deMan, J.M. 1980b. Flavor. In *Principles of Food Chemistry*. J.M. deMan, ed., pp. 227–274. Westport, CT: AVI.

Depres, J.P. 1993. Metabolic dysfunction and exercise. In *Fitness and Fatness* A.P. Hills and M.L. Whalqvist, eds. (in press).

Donelly, K., and McNaughton, L. 1992. Effects of two levels of caffeine ingestion on exercise post-oxygen consumption in untrained women. *European Journal of Applied Physiology* 65:459–463.

Drewnowski, A. 1990. The new fat replacements: A strategy for reducing fat consumption. *Post Graduate Medicine* vol. 8:111–14, 117–121.

Dubey, A.K.; Hanukoglu, A.; Hansen, B.C.; and Kowarski, A.A. 1988. Metabolic clearance rates of synthetic human growth hormone in lean and obese male rhesus monkeys. *Journal of clinical endocrinology and metabolism* 67(5):1064–1067.

Evans, H.F., and Shine, J. 1991. Human galanin: Molecular cloning reveals a unique structure. *Endocrinology* 129(3):1682–1684.

Foltin, R.W.; Fischman, M.W.; Moran, T.H.; Rolls, B.J.; and Kelly, T.H. 1990. Caloric compensation for lunches varying in fat and carbohydrate content by humans in a residential laboratory. *American Journal of Clinical Nutrition* 52:969–980.

Foltin, R.W.; Rolls, B.J.; Moran, T.H.; Kelly, T.H.; McNelis, A.L.; and Fischman, M.W. 1992. Caloric, but not macronutrient, compensation by humans for required-eating occasions with meals and snack varying in fat and carbohydrate. *American Journal of Clinical Nutrition* 55:331–342.

Ganong, W.F. 1989. *Review of Medical Physiology*. 14th edition, p. 419. London: Prentice-Hall.

Gollaher, C.J.; Swenson, E.S.; Mascioli, E.A.; Babayan, V.K.; Blackburn, G.L.; and Bistrian, B.R. 1992. Dietary fat level as determinant of protein-sparing actions of structured triglycerides. *Nutrition* 8(5):348–53.

Guistina, A.; Bocini, C.; Doga, M.; Schettino, M.; Pizzocolo, G.; and Giustina, G. 1992a. Galinin decreases circulating growth hormone level in acromegaly. *J. Clin. Endocrinol. Metab.* 74(6):1296–1300.

Guistina, A.; Girelli, A.; Bossoni, S.; Legati, F.; Schettino, M.; and Wehrenberg, W.B. 1992b. Effect of galanin on growth hormone-releasing hormone-stimulated growth hormone secretion in adult patients with non-endocrine diseases on long-term daily glucocorticoid treatment. *Metabolism* 41(5):548–51.

Hagen, E.; Istad, H.; Ose, L.; Christophersen, B.; and Fruchart, J.C. 1991. Comparison of quantification of apo A- and apo B by two different methods. In *Molecular Biology of Atherosclerosis*. Proceedings of the 17th Atherosclerosis Society Meeting, Lisbon, Portugal, 22–25 May 1991.

Hannah, J.S.; Dubey, A.K.; and Hansen, B.C. 1990. Postingestional effects of a high protein diet on the regulation of food intake in monkeys. *American Journal of Clinical Nutrition* 52:320–325.

Hansky, J. 1988. Neuroendocrine factors in food. In *Current Problems in Nutrition, Pharmacology and Toxicology*, A.J. McLean and M.L. Wahlqvist, eds., pp. 77–80. London: John Libbey.

Health and Welfare, Canada. 1992. Novel foods and novel food processes. Information letter (Health Protection Branch). Dept. of Health and Welfare, Canadian Govt. August 5, 1992, IL No. 806, pp. 1–9.

Henry, C.J.K., and Emery, B. 1986. Effect of spiced food on metabolic rate. *Human Nutrition: Clinical Nutrition* 400:165–168.

Hills, A.P., and Wahlqvist, M.L. 1993. What is overfatness? In *Fitness and Fatness*, A.P. Hills and M.L. Wahlqvist, eds. (in press).

Jones, P.J.H., and Schoeller, D.A. 1988. Polyunsaturated: Saturated ratio of diet fat influences energy substrate utilization in the human. *Metabolism* 37:145–151.

Kaldas, R.S., and Hughes, C.L., Jr. 1989. Reproductive and general metabolic effects of phytoestrogens in mammals. *Reprod. Toxicol.* 3(2):81–89.

Kanarek, R.B.; Marks-Kaufman, R.; and Lipeles, B.J. 1980. Increased carbohydrate intake as a function of insulin administration in rats. *Physiol. Behv.* 25:779–782.

Kendall, A.; Levitsky, D.A.; Strupp, B.J.; and Lissner, L. 1991. Weight loss on a low-fat diet: consequences of the imprecision of the control of food intake in humans. *American Journal of Clinical Nutrition* 53:1124–1129.

Kromhout, D.; Bosschieter, E.B.; and de Lezenne Coulander, C. 1984. Dietary fibre and 10-year mortality for coronary heart disease, cancer and all causes. *Lancet* 2:518–521.

———. 1985. The inverse relation between fish consumption and 20 year mortality from coronary heart disease. *New Engl. J. Med.* 312:1205–1209.

Kushi, L.; Lew, R.A.; Stare, F.J.; Ellison, C.R.; Lozy, M.E.; Bourke, G.; Daly, L.; Graham, I.; Hickey, N.; Mulcany, R.; and Kevaney, J. 1985. Diet and 20 year mortality from coronary heart disease. The Ireland-Boston Diet-Heart Study. *New Engl. J. Med.* 312:811–818.

Lapidus, L., and Bengtsson, C. 1986. Socioeconomic factors and physical activity in relation to cardiovascular disease and health. A 12 year follow up of participants in a population study of women in Gothenburg, Sweden. *Br. Heart* 55:295–301.

Lissner, L.; Levitsky, D.A.; Strupp, B.J.; Kalkwarf, H.J.; and Roe, D.A. 1987. Dietary fat and the regulation of energy intake in human subjects. *Am. J. Clin. Nutr.* 46:886–92.

Mascioli, E.A.; Randall, S.; Porter, K.A.; Kater, G.; Lopes, S.; Babayan, V.K.; Blackburn, G.L.; Bistrian, B.R. 1991. Thermogenesis from intravenous medium-chain triglycerides. *J. Parenter. Enteral Nutr.* 15(1):27–31.

Moreel, J.F.; Roizens, G.; Evans, A.E.; Arveiler, D.; Cambou, J.P.; Souriau, C.; Parra, H.J.; Desmarais, E.; Fruchart, J.C.; and Ducimetiete, P. 1992. The polymorphism ApoB/4311 in patients with myocardial infarction and controls: The ECTIM study. *Hum. Genet.* 89(2):169–175.

Morely, J.E.; Levine, A.S.; Yamada, T.; and Gebhard, R.L. 1983. Effect of exorphins on gastrointestinal function, hormonal release and appetite. *Gastroenterology.* 83:1517–1523.

Morris, J.N.; Marr, J.W.; and Clayton, D.G. 1977. Diet and heart: a postscript. *Br. Med. J.* 2:1307–1314.

Natarajan, M.K.; Fong, B.S.; and Angel, A. 1990. Enhanced binding of phospholipase-A2-modified low density lipoprotein by human adipocytes. *Biochem. Cell Biol.* 68(11):1243–1249.

Nishijo, H., and Ono, T. 1992. Food memory: Neural involvement in food recognition. *Asia Pacific Journal of Clinical Nutrition* 1:3–12.

Parrish, C.C.; Pathy, D.A.; and Angel, A. 1990. Dietary fish oils limit adipose tissue hypertrophy in rats. *Metabolism* 39(3):217–219.

Parrish, C.C.; Pathy, D.A.; Parkes, J.G.; and Angel, A. 1991. Dietary fish oils modity adipocyte structure and function. *J. Cell Physiol.* 148(3):493–502.

Prewitt, T.E.; Schmeisser, D.; Bowen, P.E.; Aye, P.; Dolecek, T.A.; Langenberg, P.; Cole, T. Ɛace, L. 1991. Changes in body weight, body composition, and energy intake in women fed high- and low-fat diets. *Am. J. Clin. Nutr.* 54:304–10.

Pscheidl, E.M.; Wan, J.M.; Blackburn, G.L.; Bistrian, B.R.; and Istfan, N.W. 1992. Influence of omega-3 fatty acids on splanchnic blood flow and lactate metabolism in an endotoxemic rat model. *Metabolism* 41(7):698–705.

Richman, R.M.; Steinbeck, K.S.; and Caterson, I.D. 1992. Severe obesity: The use of very low energy diets or standard kilojoule restriction diets. *Medical Journal of Australia* 156:768–770.

Risk Factor Prevalence Study Management Committee, 1985. *Risk Factor Prevalence Survey: Survey No. 2, 1983.* Canberra: National Heart Foundation of Australia and Australian Institute of Health.

———. 1990. *Risk Factor Prevalence Survey: Survey No. 3, 1989.* Canberra: National Heart Foundation of Australia and Australian Institute of Health.

Rolls, B.J.; Kim, S.; and Federoff, I.C. 1990. Effects of drinks sweetened with sucrose or aspartame on hunger, thirst and food intake in men. *Physiol. Behav.* 48(1):19–26.

Rolls, B.J.; Pirraglia, P.A.; Jones, M.B.; and Peters, J.C. 1992. Effects of olestra, a noncaloric fat substitute, on daily energy and fat intakes in lean men. *Am. J. Clin. Nutr.* 56:84–92.

Rolls, B.J.; and Shide, D.J. 1992. The influence of dietary fat on food intake and body weight. *Nutrition Reviews* 50:283–290.

Rose, D.P. 1992. Dietary fiber, phytoestrogens, and breast cancer. *Nutrition* 8(1):47–51.

Rudman, D.; Feller, A.G.; Nagraj, H.S.; Gergans, G.A.; Lalitha, P.Y.; Goldberg, A.F.; Schlenker, R.A.; Cohn, L.; Rudman, I.W.; and Mattson, D.E. 1990. Effects of human growth hormone in men over 60 years old. *N. Engl. J. Med.* 323(1): 52–4.

Schiffman, S.S. 1987. Natural and artificial sweetners. In Food and Health: Issues and Directions. M.L. Wahlqvist, R.W.F. King, J.J. McNeil, R. Swell, eds., pp. 42–48. London: John Libbey.

Seamark, R.F. 1991. The use of transgenic animals to study the role of growth factors in endocrinology. *Baillieres Clin. Endocrinol. Metab.* 5(4):833–845.

Sheppard, L.; Kristal, A.R.; and Kushi, L. 1991. Weight loss in women participating in a randomized trial of low-fat diets. *Am. J. Clin. Nutr.* 54:821–828.

Shide, D.J., and Rolls, B.J. 1993. Dietary compensation for fat reduction and fat substitutes. *Nutrition and the MD.* Vol. 19, No. 3:1–3.

Stewart, A.J.; Wahlqvist, M.L.; Stewart, B.J.; Oliphant, R.C.; and Ireland, P.D. 1990. Fatty acid pattern outcomes of a nutritional programme for overweight and hyperlipidemic Australian men. *Journal of American College of Nutrition* 9(2):107–113.

Storlien, L.H., and Bruce, D.G. 1989. Mind over metabolism: The cephalic phase in relation to non-insulin-dependent diabetes and obesity. *Biol. Psychol.* 28(1):3–23.

Swenson, E.S.; Selleck, K.M.; Babayan, V.K.; Blackburn, G.L.; and Bistrian, B.R. 1991. Persistence of metabolic effects after long-term oral feeding of a structured triglyceride derived from medium-chain triglyceride and fish oil in burned and normal rats. *Metabolism* 40(5):484–90.

Tuomilehto, J.; Tuomilehto-Wolf, E.; and Zimmit, P. 1992. Primary prevention of diabetes mellitus. In *International Textbook of Diabetes Mellitus*, K.G.M.M. Albert, R.A. DeFronzo, H. Keen, and P. Zimmet, eds., pp. 1655–73. Chicester: J. Wiley and Sons.

Ulman, L.G.; Evans, H.F.; Lismaa, T.P.; Potter, E.K.; McCloskey, D.I.; and Shine, J. 1992. Effects of human, rat and porcine galanins on cardiac vagal action and blood pressure in the anaesthetised cat. *Neurosci. Lett.* 136(1):105–108.

Vague, J. 1991. *Obesities.* London: John Libbey and Co.

Vance, M.L. 1990. Growth hormone for the elderly? (editorial) *N. Engl. J. Med.* 323(1):52–54.

Wahlqvist, M.L. 1988. Food as therapy. In *Current Problems in Nutrition, Pharmacology and Toxicolgy*, A. McLean and M.L. Wahlqvist, eds. London: John Libbey.

Wahlqvist, M.L. 1992a. Options in obesity management. *Asia Pacific Journal of Clinical Nutrition* 1:183–190.

Wahlqvist, M.L. 1992b. Non-nutrients in foods: Implications for food industry. *Food Australia* 44(12):558–560.

Wahlqvist, M.L.; Jones, G.P.; Hansky, J.; Duncan, S; Coles-Rutishauser, I.H.E.; and Littlejohn, G.O. 1981. The role of dietary fibre in human health. *Food Technology in Australia* 33(2):50–54.

Wahlqvist, M.L., and Marks, S. 1991. Diet and Obesity. In *Sugars in Nutrition*, M. Gracey, N. Kretchmer, and E. Rossi, eds. Nestle Nutrition Workshop Series Vol. 25, New York: Raven Press.

Wahlqvist, M.L.; Welborn, T.A.; and Newgreen, D.B. 1992. Very low energy diets. *Medical Journal of Australia* 156(11):752–753.

Weisinger, R.S.; Blair-West, J.R.; Denton, D.A.; Dinicolantonio, R.; McKinley, M.K.; Osborne, P.G.; and Tarjan, E. 1989. The role of brain ECF sodium and angiotensin II in sodium appetite. *Acta Physiol. Pol.* 40(3):293–300.

Welborn, T.A., and Wahlqvist, M.L. 1989. Scientific conference on "very-low-calorie diets." *Medical Journal of Australia* 151:457–458.

Williamson, D.F.; Modans, J.; Andres, R.F.; Kleinman, J.C.; Giovino, G.A.; and Byers, T. 1991. Smoking Cessation and Severity of Weight Gain in a National Cohort. *N. Engl. J. Med.* 324(ii):739–745.

Wilcox, G.; Wahlqvist, M.L.; Burger, H.G.; and Medley, G. 1990. Oestrogenic effects of plant-derived foods in postmenopausal women. *Br. Med. J.* 301:905–906.

World Health Organization Study Group. 1990. *Diet, Nutrition, and the Prevention of Chronic Diseases*. pp. 30–39. WHO Technical Report Series 797. Geneva: WHO.

Wynne, K.N.; Familari, M.; Boubiik, J.H.; Drummer, O.H.; Rae, I.D.; and Funder, J.W. 1987. Isolation of opiate receptor ligands in coffee. *Clin. Exper. Pharmacol. Physiol.* 14:785–790.

Zimmet, P. 1988. Primary prevention of diabetes mellitus. *Diabetes Care* 11:258–62.

Zimmet, P.; Dowse, G.; and Sergeanston, S. 1990. The epidemiology and natural history of NIDDM—lessons from the South Pacific. *Diabetes Metab. Rev.* 6:91–124.

Zioudrou, C.; Streaty, R.A.; and Klee, W.A. 1979. Opioid peptides derived from food protein. The exorphins. *J. Biol. Chem.* 254:2446–2449.

Chapter 5

Nutrient Control of Immune Function

Jeffrey B. Blumberg

INTRODUCTION

Nutrition scientists are approaching a goal of being able to recommend diets and provide functional foods that can promote optimal immunity in healthy individuals and maintain normal immune defenses in compromised patients. Although immunologic tests are increasingly recognized as a useful complement to traditional methods of nutritional assessment, correcting malnutrition and enhancing even apparently adequate nutritional status may prove to be effective in reducing the rate and severity of infectious diseases, increasing the responsiveness to immunization, and protecting against several chronic diseases associated with impaired immunity. These interventions involve provision of the nutrient substrates required by the immune system under stimulatory conditions in response to infectious disease or other traumas as well as under quiescent states.

Scrimshaw, Taylor, and Gordon (1959) were the first to review the early scientific work in this field and present the concept of a synergistic interaction between nutrition and the immune response to infectious disease. This interaction is now appreciated to be bidirectional, i.e.; nutritional status influences host immune responsiveness, and infectious disease has a detrimental influence on nutritional status. Interestingly, conditions associated with chronic overnutrition, e.g., obesity, cardiovascular disease, and adult-onset diabetes, significantly modulate immune function; the diets employed to treat these afflictions appear to have a far-ranging influence on host defenses.

Our growing understanding about the mechanisms by which nutritional factors affect immune function and the ontogeny of the immune response is critical to the development of functional foods targeted to this system. The general nature of the influence of diet on immunity has been shown to be both qualitative and quantitative varying with the individual nutrients. Diet may act specifically upon the lymphoid system and immune cell function or nonspecifically upon associated factors. Nutrient availability can also impact metabolic, neurologic, or endocrine functions that influence immune function. Further, nutrients are involved in the stability of the plasma membrane and the differentiation and expression of its cell surface characteristics such as antigenic determinants. Thus, nutritional factors can affect the development and maintenance of immunocompetence through many pathways; for example, nutritional factors act at multiple sites within the immune system by modulating metabolic processes. These metabolic changes may include the activation or inhibition of key enzymes, immunoregulatory mediators, and/or products of the major histocompatibility gene complex. These nutritionally induced changes can result in altered cellular immune functions, particularly in cells of T lymphocyte lineage.

Experiments with nutritional intervention in animal and cell culture models are essential to our understanding of the roles and mechanisms of individual nutrients. However, these studies may or may not be directly applicable or relevant to clinical or public health conditions. Thus, this review emphasizes clinical studies in which beneficial effects of selected nutrient interventions have been reported.

VITAMINS

Water Soluble Vitamins

B Complex Vitamins

Specific and nonspecific immune responses can be severely compromised when the B complex vitamins are deficient as a result of inadequate dietary intake or disease processes. The involvement of the B complex vitamins in immune function is not unexpected as they are essential to many aspects of cellular metabolism including sugar, protein, lipid, and nucleic acid synthesis and degradation. Although the relative importance of each of the B vitamins is not fully characterized, some human studies have been completed that suggest certain of the B vitamins have a significant influence on immunity. It is, of course, important to appreciate both the value and the confounding nature of such nutritional studies, which employ individuals with an inherited disorder, who are severely ill, or are in an advanced state of malnutrition, in determining the requirement of nutrients for optimal immunocompetence.

Humoral and cell-mediated immunity are affected by vitamin B_6 (pyridoxine) deficiency and supradietary intakes. Hodges et al. (1962) and others reported that antibody production is decreased and lymphocytopenia is produced in adults depleted of vitamin B_6. Talbott, Miller, and Kerkvleit (1987) examined the effect of a supradietary 50 mg/day vitamin B_6 supplement for two months in healthy older adults and noted increases in mitogen-stimulated lymphocyte proliferation and helper T cells. Meydani et al. (1991b), employing a vitamin B_6 depletion-repletion protocol with healthy older adults, showed that interleukin (IL)-2 production and responses to T and B cell mitogens are adversely affected by low vitamin B_6 status and that short-term supplementation with 50 mg/day of pyridoxal phosphate increased these indices of immune function above baseline values in some but not all subjects. Casciato et al. (1984) have reported that the reduced immunocompetence of renal dialysis patients, evidenced by low levels of lymphocytes with E rosette or aggregated IgG markers and decreased mitogen-stimulated lymphocyte proliferation, is reversed by 210–600 mg/week of pyridoxine.

The administration of vitamin B_{12} (cyanocobalamin) reverses the immunologic deficits in patients with pernicious anemia, perhaps via correction of a biochemical lesion depressing nucleic acid synthesis in lymphocytes (Beisel 1982). Skacel and Chanarin (1983) observed a reduction in bactericidal activity and measures of oxidative burst in patients with megaloblastic anemia and low serum vitamin B_{12}. Crist et al. (1980) noted neutropenia and/or leukopenia and related white blood cell abnormalities in children with low vitamin B_{12} levels and corrected them with cyanocobalamin supplementation of 1,000 μg/month administered intramuscularly. Seger et al. (1980) temporarily corrected the abnormal neutrophil profile in a pediatric patient with congenital transcobalamin II deficiency by the transfusion of 3,500 μg cobalamin over several days. Although Katka (1984) could not detect any difference in neutrophil function between normal subjects and patients with pernicious anemia, vitamin B_{12} therapy was associated with improvements in tuberculin skin test responses and lymphocyte proliferation to phytohaemagglutinin (PHA).

Several *ex vivo* studies indicate that individuals with low folate levels have impairments in neutrophil function that can be corrected by improved nutrient status. Youinou et al. (1982) found that the dysfunctions in neutrophil phagocytosis associated with low folic acid levels in patients were corrected with 10 mg/day folic acid although deficits in bactericidal activity remained unchanged; the *in vitro* addition of folic acid to serum from these subjects also increased phagocytic activity. The observation that mitogen stimulation of lymphocytes increases the intracellular concentration of a folate-binding protein (which may modulate nucleic acid synthesis and free folate concentration) is consistent with this evidence that folic acid deficiency adversely affects certain immune responses (DaCosta and Sharon 1980).

Biotin is a coenzyme for several enzyme-catalyzed carboxylation reactions. Fischer et al. (1982) have noted that multiple carboxylase deficiency is associated with reduced lymphocyte-mediated suppressor activity, which they corrected in a pediatric patient with 10 mg/day biotin.

Vitamin C

The involvement of vitamin C (ascorbic acid) in the maintenance of immune function has been implicated through a variety of experimental strategies. Vitamin C deficiency states have been found to be associated with decreases in the bactericidal activity and locomotion of neutrophils and macrophages, decreases in resistance to microbial infections, anergy to mycobacterial antigens, and other impairments in cell-mediated immune mechanisms (Goldschmidt et al. 1988). Weening et al. (1981) found that supraphysiological doses of vitamin C administered to patients with Chediak-Higashi syndrome and other intrinsic defects of phagocyte function resulted in less frequent and severe bacterial infections and improved neutrophil locomotion and antimicrobial activity. Animal and human studies have suggested that dietary requirements for vitamin C are increased in wound repair, surgical trauma, infectious disease, and cancer.

Several molecular and biochemical mechanisms of ascorbate-mediated immunostimulation have been proposed including: (a) modulation of intracellular cyclic nucleotide levels, (b) modulation of prostaglandin (PG) synthesis, (c) protection of 5′ lipoxygenase, (d) enhancement of cytokine production, (e) antagonism of the immunosuppressive interactions of histamine and leukocytes, and (f) neutralization of phagocyte-derived autoreactive and immunosuppressive oxidants (Anderson et al. 1990). Mononuclear leukocytes normally contain the highest levels of ascorbate among the various cellular components of blood, regardless of variations in the plasma ascorbate content. *In vitro* data indicate that vitamin C serves as the first-line plasma antioxidant in the defense against phagocyte-derived reactive oxidants; only when ascorbate is depleted does detectable free radical damage, measured by the appearance of lipid peroxides, ensue (Frei, Stocker, and Ames 1988).

Although sufficiently large and rigorously controlled intervention trials have not yet been conducted, several studies have demonstrated that enhancing vitamin C status contributes to increased immunological vigor. Anderson et al. (1980) gave 1–3 g/day vitamin C for three weeks to healthy young adults and found increases in neutrophil motility to endotoxin-activated autologous serum and in lymphocyte proliferation to concanavalin A (Con A) and PHA. Keenes et al. (1983) administered 500 mg/day vitamin C via intramuscular injection to healthy older adults and noted increases in Con A and PHA proliferative responses and delayed cutaneous hypersensitivity (DCH) to tuberculin relative to placebo controls. Panush et al. (1982) reported that 10 g/day vitamin C in young adults enhanced lymphocyte proliferative and antibody responses and DCH responses. Ziemlanski et al. (1986) noted significant increases in serum IgG, IgM, and complement C3 levels in elderly women receiving supplements of 400 mg/day vitamin C for 4 to 12 months. In their community-based survey, Goodwin and Garry (1983) observed that older adults within the top 10 percent for plasma vitamin C concentration had significantly fewer anergic responses to four different antigens and higher mean DCH scores. Watson et al. (1990) found that daily combined vitamin C (1 g) and vitamin E (800 IU) supplementation for 30 days increased lymphocyte proliferation, particularly in less physically active subjects, and PGE_1 production.

Fat Soluble Vitamins

Beta(β)-Carotene and Vitamin A

In addition to serving as a precursor of vitamin A, β-carotene is a potent quencher of singlet oxygen and an antioxidant. Evidence derived from animal models indicates that β-carotene (independent of its provitamin A activity) can protect phagocytic cells from autooxidative damage, enhance T and B lymphocyte proliferative responses, stimulate effector T cell functions, promote the production of cytokines, and increase macrophage, cytotoxic T cell, and natural killer cell tumoricidal capacities (Bendich 1991). Alexander, Newmark, and Miller (1985) reported that dosages of 180 mg/day β-carotene for two weeks in healthy adults increased the frequency of helper/inducer T lymphocytes. Preliminary reports from clinical trials with supplements of β-carotene indicate that intakes of 30–180 mg/day for a period of several weeks elevate circulating T helper and natural killer cells and increase the number of cells bearing transferrin, HLA-Dr, and IL-2 receptor markers. Malter, Schriever, and Eilber (1989) reported that a group of vegetarians had two-fold greater serum β-carotene levels than a matched nonvegetarian population; natural killer cells from the vegetarian cohort lysed twice the number of tumor cells as compared to cells from the omnivore group. The potential interactions between β-carotene and related carotenoids with antioxidant properties such as canthaxanthin have not been explored with regard to immune function.

The role of vitamin A in resistance to infection is well established, whereby its deficiency significantly increases risk of systemic morbidity and mortality, especially in children (West, Howard, and Sommer 1989). The cellular dynamics mediating this relationship have not been fully elucidated, but appear to involve vitamin A actions modifying epithelial integrity and function, lymphoid mass, specific immunity, and nonspecific mechanisms of host resistance (Chew 1987). Milton, Reddy, and Naidu (1987) and Sommer, Katz, and Tarwotjo (1984) in epidemiological studies in Indonesia and India, respectively, have shown that preexisting vitamin A deficiency increases a child's risk of developing respiratory infection. Randomized, double blind clinical trials have also demonstrated that vitamin A supplementation in physiologic doses reduces morbidity. Pinnock, Douglas, and Badcock (1986) provided 450 μg retinol equivalents (RE) daily to presumably well-nourished preschool children with a history of recurrent respiratory problems and observed a significant reduction in the frequency of respiratory illness. Shenai et al. (1987) reported that 600 μg RE administered intramuscularly every other day for a month markedly reduced the incidence of bronchopulmonary dysplasia in preterm infants at high risk of vitamin A deficiency. Infection also directly influences vitamin A status by interfering with its absorption, utilization, and excretion.

Vitamin D

Vitamin D (1,25-dihydroxyvitamin D_3) has recently been recognized as an immunoregulatory hormone serving as an immunostimulatory agent of nonspecific immunity and exerting both stimulatory and inhibitory effects on specific immune responses (Yoder and Manolagas 1991). Animal and *in vitro* evidence suggest that vitamin D directly influences all members of the mononuclear phagocyte cell lineage. Vitamin D is bound by specific receptors expressed by mononuclear phagocytes and preferentially promotes myelomonocytic progenitors to mature mononuclear phagocytes (Provvedini et al. 1983). Vitamin D also stimulates undifferentiated neoplastic myelomonocytic cells to differentiate into macrophages (McCarthy et al. 1983). Interestingly, vitamin D is actually synthesized by activated macrophages, but not other members of the mononuclear phagocyte system (Reichel et al. 1987). Activated T and B lymphocytes residing in the human thymus and tonsils as well as activated circulating peripheral blood lymphocytes from patients with rheumatoid arthritis have been found to express vitamin D receptors (Manolagas, Wernts, and Tsoukas 1986). Vitamin D also affects both lymphocyte proliferation and cytokine production and decreases immunoglobulin production (Manolagas, Hustmyer, and Yu 1990).

Clinical applications of vitamin D will require a better understanding of its precise role in immunoregulation. However, *in vitro* experiments by Crowle and Ross (1990) have demonstrated that incubation of vitamin D with human monocytes or macrophages inhibits the growth of virulent *Mycobacterium tuberculosis*. The beneficial effect of vitamin D in the treatment of tuberculosis has been documented in the past, and recent epidemiological studies further support the association between low vitamin D status and the risk of tuberculosis (Davies 1988). Although it has not been determined whether vitamin D-stimulated macrophages can actually kill the bacilli, macrophages and T lymphocytes from patients with tuberculosis do synthesize vitamin D.

Limited evidence suggests that vitamin D may prove of value in enhancing the host immune defense of patients with chronic renal failure who require long-term hemodialysis. Vitamin D has been reported by Quesada et al. (1989) to reverse the impaired natural killer cell function found in patients with end-stage renal failure and increase lymphocyte proliferation and IL-2 production. There may be further therapeutic potential for vitamin D based upon its capacity for inducing differentiation and inhibition of clonal proliferation of human leukemic cells *in vitro*, particularly for vitamin D analogs with more potent inhibitory activity and less hypercalcemic toxicity. Experimental evidence suggests a potential application for vitamin D in modulating the course of autoimmune disorders. Beneficial effects of vitamin D administration have been reported in murine models of autoimmune thyroiditis, multiple sclerosis, encephalomyelitis, rheumatoid arthritis, and systemic lupus erythematosus.

Vitamin E

Studies involving different species of experimental animals and using deficient as well as supradietary levels of vitamin E indicate that α-tocopherol is involved in the maintenance of immune function. Vitamin E deficiency states are associated with depressed plaque-forming cell, mitogenic, and mixed lymphocyte responses and alterations in macrophage membrane receptors. Vitamin E supplementation to diets has been reported to increase humoral and cell-mediated immune responses and phagocytic functions in laboratory and farm animals as well as in healthy adults and hospitalized patients (Meydani and Blumberg 1989). The association between high intakes of vitamin E and a reduced risk of infection and chronic conditions such as cancer and heart disease among the elderly may be related to an optimization of immune responsiveness.

Vitamin E can affect both the lipoxygenase and cyclooxygenase pathways of arachidonic acid metabolism. Vitamin E may exert its immunostimulatory effect by inhibiting PG synthesis and/or decreasing free radical formation. Oxygen metabolites, especially hydrogen peroxide, produced by activated macrophages depress lymphocyte proliferation; α-tocopherol has been shown to decrease hydrogen peroxide formation by peripheral blood mononuclear (PMN) cells. Immunostimulation by supplemental vitamin E is associated with increases in α-tocopherol concentration in PMN cells and decreases in plasma lipid peroxides (Meydani et al. 1990).

Several clinical trials have demonstrated the potential benefit of increased vitamin E status upon immune function. Baehner et al. (1977) reported that 1,600 IU vitamin E dosages daily for one week reduced PMN autotoxicity leading to improved phagocytosis. Chirico et al. (1983) found that 120 mg/kg of vitamin E administered intramuscularly to premature infants accelerated the normalization of phagocytic function during their first week of life. Ziemlanski et al. (1986) noted that 200 mg vitamin E daily for four months increased α_2- and β_2-globulin fractions and total serum protein. Meydani et al. (1990) conducted a one-month, double blind, placebo controlled trial in healthy older adults with 800 IU vitamin E daily and observed significant increases in DCH responses, IL-2 production, and lymphocyte proliferation to Con A in association with decreases in plasma lipid peroxides and PGE_2 synthesis. A community-based survey by Chavance et al. (1985) found that plasma vitamin E levels were positively correlated with DCH responses and high helper-inducer/cytotoxic-suppressor ratios and negatively correlated with infectious disease episodes. Chandra (1992) conducted a year-long prospective clinical trial with healthy older adults employing a multivitamin and multimineral supplement containing doses equivalent to recommended nutrient allowances except for higher levels of vitamin E (44 mg) and β-carotene (16 mg). The supplemented group showed significant improvement in several parameters of immune responsiveness and a marked reduction in illness due to infections; moreover, supplemented subjects presenting with illness were observed to require shorter antibiotic treatment regimens.

LIPIDS

n-6 Polyunsaturated Fatty Acids

Essential fatty acids and dietary fats have numerous important influences upon immune responses (Hwang 1989). Although the effect of dietary fat on immunity may be mediated in part by alterations in cell membrane composition and fluidity, serum lipoproteins, or hormone status, recent attention has been focused on lipids' capacity as substrates for eicosanoid biosynthesis. Contrasting reports in the literature concerning the immunomodulatory effects of fats and fatty acids often result from improper dietary controls or inappropriately constituted *in vitro* models (Johnston 1988). Nonetheless, it is well established that essential fatty acid deficiency results in lymphoid atrophy and depressed antibody responses both to T cell-dependent and T cell-independent antigens. Small dietary amounts of linoleic acid, an n-6 fatty acid substrate for PG and leukotriene (LT) synthesis, are required for the normal propagation and maturation of a cell-mediated immune response. High intakes of linoleic and/or arachidonic acid increase synthesis of PGE_2 by macrophages, suppress DCH and lymphocyte proliferation, and impair the rejection response of an allograft. However, this situation is complex, as linoleic acid also serves as a substrate for the synthesis of LTB_4, which has both agonist and antagonist effects on immunopotent cells.

Plant seed oils enriched in gamma(γ)-linolenic acid (GLA) have been shown to suppress inflammation and joint tissue injury in a limited number of placebo-controlled human studies. Hansen et al. (1983) and Belch et al. (1988) employed primrose seed oil (360–540 mg/day GLA) in patients with rheumatoid arthritis and suggested that such supplementation could effectively substitute for the use of nonsteroidal anti-inflammatory drugs. Pullman-Mooar et al. (1990) employed borage seed oil (1.1 g/day GLA) supplements in rheumatoid arthritis patients and reported PGE_2 and LTB_4 production by stimulated leukocytes was reduced markedly as were symptoms of joint pain and stiffness. Black currant seed oil is another rich source of GLA and appears to possess similar immunomodulatory properties.

n-3 Polyunsaturated Fatty Acids

Eicosapentaenoic (EPA) and docosahexaenoic (DHA) acids are n-3 polyunsaturated fatty acids that lead to the synthesis of PGE_3, a significantly less immunosuppressive eicosanoid than PGE_2. Alpha(α)-linolenic acid, an n-3 fatty acid present in some vegetable oils, is the essential fatty acid precursor of EPA and DHA; however, in man, delta(δ)-6 desaturase activity is quite low and thus limits this conversion. Therefore, efforts to manipulate immune responsiveness with n-3 fatty acids have required dietary sources of EPA and DHA, particularly fish oils. n-3 fatty acid supplementation decreases the release of arachidonic acid upon T lymphocyte stimulation and interferes with this critical event in cell activation. Thus, altering the n-3/n-6 intake ratio affects immune responses.

The feasibility of modifying eicosanoid synthesis in immune and accessory cells by manipulation of dietary fatty acids has been tested.

n-3 dietary supplementation has been demonstrated to decrease neutrophil chemotaxis and inhibit LTB_4 production in healthy volunteers (Lee et al. 1985) and in patients with asthma (Payan et al. 1986). In other clinical studies, dietary supplementation with n-3 fatty acids has led to improvements in patients with rheumatoid arthritis (Kremer et al. 1987) and psoriasis (Bittiner et al. 1988). Endres et al. (1989) reported that 18 g/day fish oil supplementation for six weeks in healthy adults suppressed the synthesis of the proinflammatory mediators IL-1 and IL-1β, and tumor necrosis factor, possibly via decreased production of LTB_4 and generation of the biologically less active metabolite LTB_5 from EPA. These actions are consistent with the decreased inflammatory responses reported in patients receiving n-3 supplementation and the lower incidence of inflammatory diseases such as asthma and Type I diabetes mellitus in populations with high intakes of n-3 fatty acids.

Administration of increasing doses of n-3 fatty acids has been reported to substantially improve cell-mediated immune functions, probably through alteration of PG and LT synthesis pathways (Alexander et al. 1986). However, recent reports indicate a delicate balance may exist between the immunostimulatory, antiinflammatory, and immunosuppressive actions of n-3 supplementation. Virella et al. (1989) reported depressions of B cell responses and neutrophil functions in a subject receiving 6 g/day fish oil for six weeks. Meydani et al. (1991a) found that volunteers consuming 2.4 g/day n-3 fatty acids for three months showed reduced mitogenic responses and depressed synthesis of IL-1, IL-2, and IL-6. Thus, some studies suggest n-3 fatty acid-induced decreases in cytokine production and lymphocyte proliferation may be beneficial in inflammatory and autoimmune diseases, but could also compromise cell-mediated immunity. It is interesting to note that one immunomodulatory enteral nutrient formulation containing fish oil, EPA, and DHA is now being utilized in patients (Cerra et al. 1990; Lieberman et al. 1990a). Attention is now being focused on plant sources of n-3 fatty acids such as common purslane leaves, which provide a rich source of α-linolenic acid, EPA, DHA, docosapentanenoic acid, and the antioxidants vitamins C and E and glutathione (Simopoulous et al. 1992).

AMINO ACIDS

Arginine

Gourmelen et al. (1972) first reported that 30 g/day arginine significantly increased peripheral blood lymphocyte blastogenesis in response to Con A and PHA. Barbul et al. (1981) also administered 30 g/day arginine to healthy adult volunteers and found the blastogenic response of peripheral blood lymphocytes to PHA and Con A increased several-fold over presupplementation levels. Barbul (1993) later reported that the same regimen was associated with a decrease in T suppressor subsets and increases in T helper to T suppressor ratios.

Arginine appears to improve immune parameters during physiological stresses. Injured rats, mice, guinea pigs, and humans are all reported to respond favorably to arginine supplementation at a level of approximately 3% (w/w). In addition, thymocyte proliferative response to T cell mitogens is enhanced in association with changes in either the production of IL-2 or the kinetics of IL-2 receptor expression on activated T cells. Arginine has been employed clinically in immunocompromised patients and found to increase mitogen stimulated lymphocyte proliferation. Daly, Reynolds, and Thom (1988) reported that surgical patients whose diets were supplemented with 25 g/day arginine exhibited increased T lymphocyte activation relative to a group fed glycine. Dietary arginine has also been reported to be important for the maintenance of lymphokine-activated killer cell activity (Lieberman et al. 1990b). In cancer patients undergoing major surgery, peripheral blood lymphocyte responses to Con A and PHA were higher in a group receiving 25 g/day arginine for seven days postoperatively relative to controls. The supplemented group also had increased levels of cells expressing the CD4+ (T helper) phenotype. Some evidence suggests that arginine supplementation is associated with a reduced length of hospital stay after major cancer surgery (Cerra 1992).

The mechanism by which supplemental arginine enhances the immune function of stressed animals and humans is not known. Increased growth hormone secretion or other unidentified peptides may be produced in response to supplemental arginine. Arginine may also be acting as a precursor of polyamines such as putrescine and spermidine, which are important in cell growth and differentiation; an increase in polyamine biosynthesis in activated lymphocytes is critical in the regulation of mitogenesis. It has been noted that arginine concentration in the extracellular space surrounding wounds, tumors, and other inflammatory sites is very low, but activated macrophages contain arginase, which is capable of converting arginine to urea and ornithine. Locally produced ornithine thus could be converted to polyamines while the urea could produce a local pH conducive to collagen synthesis. Substitution of polyamines or their precursors, e.g., citrulline, does not appear to duplicate the immunomodulatory effects of arginine. Arginine also serves as a source of nitrous and nitric oxides, which may act as mediators of immune function.

Glutamine

Glutamine is the most abundant plasma free amino acid in the body and functions in a wide variety of metabolic reactions including its provision as fuel to lymphocytes and other rapidly dividing cells such as mucosal cells of the gastrointestinal tract. Like arginine, glutamine is classified as a nutritionally dispensable (nonessential) amino acid, but evidence has accumulated suggesting it serves as a conditionally essential nutrient; i.e., its endogenous synthesis is inadequate to meet the body's needs under certain clinical conditions (Lacey and Wilmore 1990). Recent studies suggest the rate of glutamine utilization by macrophages and lymphocytes (and possibly for all immune cells) is either similar to or greater than that of glucose even in a quiescent state (Newsholme and Parry-Billings 1990). This high rate of glutamine utilization may provide optimal conditions for regulation of the use of intermediates for synthesis of purine and pyrimidine nucleotides during the cell cycle. Any significant decrease in the availability of glutamine or its rate of utilization by lymphocytes would be expected to decrease their rate of proliferation and their ability for rapid response to immune challenges. Glutamine has also been identified as a critical nutrient for the maintenance of the intestinal immune system (gut-associated lymphatic tissue) and secretory IgA synthesis as well as for the prevention of bacterial translocation from the gut following the stress of burns, surgery, or other trauma (Alverdy 1990).

PURINE AND PYRIMIDINE NUCLEOTIDES

A dietary requirement for a source of preformed purine and/or pyrimidine bases has been identified as necessary for normal development of cellular immune responses. Nucleotides are precursors of DNA and RNA and also serve a multitude of other important functions in cellular energetics and metabolism. Genetic defects in nucleotide metabolism lead to a number of disease states with many effects on the immune system. Early assumptions about the non-essentiality of nucleotides for optimal growth and function of metabolically active cells such as lymphocytes, macrophages, and intestinal cells were based on studies with healthy individuals. However, nucleotide requirements appear evident with a stress like an immune challenge.

Cellular and humoral immunity are both impaired in patients with adenosine deaminase deficiency (Giblett et al. 1972). Defective T cell function has also been found in patients with purine nucleoside phosphorylase deficiency. Cells treated with deoxycoformycin, an inhibitor of adenosine deaminase, do not respond to mitogen stimulation because of a block related to nucleotide levels prior to entry of the cells into the S phase of the cell cycle (Thuller et al. 1985). A role for dietary nucleotides in the modulation of cellular immunity is also suggested by the observation that renal allograft patients supported with hyperalimentation, which contains no nucleotide sources, display a suppressed immune response to the allograft despite reduction in pharmacoimmunosuppression (Van Buren, Kulkarni, and Rudolph 1983). Uracil administration can restore DCH and T cell proliferative responses to specific antigens and reduce abscess formation to gram-positive organisms in mice (Kulkarni et al. 1986). Uracil has also been reported to reverse the immunosuppression associated with blood transfusion in experimental settings. Dietary nucleotides may also be effective in macrophage activation of the T helper/inducer populations. Studies in animal models have demonstrated that use of a defined diet without a source of nucleotides decreases cellular immune function and resistance to infection and increases allograft survival (Rudolph et al. 1990). These changes appear attributable to a higher portion of lymphocytic cells that are incapable of making the transition to the S phase of the cell cycle.

GLUTATHIONE

Glutathione (GSH) is the most abundant low molecular weight thiol-containing compound in cells and a strong free radical scavenger. In its reduced form GSH protects against various oxidants, free radicals, and cytotoxic agents. Adequate intracellular levels of GSH are necessary for lymphocyte activation; lymphocytes exposed to sulfhydryl oxidizing agents have a decreased proliferative response to mitogens (Noelle and Lawrence 1981). The depletion of GSH lowers mitogenic responses, and the *in vitro* addition of this tripeptide into culture medium reverses it. Furukawa, Meydani, and Blumberg

(1987) reported that dietary GSH supplementation in old mice significantly enhanced DCH response and lymphocyte proliferation. It has been suggested that the targets for GSH and thiol oxidizing agents could include membrane and cytoplasmic proteins that regulate lymphocyte proliferation. Clinical trials examining the efficacy of GSH supplementation on immunity are lacking.

MINERALS

Selenium

Selenium, presumably via its incorporation into cytosolic glutathione peroxidase (GSHPx) and biomembrane-associated phospholipid hydroperoxide GSHPx as a selenocysteine residue, has been associated with the expression of specific and nonspecific and cell-mediated immune responses (Spallholz 1990). Selenium and/or GSHPx is incorporated and avidly retained in cells derived from undifferentiated bone marrow stem cells and cells modified by the immune system. Selenium-dependent GSHPx appears to control excessive production of peroxidative substrates such as hydrogen peroxide in cells that elicit phagocytic respiratory bursts. Cells deficient in GSHPx have been noted to lose their ability to assemble microtubules and mount a cytotoxic attack. GSHPx has also been identified as a modulator of the arachidonic acid cascade because lowered selenium or GSHPx status depresses the reduction of PGG_2 and the hydroperoxy-eicosatetraenoic acids and decreases eicosanoid biosynthesis (Schoene, Moris, and Levander 1986).

Selenium deficiency in animals depresses components of the immune response, while selenium adequacy or moderate supplementation promotes viable immune responses and selenium toxicity produces immunosuppression. Dietary selenium sufficient or in excess of that necessary for GSHPx synthesis also reduces the frequency, growth, and size of tumors in animals and in tissue culture. Dimitrov et al. (1986) supplemented subjects with 400 μg/day sodium selenite and noted increased plasma selenium and enhanced natural killer cell activity *in vitro*. Frank selenium deficiency is unlikely to occur except during unsupplemented total parenteral nutrition; however, inherited deficiencies of GSHPx result in a syndrome of recurrent and severe bacterial infections secondary to defective intracellular killing by leukocytes, further indicating the important role of selenium in immune function (Rutenberg et al.

1977). Interestingly, Dworkin et al. (1988) reported that patients with AIDS and AIDS-related complex possess less plasma and erythrocyte selenium and GSHPx activity.

Other Minerals

Similar to selenium, several other metalloenzymes are directly involved in immunity and antioxidant defenses including copper/zinc superoxide dismutase and iron-catalase. It is assumed that the efficacy of these minerals is dependent upon their incorporation into these antioxidant enzymes; an increase in mineral concentrations without a concomitant increase in enzyme activity would not necessarily be expected to result in increased immune function or antioxidant status. However, these minerals are also significantly involved in immune responsiveness beyond their function in antioxidant enzymes; for example, thymulin, a key thymic factor, and δ-6 desaturase, which converts linoleic acid to γ-linolenic acid, are zinc dependent.

An impairment of immunity has been firmly established in experimental and clinical conditions in which a deficiency in copper, iron, manganese, or zinc occurs (Fletcher et al. 1988). Further, repletion to nutritional sufficiency is associated with restoration of immune function. As noted, it is not always possible to extrapolate immunologic observations of nutritional effects in experimental animals to studies of human populations; relevant examples have been demonstrated. For example, iron supplementation in populations with a high incidence of deficiency anemia has been shown to decrease the morbidity from infectious and diarrheal diseases (Scrimshaw, Taylor, and Gordon 1988). Individual supplementation at moderate levels above apparent dietary requirements for these minerals have been reported to enhance some aspects of immune function. However, it is important to recognize that mineral overloads are associated with toxicity and, in some cases, lowered host defenses.

CONCLUSION

This review has focused on human studies reporting beneficial effects on immunity of nutrients capable of being formulated into functional foods. Those studies that offered equivocal or no positive results have not been discussed. As noted above, nutritional interventions in animal and cell culture models may or may not have clinical and public health relevance. The increasing recognition of the potential impact of nutrient-nutrient interactions also requires some caution in initiating and interpreting many studies, particularly those focusing on a single nutrient. Although there are still significant gaps and discrepancies in our understanding and application of nutritional immunology, recent studies on macro- and micronutrients have provided us with substantial progress in this regard. Importantly, preliminary data from studies employing phytochemicals derived from foods also suggest great promise in their immunomodulatory capacity. For example, catechins and flavonoids from green tea and berries; γ-glutamyl allylic cysteines from aged garlic extract; triterpenoids from licorice; and polyacetylenes from parsley, carrots, and celery appear to influence immunologic activity and could prove to be useful components of functional foods.

Dietary manipulation can now be used to beneficially affect immunity in patients and can potentially also serve in general populations to decrease the risk of chronic diseases associated with age-related immune deficits. Effective nutritional interventions in the immune system may find value not only in therapeutic applications, but also in the prophylactic treatment of subjects at risk of immunoincompetence because of illness or prior to immunodepressive drug and surgical regimens. Further research on nutrients and phytochemicals is required to define efficacious interventions appropriate to the goals of developing functional foods that can maintain normal immune defenses in compromised patients and promote optimal immune function in healthy individuals to reduce their risk of infectious and chronic diseases.

References

Alexander, M.; Newmark, H.; and Miller, R.G. 1985. Oral beta-carotene can increase the number of OKT4+ cells in human blood. *Immunol. Lett.* 9:221–225.

Alexander, J.W.; Saito, H.; Ogle, C.; and Tracki, O. 1986. The importance of lipid type in the diet after burn injury. *Ann. Surg.* 204:1–8.

Alverdy, J.C. 1990. Effects of glutamine-supplemented diets on immunology of the gut. *JPEN* 14:109S–113S.

Anderson, R.; Oosthuigen, R.; Maritz, R.; Theron, A.; and Van Rensburg, A. 1980. The effect of increasing weekly doses of ascorbate on certain cellular and humoral immune functions in normal volunteers. *Am. J. Clin. Nutr.* 33:71–76.

Anderson, R.; Smit, M.J.; Joone, G.K.; and Van Staden, A.M. 1990. Vitamin C and cellular immune functions. Protection against hypochlorous acid-mediated inactivation of glyceraldehyde-3-phosphate dehydrogenase and ATP generation in human leukocytes as a possible mechanism of ascorbate-mediated immunostimulation. *Ann. N.Y. Acad. Sci.* 587:34–48.

Baehner, R.L.; Boxer, L.A.; Allen, J.M.; and Davis, J. 1977. Autooxidation as a basis for altered function by polymorphonuclear leukocytes. *Blood* 50:327–335.

Barbul, A. 1993. The role of arginine as an immune modulator. In *Nutrient Modulation of the Immune Response,* S. Cunningham-Rundles, ed., pp. 47–61. New York: Marcel Dekker, Inc.

Barbul, A.; Sisto, D.A.; Wasserkrug, H.L.; and Efron, G. 1981. Arginine stimulates lymphocyte immune response in healthy human beings. *Surgery* 84:244–251.

Beisel, W.R. 1982. Single nutrients and immunity. *Am. J. Clin. Nutr.* 35(Suppl):417–468.

Belch, J.J.F.; Ansel, D.; Madhok, R.; and O'Dowd, A. 1988. Effects of altering dietary essential fatty acids on requirements for nonsteroidal antiinflammatory drugs in patients with rheumatoid arthritis: A double blind placebo controlled study. *Ann. Rheum. Dis.* 47:96–102.

Bendich, A. 1991. Carotenoids and immunity. *Clin. Appl. Nutr.* 1:45–51.

Bittiner, S.B.; Tucker, W.F.; Cartwright, I.; and Bleehen, S.S. 1988. A double blind, randomized, placebo-controlled trial of fish oil in psoriasis. *Lancet* 1:378–380.

Casciato, D.A.; McAdam, L.P.; Kopple, J.D.; Bluestone, R.; Goldberg, L.S.; Clements, P.J.; and Knutson, D.W. 1984. Immunologic abnormalities in hemodialysis patients: Improvement after pyridoxine therapy. *Nephron* 38:9–16.

Cerra, F.B. 1992. Role of nutrition in the management of malnutrition and immune dysfunction of trauma. *J. Am. Coll. Nutr.* 11:512–518.

Cerra, F.B.; Lehman, S.; Konstantinides, N.; Konstantinides, F.; Shronts, E.P.; and Holman, R. 1990. Effect of enteral nutrient on in vitro tests of immune function in ICU patients: A preliminary report. *Nutrition* 6:84–87.

Chandra, R.K. 1992. Effect of vitamin and trace-element supplementation on immune responses and infection in elderly subjects. *Lancet* 340:1124–1127.

Chavance, M.; Brubacher, G.; Herbeth, B.; Vernes, G.; Mistacki, T.; Deti, F.; Fournier, C.; and Janot, C. 1985. Immunological and nutritional status among the elderly, In *Nutrition, Immunity, and Illness in the Elderly,* R.K. Chandra, ed., pp. 137–145. New York: Pergamon Press.

Chew, B.P. 1987. Vitamin A and beta-carotene on host defense. *J. Dairy Sci.* 70:2732–2743.

Chirico, G.; Marconi, M.; Colombo, A.; Chiara, A.; Randini, G.; and Ugazio, A.G. 1983. Deficiency of neutrophil phagocytosis in premature infants: Effect of vitamin E supplementation. *Acta Paediatr. Scand.* 72:521–524.

Crist, W.M.; Parmley, R.T.; Holbrook, C.T.; Castleberry, R.P.; Denys, F.R.; and Malluh, A. 1980. Dysgranulopoietic neutropenia and abnormal monocytes in childhood vitamin B_{12} deficiency. *Am. J. Hematol.* 9:89–107.

Crowle, A.J., and Ross, E.J. 1990. Comparative abilities of various metabolites of vitamin D to protect cultured human macrophages against Tubercle Bacilli. *J. Leuk. Bio.* 47:545–550.

DaCosta, M., and Sharon, M. 1980. The synthesis of folate-binding in lymphocytes during transformation. *Br. J. Haematol.* 46:575–579.

Daly, J.M.; Reynolds, J.; and Thom, A. 1988. Immune and metabolic effects of arginine in the surgical patient. *Ann. Surg.* 208:512–523.

Davies, P.D. 1988. A possible link between vitamin D deficiency and impaired host defense to Mycobacterium Tuberculosis. *Tubercle* 66:301–306.

Dimitrov, N.V.; Charamella, L.J.; and Meyer, C.H. 1986. Modulation of natural killer cell activity by selenium in humans. *J. Nutr. Growth Cancer* 3:193–198.

Dworkin, B.M.; Rosenthal, W.S.; Wormser, G.P.; Weiss, L.; Nunez, M.; Joline, C. "d Herp, A. 1988. Abnormalities of blood selenium and glutathione peroxidase activity in patients with Acquired Immunodeficiency Syndrome and AIDS-Related Complex. *Biol. Trace Elm. Res.* 15:167–177.

Endres, S.; Ghorbani, B.S.; Kelley, V.E.; Georgilis, K.; Lannemann, G.; Vander Meer, J.W.; Cannon, J.G.; Rogers, T.S.; Klempner, M.S.; Weber, P.C.; Schaefer, E.J. wlff, S.M.; and Dinarello, C.A. 1989. The effect of dietary supplementation with n-3 polyunsaturated fatty acids on the synthesis of interleukin-1 and tumor necrosis factor by mononuclear cells. *N. Engl. J. Med.* 320:265–271.

Fischer, A.; Munnich, A.; Saudubray, J.M.; Mamas, S.; Coudie, F.X.; Charpentier, C.; Dray, F.; Friezal, J.; and Griscelli, C. 1982. Biotin-responsive immunoregulatory dysfunction in multiple carboxylase deficiency. *J. Clin. Immunol.* 2:35–38.

Fletcher, M.P.; Gershwin, M.E.; Keen, C.L.; and Hurley, L. 1988. Trace element deficiencies and immune responsiveness in humans and animal models. In *Nutrition and Immunology*, R.K. Chandra, ed., p. 215. New York: Alan R. Liss.

Frei, B.; Stocker, R.; and Ames, B.N. 1988. Antioxidant defenses and lipid peroxidation in human blood plasma. *Proc. Natl. Acad. Sci.* 85:9748–9752.

Furukawa, T.; Meydani, S.N.; and Blumberg, J.B. 1987. Reversal of age-associated decline in immune responsiveness by dietary glutathione supplementation in mice. *Mech. Age. Dev.* 38:107–117.

Giblett, E.R.; Anderson, J.E.; Cohen, F.; Pollara, B.; and Meuwissen, H.J. 1972. Adenosine-deaminase deficiency in two patients with severely impaired cellular immunity. *Lancet* 2:1067–1070.

Goldschmidt, M.C.; Masin, W.J.; Brown, L.R.; and Wyde, P.R. 1988. The effect of ascorbic acid deficiency on leukocyte phagocytosis and killing of *Actinomyces viscosus*. *Int. J. Vit. Nutr.* 58:326–33.

Goodwin, J.S., and Garry, P.J. 1983. Relationship between megadose vitamin supplementation and immunological function in a healthy elderly population. *Clin. Exp. Immunol.* 51:647–653.

Gourmelen, M.; Donnadieu, M.; Schimpff, R.M.; Lestradet H.; and Girard, F. 1972. Effect of ornithine hydroxhloride on growth hormone (HGH) plasma levels. *Ann. Endocrinol.* 33:526–528.

Hansen, T.M.; Lerch, A.; Kassis, V.; Lorenzen, I.; and Sondergard, J. 1983. Treatment of rheumatoid arthritis with PGE_1 precursors cis-linoleic acid and gamma linolenic acid. *Scand. J. Rheumatol.* 12:85–89.

Hodges, R.E.; Bean, W.B.; Ohlson, M.A.; and Bleiler, R.E. 1962. Factors affecting human antibody response, IV. Pyridoxine deficiency. *Am. J. Clin. Nutr.* 11:180–186.

Hwang, D. 1989. Essential fatty acids and immune response. *FASEB J.* 3:2052–2061.

Johnston, P.V. 1988. Lipid modulation of immune responses. *Contemp. Issues Clin. Nutr.* 11:37–86.

Katka, K. 1984. Immune functions in pernicious anaemia before and during treatment with vitamin B_{12}. *Scand. J. Haematol.* 32:76–82.

Keenes, B.; Dumont, I.; Brohee, D.; Hubert, C.; and Neve, P. 1983. Effect of vitamin C supplementation on cell-mediated immunity in old people. *Gerontology* 29:305–310.

Kremer, J.M.; Jubiz, W.; Michalek, A.; Rynes, R.I.; Bartholomew, L.W.; Bigaouette, J.; Timchalk, M.; Beeler, D.; and Lininger, L. 1987. Fish-oil fatty acid supplementation in active rheumatoid arthritis: A double-blinded, controlled, crossover study. *Ann. Intern. Med.* 106:497–503.

Kulkarni, A.D.; Fanslow, W.C.; Rudolph, F.B.; and Van Buren, C.T. 1986. Effect of dietary nucleotides on response to bacterial infections. *JPEN* 10:169–171.

Lacey, J.M., and Wilmore, D.W. 1990. Is glutamine a conditionally essential amino acid? *Nutr. Rev.* 48:297–309.

Lee, T.H.; Hoover, R.I.; Williams, J.D.; Sperling, R.I.; Ravalese, J.; Spur, B.W.; Robinson, D.R.; Corey, E.J.; Lewis, R.A.; and Austen, K.F. 1985. Effect of dietary enrichment with ecosapentaenoic and docosahexaenoic acids on in vitro neutrophil and monocyte leukotriene generation and neutrophil function. *N. Engl. J. Med.* 312:1217–1224.

Lieberman, M.D.; Shou, J.; Torres, A.S.; Weintraub, F.; Goldfine, J.; Sigal, R.; and Daly, J.M. 1990a. Effects of nutrient substrates on immune function. *Nutrition* 6:88–91.

Lieberman, M.D.; Sigal, R.K.; Evantash, E.; and Daly, J.M. 1990b. L-arginine potentiates lymphokine-activated killer cell (LAK) cytotoxicity. *FASEB J.* 4:A1041.

Malter, M.; Schriever, G.; and Eilber, U. 1989. Natural killer cells, vitamins, and other blood components of vegetarian and omnivorous men. *Nutr. Cancer* 12:271–278.

Manolagas, S.C.; Hustmyer, F.G.; and Yu, X. 1990. Immunomodulating properties of 1,25-dihydroxyvitamin D_3. *Kidney International* 38:S9–S16.

Manolagas, S.C.; Wernts, D.A.; and Tsoukas, C.D. 1986. 1,25-Dihydroxyvitamin D_3 receptors in lymphocytes from patients with rheumatoid arthritis. *J. Lab. Clin. Med.* 108:596–600.

McCarthy, D.M.; San Miguel, J.F.; Freake, H.C.; Green, P.M.; Zola, H.; Catovsky, D.; and Goldman, J.M. 1983. 1,25-Dihydroxyvitamin D_3 inhibits proliferation of human promyelocytic leukaemic (HL-60) cells and induced monocyte-macrophage differentiation in HL-60 and normal human bone marrow cells. *Leukemic Res.* 7:51–55.

Meydani, S.N.; Barklund, P.; Liu S.; Meydani, M.; Miller, R.A.; Cannon, J.G.; Morrow, F.D.; Rocklin, R.; and Blumberg, J.B. 1990. Vitamin E supplementation enhances cell-mediated immunity in healthy elderly subjects. *Am. J. Clin. Nutr.* 52:557–563.

Meydani, S.N., and Blumberg, J.B. 1989. Nutrition and immune function in the elderly. In *Nutrition, Aging, and the Elderly*, H.N. Munro and D.E. Danford, eds., pp. 61–87. New York: Plenum Publishing.

Meydani, S.N.; Endres, S.; Woods, M.M.; Goldin, B.R.; Soo, C.; Morrill-Labrode, A.; Dinarello, C.A.; and Gorbach, S.L. 1991a. Oral n-3 fatty acid supplementation suppresses cytokine production and lymphocyte proliferation: Comparison of young and older women. *J. Nutr.* 121:547–555.

Meydani, S.N.; Ribaya-Mercado, J.D.; Russell, R.M.; Sahyoun, N.; Morrow, F.D.; and Gershoff, S.N. 1991b. Vitamin B_6 deficiency impairs interleukin-2 production and lymphocyte proliferation in elderly subjects. *Am. J. Clin. Nutr.* 53:1275–1280.

Milton, R.C.; Reddy, V.; and Naidu, A.N. 1987. Mild vitamin A deficiency and childhood morbidity—an Indian experience. *Am. J. Clin. Nutr.* 46:827–829.

Newsholme, E.A., and Parry-Billings, M. 1990. Properties of glutamine release from muscle and its importance for the immune system. *JPEN* 14:63S–67S.

Noelle, R.J., and Lawrence, D.A. 1981. Determination of glutathione in lymphocytes and possible association of redox state and proliferative capacity of lymphocytes. *Biochem. J.* 198:571–579.

Panush, R.S.; Delafuente, J.C.; Katz, P.; and Johnson, J. 1982. Modulation of certain immunologic responses by vitamin C. III. Potentiation of *in vitro* and *in vivo* lymphocyte responses. *Int. J. Vit. Nutr. Res.* 23(Suppl.):35–47.

Payan, D.G.; Wong, M.Y.; Chernov-Rogan, T.; Valone, F.H.; Pickett, W.C.; Blake, V.A.; Gold, W.M.; and Goetzl, E.J. 1986. Alterations in human leukocyte function induced by ingestion of eicosapentaenoic acid. *J. Clin. Immunol.* 6:402–410.

Pinnock, C.B.; Douglas, R.M.; and Badcock, N.R. 1986. Vitamin A status in children who are prone to respiratory tract infections. *Aust. Paediatr. J.* 22:95–99.

Provvedini, D.M.; Tsoukas, C.D.; Deftos, J.L.; and Manolagas, S.C. 1983. 1,25-Dihydroxyvitamin D_3 receptors in human leukocytes. *Science* 221:1181–1183.

Pullman-Mooar, S.; Laposata, M.; Lem, D.; Holman, R.T.; Leventhal, L.J.; DeMarco, D.; and Zurier, R.B. 1990. Alteration of the cellular fatty acid profile and the production of eicosanoids in human monocytes by gamma linolenic acid. *Arthritis Rheum.* 33:1526–32.

Quesada, J.M.; Solana, R.; Martin, A.; Santamaria, M.; Serrano, I.; Martinez, M.E. jama, P.; and Pena, J. 1989. The effect of calcitriol on natural killer cell activity in hemodialyzed patients. *J. Steroid Biochem.* 34:423–425.

Reichel, H.; Koeffler, H.P.; Barbes, R.; and Norman, A.W. 1987. Regulation of 1,25 dihydroxyvitamin D_3 production by cultured alveolar macrophages from normal human donors and from patients with pulmonary sarcoidosis. *J. Clin. Endocrinol. Metab.* 65:1201–1209.

Rudolph, F.B.; Kulkarni, A.D.; Fanslow, W.C.; Pizzini, R.P.; Kumar, S.; and Van Buren, C.T. 1990. Role of RNA as a dietary source of pyrimidines and purines in immune function. *Nutrition* 6:45–52.

Rutenberg, W.D.; Yang, M.C.; Doberstyn, E.B.; and Bellanit, J.A. 1977. Multiple leukocyte abnormalities in Chronic Granulomatosis Disease: A familial study. *Pediat. Res.* 11:58–63.

Schoene, N.W.; Moris, V.C.; and Levander, O.A. 1986. Altered arachidonic acid metabolism in platelets and aortas from selenium-deficient rats. *Nutr. Res.* 6:75–83.

Scrimshaw, N.S.; Taylor, C.E.; and Gordon, J.E. 1959. Interactions of nutrition and infection. *Am. J. Med. Sci.* 237:367–403.

———. 1988. Interactions of nutrition and infection. *Nutrition* 4:13–50.

Seger, R.; Frater-Schroder, M.; Hitzig, W.H.; Wildeuer, A.; and Linnell, J.C. 1980 Granulocyte dysfunction in transcobalamin II deficiency responding to leucovorin or hydroxycobalamin-plasma transfusion. *J. Inherit. Metab. Dis.* 3:3–9.

Shenai, J.P.; Kennedy, K.A.; Chytil, F.; and Stahlman, M.T. 1987. Clinical trial of vitamin A supplementation in infants susceptible to bronchopulmonary dysplasia. *J. Pediatr.* 111:269–277.

Simopoulous, A.P.; Norman, H.A.; Gillaspy, J.E.; and Duke, J.A. 1992. Common purslane: A source of omega-3 fatty acids and antioxidants. *J. Am. Coll. Nutr.* 11:374–382.

Skacel, P.O., and Chanarin, I. 1983. Impaired chemiluminescence and bactericidal killing neutrophils from patients with severe cobalamin deficiency. *Br. J. Haematol.* 44:203–215.

Sommer, A.; Katz, J.; and Tarwotjo, I. 1984. Increased risk of respiratory disease and diarrhea in children with pre-existing mild vitamin A deficiency. *Am. J. Clin. Nutr.* 40:1090–1095.

Spallholz, J.E. 1990. Selenium and glutathione peroxidase: Essential nutrient and antioxidant component of the immune system. *Adv. Exper. Med. Biol.* 262:145–158.

Talbott, M.C.; Miller, L.T.; and Kerkvleit, N. 1987. Pyridoxine supplementation: Effect of lymphocyte response in elderly persons. *Am. J. Clin. Nutr.* 46:659–664.

Thuller, L.; Arenzana-Seisdedos, F.; Perignon, J.L.; Munier, A.; Hamet, M.; and Cartier, P.H. 1985. Interleukin 1 liberation and absorption capacities of rat T lymphocytes in conditions of severe adenylic nucleotide pool depletion due to adenosine deaminase deficiency. *Immunopharmacol.* 10:89–95.

Van Buren, C.T.; Kulkarni, A.D.; and Rudolph, F. 1983. Synergistic effect of a nucleotide-free diet and cyclosporine on allograft survival. *Transplant Proc.* Suppl. 1–2:2967–2971.

Virella, G.; Kilpatrick, J.M.; Rugeles, M.T.; Hyman, B.; and Russell, R. 1989. Depression of humoral responses and phagocytic functions in vivo and in vitro by fish oil and eicosapentanoic acid. *Clin. Immun. Immunopath.* 52:257–270.

Watson, R.R.; Benedict, J.; Mayberry, J.C.; Hicks, M.J., and Moriguchi, S. 1990. Supplementation of vitamins C and E and cellular immune function in young and aging men. *Ann. N.Y. Acad. Sci.* 498:530–533.

Weening, R.S.; Schoorel, E.P.; Roos, D.; Van Schaik, M.L.; Voetman, A.A.; Bot, A.A.; Batenburg-Plenke, A.M.; Willems, C.; and Zeijlemaker, W.P. 1981. Effect of ascorbate on abnormal neutrophil, platelet and lymphocytic function in a patient with the Chediak-Higashi Syndrome. *Blood* 57:856–865.

West, K.P.; Howard, G.R.; and Sommer, A. 1989. Vitamin A and infection: Public health implications. *Ann. Rev. Nutr.* 9:63–86.

Yoder, M.C., and Manolagas, S.C. 1991. Vitamin D and its role in immune function. *Clin. Appl. Nutr.* 1:35–44.

Youinou, P.Y.; Garre, M.A.; Menez, J.F.; Boles, J.M.; Morin, J.F.; Pennec, Y.; Miossec P.J.; Morin, P.P.; and LeMenn, G. 1982. Folic acid deficiency and neutrophil dysfunction. *Am. J. Med.* 73:652–567.

Ziemlanski, S.; Wartanowicz, M.; Kios, A.; Raczka, A.; and Kios, M. 1986. The effects of ascorbic acid and alpha-tocopherol supplementation on serum proteins and immunoglobulin concentrations in the elderly. *Nutr. Internatl.* 2:1–5.

Chapter 6

Dietary Factors Modulating the Rate of Aging

Huber R. Warner and
Sooja K. Kim

INTRODUCTION

There is general agreement that diet has important implications for health and susceptibility to disease, but there is little direct evidence that dietary factors alter the progress of aging per se. One difficulty is that it is not easy to distinguish changes due to "true aging" from changes associated with age-related disease. There are numerous theories of aging (Warner et al. 1987), many of which are not mutually exclusive, but share common elements. However, it seems clear that the longevity of an individual within a population depends on his/her genetic constitution and environmental experience. Thus, one can ask how dietary factors might modulate age-related changes in gene expression and/or macromolecular damage due to environmental insults, both of which presumably determine the rate of aging of any particular individual of a species.

The purpose of this chapter is to discuss some of the possibilities and to provide evidence that might eventually lead to successful dietary interventions to reduce the rate of aging. In this chapter we assume that there is substantial overlap between risk factors for aging and risk factors for age-related pathology. Therefore, an intervention to reduce degenerative processes leading to pathological changes would be expected to slow aging as well.

CALORIC RESTRICTION

The restriction of caloric intake, while maintaining an appropriate intake of essential nutrients, is the most universal intervention known for the extension of life span in animals. Although most life span extension experiments have utilized mice and rats, the intervention also works in protozoa, rotifers, nematodes, insects, spiders, and fish. Weindruch and Walford (1988) have recently published an extremely comprehensive discussion of this area of research, and this volume is recommended to the reader interested in the details of this phenomenon. A brief summary, which also includes a discussion of the significance of caloric restriction for human aging (Masoro 1989), is also recommended. The purpose here is to briefly discuss the salient features of this phenomenon and to present some more recent findings.

Nature of the Phenomenon

The usual procedure in mice or rats is to begin the restriction at about 1 to 3 months of age. When Weindruch et al. (1986) fed mice 40 kcal per week instead of the 85 kcal consumed per week by the control mice, the average and the maximum life span were increased 38% and 34%, respectively. This extension of maximum life span indicates that caloric restriction slows some critical process(es), which is a major factor in determining the rate of aging of the entire organism.

The age of initiation of the caloric restriction has also been investigated. Whereas the magnitude of the life-extending effect is reduced to about 10% when the restriction is started late in life, e.g., 12 months of age, this still represents substantial extension of both average and maximum life span (Weindruch and Walford 1982). Comparable results have been reported by Yu, Masoro, and McMahan (1985) for rats.

Despite the large amount of work that has been done in this area, it is still not clear how caloric restriction produces life span extension. Nevertheless, it has been shown that many, but not all, age-related changes are in fact delayed by caloric restriction. These changes include the amount of cellular damage sustained with increasing age, as well as changes in the expression of many genes. Some of these are discussed below, but none of these changes can be directly or unequivocally related to the extension of life span observed.

Effect of Caloric Restriction on Gene Expression

Weindruch and Walford (1988) have compiled a comprehensive list of changes in enzyme activities in a variety of tissues that accompany caloric restriction. In general, activities either stay the same or increase. Particularly relevant are the increases in liver catalase and, in some cases, superoxide dismutase in rodents, because of their role in preventing oxidative damage.

However, a better question to ask might be: What is the effect of caloric restriction on age-related changes in gene expression? Here the data are more sparse. Richardson et al. (1987) have shown that caloric restriction reverses the age-related decline in the transcription of the gene for α_{2u}-globulin, and Waggoner et al. (1990) have shown that the age-related increase in apolipoprotein A1 transcription is also reversed, whereas the age-related changes in expression of the genes for c-myc and apolipoprotein B are unaffected by caloric restriction. Chatterjee et al. (1989) obtained similar results for expression of α_{2u}-globulin, and also found that caloric restriction delays the age-dependent appearance of the messenger RNA for senescence marker protein 2 (SMP-2) in liver. SMP-2 is a sulfotransferase specific for dehydroepiandrosterone (DHEA); therefore this enzyme may be important in the metabolism and transport of DHEA, a compound that has been shown to have remarkable effects on restoring immune function in old mice (Daynes and Araneo 1992).

These results indicate that caloric restriction can delay age-related changes in the transcription of some, but not all, genes. What remains far from clear is which, if any, of these changes play a central role in extending life span and/or delaying the onset of age-related disease.

Effect of Caloric Restriction on Macromolecular Damage

One of the most prevalent insults faced by cells is oxidative damage to macromolecules, including nucleic acids, proteins, and lipids. This damage presumably results primarily from the production of reactive oxygen species (ROS) by mitochondria during normal metabolism. This diversion of oxygen to the production of ROS in mitochondria has been estimated to be in the range of about 1 to 4% (Wallace 1992). ROS are also produced by specific enzymatic reactions such as xanthine oxidase, NADPH oxidase, and a wide variety of mixed function oxidases, and are generated non-enzymatically in the presence of iron, as well as by metabolism of organic solvents, industrial pollutants, components of tobacco smoke, and by exposure to radiation (Machlin and Bendich 1987).

DNA Damage and Repair

Oxidative damage to DNA has been demonstrated to occur in rodents, and increases with age. Most recent published work has focused on the presence of 8-hydroxyguanine in DNA because of the development of a good assay system for this damaged base (Park et al. 1992). Fraga et al. (1990) have shown that in the rat 8-hydroxyguanine accumulates in DNA with age in the liver, kidney, and intestine, but not in the brain or testes. They also observed that the excretion of 8-hydroxy deoxyguanosine in the urine decreased with increasing age, indicating that repair of damaged DNA decreases with increasing age. Thus, it would be important to determine whether caloric restriction attenuates the amount of DNA damage occurring *in vivo*, and/or increases the rate of DNA repair. Data on this are limited and somewhat contradictory. Both Licastro et al. (1988) and Weraarchakul et al. (1989) have shown that repair of UV-induced damage is increased by caloric restriction, but Saul, Gee, and Ames (1987) failed to demonstrate a significant effect of caloric restriction on urinary excretion of oxidized thymine or thymidine. Finally, Randerath, Reddy, and Disher (1986) found that rodent DNA contains very low levels of unidentified modified bases (called I-compounds) that increase with age. However, caloric restriction not only did not reverse this increase, but actually increased the rate of I-compound content of the DNA (Randerath et al. 1993). Vitamin E also does not prevent the accumulation of I-compounds, but also increases it (Li et al. 1991). Thus, whereas these and other studies show that the content of I-compounds in the DNA is influenced by diet, it appears that I-compounds in DNA do not play a major role in aging in rodents.

Protein Damage and Repair

The significance of oxidative damage to proteins and its role in aging has only fairly recently come to be appreciated (Stadtman 1992). Oxidized proteins are found both in tissues and in cells grown in culture, and have been shown to increase with increasing age. The rate of accumulation of oxidized proteins is much greater in the rat than in humans, and inversely correlates with the life span of these two species. The oxidation occurs primarily in the side chains of metal-binding amino acids such as lysine, arginine, and histidine, or in the side chains of aromatic amino acids such as phenylalanine, tyrosine, or tryptophan (Stadtman 1992). Much of this protein oxidation is thought to occur as a "caged" reaction at or near the iron binding sites (Stadtman 1991). Starke-Reed and Oliver (1989) have estimated that as much as 50–70% of the total cellular protein may be oxidized, especially in senescent cells.

A single report on the effect of caloric restriction in rats indicates that 40% caloric restriction reduces the accumulation of oxidized protein between 9 and 15 weeks of age (Youngman, Park, and Ames 1992). A similar result was achieved by protein restriction. This suggests that caloric restriction may in fact attenuate protein oxidation *in vivo*, but longer term experiments in older animals are needed to confirm this. It is also known that protein degradation decreases in the brain with increasing age (Starke-Reed and Oliver 1989), and it would be of interest to know the effect of caloric restriction on protein turnover in brain and other tissues.

Lipid Peroxidation and Repair

Laganiere and Yu (1987) have shown that peroxidation of lipids in both microsomal and mitochondrial membranes increases with age, but this peroxidation is reduced about 50% by caloric restriction. This difference could not be attributed to the presence of vitamin E because the levels in these membranes were higher in *ad libitum*-fed animals than in calorically restricted animals. A similar conclusion was drawn from the amount of malondialdehyde produced when these membranes were incubated with ferrous ion in the presence of ADP and NADPH. In these experiments caloric restriction also altered the fatty acid composition of the membranes, both increasing the linoleic acid content and lowering the docosapentaenoic acid (22:5) content by 50%.

The above results on oxidative damage to DNA, protein, and lipids provide some support for the idea that caloric restriction may slow the rate of aging by either reducing the amount of damage to macromolecules or increasing the repair activity in the cell.

How Does Caloric Restriction Work?

Because so many age-related changes are reversed by caloric restriction, it is difficult to pinpoint the critical event among all of these changes. A useful starting point may be to at least understand the relationship between glucose intake and the use of that glucose to produce metabolic energy. Masoro et al. (1992) have shown that plasma glucose levels are lower at all times of the day (usually in the range of 10–50% lower) in calorically restricted rats than in *ad libitum*-fed rats at all ages, and that insulin levels are even more (40–80%) reduced. Nevertheless, the calorically restricted rats used the same amount of carbohydrate per gram of lean body mass, in spite of the lower blood glucose and insulin levels. These authors refer to this as altering the "characteristics" of glucose fuel use, but we are little closer to understanding why this extends life span or delays the onset of age-related diseases such as cancer. It is consistent with the theory that nonenzymatic glycosylation (glycation) is a contributing factor in aging (Cerami 1985), and in fact hemoglobin glycation is reduced significantly by caloric restriction (Masoro, Katz, and McMahan 1989), but there is no reason to believe that glycation per se is sufficient to explain aging.

Does Caloric Restriction Work in Primates?

Nonhuman Primates

Whereas the extensive findings in rodents are promising, the question remains whether caloric restriction will extend life and/or improve health in humans. At least two research groups, one at the Gerontology Research Center in Baltimore, MD, and another at the

University of Wisconsin, have begun calorically restricting rhesus monkeys to determine whether life span and health span are extended. Unfortunately these studies have not yet progressed long enough to provide any unequivocal evidence that caloric restriction will provide any long-term health benefits to these nonhuman primates, although it is clear that 30% reduction in calories can be tolerated in rhesus monkeys without adverse effects (Kemnitz et al. 1993).

Humans

Weindruch and Walford (1988) have discussed several human populations where caloric restriction is being naturally practiced because of cultural traditions. The best example is provided by the residents of Okinawa who ingest a standard diet consisting mainly of fish, cereals, and vegetables, and who are reported to have an average longevity substantially greater than that of residents of any other Japanese island (Kagawa 1978). There are plenty of examples of other populations that are subjected to a calorically restricted diet without increasing longevity, but in most, if not all, of these cases the diet may be also nutritionally deficient.

A more controlled human experiment is currently underway in Arizona. Walford, Harris, and Gunion (1992) have described the results of a one-year study in which eight subjects (including Dr. Walford) have been consuming a nutrient-dense diet low in both calories (about 1,800 kcal per day) and fat (10% of the total calories). These four men and four women have experienced not only a 11–16% weight loss, but also significant reductions in total serum cholesterol levels (from 191 mg per dl to 124 mg per dl), fasting glucose (from 92 to 74 mg per dl) and blood pressure (from 109/74 to 89/58 mm mercury). These reductions may be promising indicators of improved health in these eight individuals, but certainly do not yet indicate whether caloric restriction in humans will eventually be shown to extend life span.

Calorie Restriction and Delay of Age-Related Disease

An early observation was that caloric restriction not only extends the maximum life span of rodents, but also delays the induction of tumors (Weindruch and Walford 1988). Our earlier discussion about the reduction of oxidative damage by caloric restriction provides at least one potential explanation for this observation. Of even more importance is the effect of caloric restriction on other forms of age-related pathological change. In the most comprehensive study yet performed, Bronson and Lipman (1991) found that caloric restriction not only reduced tumor incidence in mice, but also reduced total pathological lesions and lymphoid nodules. The mice studied included four different genotypes and both sexes, and 135 different lesions were observed. In general the calorically restricted mice developed the same pattern of lesions as the *ad libitum*-fed mice, suggesting that some common process critical in the development of most age-related pathologies may be operating, but attenuated in the calorically restricted mice. Similar but less extensive results have been reported earlier for renal disease (Fernandes et al. 1978) and a variety of other diseases (Weindruch and Walford 1988).

DIETARY LIPIDS

There is a vast amount of literature on the role of lipids and lipid metabolism in cardiovascular disease, and there is no need to cover that ground here (see Chapter 2). However, there are data available on the effect of dietary lipid composition on life span in rodents. Fernandes (1989) has found that the substitution of fish oil for corn oil in the diet of autoimmune disease-prone B/W mice has a profound effect on increasing the maximum life span, and that caloric restriction increases the life span even further. It is not clear from this experiment whether the effects observed are due to the substantially different fatty acid compositions of these two oils, or whether the presence of the omega(ω)-3 fatty acids in fish oils is responsible.

Fish oils, and the role of ω-3 fatty acids in reducing aging-related diseases such as cancer and cardiovascular disease, have been widely studied (Lands 1986). Replacement of corn oil by fish oil reduces serum cholesterol (Fernandes 1989), reduces tumor growth (Fernandes and Venkatraman 1991), and improves immune function (Fernandes 1990). The role of fish oil in improved health might be an important factor in the apparent increased longevity of the Okinawan population referred to above.

DIETARY INTAKE OF ESSENTIAL MICRONUTRIENTS

Oxidative damage has long been assumed to be a major factor in mammalian aging (Harman 1956), but while the suggestion remains promising (Warner 1992), unequivocal data in support of this theory have been difficult to obtain. One approach has been to look for correlations among aging, life span, and oxidative damage and repair. Cutler (1991) has attempted many such correlations with mixed results. He has shown that serum carotenoid levels show a complex positive correlation with maximum life span in both primates and nonprimates, and that when divided by a quantity known as specific metabolic rate (calories consumed per lifetime per gram body weight per day), serum levels of vitamin E, urate, and ascorbate also positively correlate with maximum life span. Antioxidant enzymes such as superoxide dismutase, catalase, and glutathione peroxidase show both positive and negative correlations with maximum life span. The correlation approach is also complicated by the fact that correlations may also be tissue and gender specific (Warner 1992). However, there is a strong inverse relationship between maximum life span and steady state levels of 8-hydroxyguanine in liver DNA (Cutler 1991).

It is generally assumed that nutrient antioxidants play an important role in protecting against oxidative damage *in vivo* (Machlin and Bendich 1987). One approach to evaluate the importance of these micronutrients in preventing oxidative damage is to simply ask how effective micronutrients are in responding to an oxidative challenge. Frei, Stocker, and Ames (1988) have exposed human plasma to a water soluble peroxyl radical generator at 37° C and then determined the rate of disappearance of various antioxidant components of the plasma. Vitamin C disappears very rapidly, followed by bilirubin, urate, and vitamin E, in that order; lipid peroxides do not begin to appear until all the vitamin C is consumed, but well before the other antioxidants are even 50% consumed. These results suggest that vitamin C may play the major role in attenuating oxidative damage of water soluble compounds, at least in plasma.

Zamora, Hidalgo, and Tappel (1991) have tested the importance of dietary vitamin E, β-carotene, selenium, and coenzyme Q_{10} in protecting rat liver and isolated erythrocytes from oxidative damage. Vitamin E and selenium were the most effective at preventing liver damage (as assayed by release of transaminases), whereas coenzyme Q_{10} was not effective. All four were able to reduce the amount of lipid peroxidation by about 50% in unchallenged erythrocytes, but were ineffective in protecting the erythrocytes against a subsequent peroxidative challenge at the levels employed.

The above discussion is not intended to be a comprehensive review of the literature on dietary antioxidants, but rather to provide some example of approaches being used to study the role of dietary antioxidants in preventing oxidative damage, and by implication, in slowing the rate of aging. What follows is a more detailed discussion of how some of these micronutrients may impact aging-related changes.

Vitamin C

Although vitamin C (ascorbic acid) has received considerable publicity as a dietary intervention to reduce the incidence of colds (Pauling 1970), few data exist to demonstrate how vitamin C might be helpful in preventing other diseases. Recently Fraga et al. (1991) found that oxidative damage to sperm DNA shows a strong inverse correlation to vitamin C levels in human seminal fluid. The average level of 8-hydroxyguanine in human sperm DNA is about 13 fmol per μg DNA when vitamin C levels in seminal fluid exceed 25 μmolar, but when vitamin C levels fall below 25 μmolar the 8-hydroxyguanine level increases dramatically to as much as 80 fmol per μg DNA. Varying the dietary intake of vitamin C strongly influenced the vitamin C levels in seminal fluid. Vitamin C intake may be especially critical in smokers because lower serum levels of vitamin C are found in smokers (Schechtman, Byrd, and Gruchow 1989), and these low levels correlate with increased incidence of cancer in offspring of male smokers (John, Savitz, and Sandler 1991).

Vitamin E

Packer (1991) states that "Vitamin E is well accepted as nature's most effective lipid-soluble, chain-breaking anti-oxidant," but the question remains as to what the relevance of this is to aging. Vitamin E supplementation has been implicated in an improvement in mental well-being in Finnish nursing home patients (Tolonen, Halme, and Sarna 1985), reduction in blood lipid peroxides in elderly people in Poland (Wartanowicz et al. 1984) and Finland (Tolonen et al. 1988), reduction of cataracts (Robertson, Donner, and Trevithick 1989), reduction of osteoarthritis-associated pain (Machtey and Ouaknine 1978; Blankenhorn 1986), and reduction of oral cancer (Gridley et al. 1992).

Vitamin E has also been implicated in maintenance of immune function in elderly people (Meydani et al. 1990). These authors have shown that vitamin E supplementation in the diet of healthy older subjects is accompanied by increased plasma vitamin E levels, reduced plasma lipid peroxides, increased interleukin-2 production, and a slight increase in Con A-stimulated lymphocyte proliferation. Similar results have been obtained in mice (Meydani et al. 1986). Vitamin E deficiency in rats also leads to premature erythrocyte aging as measured by degradation of band 3 protein, which leads to the appearance of "senescent-cell antigen" on the cell surface (Kay et al. 1986). Thus, it is clear that the antioxidant potential of vitamin E is important in maintaining cellular function.

Iron

If one accepts the hypothesis that oxidative damage is an important risk factor for both aging and age-related disease, it is relevant to ask what dietary factors may exacerbate the problem. Because of its role in the Fenton reaction, where it catalyses the conversion of hydrogen peroxide to hydroxyl radical, iron is an obvious candidate. Sullivan (1992) has recently summarized the case for the role of stored iron in ischemic heart disease. He particularly cites the work of Salonen et al. (1992), which suggests that both high serum ferritin concentration and high dietary iron intake correlate with increased risk of myocardial infarction in Finnish men. These results suggest the possibility that iron intake may be a critical factor not only in ischemic heart disease, but also in age-related degeneration due to oxidative damage of proteins (because most of the iron in the body is bound to proteins).

Iron in the body is mostly tightly bound to proteins such as hemoglobin, cytochromes, transferrin, and ferritin. Because of this tight binding it has not been possible to rapidly deplete iron stores by simply decreasing dietary intake (Sullivan 1992). Therefore, to be effective, attempts to decrease body iron stores by decreasing dietary intake should be accompanied either by interference with iron absorption (Thomas, Salsburey, and Greenberger 1972) or by use of compounds such as desferrioxamine to chelate the iron (Williams, Zweier, and Flaherty 1991).

Other Trace Metal Micronutrients

Copper may also play dual and contrasting roles in tissue damage because of oxidative damage. Salonen et al. (1991) have shown that high serum levels of copper correlate with increased risk of myocardial infarction in Finnish men. By analogy, one could extrapolate this to mean that high levels of copper would also contribute to the degenerative processes associated with aging. However, copper is also required for the cytoplasmic form of superoxide dismutase, the so-called Cu/Zn-superoxide dismutase.

Another antioxidant defense enzyme, glutathione peroxidase, requires selenium in the form of selenocysteine for activity. Zamora et al. (1991) have shown that supplementing a selenium-deficient diet with 1 mg selenium per kg diet increases the activity of glutathione peroxidase in the plasma of rats by over ten-fold. It is reasonable to assume that the relationship between selenium intake and glutathione peroxidase activity is responsible for the beneficial effects of selenium in reducing the risk of cancer (Salonen et al. 1985) and cardiovascular disease (Salonen et al. 1982). Selenium also appears to inhibit the peroxidation-induced increase in the synaptosomal uptake of dopamine associated with a low hematocrit, and a low hematocrit is one of the more common findings in elderly persons (Lipschitz 1986).

Both vitamin E and selenium have antioxidant effects when added to normal diets, but these effects differ in different organ systems and in prolongation of life. Although dietary selenium may regulate the activity of glutathione peroxidase and protect cells from lipid peroxidation, varying dietary selenium had no effect on either the accumulation or fluorescence characteristics of lipid-derived pigments in any type of peripheral neurons, whereas vitamin E deficiency induced an increase in pigment formation (Koistinaho, Alho, and Hervonen 1990). Vitamin C and vitamin E interact to affect serum levels of iron, total iron-binding capacity, and the distribution of iron in the rat liver and spleen. Iron uptake by cultured rat myocardial cells and cell membranes is affected by both vitamins C and E (Hershko, Link, and Prinson 1987; Link et al. 1989). These investigators also showed that accumulation of iron resulted in increased cellular concentrations of malonyldialdehyde (MDA), a breakdown product of lipids indicating increased peroxidation, and vitamin E inhibited MDA formation in iron-loaded heart cells. However, vitamin C accelerated malondialdehyde production, even though the addition of vitamin C resulted in a significant reduction in iron uptake. The protective effect of vitamin E against lipid peroxidation in myocardial cells was also directly demonstrated.

Thus, excess dietary copper or iron represents a risk to mature organisms including man, but the level of exposure necessary to increase age-dependent degenerative changes is at present unknown. Dietary manipulation of both iron and copper intake may prove to be a simple and straightforward means of attenuating the aging process in general. In understanding the role of micronutrients such as iron, copper, and selenium in age-related pathology it is important to develop a broader definition of "normal status." Sullivan (1992) suggests that a physiologically normal iron status should be defined as "the absence of iron stores without anemia." Assuming it is possible to determine such a status, comparable definitions might apply to the status of copper, selenium, and other micronutrients.

SUMMARY

From the above discussion three potential dietary strategies for affecting the rate of aging in humans emerge: (1) reduction of caloric intake, (2) prevention and/or repair of oxidative damage, and (3) intake of ω-3 fatty acids. These strategies may not be mutually exclusive, e.g., caloric restriction may, in fact, decrease net oxidative damage to macromolecules. Central to the discussion was the assumption that risk factors that lead to poor health in older individuals are also risk factors for the development of age-related pathology, and that oxidative damage is intimately linked to both. Finally, implicit in the discussion, but not directly addressed, is the importance of taking a moderate approach to dietary supplementation and depletion, which requires a more thorough knowledge of contrasting roles of macro- and micronutrients.

References

Blankenhorn, G. 1986. Clinical efficiency of spondyvit (vitamin E) in activated arthroses. A multicenter, placebo-controlled, double-blind study. *Z. Orthop.* 124:340–343.

Bronson, R.T., and Lipman, R.D. 1991. Reduction in rate of occurrence of age related lesions in dietary restricted laboratory mice. *Growth, Develop Aging* 55:169–184.

Cerami, A. 1985. Hypothesis: Glucose as a mediator of aging. *J. Am. Geriatr. Soc.* 33:626–634.

Chatterjee, B.; Fernandes, G.; Yu, B.P.; Song, C.; Kim, J.M.; Demyan, W.; and Roy, A.K. 1989. Calorie restriction delays age-dependent loss in androgen responsiveness of the rat liver. *FASEB J.* 3:169–173.

Cutler, R.G. 1991. Antioxidants and aging. *Am. J. Clin. Nutr.* 53:373S–379S.

Daynes, R.A., and Araneo, B.A. 1992. Prevention and reversal of some age-associated changes in immunologic responses by supplemental dehydroepiandrosterone sulfate therapy. *Aging: Immunol. Infect. Dis.* 3:135–154.

Fernandes, G. 1989. Effect of dietary fish oil supplement on autoimmune disease: Changes in lymphoid cell subsets, oncogene MRNA expression and neuroendocrine hormones. In *Health Effects of Fish and Fish Oils*, R. Chandra, ed., pp. 409–433. St. Johns, Newfoundland: ARTS Biomedical Publishers.

———. 1990. Modulation of immune functions and aging by omega-3 fatty acids and/or calorie restriction. In *The Potential for Nutritional Modulation of Aging Processes*, D.K. Ingram, G.T. Baker, and N.W. Shock, eds., pp. 263–287. Trumbull, CT: Food and Nutrition Press.

Fernandes, G.; Friend, P.; Yunis, E.J.; and Good, R.A. 1978. Influence of dietary restriction on immunological function and renal disease in (NZB × NZW) F1 mice. *Proc. Natl. Acad. Sci. USA* 76:1500–1504.

Fernandes, G., and Venkatraman, J.T. 1991. Modulation of breast cancer growth in nude mice by omega-3 lipids. *World Rev. Nutr. Diet.* 66:488–503.

Fraga, C.G.; Shigenaga, M.K.; Park, J.-W.; Degan, P.; and Ames, B.N. 1990. Oxidative damage to DNA during aging: 8-hydroxy-2′-deoxyguanosine in rat organ DNA and urine. *Proc. Natl. Acad. Sci. USA* 87:4533–4537.

Fraga, C.G.; Motchnik, P.A.; Shigenaga, M.K.; Helbrock, H.J.; Jacob, R.A.; and Ames, B. 1991. Ascorbic acid protects against endogenous oxidative DNA damage in human sperm. *Proc. Natl. Acad. Sci. USA* 88:11003–11006.

Frei, B.; Stocker, R.; and Ames, B.N. 1988. Antioxidant defense and lipid peroxidation in human blood plasma. *Proc. Natl. Acad. Sci. USA.* 85:9748–9752.

Gridley, G.; McLaughlin, J.K.; Block, G.; Blot, W.J.; Gluch, M.; and Fraumeni, J.F., Jr. 1992. Vitamin supplement use and reduced risk of oral and pharyngeal cancer. *Am. J. Epidemiol.* 135 (10):1083–1092.

Harman, D. 1956. Aging: a theory based on free radical and radiation chemistry. *J. Gerontol.* 11:298–300.

Hershko, C.; Link, G.; and Prinson, A. 1987. Modification of iron uptake and lipid peroxidation by hypoxia, ascorbic acid, and α-tocopherol in iron-loaded rat myocardial cell cultures. *J. Lab. Clin. Med.* 110:355–361.

John, E.M.; Savitz, D.A.; and Sandler, D.P. 1991. Parental exposure to parents' smoking and childhood cancer. *Am. J. Epidemiol.* 133:123–132.

Kagawa, Y. 1978. Impact of westernization on the nutrition of Japanese: Changes in physique, cancer, longevity and centenarians. *Prev. Med.* 7:205–217.

Kay, M.M.B.; Bosman, G.J.C.G.M.; Shapiro, S.S.; Bendich, A. ˜d Bassel, P.S. 1986. Oxidation as a possible mechanism of cellular aging: Vitamin E deficiency causes premature aging and IgG binding to erythrocytes. *Proc. Natl. Acad. Sci. USA* 83:2463–2467.

Kemnitz, J.W.; Weindruch, R.; Roecker, E.B.; Crawford, K.; Kaufman, P.L.; and Ershler, W.B. 1993. Dietary restriction of adult male rhesus monkeys: Design, methodology and preliminary findings from the first year of study. *J. Gerontol.* 48 (1):B17–B26.

Koistinaho, J.; Alho, H.; and Hervonen, A. 1990. Effect of Vitamin E and selenium supplement on the aging peripheral neurons of the male Sprague-Dawley rat. *Mech. Ageing Dev.* 51:63–72.

Lands, W.E.M. 1986. *Fish and Human Health.* Orlando, FL: Academic Press.

Langaniere, S., and Yu, B.P. 1987. Anti-lipoperoxidation action of food restriction. *Biochem. Biophys. Res. Commun.* 145:1185–1191.

Li, D.; Wang, Y.-M.; Nath, R.G.; Mistry, S.; and Randerath, K. 1991. Modulation by dietary vitamin E of I-compounds (putative indigenous DNA modifications) in rat liver and kidney. *J. Nutr.* 121:65–71.

Licastro, F.; Weindruch, R.; Davis, L.J.; and Walford, R.L. 1988. Effect of dietary restriction upon the age-associated decline of lymphocyte DNA-repair activity in mice. *Age* 11:48–52.

Link, G.; Prinson, A.; Kahane, I.; and Hershko, C. 1989. Iron binding modifies the fatty acid composition of cultured rat myocardial cells and liposomal vesicles: Effect of ascorbate and α-tocopherol on myocardial lipid peroxidation. *J. Lab. Clin. Med.* 114:243–249.

Lipschitz, D.A. 1986. Nutrition and aging of the hematopoietic system. In *Nutrition and Aging* M.L. Hutchinson and H.N. Munro., eds., pp. 251–262. Orlando, FL: Academic Press.

Machlin, L.J., and Bendich, A. 1987. Free radical tissue damage: Protective role of antioxidant nutrients. *FASEB J.* 1:441–445.

Machtey, I., and Ouaknine, L. 1978. Tocopherol in osteoarthritis: A controlled pilot study. *J. Am. Geriatr. Soc.* 26:328–330.

Masoro, E. 1989. Food restriction research: Its significance for human aging. *Am. J. Human Biol.* 1:339–345.

Masoro, E.J.; Katz, M.S.; and McMahan, C.A. 1989. Evidence for the glycation hypothesis of aging from the food-restricted rodent model. *J. Gerontol.* 44(1):B20–B22.

Masoro, E.J.; McCarter, R.I.M.; Katz, M.S.; and McMahan, C.A. 1992. Dietary restriction alters characteristics of glucose fuel use. *J. Gerontol.* 47 (6):B202–B208.

Meydani, S.N.; Meydani, M.; Verdon, C.P.; Shapiro, A.C. Blumberg, J.B.; and Hayes, K.C. 1986. Vitamin E supplementation suppresses prostaglandin E_2 synthesis and enhances immune response of aged mice. *Mech. Ageing Dev.* 34:191–201.

Meydani, S.N.; Barklund, M.P.; Liu, S.; Meydani, M.; Miller, R.A.; Cannon, J.G.; Morrow, F.D.; Rocklin, R.; and Blumberg, J.B. 1990. Vitamin E supplementation enhances cell-mediated immunity in healthy elderly subjects. *Am. J. Clin. Nutr.* 52:557–563.

Packer, L. 1991. Protective role of vitamin E in biological systems. *Am. J. Clin. Nutr.* 53:1050S–1055S.

Park, E.M.; Shigenaga, M.K.; Degan, P.; Korn, T.S.; Kitzler, J.W.; Wehr, C.M.; Kolachan, P.; and Ames, B.N. 1992. Assay of excised oxidative DNA lesions: Isolation of 8-oxoguanine and its nucleoside derivatives from biological fluids with a monoclonal antibody column. *Proc. Natl. Acad. Sci. USA* 89:3375–3379.

Pauling, L. 1970. *Vitamin C and the Common Cold.* San Francisco: W.H. Freeman.

Randerath, K.; Hart, R.; Zhou, G.-D.; Reddy, R.; Danna, T.F. ˜d Randerath, E. 1993. Enhancement of age-related increases in DNA I-compound levels by calorie restriction: Comparison of male B-N and F-344 rats. *Mutation Res.* 295:31–46.

Randerath, K.; Reddy, M.V.; and Disher, R.M. 1986. Age and tissue related DNA modifications in untreated rats: Detection by ^{32}P-postlabeling assay and possible significance for spontaneous tumor induction and aging. *Carcinogenesis* 7:1615–1617.

Richardson, A.; Butler, J.A.; Rutherford, M.S.; Semsei, I. θ, M.-Z.; Fernandes, G.; and Chiang, W.-H. 1987. Effect of age and dietary restriction on the expression of α_{2u}-globulin. *J. Biol. Chem.* 262 (26):12851–12825.

Robertson, J.; Donner, A.P.; and Trevithick, J.R. 1989. Vitamin E intake and risk of cataracts in humans. *Ann. NY Acad. Sci.* 570:372–382.

Salonen, J.T.; Alfthan, G.; Huttunen, J.K.; Pikkarainen, J.; and Puska, P. 1982. Association between cardiovascular death and myocardial infarction and serum selenium in a matched-pair longitudinal study. *Lancet* 2:175–179.

Salonen, J.T.; Salonen, R.; Lappeteläinen, R.; Mäenpää, P.; Alfthan, G.; and Puska, P. 1985. Risk of cancer in relation to serum concentrations of selenium and vitamins A and E: Matched case-control analysis of prospective data. *Br. Med. J.* 290:417–420.

Salonen, J.T.; Salonen, R.; Korpela, H.; Suntioinen, S.; and Tuomilehto, J. 1991. Serum copper and the risk of acute myocardial infarction: A prospective population study in men in eastern Finland. *Am. J. Epidemiol.* 134:268–276.

Salonen, J.T.; Nyyssönen, K.; Korpela, H.; Tuomileho, J.; Seppänen, R.; and Salonen, R. 1992. High stored iron levels are associated with excess risk of myocardial infarction in eastern Finnish men. *Circulation* 86:803–811.

Saul, R.L.; Gee, P.; and Ames, B.N. 1987. In *Modern Biological Theories of Aging*, H.R. Warner, R.N. Butler, R.L. Sprott, and E.L. Schneider, eds., pp. 113–129. New York: Raven Press.

Schechtman, G.; Byrd, J.C.; and Gruchow, H.W. 1989. The influence of smoking on vitamin C status in adults. *Am. J. Publ. Health* 79:158–162.

Stadtman, E.R. 1991. Metal catalysed oxidation of proteins. *J. Biol. Chem.* 266:2005–2008.

———. 1992. Protein oxidation and aging. *Science* 257:1220–1224.

Starke-Reed, P.E., and Oliver, C.N. 1989. Protein oxidation and proteolysis during aging and oxidative stress. *Arch. Biochem. Biophys.* 275:559–567.

Sullivan, J.L. 1992. Stored iron and ischemic heart disease: Empirical support for a new paradigm. *Circulation* 86:1036–1037.

Thomas, F.B.; Salsburey, D.; and Greenberger, N.J. 1972. Inhibition of iron absorption by cholestyramine: Demonstration of diminished iron stores following prolonged administration. *Am. J. Dig. Dis.* 17:263–269.

Tolonen, M.; Halme, M.; and Sarna, S. 1985. Vitamin E and selenium supplementation in geriatric patients. *Biol. Trace Element Res.* 7:161–168.

Tolonen, M.; Sarna, S.; Halme, M.; Tuominen, S.E.J.; Westermarck, T.; Nordberg, U.-R.; Keinonen, M.; and Schrijver, J. 1988. Antioxidant supplementation decreases TBA reactants in serum of elderly. *Biol. Trace Element Res.* 17:221–228.

Waggoner, S.; Gu, M.-Z.; Chiang, W.-H.; and Richardson, A. 1990. The effect of dietary restriction on the expression of a variety of genes. In *Genetic Effects on Aging II*, David E. Harrison, ed., pp. 255–273. Caldwell, NJ: The Telford Press.

Walford, R.L.; Harris, S.B.; and Gunion, M.W. 1992. The calorically restricted, low-fat, nutrient dense diet in Biosphere 2 significantly lowers blood glucose, total white cell count, cholesterol, and blood pressure in humans. *Proc. Natl. Acad. Sci. USA* 89:11533–11537.

Wallace, D.C. 1992. Mitochondrial genetics: A paradigm for aging and degenerative diseases? *Science* 256:628–632.

Warner, H.R. 1992. Overview: Mechanisms of antioxidant action on life span. In *Antioxidants: Chemical, Physiological, Nutritional and Toxicological Aspects*, Helmut Sies, John Erdman, Jr., George Baker III, Carol Henry, and Gary Williams, eds., pp. 151–161. Princeton, NJ:Princeton Scientific.

Warner, H.R.; Butler, R.N.; Sprott, R.L.; and Schneider, E.L. 1987. *Modern Biological Theories of Aging*. New York:Raven Press.

Wartanowicz, M.; Panczenko-Kresowka, B.; Ziemlanski, S.; Kowalska, M.; and Okolska, G. 1984. The effect of α-tocopherol and ascorbic acid on the serum lipid peroxide level in elderly people. *Ann. Nutr. Metab.* 28:186–191.

Weindruch, R., and Walford, R.L. 1982. Dietary restriction in mice beginning at one year of age: Effects on life span and spontaneous cancer incidence. *Science* 215:1415–1418.

———. 1988. *The retardation of aging and disease by dietary restriction*. Springfield, IL: Charles C. Thomas.

Weindruch, R.; Walford, R.L.; Fligiel, S.; and Guthrie, D. 1986. The retardation of aging by dietary restriction: Longevity, cancer, immunity, and lifetime energy intake. *J. Nutrition* 116:641–654.

Weraarchakul, N.; Strong, R.; Wood, W.G.; and Richardson, A. 1989. The effect of aging and dietary restriction on DNA repair. *Exp. Cell Res.* 181:197–204.

Williams, R.E.; Zweier, J.L.; and Flaherty. 1991. Treatment with deferoxamine during ischemia improves functional and metabolic recovery and reduces reperfusion-induced oxygen radical generation in rabbit hearts. *Circulation* 83:1006–1014.

Youngman, L.D.; Park, J.-Y.K.; and Ames, B.N. 1992. Protein oxidation associated with aging is reduced by dietary restriction of protein or calories. *Proc. Natl. Acad. Sci. USA* 89:9112–9116.

Yu, B.P.; Masoro, E.J.; and McMahan, C.A. 1985. Nutritional influence on aging of Fischer 344 rats. I. Physical, metabolic and longevity characteristics. *J. Gerontol.* 40:657–670.

Zamora, R.; Hidalgo, F.J.; and Tappel, A.L. 1991. Comparative antioxidant effectiveness of dietary β-carotene, vitamin E, selenium, and coenzyme Q_{10} in rat erythrocytes and plasma. *J. Nutr.* 121:50–56.

Chapter 7

Mood and Performance Foods

Herbert L. Meiselman and
Harris R. Lieberman

INTRODUCTION

Over the past two decades a substantial body of research has developed demonstrating that foods can have an impact on behavior and mood. This phenomenon is in contrast with other functions of foods covered in this volume, because many of the behavioral effects of foods are rapid and short lived rather than slowly accumulating over time. General reviews of the topic of foods that have an impact on mood and performance can be found in Kruesi and Rapaport (1986), Kanarek and Marks-Kaufman (1991), Lieberman (1989), and the National Research Council (in press).

Despite some sensational claims and a growing body of solid research, there have been few attempts to apply the mood-altering and performance-altering potential of foods to real world problems. Part of the reason for this relates to the need for more maturation and growth of the field. Part relates to difficulties of conducting what is, inherently, interdisciplinary research involving chemistry and biochemistry, physiology, pharmacology, psychology, and medicine. However, a major reason relates to the vast range of methods that has been applied to this research area. This diversity has raised important methodological questions pertaining not only to mood/ performance foods, but also to general issues relating to the assessment of human behavior. Unfortunately, the wide variety of methods has yielded an assortment of results, making simple predictions difficult. Perhaps the alteration of mood and performance by foods is inherently complex; perhaps that apparent complexity can be reduced with further research and understanding.

This chapter will review some of the important methodological issues pertaining to mood and performance and then will discuss recent mood and performance research on caffeine, amino acids, and carbohydrates.

METHODOLOGICAL ISSUES

The growth of the field of nutrition and behavior has resulted in a growth in related methodological studies. Various methodological issues have been discussed in a number of references (Kruesi and Rapaport 1986; Christensen 1991; Meiselman and Kramer in press; Lieberman, Spring, and Garfield 1986b; Lieberman 1989). Some key questions relate to the determination of baseline, selection of subjects, selection of test foods, design and delivery of treatment (drug) conditions, and selection of dependent measures of mood and performance.

Baseline

The period prior to testing is important for two reasons. It is generally used as a baseline against which to compare the treatment. Additionally, it provides a more uniform context in which the treatment can work. There are a number of alternative ways to control the period before testing (see Christensen 1991; Meiselman and Kramer in press). The most demanding is a lengthy elimination or washout phase comparable to those sometimes used in allergy research. A washout phase allows for baseline measures without any of the treatment material in the system, and it also allows for the treatment to manifest itself fully upon introduction. However, a washout phase is quite intrusive to the subject's normal eating, and one might question whether one is observing a baseline mood and performance or merely another demanding treatment.

Less intrusive than a complete washout phase is abstinence from eating for a brief period, usually after dinner the previous day. Alternatively, the experimenter can provide or recommend one or more standardized meals before the treatment. Standard meals can be consumed at home (but without supervision) or at the laboratory. Both abstinence from eating and standardized meals still change the context of the treatment, as noted by Meiselman and Kramer (in press).

In addition to concerns regarding baseline eating, one must also consider the baseline mood and behavior of subjects. If one is interested in anxiety, should one reduce anxiety through relaxing conditions, or exaggerate anxiety through stressful conditions, or merely measure what anxiety exists? The problem with the latter may be the wide range in baseline mood that could result. This range could make it more difficult to observe treatment effects if those effects are present differentially as a function of pre-existing mood state.

Expectancy and Placebo

Subjects' expectations about what might occur and/or about what they "should do" can have dramatic effects on behavior. There is substantial evidence of expectancy effects for painkillers, alcohol, and caffeine. The solution to this problem has been the use of a placebo, which is designed to deliver zero level of the treatment (or drug) presented in an identical form as the treatment. A placebo should look, taste and smell the same as the treatment. The instructions to the subject should be the same. When placebos are combined with double blind conditions in which both the experimenter and the subject are not informed of the treatment condition, then the outcomes of treatment can be distinguished from the outcomes of placebos. The latter is the effect of expectancy.

Unfortunately, a placebo is not always strictly possible. It is possible to substitute decaffeinated coffee for caffeinated coffee (if one demonstrated the same sensory properties). But it is somewhat more difficult to provide a long-term caffeine-free diet. It is possible to substitute sugar in a single product, but very difficult to provide carbohydrate-free diets.

The way that instructions are worded can affect subjects even under what appears to be appropriate experimental design. Kirsch and

Weixel (1988) gave subjects different levels of a placebo of decaffeinated coffee with either double blind or deceptive instructions. With deceptive instructions, subjects expected an active drug (caffeine); with double blind instructions, subjects expected either a placebo or an active drug. The two instruction conditions produced different, sometimes opposite, results. Kirsch and Weixel (1988) recommend a balanced placebo design (Marlatt and Rohsenow 1980) that includes a condition in which subjects are told they will not get a drug, but do receive it; however, this may present practical and ethical dilemmas.

Selecting Subjects

When conducting research on effects of foods on mood and performance, or considering application of that research, subject selection and variability must be considered. One cannot overestimate the problem of individual differences, although Dews (1984) has correctly cautioned not to attribute all variability to inherent subject variability.

First, when considering foods, one must determine whether an individual being tested avoids the food, uses the food moderately, or uses the food heavily. People who avoid specific foods are probably not good subjects in studies designed to demonstrate eventual uses of those mood/behavior altering foods. Most people have a surprisingly narrow food repertoire, and many individuals and cultures exclude large numbers of foods (Rozin and Vollmecke 1986). However, many individuals who attempt to completely exclude common food constituents such as caffeine from their diet are not successful. Studies of some mood/behavior enhancing foods limit themselves to users or moderate users, but there are no standards for how to determine who serves as subjects. Generalizing from moderate users to other groups may not be appropriate.

Another key issue in mood/behavior research is in the area of psychological or psychiatric disorders. Many of the mood studies have used hospitalized patients (Sawyer, Julia, and Turin 1982). What is the relationship between decreased anxiety in a clinical patient to mood changes in a nonclinical ("normal") subject? Also, when considering overall dietary influences on mood/behavior, one must consider that a number of mood states might depress or increase eating, thereby reducing or increasing the potential mood-altering effects of food.

When considering the possible effects of treatments, individual differences are also a factor either because the dose-response relationship varies across subjects, or because some subjects do not respond at all. The latter issue has been shown to be a key factor with monosodium glutamate studies; i.e., many people do not have any allergic reaction to it, and simply asking people whether they do is not a good predictor (Meiselman 1987).

Drugs/Foods/Meals

The delivery of the treatment to the subject involves one of the following: delivery of a drug by itself; of a drug within a food, perhaps as a natural constituent; or of a drug within an entire meal. For example, caffeine can be delivered in pure form in a capsule, in a solid food, or in a cup of coffee. Carbohydrate can be delivered as sugar water or as an entire carbohydrate-rich meal. Meals can be extended to form diets. The issue of relevance of stimuli in food research has been discussed (Meiselman 1992; Meiselman and Kramer in press; Spitzer and Rodin, 1981). The context in which a drug is delivered (drug/food/meal) can have a great influence on whether that drug is consumed (Hirsch and Kramer in press) and could influence the behavior/mood effects of the drug.

Measures of Mood and Performance

One of the greatest methodological stumbling blocks in food/behavior research is the selection of specific tests of mood and behavior performance (Lieberman 1989, 1992; Lieberman, Spring, and Garfield 1986). A very wide range of tests is available to measure moods/attitudes and to measure various aspects of performance. The latter is sometimes divided into sensory, physical, and cognitive performance. Sensory tests include simple thresholds and more complex signal detection tests. Physical measures can include everything from simple motor tests (e.g., reaction time) to complex marathon runs through obstacle courses. Cognitive tests include everything from simple tests of choice reaction time to highly complex problem-solving, learning, and memory tests. Many of these tests depend on multiple factors within the individual, and a drug effect cannot be simply tied to one factor. Further, performance on these tests can show high intrasubject and intersubject variation. Many tests will require a long practice period in order to reach a stable performance level against which to measure treatment. Often, factors such as fatigue, practice, time of day, and other situational factors will be more powerful than the treatment, thereby making it difficult to observe treatment effects. Experimenters often include a number of different mood/behavior measures; this increased number of measures requires different statistical treatment to control for the increased number of opportunities to observe effects (Christensen, Miller, and Johnson 1991). This statistical correction is often not made, increasing the chances for drawing incorrect conclusions.

Not only has there been difficulty in selecting the best measure of mood or vigilance or some other mood/behavior, but there is a growing recognition that we need more integrated measures. Such measures would reflect sensory, physical, and cognitive components working together. Overall job performance might be such a measure. Do drugs or diet affect the way people function? Do they affect our overall cognitive functioning, not just vigilance or memory? The need for such integrative measures will eventually result in development and testing of such techniques.

CAFFEINE

Caffeine is a member of the class of naturally occurring substances called methylxanthines, whose discovery and identification have been reviewed by Arnaud (1984). The caffeine compound 1,3,7-trimethyl xanthine was discovered as early as 1820 and its chemical structure determined as early as 1875. Other methylxanthines may have behavioral effects, but caffeine has received the most attention, probably because it is found in much greater quantities in foods than other xanthines.

Unfortunately, despite knowledge about its chemistry, its usage, and its acute physiological effects, there is still disagreement regarding its behavioral effects in man (Lieberman 1992; Dews 1984; Sawyer, Julia, and Turin 1982; Kruesi and Rapaport 1986).

Caffeine Use

The amount of caffeine in foods varies widely. Coffee has both the highest and the most variable caffeine content (Barone and Roberts 1984). In general, freshly brewed coffee has a higher caffeine content than instant coffee. Both contain more caffeine than tea, cola beverages, or other soft drinks. Decaffeinated coffee, cocoa, and chocolate milk tend to have a very low caffeine content (Barone and Roberts 1984; Lieberman 1992). Coffee (75%) and tea (15%) accounted for 90% of caffeine consumption in the United States approximately ten years ago; in Canada, coffee accounted for less (60%) and tea more (30%) (Barone and Roberts 1984). The extent to which different foods contribute to caffeine consumption varies considerably throughout the world. Current trends to drink less coffee are well established in the United States. In addition to beverages, a number of nonprescription preparations contain caffeine (Sawyer, Julia, and Turin 1982) including painkillers, cold medicines, and over-the-counter stimulants.

A key issue relative to the behavioral effects of caffeine is the distinction between users and nonusers. This distinction is often used as a criterion of subject selection. In their detailed analysis of standard consumption estimates, Barone and Roberts (1984) deal mainly with coffee consumers. As more people make what they believe to be health-oriented dietary decisions, there needs to be more detailed data collected on consumers/nonconsumers of caffeine. As Sawyer, Julia, and Turin (1982) have noted, brief abstention from caffeine does not rid the body of caffeine in heavy caffeine users. Caffeine's behavioral effects need to be examined in individuals who are consuming their normal levels of caffeine.

Caffeine Physiology and Effects

The physiological effects of caffeine appear to depend on the mediating role of adenosine, a neuromodulator of similar chemical structure (Snyder 1984). The methylxanthines, including caffeine, block adenosine at its receptor sites and produce an increase in central nervous system activity (Snyder 1984; Lieberman 1992). Sawyer, Julia, and Turin (1982) and Hirsch (1984) summarize the physiological effects of caffeine. They are also discussed in various chapters of the volume edited by Dews (1984).

Dews (1984) has argued against the widely held belief that caffeine's effects vary greatly from person to person. One must consider dosage factors to equate for body weight and rates of stomach emptying and intestinal absorption, all leading to different plasma levels of caffeine. One must also consider different foods normally consumed with caffeine (milk and sugar, fatty foods, etc.), the effects of tolerance, the effects of the normal eating situation, and certain personality factors. Given all these variables, and perhaps others as yet unknown, it is not surprising that the effects of caffeine are seen as highly individual, although it is not at all clear that caffeine by itself produces such idiosyncratic effects.

Behavioral Effects: Physical

Comprehensive data on caffeine's effects on simple sensory or motor functions are not available. Although moderate caffeine doses have not been shown to affect simple sensory tasks, there has been only limited testing of a few sensory modalities (Dews 1984; Lieberman 1992).

Caffeine appears to reduce simple and complex reaction time, although Lieberman (1992) points out that "such effects are difficult to detect consistently because of their modest size, nonmonotonic nature, and what may be a complex relationship between dose administered and the caffeine consumption history of the subjects."

Behavioral Effects: Cognitive

Caffeine in moderate doses appears to increase vigilance. Lieberman (1992), in a series of studies, adapted the Wilkinson vigilance test and consistently found that caffeine affects the ability of subjects to distinguish shorter from longer tones. In the first study, Lieberman utilized a double blind crossover design and excluded smokers and heavy caffeine consumers. Caffeine, in capsule form, at 32, 64, 128, and 256 mg was provided at 0800 hours followed by morning testing. Vigilance detection rate increased for all four caffeine doses. This result has been replicated for 64, 128, and 256 mg (Lieberman et al. 1987b, 1988). Although it is surprising that the four original caffeine doses did not result in significant differences in vigilance scores, clearer dose response relationships were seen in a follow-up study (Lieberman et al., 1987a,b; Lieberman 1992).

Sawyer, Julia, and Turin (1982) in an earlier review noted the mixed results for vigilance testing, with some studies showing increased vigilance and some studies showing no increase (one visual vigilance study showed a decrease). Loke and Meliska (1984) failed to observe a vigilance effect of caffeine using a visual vigilance task. Fine et al. (in press) have found robust effects of caffeine on visual

Table 7-1. The Effect of Caffeine on Alertness.

	Dependent variable	Effect	Dose (mg)
Clubley et al. 1979	Mental sedation	+	75, 150, 300
Leathwood and Pollet 1982	Bipolar scales of vigor, awake, full-of-go	+	100
Griffiths et al. 1990	Alertness, etc.	+	100
Roache and Griffiths 1987	Alertness, vigor	+	200, 400, 600
	Fatigue	−	200, 400, 600
Loke 1988	Alert, energetic	o	0, 200, 400
Svensson, Persson, and Sjoberg 1980	Activation	o	0, 100
Swift and Tiplady 1988	Alert	+[a]	0, 200
Lieberman et al. 1987b	Alert, vigor	+	64, 128

[a]Young volunteers only.

vigilance using a test constructed to incorporate key characteristics of the modified Wilkinson vigilance task used by Lieberman et al. (1987b). Loke, Hinrichs, and Ghoneim (1985) had also failed to find effects of caffeine on a variety of cognitive and other tasks.

Given the difficulty of reliably showing caffeine effects on reaction time on vigilance, it is not surprising that more complex cognitive tasks such as memory or learning show inconclusive results. Lieberman (1992) catalogs a number of failures to observe such effects. Sawyer, Julia, and Turin (1982) and Dews (1984) also review the studies in this area. Lieberman (1992) has noted the high intersubject variability of such complex cognitive tasks and suggested the need for greater sample sizes. More complex cognitive tasks also bring into play more of the other variables of the subject, the task, and the situations; thus, it is more difficult for a caffeine effect, if any, to show through the noise.

Behavioral Effects: Mood

Caffeine appears to enhance alertness and reduce fatigue; it may reduce anxiety at lower dosage and increase anxiety at higher dosage. A large number of studies confirm that caffeine in a wide variety of dosages increases alertness and related moods and reduces fatigue and related moods (Table 7-1.). Several things do not appear to have been investigated to date: whether caffeine exerts a discriminable alerting effect for long-term users, as opposed to brief laboratory tests; whether caffeine added to a baseline of caffeine consumption produces any increased alertness; and how caffeine's alerting properties differ in users and nonusers. These types of gaps in the data make it difficult to fully utilize caffeine to promote alertness.

Sawyer, Julia, and Turin (1982) ascribe a number of anxiety reactions to caffeine, including restlessness and depression, while Lieberman (1992) concludes that "caffeine can increase anxiety when administered in single bolus doses of 300 mg or higher, which is many times greater than the amount present in a single serving of a typical caffeine-containing beverage."

The results of a number of studies are shown in Table 7-2.

In their review, Sawyer, Julia, and Turin (1982) report no causal relationship between caffeine and depression, but report a caffeine-depression association in studies that also report a caffeine-anxiety association. They go on to discuss the possibility that caffeine relieves anxiety, but later enhances depression, thus leading to more caffeine to reduce anxiety. Lieberman (1988) noted reduced depression following moderate caffeine doses.

Simulator Studies

As noted above, caffeine studies to date generally do not predict how to deal with real-life living and working environments. Two studies address such concerns and demonstrate that such practical studies are possible. Regina et al. (1974) found significant positive effects of 200 mg caffeine, as compared to a placebo, in a 90-minute simulated driving test by male drivers. After 90 minutes, a second 200-mg dose yielded limited additional effects. Johnson (1991) also found significant positive effects of 200 mg caffeine, as compared to a placebo, this time for three hours of simulated rifle firing by male soldiers. A common element in these very different kinds of studies may be the inclusion of a substantial vigilance component in the simulated task. Such findings reinforce the results of studies that employed laboratory vigilance tasks and found behavioral effects of caffeine (Lieberman 1992).

Caffeine: Summary

The most robust behavioral consequence of caffeine administration appears to be improved performance on tasks that require sustained

Table 7-2. The Effect of Caffeine on Anxiety.

	Subject	Variable	Effect	Dose
Gilliland and Andress 1981	College students	State Trait Anxiety Inventory	+	1–5 cups/day, >5 cups/day
		Beck Depression Scale[a]	+	
Winstead 1976	Patients	State Trait Anxiety Inventory	+	>5 cups/day
Greden et al. 1978	Patients	State Trait Anxiety Inventory	+	>5 cups/day
Cole et al. 1978	Drug users	State Trait Anxiety Inventory	+	300 mg
Loke 1988	College students	20 6-point scale S-nervous, tense	+	0[b], 200[b], 400
Roache and Griffiths 1987		Tension	o	200, 400, 600
Svensson, Persson, and Sjoberg 1980	College students	Calmness scale	o	0, 100
Lieberman et al. 1987b	Young males	Profile of Mood Scale	–	64, 128

[a]For high coffee (>5 cups per day) users.
[b]No effect.

vigilance. Caffeine also appears to have effects on mood states, such as alertness and fatigue, that are consistent with its effects on vigilance. The effects of caffeine on more complex cognitive tasks remains unresolved, but it is unlikely that this substance has direct, robust effects on complex cognitive function. Because caffeine has been shown to improve vigilance in laboratory tasks and simulations of tasks with substantial vigilance components, its use in situations where the maintenance of vigilance is critical, such as vehicle operation, may be beneficial.

LARGE NEUTRAL AMINO ACIDS

Amino acids have only recently been evaluated for possible effects on behavior, unlike caffeine which has been studied for many years. In those cases where effects have been observed, they are of a modest nature, as would be expected from compounds that are found in significant quantities in common foods.

The amino acids most likely to alter behavior directly are those that cross the blood-brain barrier. Amino acids typically lack the physicochemical properties that permit access to the brain; thus, specific transport mechanisms must exist to facilitate their transport into the central nervous system. Such a mechanism exists for the large neutral amino acids (LNAA) and allows these compounds to readily cross the blood-brain barrier. Several LNAA are of special interest because they are precursors of critical brain neurotransmitters. Furthermore, under certain conditions their availability may alter the concentration of the central neurotransmitters that are synthesized from them (Wurtman, Hefti, and Melamed 1981; Fernstrom 1983). Therefore, a rational basis exists for examining the consequences of administration of these amino acids on behavior.

The amino acid that has been shown to affect brain neurotransmission most readily is tryptophan, an essential amino acid and the precursor of the neurotransmitter serotonin and the hormone melatonin. Administration of tryptophan in sufficient quantities increases both brain tryptophan and serotonin concentration. Another large neutral amino acid, tyrosine, is the precursor of a class of several neurotransmitters, the catecholamines. Dopamine, norepinephrine, and epinephrine are catecholamines that have a variety of central and peripheral functions. Catecholaminergic neurons only appear to be precursor sensitive under certain conditions. As a consequence, it has been more difficult to demonstrate unequivocal effects of tyrosine compared to tryptophan on brain function (Lieberman in press). It has been suggested that protein and carbohydrate foods may affect behavior because of the effects these foods have on the large neutral amino acids and therefore, perhaps, brain neurotransmitters.

Tryptophan

Initial reports that L-tryptophan had behavioral effects appeared in the 1960s (for a review see Hartmann 1982). The currently available literature indicates that when tryptophan is administered in sufficient doses it has sedative-like effects on humans. A compound that has such effects may or may not promote sleep; if it does, it is also a hypnotic. It appears that substantial sedative-like effects are only present when doses of three grams or higher are administered, although clear dose-response relationships have not been documented (Borbely and Youmbi-Balder 1986). Daily tryptophan intake in the human diet ranges from 1 to 1.5 grams, so tryptophan's behavioral effects do not appear to be present when it is consumed in doses found in foods. Furthermore, considerable controversy exists regarding the nature of tryptophan's effects on arousal level and sleep. Many human studies have demonstrated that this amino acid reduces alertness, as measured by self-report mood questionnaires, and sleep latency as measured by EEG recordings (Hartmann 1982; Lieberman et al. 1982). Several negative studies have also appeared, but they administered doses below three grams or only tested a few subjects (Nicholson and Stone 1979; Borbely and Youmbi-Balder 1986). Although reviewers generally agree that tryptophan increases sleepiness and reduces sleep latency in doses of three grams or more, they fail to reach definite conclusions on the value of tryptophan as a clinical hypnotic (Borbely and Youmbi-Balder 1986; Lieberman 1989). Although tryptophan does not appear to be as potent as prescription drugs (that are usually employed as hypnotic drugs) such as the benzodiazepines, it may have some clinical utility as a treatment for mild insomnia. However, this is a controversial area and further clinical trials are essential if tryptophan is to be employed in this manner.

Currently, the issue of the clinical utility of tryptophan is largely moot, because ingestion of this amino acid has recently been implicated in an outbreak of a rare, sometimes fatal, syndrome called Eosinophilia-myalgia syndrome (EMS). This disease, as the name implies, is characterized by generalized myalgia and an elevated peripheral eosinophil count (Hertzman et al. 1990). Due to the association of tryptophan and tryptophan-containing supplements with EMS, these compounds have been withdrawn from the U.S. market by order of the U.S. Food and Drug Administration. However, it now appears that EMS is caused by a contaminant found in the tryptophan produced by one manufacturer (Belongia et al. 1990), although this is still a controversial issue.

The sedation observed in humans after tryptophan administration has been attributed by Spinweber et al. (1983) and others to a postulated role for serotonin neurons in the regulation of sleep (Jouvet 1973, 1983). There is considerable controversy over the role of serotonin in the regulation of sleep; other mechanisms to account for the action of tryptophan, such as indirect effects on dopaminergic function (Hartmann 1982) on the central nervous system, have been advanced.

Central serotoninergic systems are also involved in the regulation of other mood states, including depression. Many antidepressants are potent serotoninergic agonists. A number of clinical studies have evaluated tryptophan for possible antidepressant activity with variable results (for a review, see Young 1986). However, a single tryptophan-deficient meal has been clearly shown to produce transient depression in normal humans (Young et al. 1985; Smith et al. 1987).

It should be noted that doses of tryptophan that are psychopharmacologically active probably produce changes in brain tryptophan larger than those produced by any single meal. Therefore, although tryptophan is a common food constituent, it must be given in pure form for it to have substantial hypnotic-like effects; accordingly, it could be considered a drug, not a food. The putative changes in behavior that foods such as carbohydrates or proteins may produce because they also can alter brain tryptophan and serotonin concentration are more difficult to demonstrate, as discussed below.

Tyrosine

Tyrosine, another LNAA, is the precursor of three catecholamines found in the brain: dopamine, norepinephrine, and epinephrine. Although it has been demonstrated that increased tyrosine availability can increase catecholaminergic neurotransmission, this only occurs in certain specific situations. It appears that catecholaminergic neurons are only precursor sensitive when they are firing frequently (Wurtman, Hefti, and Melamed 1981; Milner and Wurtman 1986). In the brain, as in the periphery, increased release of catecholamines is associated with exposure to acute stress, and the catecholamine most closely associated with acute stress is norepinephrine (Stone 1975). Although tyrosine appears to have little effect on normal animals and humans (Lieberman et al. 1982), when it is administered to stressed subjects it appears to have some beneficial effects. Presumably, under conditions of severe acute stress, noradrenergic neurons fire more frequently and expend available transmitter. The availability of extra precursor due to the administration of tyrosine appears to increase catecholamine release and improve various aspects of behavior.

In animals, tyrosine pretreatment has been shown to reduce the adverse effects of tail shock, cold stress, and hypoxia (Brady, Brown, and Thurmond 1980; Lehnert et al. 1984a,b; Rauch and Lieberman 1990; Luo et al. 1992; Lieberman et al. 1992). In humans, tyrosine reduces some of the adverse behavioral consequences of exposure to a combination of stressors—hypobaric hypoxia and cold. In two similar studies (Banderet and Lieberman 1989; Lieberman, Banderet and Shukitt-Hale 1990), administration of tyrosine in doses of 85 to 170 mg/kg restored performance and mood and relieved symptoms among those individuals who were most affected by exposure to the stressors. Complex cognitive performance such as learning and memory appeared to improve, as did reaction time and vigilance, when subjects received tyrosine. Mood states, such as fatigue and depression, also improved, and a variety of symptoms associated with exposure to hypoxia, such as headache, were partially relieved. Tyrosine also appears to prevent some of the adverse effects of cardiovascular stressors on animals and humans (Scott et al. 1981; Conlay, Maher and Wurtman 1981, 1985; Banderet and Lieberman 1989; Dollins 1990).

Phenylalanine

The artificial sweetener aspartame consists of about 50% phenylalanine. Like tyrosine, phenylalanine is a precursor of the catecholamines, and because it is also an LNAA its plasma availability influences the access of other LNAA to the brain. Several studies have been conducted to examine the effects of high doses of aspartame on humans, which greatly elevate plasma phenylalanine levels. No effects on behavior were detected (Ryan-Harshman, Leiter, and Anderson 1987; Lieberman et al. 1988). However, chronically elevated levels of plasma phenylalanine that occur as a consequence of phenylketonuria do have a substantial adverse impact on brain and development.

Amino Acids: Summary

Acute administration of certain amino acids in sufficient doses (i.e., in this context they could be considered to be drugs, not foods) can alter brain neurotransmission; however, the effects of these food constituents on behavioral functions remain controversial. It is likely that LNAA tryptophan, when given in doses of at least three grams, is a sedative, although the efficacy of this amino acid as a clinically useful hypnotic remains to be established. Tyrosine, another LNAA, has been studied for possible beneficial behavioral effects on laboratory animals and humans exposed to acute stress. To date, only a small number of studies have been conducted and definitive conclusions are not possible, although improvements in performance and mood have been observed. Phenylalanine, which has been examined in a few studies because it is a major constituent of aspartame, has not been found to have any behavioral effects as yet. Considerable future research to define the mechanisms of action of these compounds and determine their behavioral or physiological effects is necessary.

PROTEIN AND CARBOHYDRATE FOODS

Scientific interest in possible behavioral effects of protein and carbohydrate foods can be largely attributed to the demonstration in the early 1970s that these foods can have opposite effects on brain concentration of tryptophan and its neurotransmitter product serotonin (Fernstrom and Wurtman 1972; for reviews, see Wurtman, Hefti, and Melamed 1981; Fernstrom 1983). Intuitively, it would be expected that protein foods would increase brain tryptophan because they contain tryptophan. Also, because amino acids are not present in pure carbohydrate, it would be anticipated that such foods would not affect brain levels of tryptophan or any other amino acid. These assumptions have been proven to be incorrect. Protein meals actually lower brain tryptophan while carbohydrate meals substantially elevate brain tryptophan and increase brain serotonin concentration. Carbohydrate consumption does increase peripheral plasma tryptophan concentration, but the transport of tryptophan across the blood-brain barrier is determined by another critical factor. Tryptophan's access to the brain is determined by the ratio of its plasma concentration to that of the other LNAAs, not by its absolute plasma concentration (Pardridge 1986; Wurtman, Hefti, and Melamed 1981; Fernstrom 1983). Because tryptophan is the scarcest of the LNAAs with regard to its quantity in most protein foods, its plasma ratio, as opposed to absolute quantity, declines after such foods are ingested (Fernstrom and Wurtman 1972; Lieberman, Spring, and Garfield, 1986b). Protein meals actually lower plasma tryptophan ratio and, consequently, less tryptophan enters the brain and is available for the synthesis of serotonin.

Pure carbohydrate meals, even though they contain no amino acids, also affect the plasma levels of the large neutral amino acids. Such meals actually increase the ratio of plasma tryptophan to the other large neutral amino acids, the opposite outcome produced by eating protein (Wurtman, Hefti and Melamed 1981; Fernstrom 1983). This unanticipated consequence results because of the secretion of insulin that occurs when carbohydrates are consumed. Insulin lowers the plasma levels of the other LNAAs relative to tryptophan. Because more tryptophan is available for transport into the brain, carbohydrate meals increase brain serotonin. It appears that carbohydrate meals generally increase the tryptophan/LNAA ratio regardless of the type of carbohydrate ingested (Lieberman, Caballero, and Finer 1986a). The effects of a carbohydrate meal resemble those following ingestion of pure tryptophan, which is behaviorally active doses produces a larger rise in the plasma tryptophan ratio than even the largest carbohydrate meal (Lieberman 1989). A number of factors, including interval between meals, prefeeding metabolic state and type of carbohydrate, influence the magnitude of the rise in tryptophan ratio produced by carbohydrate meals (Ashley, Liardon, and Leathwood 1985).

The discovery that protein and carbohydrate meals had very different effects on a key brain neurotransmitter, serotonin, provided a rational basis for investigating the effects of such foods on behavior (Lieberman, Spring, and Garfield 1986). Based on anecdotal information, many individuals have associated sucrose or other sugars with increased motor activity, probably because such foods provide rapidly available metabolic energy when they are consumed. In spite of convincing scientific evidence to the contrary, many lay people still believe that sugar causes hyperactivity in children (Rapaport 1986; Lieberman 1985). However, because carbohydrate increases brain tryptophan, and tryptophan in nonphysiological doses has sedative-like effects, it would seem more likely that carbohydrate meals have similar effects. To date, only a few studies have addressed this question with variable results.

One of the first studies to examine the effects of protein and carbohydrate foods on human mood and performance was conducted by Spring et al. (1982). They employed a parallel design to study a large heterogenous group of adults who received either high protein or high carbohydrate meals for breakfast or lunch. There were differences in response to the meals as a function of age and gender. Females reported greater sleepiness after the carbohydrate meal, and males reported greater calmness after carbohydrates. Performance on an auditory attention task was impaired among older subjects following the high carbohydrate lunch. In a crossover study of 40 young males, protein versus carbohydrate meals had no significant effects on mood state. However, auditory reaction time was significantly slower when subjects consumed a carbohydrate lunch (Lieberman, Wurtman, and Teicher 1989). In another study by Spring et al. (1989), the effects of a carbohydrate, protein, and balanced lunch were examined in seven normal young women. The carbohydrate produced significantly greater fatigue two hours after its consumption, but did not affect other aspects of mood or performance.

Smith and colleagues (1988) also examined the effects of protein and carbohydrate lunch meals on mood and performance. Their subjects were 11 young males and females. This study failed to detect any effects of carbohydrate versus protein meals on mood state, but observed slower response to peripheral visual stimuli on a search task (Smith et al. 1988).

In one of the largest studies conducted to date on the effects of protein versus carbohydrate meals on mood and performance, Lieberman, Wurtman, and Teicher (1989) administered breakfast and dinner meals to 86 subjects. Four groups were tested: 24 elderly females, 21 elderly males, 21 young males, and 20 young females, and each subject received every meal. The subjects in this study were inpatients in a clinical research center, so that many nutritional and other variables that might affect day-to-day mood and performance were controlled. No general effects of meals were found in this study, nor were there differential effects of the meals on young versus elderly subjects. There were significant interactions between meal type and time of administration. The protein meal at breakfast induced more self-reported fatigue than the isocaloric carbohydrate meal. At dinner a carbohydrate meal produced more fatigue, especially among the young subjects. The meals had no effect on performance even though a large battery of performance tasks that had previously been employed to assess the effects of nutrients on behavior were employed. The complex effects of meal composition on behavior and the lack of generalized effects of such meals observed in this and other studies suggest that the mechanisms underlying the effects of these foods on behavior are not yet recognized.

There does appear to be at least one population that may be more sensitive to consumption of protein versus carbohydrate meals than the general population. In one study, substantial changes in the mood state of obese individuals following consumption of carbohydrate-rich meals were observed (Lieberman, Wurtman, and Chew 1986c). In that study, obese volunteers were initially categorized as either carbohydrate cravers or noncravers. This was accomplished by objective measurement of actual meal and snack food intake during a five-day admission to a clinical research center. Those individuals who were classified as carbohydrate cravers ate normal meals, but their consumption of carbohydrate-rich snacks, which were typically consumed without protein several hours after a meal, accounted for 30% or more of their total daily calorie intake. The noncarbohydrate cravers, who appear to be a smaller proportion of the population of overweight individuals, ate carbohydrate snacks right after protein-rich meals or consumed snacks containing both protein and carbohydrate.

When both groups of subjects were given a test meal consisting of a nonsweet, high carbohydrate, protein-free lunch (104 g starch), significant differences in mood state were observed two hours later. In general, the overall consequence of consuming the high carbohydrate meal was induction of positive mood changes among the carbohydrate craving group and negative changes among the noncravers. Depression decreased among the carbohydrate cravers and increased among noncarbohydrate cravers. Sleepiness and fatigue increased in the noncarbohydrate cravers but changed little among the carbohydrate cravers.

These findings indicate that obese individuals may select or avoid specific foods because of the acute impact the macronutrients contained in the foods have on their mood state.

Protein and Carbohydrate: Summary

Definitive effects of protein and carbohydrate meals on human mood and performance have not been established. A few studies suggest that consumption of carbohydrate meals may be more calming than protein or balanced meals. However, these effects appear to depend on the time of day, gender, food preferences, weight, and age of the test population. Consistent, replicated effects have not as yet been demonstrated. Future research must demonstrate that such effects are robust enough to be replicated within and between laboratories.

References

Arnaud, M.J. 1984. Products of metabolism of caffeine. In *Caffeine*, P.B. Dews, ed., pp. 3–38, New York: Springer-Verlag.

Ashley, D.V.M.; Liardon, R.; and Leathwood, P.D. 1985. Breakfast meal composition influences plasma tryptophan to large neutral amino acid ratios of healthy lean young men. *Journal of Neural Transmission* 63:271–283.

Banderet, L.E., and Lieberman, H.R. 1989. Treatment with tyrosine, a neurotransmitter precursor, reduces environmental stress in humans. *Brain Research Bulletin* 22:759–762.

Barone, J.J., and Roberts, H. 1984. Human consumption of caffeine. In *Caffeine*, P.B. Dews, ed., pp. 59–76, New York:Springer-Verlag.

Belongia, E.A.; Hedberg, C.W.; Gleich, G.J.; White, K.E.; Mayeno, A.N.; Loegering, D.A.; Dunnette, S.L.; Pirie, P.L. δcDonald, K.L.; and Osterholm, M.T. 1990. An investigation of the cause of the eosinophilia-myalgia syndrome associated with tryptophan use. *New England Journal of Medicine* 323(6):357–365.

Borbely, A.A., and Youmbi-Balderer, G. 1986. Effects of tryptophan on human sleep. In *Sleep, aging and related disorders*. Interdisciplinary Topics in Gerontology, vol. 22, W. Emser, D. Kurtz, and W.B. Webb, eds., pp. 17–32, Basel: Varger.

Brady, K.; Brown, J.W.; and Thurmond, J.B. 1980. Behavioral and neurochemical effects of dietary tyrosine in young and aged mice following cold swim stress. *Pharmacology, Biochemistry and Behavior* 12:667–674.

Christensen, L. 1991. Issues in the Design of Studies Investigating the Behavioral Concomitants of Foods. *Journal of Consulting and Clinical Psychology* 59(6):874–882.

Christensen, L.; Miller, J.; and Johnson, D. 1991. Efficacy of caffeine versus expectancy in altering caffeine-related symptoms. *Journal of General Psychology* 118(1):5–12.

Clubley, M.; Bye, C.E.; Henson, T.A.; Peck, A.W.; and Riddington, C.J. 1979. Effects of caffeine and cyclizine alone and in combination on human performance, subjective effects and EEG activity. *British Journal of Clinical Pharmacology* 7:157–163.

Cole, J.O.; Pope, H.G.; LaBrie, R.; and Ionescu-Pioggia, M. 1978. Assessing the subjective effects of stimulants in casual users. *Clinical Pharmacology and Therapeutics* 24:243–252.

Conlay, L.A.; Maher, T.J.; and Wurtman, R.J. (1981). Tyrosine increases blood pressure in hypotensive rats. *Science* 212:559–560.

———. 1985. Tyrosine accelerates catecholamine synthesis in hemorrhaged hypotensive rats. *Brain Research* 333:81–84.

Dews, P.B. 1984. Behavioral effects of caffeine. In *Caffeine*, P.B. Dews, ed., pp. 86–106, New York: Springer-Verlag.

Dollins, A.B.; Krock, L.P.; Storm, W.F.; and Lieberman, H.R. 1990. Tyrosine decreases physiological stress caused by lower body negative pressure (LBNP). *Aviation, Space and Environ. Med.* 61(5):491.

Fernstrom, J.D. 1983. Role of precursor availability in the control of monoamine biosynthesis in the brain. *Physiology Review* 63:484–546.

Fernstrom, J.D., and Wurtman, R.W. 1972. Brain serotonin content: Physiological regulation by plasma neutral amino acids. *Science* 178:414–416.

Fine, B.J.; Kobrick, J.L.; Lieberman, H.R.; Marlowe, B.; Riley, R.H.; and Tharion, W.J. (in press). Changes in Human Vigilance Following Treatments with Caffeine or Diphenhydramine. *Psychopharmacology*.

Gilliland, K., and Andress, D. 1981. Ad lib caffeine consumption, symptoms of caffeinism and academic performance. *American Journal of Psychiatry* 138:512–514.

Greden, J.F.; Fontaine, P.; Lubetsky, M.; and Chamberlin, K. 1978. Anxiety and depression associated with caffeinism among psychiatric inmates. *American Journal of Psychiatry* 13:963–966.

Griffiths, R.R.; Evans, S.M.; Heisman, S.J.; Preston, K.L.; Sannerud, C.A.; Wolf, B.; and Woodson, P.P. 1990. Low-dose caffeine discrimination in humans. *Journal of Pharmacology and Experimental Therapeutics* 252:970–978.

Hartmann, E. 1982. Effects of L-tryptophan on sleepiness and on sleep. *Journal of Psychiatric Research* 17:107–113.

Hertzman, P.A.; Blevins, W.L.; Mayer, J.; Greenfield, B.; Ting, M.; and Gleich, G.J. 1990. Association of the eosinophilia-myalgia syndrome with the ingestion of tryptophan. *New England Journal of Medicine* 322:869–873.

Hirsch, K. 1984. Central nervous system pharmacology of the dietary methylxanthines. In *The Methylxanthine Beverages and Foods: Chemistry, Consumption, and Health Effect*, G.A. Spiller, ed., New York: Allan R. Liss, Inc.

Hirsch, E., and Kramer, F.M. (1993). Situational influences on food intake. In *Nutritional Needs in Hot Environments*, Bernadette M. Marriott, ed., pp. 215–243, Washington, DC: National Academy Press.

Johnson, R.F. 1991. Rifle Firing Simulation: Effects of MOPP, heat, and medications on marksmanship. In *Proceedings of the Thirty-Third Annual Military Testing Association*, pp. 530–535. San Antonio, TX.

Jouvet, M. 1973. Serotonin and sleep in the cat. In *Serotonin and Behaviour*, J. Barchas and E. Usdin, eds., pp. 385–400. New York: Academic Press.

———. 1983. Serotonergic and nonserotonergic mechanism in sleep. In *Sleep Disorders Basic and Clinical Research*, M. Chase and E. Weitzman, eds., New York: Spectrum.

Kanarek, R.B., and Marks-Kaufman, R. 1991. *Nutrition and Behavior: New Perspectives*. New York: AVI.

Kirsch, I., and Weixel, L.J. 1988. Double-Blind Versus Deceptive Administration of a Placebo. *Behavioral Neuroscience* 102(2):319–323.

Kruesi, M.J., and Rapaport, J.L. 1986. Diet and human behavior: How much do they affect each other? In *Annual Review of Nutrition*, R.E. Olson, E. Beutler, and H.P. Broquist, eds., pp. 113–130. Palo Alto, California: Annual Reviews Inc.

Leathwood, P., and Pollet, P. 1982. Diet-induced mood changes in normal populations. *Journal of Psychiatric Research* 17:147–154.

Lehnert, H.; Reinstein, D.K.; Strowbridge, B.W.; and Wurtman, R.J. 1984a. Neurochemical and behavioral consequences of acute uncontrollable stress: Effects of dietary tyrosine. *Brain Research* 303:215–223.

Lehnert, H.; Reinstein, D.K.; and Wurtman, R.J. 1984b. Earl Usdin, ed. Tyrosine reverses the depletion of brain norepinephrine and the behavioral deficits caused by tail-shock stress in rats. In *Stress: The Role of the Catecholamines and Other Neurotransmitters*, pp. 81–91. New York: Gordon and Breach.

Lieberman, H.R. 1985. Sugars and behavior. *Clinical Nutrition* 4(6):195–199.

———. 1988. Beneficial effects of caffeine. In *Twelfth International Scientific Colloquium on Coffee*, pp. 102–107. Paris: ASIC.

———. 1989. Cognitive effects of various food constituents. In *Handbook of the Psychophysiology of Human Eating*, R. Shepherd, ed., pp. 251–270. Chichester: England U.K. Wiley.

———. 1992. Caffeine. In *Factors Affecting Human Performance Vol. II: The Physical Environment*, D. Jones and A. Smith, eds., pp. 49–72. London: Academic Press.

———. (in press). *Tyrosine and Stress: Human and Animal Studies*. Washington, D.C.: National Academy Press.

Lieberman, H.R.; Corkin, S.; Spring, B.J.; Growdon, J.H.; and Wurtman, R.J. 1982. Mood, performance and pain sensitivity: Changes induced by food constituents. *Journal of Psychiatric Research* 17:135–145.

Lieberman, H.R.; Spring, B.; and Garfield, G.S. 1986. The behavioral effects of food constituents: Strategies used in studies of amino acids, protein, carbohydrates and caffeine. *Diet and Behavior: A Multidisciplinary Evaluation, Nutrition Reviews* 44(suppl):61–70.

Lieberman, H.R.; Caballero, B.; and Finer, N. 1986a. The composition of lunch determines afternoon tryptophan ratios in humans. *Journal of Neural Transmission* 65:211–217.

Lieberman, H.R.; Spring, B.; and Garfield, G.S. 1986b. The behavioral effects of food constituents: Strategies used in studies of amino acids, protein, carbohydrate and caffeine. *Nutrition Reviews* 44(suppl):61–70.

Lieberman, H.R.; Wurtman, J.J.; and Chew, B. 1986c. Changes in mood after carbohydrate consumption among obese individuals. *American Journal of Clinical Nutrition* 44:772–778.

Lieberman, H.R.; Wurtman, R.J.; Emde, G.G.; Roberts, C.; and Coviella, I.L.G. 1987a. The effects of low doses of caffeine on human performance and mood. *Psychopharmacology*, 92:308–312.

Lieberman, H.R.; Wurtman, R.J.; Emde, G.G.; and Coviella, I.L.G. 1987b. The effects of caffeine and aspirin on mood and performance. *Journal of Clinical Psychopharmacology* 7(5):315–320.

Lieberman, H.R.; Caballero, B.; Emde, G.G.; and Bernstein, J.G. 1988. The effects of aspartame on human mood, performance and plasma amino acid levels. In

Dietary Phenylalanine and Brain Function, R.J. Wurtman and E. Ritter-Walker, eds. Boston: Birkhäuser.

Lieberman, H.R.; Wurtman, J.J.; and Teicher, M.H. 1989. Aging, nutrient choice, activity and behavioral responses to nutrients. In *Nutrition and the Chemical Senses in Aging. Annals of the New York Academy of Sciences* 561:196–208.

Lieberman, H.R.; Banderet, L.E.; and Shukitt-Hale, B. 1990. Tyrosine protects humans from the adverse effects of acute exposure to hypoxia and cold. *Soc. Neurosci. Abstr.* 16:272.

Lieberman, H.R.; Shukitt-Hale, B.; Luo, S.; Devine, J.A.; and Glenn, J.F. 1992. Tyrosine reduces the adverse effect of hypobaric hypoxia on spatial working memory of the rat. *Soc. Neurosci. Abstract* 18(1):715.

Loke, W.H. 1988. Effects of caffeine on mood and memory. *Physiology and Behavior* 44:367–372.

Loke, W.H., and Meliska, C.J. 1984. Effects of caffeine use and ingestion on a protracted visual vigilance task. *Psychopharmacology* 84:54–57.

Loke, W.H.; Hinrichs, J.V.; and Ghoneim, M.M. 1985. Caffeine and diazepam: Separate and combined effects on mood, memory, and psychomotor performance. *Psychopharmacology* 87:344–350.

Luo, S.; Villamil, L.; Watkins, C.J.; and Lieberman, H.R. 1992. Further evidence tyrosine reverses behavioral deficits caused by cold exposure. *Soc. Neurosci. Abstract* 18(1):716.

Marlatt, G.A., and Rohsenow, D.J. 1980. Cognitive processes in alcohol use: Expectancy and the balanced placebo design. In *Advances in Substance Abuse*, N.K. Mello, ed., pp. 159–199, Greenwich, CT: JAI Press.

Meiselman, H.L. 1987. Use of sensory evaluation and food habits data in product development. In *Umami: A Basic Taste*, Y. Kawamura and M.R. Kare, eds., pp. 307–326. New York: Marcel Dekker, Inc.

———. 1992. Methodology and theory in human eating research. *Appetite* 19:49–55.

Meiselman, H.L., and Kramer, F.M. (in press). The role of context in behavioral effects of foods. In *An Evaluation of Potential Performance Enhancing Food Components for Operational Rations*, Bernadette M. Marriott, ed., Washington, D.C.: National Academy Press.

Milner, J.D., and Wurtman, R.J. 1986. Catecholamine synthesis: Physiological coupling to precursor supply. *Biochemical Pharmacology* 35:875–881.

National Research Council. (in press). B. Marriott and J. Grumstrup-Scott, eds. Washington, D.C.: National Academy Press.

Nicholson, A.N., and Stone, B.M. 1979. L-tryptophan and sleep in healthy man. Electroenceph. clin. *Neurophysiology* 47: 539–545.

Pardridge, W.M. 1986. Blood-brain barrier transport of nutrients. *Nutrition Reviews* 44(suppl.):15–25.

Rapaport, J.L. 1986. Diet and hyperactivity. *Nutrition Reviews* 44(May, suppl.):158–162.

Rauch, T.M., and Lieberman, H.R. 1990. Pre-treatment with tyrosine reverses hypothermia induced behavioral depression. *Brain Research Bulletin* 24:147–150.

Regina, E.G.; Smith, G.M.; Keiper, C.G.; and McKelvey, R.K. 1974. Effects of caffeine on alertness in simulated automobile driving. *Journal of Applied Psychology* 59(4):483–489.

Roache, J.D., and Griffiths, R.R. 1987. Interactions of diazepam and caffeine: Behavioral and subjective dose effects in humans. *Pharmacology, Biochemistry and Behavior* 26:801–812.

Rozin, P., and Vollmecke, T.A. 1986. Food likes and dislikes. *Annual Review of Nutrition* 6:433–456.

Ryan-Harshman, M.; Leiter, L.A.; and Anderson, G.H. 1987. Phenylalanine and aspartame fail to alter feeding behavior, mood and arousal in men. *Physiology and Behavior* 39:247–253.

Sawyer, D.A.; Julia, H.L.; and Turin, A.C. 1982. Caffeine and Human Behavior: Arousal, Anxiety, and Performance Effects. *Journal of Behavioral Medicine* 5(4):415–439.

Scott, N.A.; DeSilva, R.A.; Lown, B.; and Wurtman, R.J. 1981. Tyrosine administration decreases vulnerability to ventricular fibrillation in the normal canine heart. *Science* 211:727–729.

Smith, A.; Leekam, S.; Ralph, A.; and McNeill, G. 1988. The influence of meal composition on post-lunch changes in performance efficiency and mood. *Appetite* 10:195–203.

Smith, S.E.; Pihl, R.O.; Young, S.N.; and Ervin, F.R. 1987. A test of possible cognitive and environmental influences on the mood-lowering effect of tryptophan depletion in normal males. *Psychopharmacology* 91:451–457.

Snyder, S.H. 1984. Adenosine as a mediator of the behavioral effects of xanthines. In *Caffeine*, P.B. Dews, ed., pp. 129–141. New York: Springer-Verlag.

Spinweber, C.L.; Ursin, R.; Hilbert, R.P.; and Hilderbrand, R.L. 1983. *Electroencephalography and Clinical Neurophysiology* 55:652–661.

Spitzer, L., and Rodin, J. 1981. Human eating behavior: A critical review of studies in normal weight and overweight individuals. *Appetite* 2:293–329.

Spring, B.; Maller, O.; Wurtman, J.J.; Digman, L.; and Cozolino, L. 1982. Effects of protein and carbohydrate meals on mood and performance: Interactions with sex and age. *Journal of Psychiatric Research* 17:155–167.

Spring, B.; Chiodo, J.; Harden, M.; Bourgeois, M.J.; Mason, J.D.; and Lutherer, L. 1989. Psychobiological effects of carbohydrates. *Journal of Clinical Psychiatry* 50 (5 suppl):27–33.

Stone, E.A. 1975. Stress and catecholamines. In *Catecholamines and Behaviour*, A.J. Freidhoff, ed., pp. 31–72. New York: Plenum Press.

Svensson, E.; Persson, L.; and Sjoberg, L. 1980. Mood effects of diazepam and caffeine. *Psychopharmacology* 67:73–80.

Swift, C.G., and Tiplady, B. 1988. The effect of age on the response to caffeine. *Psychopharmacology* 94:29–31.

Winstead, D.K. 1976. Coffee consumption among psychiatric inpatients. *American Journal of Psychiatry* 133:1447–1450.

Wurtman, R.J.; Hefti, F.; and Melamed, E. 1981. Precursor control of neurotransmitter synthesis. *Pharmacology Review* 32:315–335.

Young, S.N. 1986. The clinical psychopharmacology of tryptophan. In *Nutrition and the Brain*, vol. 7, R.J. Wurtman and J.J. Wurtman, eds., pp. 49–88. New York: Raven.

Young, S.N.; Smith, S.E.; Pihl, R.O.; and Ervin, F.R. 1985. Tryptophan depletion causes a rapid lowering of mood in normal males. *Psychopharmacology* 87:173–177.

Chapter 8

Medical Foods

Mary K. Schmidl and
Theodore P. Labuza

INTRODUCTION

Medical foods, also known as enteral formulas, used to feed hospitalized patients or to be a major dietary factor for those with unusual diseases, have been an important achievement in science and medicine during the past 40 years. These truly are critical "**functional foods**" for special dietary purposes. There are over 180 of these special products in the market produced by about 20 manufacturers (Talbot 1990; Hatten and Mackey 1990). The products that are fed by nasogastric feeding devices in sterile solution form are generally considered foods (enteral foods) to differentiate them from sterile parenteral nutrition solutions fed by vein, which are considered as drugs and which need to be precleared through the Food and Drug Administration. It has been estimated by the Federated Societies for Experimental Biology (Anon. 1992) that there are over 5 million patients on enteral medical foods in the United States. These

Adapted in part from an IFT Scientific Status Summary (1992) with permission of The Institute of Food Technologists (Chicago, IL)

foods comprise essentially sterile solutions or rehydratable powders with a value exceeding $1 billion.

Advances in medical foods have been in four main areas. First, our knowledge of dietary management of chronic disease, inborn errors of metabolism, immune responses, and injury based on human physiology and biochemistry has increased. Second, advances in food technology and the availability of certain chemical constituents have allowed for the manufacturing of targeted food products containing various specialized nutrients to supply critical nutritional needs in various formats other than sterile liquids or rehydrated powders. Third, there has been considerable improvement in medical devices, delivery and packaging systems, infusion pumps, and nonsurgical techniques that allows for convenient administration of liquid products, especially for nasogastric feeding. Finally, there is a current renewed interest in enteral nutrition support dictated by hospital cost containment measures, safety issues, and physiological benefits, including maintenance of small intestine mass and pancreatic function. All of these advances were critical in the development of these functional foods.

Medical foods are designed to function by providing nutritional support to individuals who are unable to ingest adequate amounts of food in a conventional form, or by providing specialized nutritional support to patients who have special physiological and nutritional needs. Depending on the patient's nutritional status, a medical food may either supplement a diet or be the sole source of nutrition for extended periods of time. A report, by the Johns Hopkins Center for Hospital Finance and Management (Steinburg and Anderson 1989) on the financing of nutritional support, stated that in the United States the average duration of treatment with medical foods for hospitalized patients was 12 days when on oral supplements as compared to 18.5 days for those on tube feedings (Steinberg and Anderson 1989). On the other hand, the treatment of infants who are diagnosed at birth as having phenylketonuria includes specialized medical foods, sometimes designated as "**orphan**" medical foods, in a convenient and recognizable form throughout their lives. The need for these special foods almost disappears when they become teenagers, but when females become pregnant the need returns basically to protect the fetus. These special dietary foods fall in between the category of a normal food and that of a drug and their regulatory status is unclear. Orphan medical foods can be defined as those products that are used to treat conditions for which there are a low number of new cases per year. As a point of reference, inborn errors of metabolism with respect to phenylketonuria (PKU), urea cycle disorder, and glycogen storage disorder comprise about one in 10 thousand of births or about 400 new cases per year (~20,000 total population), while maple syrup urine disease, propionic acid acidemia, and methyl malonic acidemia present in 1 in 120,000 births, resulting in 20–40 cases per year and a total population of only about 2,000 in the United States.

A key to understanding the development of nutritional products for clinical use and to maintaining individuals with metabolic disorders is to recognize that the physiological and nutritional requirements of a hospitalized or compromised person are generally different from those of an average healthy individual. For example: (a) hospitalized patients may have an increased need for calories (Kinney 1976); (b) these same patients usually have an increased need for protein as a result of trauma or sepsis (Bistrian et al. 1974); (c) those patients with a disease of, resection of, or obstructions of the gastrointestinal tract will have a decreased ability to absorb nutrients; (d) many hospitalized patients lack mobility, which can affect the biochemical requirements for specific nutrients (e.g., calcium); (e) hospitalized patients exhibit malfunction of the food/nutrient processing organs (e.g., stomach, intestine, liver, pancreas, or kidney), which can greatly alter nutrient requirements; (f) some patients such as those with AIDS have a severely depressed immune response, may become hypersensitive to normal food proteins, and may need precursors to stimulate immunological function; (g) some people have anaphylactic reactions to specific food components and need special avoidance foods and diets that are low in or devoid of the causative agents. At the same time these diets must supply all needed nutrients and calories; and (h) those with inborn errors of metabolism; e.g., phenylketonuria (PKU), maple syrup urine disease (MSUD), homocystinuria, galactosemia, fructosuria, and hyperlipoproteinaemia must either strictly avoid specific food components/nutrients to prevent illness or death, or must ingest increased amounts of certain metabolites to stimulate a specific metabolic pathway. Medical foods can be delivered in many forms, e.g., as sterile liquids that can be consumed directly or fed by a nasogastric or intestinal tube route, as dry powders which must be rehydrated, or as solid bars with the nutrients in a chewable form.

Although a few basic principles of supplying nutrients to the hospitalized patient or compromised person have been identified and established, one must remember that this field of medical nutrition is evolving rapidly. There are still many problem areas that await solution through future research. The factors that will influence future developments include demographics of the disease, advances in food processing technology, and government regulations through their impact on private industry with respect to stimulating research and development of useful products.

INTERNATIONAL AND U.S. REGULATION

Since the passage of the Pure Food and Drug Act in 1906 in the United States, there has been considerable controversy in regard to special dietary products that are at the interface of food and drugs (Merrill and Hutt 1990). With the passage of the U.S. Food, Drug and Cosmetic Act (1938), medical foods were regarded as prescription drugs in order to ensure that their use would be supervised by physicians and to prevent misuse by individuals with the disease. Manufacturers had to apply for a Investigational New Drug (IND) regulation to market these products, a difficult and costly process-inhibiting development. The first product, Lofenalac, a modified casein hydrolysate with low phenylalanine content, was approved in 1957 for infants with phenylketonuria.

In 1972 the FDA reclassified medical foods as Special Dietary Foods to enhance their research and development and availability. The FDA relaxed the nutrition labeling regulations for these products and defined them as "foods represented for use solely under medical supervision to meet nutritional requirements in specific medical conditions."

In 1988 Congress amended the Orphan Drug Act to include the first legal definition of medical foods (21 U.S. Code 360 ee (b)). According to the act, the term "**medical food**" means a food formulated to be consumed or administered enterally under the supervision of a physician and which is intended for the specific dietary management of a disease or condition for which distinctive nutritional requirements based on recognized scientific principles are established by "**medical evaluation**" (U.S. Congress 1988). This definition was intended to give credence to the concept of orphan medical foods, which were defined in the 1988 amendments as a subcategory of medical foods that include products useful in the management of "any disease or condition that occurs so infrequently in the United States that there is no reasonable expectation that a medical food for such disease or condition will be developed without assistance" (21 U.S. Code 360 ee (b)). This definition was further incorporated into the Nutrition Labeling and Education Act of 1990 (21 U.S. Code 343; November 8, 1990) and is the current authoritative definition of medical foods (U.S. Congress 1990).

The U.S. Infant Formula Act passed in 1980 has specific nutrient requirements for complete infant formulas fed by mouth because they are the sole source of feeding. This category therefore defines special medical foods for healthy persons rather than for the ill. The FDA did allow under 21 CFR 107.50 (a), for the manufacture of infant formulas without all of the required nutrients (21 CFR 107.10). These products can be regarded as medical "**Exempt Infant Formulas**" in which nutrients have been deleted so they could be used for infants with inborn errors of metabolism.

An FDA Compliance Program (CPGM 7321.002) in 1988 evaluated medical food manufacturers with respect to proper formulation, appropriate microbiological standards, and reasonable therapeutic claims for their products. The FDA found that in general, the medical food manufacturers were meeting the requirements, although there have been problems with some manufacturers of certain dietary supplements, (for example, the tryptophan-EMS crisis, Hertzman et al. 1991), as well as with an infant formula that was made with improper chloride levels.

A critical factor relating to safety and health benefit claims that seems to inhibit research and development is the fact that many of the intermediary metabolites that could be added to medical foods to help alleviate a disease condition either are drugs or might be considered as drugs, e.g., homocysteine or hydroxycobalamine. Because of the small market, the manufacturer cannot afford to get approval of the metabolite either as a drug through an IND application or as a food additive. It is possible that this might be changed in the future because exemptions for investigational use of unapproved food additives can be made if consistent with public health.

On an international level, the FAO/WHO Codex Commission (July 1991) has approved standards for foods for special medical dietary uses [CC NFDU]. This is critical, as it will instill international harmonization in this area and lead to a larger market for manufacturers. For example, as noted earlier, PKU affects only 1 in 10,000 new births, leading to under 400 new cases per year in the United States with a total affected population of under 20,000. If a product were consumed only once daily by all these people in the United States, there would be a need for only about 7 million units per year. A large food manufacturing facility making 200 units/minute would produce this in about one month, certainly not the type of operation that would be attractive to large-scale food manufacturers. An even worse case would be for products to treat maple syrup urine disease (MSUD), which affects only 1 person in 120,000, requiring about two shifts to produce U.S. needs for one year. International agreement on requirements for specific medical foods for various disorders would create much larger markets and perhaps stimulate their development (Labuza 1981).

Several groups have been formed over the past few years to promote the development of medical foods. These include the Life Science Research Office of the Federated Societies for Experimental Biology (1977) (Fisher et al. 1977); a task force of the FDA (1983); the Codex Committee on Nutrition and Foods for Special Dietary Uses, a subsidiary body of the Codex Alimentarius Commission (Cheney 1989); and the Enteral Nutrition Council (Mountford and Cristol 1989). In addition, FASEB held a specific symposium on medical foods (Anon. 1992). Some of these efforts have been reviewed in several papers published in a special issue of the *Food Drug Cosmetic Law Journal* [vol. 44 (4):461–591].

HISTORICAL

The history of medical foods and enteral liquid feedings began with very primitive techniques and has evolved to routine methods of delivery in controlled clinical settings for nutritional management of hospitalized patients as well as delivery to those with special metabolic diseases being treated at home. Before medical foods could be developed, a basic knowledge of the nutrients required to sustain life and the metabolic pathways was necessary. It was not until the 1930s and 1940s, when the crucial roles of the essential amino acids, fatty acids, minerals, and vitamins were identified, that such development could began to take place. Following these discoveries and the understanding of metabolic disorders, the application to medicine became evident.

The earliest "**elemental**" medical food products were spin-offs of the space program (Greenstein 1957; Winitz 1965). The early work, sponsored by the National Aeronautics and Space Administration (NASA), involved the use of elemental diets for in-flight space feeding to minimize the volume of human waste products. Urine presented little problem, but such was not the case for fecal material. One obvious way to eliminate bowel movements was to remove the need for them. The goal therefore became to deliver a nutrient solution to the jejunum that would not require digestion and would be completely absorbed in the upper gastrointestinal tract. This formulation would have to be made from pure amino acids, sugar, essential vitamins, minerals, and fatty acids with no undigestible residue. These diets soon found other applications as medical foods.

One of the earliest published studies employing a medical food was entitled *Use of the "Space Diet" in the Management of a Patient with Extreme Short Bowel Syndrome* (Thompson et al. 1969). A hospitalized patient without a functional gastrointestinal tract, pancreas, etc. (e.g., the short bowel syndrome) would not have the ability to digest nutrients, but might be able to absorb nutrients in their simplest form. The physician could feed the space diet above or below a fistula, inflamed bowel, obstructed bile duct, or other pathologic site. Elemental diets currently available include an array of products for the health care industry to treat nearly all types of patients with compromised guts and the 30–40 or so diseases caused by inborn errors of metabolism that can be managed by dietary means. One problem is that in an elemental form these diets generally are unpalatable and offer no variety for the patient.

Following these early studies, increased recognition of the role of nutrition in human health and disease led to increased research in total parenteral nutrition (TPN) products for specific conditions. These products contain all the necessary nutrients of an elemental diet, but are administered intravenously to patients who cannot eat enough or not at all, will not eat, should not eat, or cannot be fed by mouth or tube. These formulas are classified as drugs because their route of administration is by vein. Parenteral formulas are needed primarily to provide nutritional support especially after surgery, when the gastrointestinal tract is unable to absorb nutrients or when complete bowel rest is required. In most other cases, however, medical foods fed by mouth or tube fed are a more desirable alternative as long as there is access to and some function in the small bowel.

The importance of nutrition in the hospital setting became very clear in 1974 when a study showed that 50% of surgical patients in the U.S. suffered from moderate to severe protein-calorie malnutrition, both prior to and after surgery (Bistrian et al. 1974). The same investigators in another survey on general hospitalized patients found that about 44% were suffering from severe caloric deficiency, but fewer suffered from protein deficiencies than did the surgical patients (Bistrian et al. 1976). This has been documented by others in long-term care populations at home or nursing homes where eating is difficult or hard to manage (Shaver et al. 1980). In these cases poor nutritional support can be a critical deterrent to the recovery of the patient. In some cases the formulation is not a cure for the disease, but appears to increase the patients' tolerances for treatments of the disease, e.g.; drugs, radiation, chemotherapy, and surgery. Thus appropriate nutritional support not only corrects malnutrition, but can help in the prevention of complications and needless deaths (Twomey 1985; American Dietetic Association 1986).

CLASSIFICATION

There are numerous medical food products and classification systems (Bell 1989; Heimburger and Weinsier 1985). Available products range from nutritionally complete to nutritionally incomplete (modular) products as well as disease-specific products designed to limit or increase certain nutrients or intermediary metabolites. The FDA Compliance Program Guidance Manual (CPGM 7321.002; Food and Drug Administration 1988/1989/1990) (Anon. 1988) has identified four major categories of products (Food and Drug Administration 1989). These are described by Hattan and Mackey (1989) as follows.

Nutritionally Complete

Nutritionally complete enteral formulations supply all the required proteins, fats, carbohydrates, vitamins, and minerals in sufficient quantities to maintain nutritional status of an individual receiving no other source of nourishment for a wide variety of clinical conditions. The inventory of commercially prepared, complete formulations range from those composed of whole proteins, carbohydrates, and fats to those consisting of simple sugars, amino acids, and medium-chain triglycerides (MCT). The latter group is indicated for individuals with impaired ability to digest or absorb intact nutrients. The raw materials used in the manufacture of these formulations are supplied in such forms as calcium and sodium caseinate, soy protein isolate, hydrolyzed corn starch, and corn, canola, soy, and sunflower oils. The original formulas used in the 1960s were whole foods that had been homogenized in a blender into a liquid state. Although still available, these "**blenderized**" formulations make up a small fraction of the nutritionally complete formulations on the market today.

Nutritionally Incomplete

Enteral formulations that supply a single nutrient or combinations of nutrients in quantities insufficient to maintain the nutritional status of a normal, healthy individual are considered nutritionally incomplete. These products are known as modular components; they can be used as a supplemental source of nutrients and calories in an otherwise normal diet (e.g., MCT oil for extra calories), or can be combined with other modular components to produce a nutritionally complete formula. They permit flexibility so that the formula can be tailored to meet the special requirements of individual patients. The disadvantages of modular products are that (a) most institutions do not have adequate labor to mix modular formulations, (b) the intermixing of various modulars to produce a formulation tailored to the needs of individual patients requires considerable expertise, care to prevent microbial contamination, and expense, (c) when the formulations are prepared in the hospital or at home by those untrained in dietetics or clinical nutrition, there is an increased potential for environmental and/or microbial contamination, due to product handling, and (d) there can be an induction of metabolic disorders or deficiency states, due to lack of or excess of specific nutrients in the blended formula.

Formulas for Metabolic (Genetic) Disorders

This category of medical foods consists of those formulations manufactured specifically for individuals with inborn errors of metabolism; e.g., PKU, MSUD, urea cycle disorder (affects 1 in 25,000), glycogen storage disease (affects 1 in 25,000), propionic acidemia (affects 1 in 120,000), or methyl malonic acidemia (affects 1 in 120,000). Other diseases for which the epidemiology of incidence is much less established include homocystinuria (need to remove methionine), organic acidemia (limit isoleucine and methionine), galactosemia (provide a galactose- and lactose-free diet), fructosemia (eliminate all sucrose, fructose, and sorbitol), and histidinenemia (need to remove histidine). It should be noted that in all these metabolic disorders the histopathology may change with age, and thus the nutrient requirements change, which then elicits a modification in the medical food of usually unknown dimensions. The challenge to the food manufacturer to establish technologies to create these medical foods is thus very difficult. When the formulations are labeled for use by infants (birth through 12 months), the product falls under the requirements of the exempt infant formula regulations as noted earlier. However, many of these products are manufactured and labeled for use in the treatment of metabolic disorders in children, adolescents, and adults, and, therefore are not subject to those infant formula provisions. Although not nutritionally complete in the traditional sense, (PKU products, for example, contain a reduced level of aromatic amino acids to which the minimum amount of phenylalanine needed to meet the growth requirement of the infant is added), these formulas are designated to provide the nutrient composition necessary for the growth and development of persons afflicted with the metabolic disorder. The **NORD** (National Organization for Rare Diseases), a public organization, has been formed to help promote research into the dietary restrictions and requirements for these diseases, as well as to promote third party insurance payments for the special medical foods. Few medical insurance policies pay for the medical foods required to treat these diseases, because the products technically are not drugs.

Oral Rehydration Solutions

This category of medical foods consists of products indicated for the maintenance of replacement of water and electrolytes that have been lost due to mild to severe diarrhea such as from cholera or due to other causes of dehydration. Standard components of these formulations include sodium, chloride, potassium, citrate, dextrose, and water.

TECHNOLOGIES AND SPECIALIZED NUTRIENTS

Because of the specific clinical requirements for nutritional support, unique technologies or ingredients have been used to manufacture medical products. These are discussed below.

Protein/Amino Acid Sources

Most commercially available enteral formulas consist of whole (intact) protein as the nitrogen source in a liquid or powder and meet the minimum standard for quantity and quality of protein (FAO/WHO 1965). Formulas with nitrogen in the form of peptides or free amino acids are also available for patients with malabsorption or specific organ disease (renal or hepatic failure). These elemental diets were developed to provide "**predigested**" protein to patients with impaired mucosal absorption. These diets may induce vomiting, diarrhea, and electrolyte abnormalities because of their high osmolality, which may negate much of their potential benefits (Koretz and Meyer 1980). Work has shown that protein in the form of small peptides (dipeptides and tripeptides) is more rapidly and efficiently absorbed than protein in the form of free amino acids. Further studies have identified a specific peptide carrier system in the intestinal brush border that is noncompetitive and independent of free amino acid carriers (Matthews and Adibi 1976; Grimble et al. 1987). Partial enzymatic hydrolysis of proteins can be used to prepare a mixture composed of short-chain peptides, which can be fed as such in the formula, or they can be incorporated into a gel-like structure such as cheese or surimi, extruded into a snack-like product or freeze dried and compressed into a nutrient bar for use by those who can chew.

There is no naturally occurring protein that is low in phenylalanine, so there is no natural food on which to base a PKU diet. Thus one must use either a mixture of synthetic amino acids limiting or excluding phenylalanine completely or develop a protein hydrolysate from which the phenylalanine has been removed by absorption onto charcoal. As synthetic amino acids are expensive, hydrolysates are commonly used. Casein is the usual starting material, although some amino acids may have to be supplemented (Bichel et al. 1954). One problem with hydrolysates is their unpleasant, bitter flavor of peptides that can cause refusal to eat them, even with infants. Challenges for the future include separating out the bitter peptides and replacing those missing amino acids with nonbitter dipeptides, and the possible use of the plastein reaction to recombine the peptides into larger molecular weight proteins with little or no flavor.

Branched-Chain Amino Acids

High levels of branched-chain amino acids, such as leucine, isoleucine, and valine are often recommended to speed recovery for multiple-trauma or burn patients (Brennan et al. 1986; Alexander and Gottschlish 1990). The mechanism as to how these amino acids function physiologically is unclear. One general explanation is that under normal nutritional and physiological conditions, fuel requirements of the peripheral tissues are largely met by glucose and fatty acids, while amino acids generally contribute little toward overall fuel economy. However, during those abnormal metabolic states induced by trauma or sepsis, fat mobilization and utilization and glucose utilization are decreased as a result of hormonal changes precipitated by the trauma. Under these conditions, branched-chain amino acids derived from muscle protein catabolism become an important and significant potential source of fuel. When the branched-chain amino acid undergoes deamination, the free amino group is transaminated into other organic precursors, yielding alanine and glutamine (Cerra et al. 1983). The alanine is released from the muscle and is transported to the liver, where it is taken up and rapidly converted to glucose by the associated gluconeogenic metabolic pathways. Additionally, the oxidative decarboxylation of the ketoacids to form thioesters of coenzyme A allows this organic moiety to be readily utilized by the tricarboxylic acid cycle for energy in muscle cells (Freund et al. 1979; Echenique et al. 1984).

Glutamine

Glutamine metabolism appears to be significantly altered following injury or catabolic disease states. Conditions such as trauma and sepsis are associated with increased gastrointestinal consumption of glutamine (Alverdy et al. 1985, Alverdy 1990). Low plasma glutamine levels have been correlated with diarrhea, villous atrophy, mucosal ulceration, and intestinal necrosis (Souba et al. 1985). Mucosal degeneration and atrophy are particularly undesirable in the critically ill patient, as it predisposes them to bacterial translocation, i.e., bacterial passing from the gastrointestinal tract into the bloodstream, causing sepsis (Deitch et al. 1987). Recent animal studies suggest that elemental diets supplemented with glutamine have a tropic effect on the gastrointestinal tract with subsequent improvement in intestinal integrity (Rombeau 1990).

Because most dietary proteins contain glutamine, most medical foods will contain glutamine (Lacey and Wilmore 1990). Whether the amount found in the protein is adequate for the hospitalized patient is undetermined, and needs further research. It is unclear whether glutamine must be provided to the patient in a free form (crystalline amino acid) or bound form in order to extend the beneficial effect to the gastrointestinal tract. It seems logical that glutamine in any form (free, bound in proteins, bound in hydrolysates) would be beneficial in the gut for maintaining and restoring integrity. Further research is needed to elucidate the mechanism and verify these results.

Carnitine

Although it has always been assumed that individuals eating a diet containing sufficient levels of the essential amino acids would synthesize adequate amounts of carnitine, it has recently become apparent that there are individuals with a systemic carnitine deficiency (Borum 1983). In addition, under certain disease conditions, carnitine becomes essential and thus has been used to fortify medical food products. The daily requirement for carnitine is unknown for all mammalian species including humans, but it is known to be synthesized in the liver from the essential amino acids lysine and methionine. If the liver is impaired, the synthesis of carnitine may also be impaired. Because all the long-chain fatty acids supplied in the diet must be transported into the mitochondria via a carnitine pathway before they can be oxidized to produce energy, adequate levels of carnitine in the tissue are essential for these individuals (Fritz 1959). Testasecca (1987) also showed that added carnitine improved energy metabolism of patients under TPN support, while Bohles et al. (1984) showed improved muscle mass of hospitalized patients given supplemental carnitine.

Taurine

Taurine is important for normal retinal development and in the synthesis of bile salts (Hayes 1988). It is a beta amino acid that contains sulfur, but is unable to covalently bond with other amino acids. This unique amino acid may be essential for infants, children, and perhaps critically ill adults. Studies on taurine supplementation and its effect on fat absorption have shown conflicting results. However, some studies with cystic fibrosis patients showed improvement in fat absorption, growth, and weight gain following taurine supplementation (Belli et al. 1987). These latter beneficial effects are most likely due to the fact that cystic fibrosis is associated with an increased proportion of glycine conjugated bile acids concomitant with decreased taurine conjugated bile acids, which thus may contribute to malabsorption. Several medical food products are now supplemented with taurine.

Ribonucleic Acids (RNA)

Ribonucleic acids (RNA) in the body direct the synthesis of cellular protein. The common nucleotide bases that make up RNA include the purines, adenine and guanine, and the pyrimidines, cytosine, thymine, and uracil. These building blocks are formed in various tissues or are obtained from the diet components in the gut. They are present in large quantities in yeast hydrolysates as well as in fish. Certain rapidly growing cells, such as T lymphocytes and intestinal epithelial cells appear to lack the ability to synthesize these nucleotides and thus depend on the salvage pathway or dietary sources for sufficient amounts of nucleotides to be able to synthesize proteins and proliferate. It appears that these salvage/dietary sources are inadequate during severe metabolic stress such as trauma, burns, and sepsis, thereby contributing to the decrease in immune dysfunction under these conditions. In these cases they must be supplied in the medical food. Recent work shows that RNA may be essential for the maintenance of normal cellular immunity and for host resistance under certain conditions (Kulkarni et al. 1986). Nucleotides also may be needed in the diet for cancer and AIDs patients, who by virtue of the disease have suppressed immune function. Diets containing nucleotides have also been reported to decrease delayed hypersensitivity responses (Kulkarni et al. 1986), increased resistance to infections (Fanslow et al. 1988), and increased interleukin-2 production (van Buren et al. 1985). These immunosuppressive effects are directly related to the unique metabolic requirement of T lymphocytes for nucleotides, which appears to be necessary for T cell maturation and expression of phenotypic markers. Based on this, the addition of nucleotides to medical foods, most of which are nucleotide free, may have therapeutic applications for those who are immunocompromised due to metabolic stress or illness and at risk of developing infectious complications. On the other hand, Fanslow et al. (1988) demonstrated that for organ transplants, it is desirable to suppress the immune system in order to inhibit organ rejection, which can be accomplished by a diet free of nucleotides.

Lactose-Free Products

Although lactase is normally present in the gut of young persons of all racial groups, some loss of this intestinal oligosaccharidase during aging, causes adverse reactions when lactose-containing dairy foods are consumed. In fact, much of the world's adult population has lactase deficiency (Paige and Bayless 1981) and must limit their intake of some dairy products. Although deficiency of this enzyme may indeed be a delayed genetic condition, the high prevalence of malnutrition in small intestinal disease in many of the racial groups with adult lactase deficiency suggests that environmental factors may also be important in producing the deficiency state. Medical foods can obviously be made without lactose, but if a dairy-based product is developed, lactase can be added followed by holding at reduced temperature to hydrolyze the lactose to glucose and galactose, or lactose can be directly removed from the milk solids by means of ultrafiltration or ion exchange of the solution. In addition, one potential solution is the binding of the lactase enzyme on the inner packaging surface for liquid foods, where it can then cleave the lactose during distribution (Labuza and Breene 1989).

Special Dietary Fats

Medium-chain triglycerides (MCTs), which contain fatty acids composed of 6–10 linear carbon units, have been shown to be of use in medical foods. Unlike longer chain triglycerides, MCTs are absorbed intact by the intestinal mucosa and then the fatty acids are released by a mucosal cell microsomal lipase. They are not reesterified in the liver as are longer chain fatty acids, but are transported via the portal blood to the liver as a free fatty acid complexed with albumin. Thus they reach the liver more quickly. The majority of MCTs are retained in the liver rather than in other tissues or organs. MCTs can be used more readily by the cell because they do not require carnitine for their metabolism. The ability to synthesize carnitine is in question (as noted earlier) depending on the disease or metabolic state of the patient. MCT oil has been specifically indicated for pancreatic insufficiency, biliary atresia, chyluria, chylous fistula, celiac disease, small bowel resection, and cystic fibrosis.

MCTs are usually made from coconut oil, which naturally contains approximately 65% medium-chain fatty acids. To create MCTs, the triglycerides of the coconut oil are hydrolyzed off by sodium hydroxide, fractionated by distillation and/or super critical carbon dioxide, and then the six-, eight-, and ten-carbon units are re-esterified onto the glycerol molecule. Medium-chain triglycerides are most often used in combination with long-chain triglyceride oils in enteral formulas to provide the essential fatty acids via the LCTs, yet retain the advantage of the easy digestibility, absorption, and metabolism of the MCTs. Although MCTs and LCTs will compete for absorption, the administration of MCT in conjunction with LCT actually increases the total intestinal absorption of both over that obtained when either is administered alone (Bach and Babayan 1982). The MCT oil also provides fewer calories (8.3 kcal/g) and may be helpful in caloric reduction.

Structured lipid refers to a synthesized triglyceride that contains both MCTs and LCTs of specific distribution. The structured lipid is made by hydrolyzing a mixture of long- and medium-chain triglycerides and allowing them to re-esterify randomly forming a mixed triglyceride molecule that is chemically distinct from the physical mixtures of medium-chain and long-chain triglycerides, and may be more optimal from a digestion and metabolic standpoint. Long-chain fatty acids, preferably linoleic, can be added to meet the essential fatty acid requirement (Recommended Dietary Allowances 1989; Babayan 1987). Fish oils may also be used to provide new triglycerides containing eicosapentanoic acid or docosahexaenoic acid for patients who may benefit from ω-3 fatty acids.

Short-chain fatty acids such as acetate, butyrate, and propionate produced by bacterial fermentation in the colon and cecum have been shown to contribute up to 30% of the energy requirement for humans on high fiber diets (Slaven 1987). Short-chain fatty acids are readily absorbed by the colonic mucosa and those not metabolized there are transported to the liver for conversion to ketone bodies and other lipids. It has been suggested that under stress conditions the body may prefer to use ketone bodies, though exercise should be limited (Birkhahn and Border 1981). Medical food products containing fermentable fibers as precursors of short-chain fatty acids are a distinct possibility for the future.

Fiber

Since the mid 1980s, fiber has been added to some medical foods to improve gastrointestinal function by regulating transit time and facilitating absorption of fluid and electrolytes from the gut lumen

(Anderson 1989). Soy polysaccharide fiber is commonly used due to its intrinsic low viscosity and ability to flow through feeding tubes for oral tube feeding. Soy polysaccharide fiber is 94% insoluble dietary fiber and 6% soluble dietary fiber as measured by the AOAC method (Lo 1990). Insoluble fibers generally are not as fermentable by colonic bacteria as are soluble fibers and tend to increase stool wet weight by binding water. In hyperlipidemic patients, soy polysaccharide has been shown to decrease cholesterol levels and to lower serum glucose levels (Lo et al. 1986). Studies on gastrointestinal effects of enteral formulas with added soy polysaccharide fiber in critically ill patients have shown variable results. Based on current data, high fiber formulas (10–30 g per daily intake) are most likely beneficial to patients on long-term enteral feeding, particularly if the fibers are fermented to volatile fatty acids, which may be a preferred fuel for the mucosal cell of the gastrointestinal tract (Klurfeld 1987). Other fiber sources such as oat, pea, and natural gum sources including carrageenin and xanthan gum could be used in solid chewable products formed by extrusion, drying, or gelation techniques.

DELIVERY SYSTEMS

Medical foods are currently manufactured and packaged in two basic forms: (1) dry powders that are reconstituted into liquids, and (2) liquid products that are packaged in steel cans, glass containers, or multilaminated (plastic/foil/paperboard) containers. Both dietary forms tend to produce a monotonous diet because the product is fed either through a tube or drunk as a liquid, thus producing little oral stimulation. Of additional consideration is that all the components of a liquid product are heat sterilized together and remain mixed during distribution from the manufacturer, thereby increasing the potential for adverse chemical reactions that degrade essential nutrients or produce undesirable end-products, including off-flavors, bitterness, and toxic substances. Labuza and Massaro (1990) and Schmidl et al. (1988) studied and reviewed some of the storage problems of parenteral and enteral liquid medical food systems. Schmidl (1983) showed that one could prevent browning and emulsion destabilization of a fat containing TPN by acidifying to pH 3.4–4 and using a succinylated starch as an emulsifier.

Glass bottles are used for liquid products because they are transparent, which makes following the amount fed easy to read. However, bottles are also heavy and breakable, and the intake of air that is required to empty the container (if the product is tube fed) is sometimes thought to increase the risk of contaminating the commercially sterile product. Multilayer films composed of laminated polyethylene and/or polyvinyl chloride can be made into transparent collapsible pouches for tube feeding. These are impact resistant, easy to transport and handle, and flexible enough to allow easy administration without air intake. The bag can be squeezed with pressure cuffs for rapid administration of large quantities of liquids for bolus feeding. However, from a manufacturing point of view, the production speeds and seal integrity of the pouches remain of concern. Semi-rigid multilaminate polypropylene/ethylene vinyl alcohol containers are also impact resistant at room temperature, but because they are not fully collapsible, they generally require air priming in the same way that glass bottles do. These prefilled one-liter containers have become very popular for enteral liquids because they are easier to handle than a flexible bag, save nursing and staff time (no mixing of powders), reduce risk of formula contamination (Anderton 1985), generally are less expensive, and are particularly desirable for a long-term stable patient where the care may be at home. Closed-system containers may be a disadvantage for unstable patients where there may be a need to modify the formulas frequently (add water, protein, electrolytes).

More recently, aseptic sterilized product/packages systems have become more widely used in the medical food market. These containers are laminated aluminum foil, with layers of polyethylene, polypropylene, and paper board, the traditional "**juice box**" type brick-shaped containers. The rectangular shape occupies less space than round containers, and storage space is limited in nursing homes and hospitals. These products require that the patient be able to feed himself/herself by sipping.

Powdered products require reconstitution, which increases the risk of microbial contamination (Krey 1989). However, these products are generally less expensive than equivalent ready-to-feed liquids and require less storage space. Another advantage of powdered formulas is the ability to feed them in a highly concentrated form (partially hydrated). This is a useful option for fluid restricted patients. Finally, if properly dehydrated and mixed under conditions to keep the moisture content low enough (generally at the monolayer value), chemical reactions during distribution are minimized (Labuza 1980).

PROCESSING OPPORTUNITIES

As discussed at the Federated Societies for Experimental Biology symposium on medical foods (Anon. 1992), little thought has been given to delivery systems and/or technologies other than powders or liquids to help produce medical food products. Levine et al. (1985) noted the lack of understanding by the medical field of food technologies, which has perhaps contributed to a more drug-like approach in medical food development. This is surprising, as the food industry has certainly made significant progress in designing lower salt, lower calorie, lower cholesterol, and higher fiber- and calcium-containing foods, using new food ingredients, such as artificial sweeteners and protein or carbohydrate based fat substitutes, new methods of separation such as super critical carbon dioxide extraction, and newer methods of processing such as aseptic processing, and hot and cold extrusion. (Schmidl and Labuza 1985, 1990; Labuza 1985).

Labuza (1977, 1986) explored approaches the food industry could pursue to create foods specifically for the renal deficient patient, but few companies have entered this field, probably because of the very small market. Thus those who need specialty products to manage their medical condition are faced with few product options, and because of diet monotony, often break their diets, leading to medical problems.

One possible route for new products is that of intermediate moisture food technology derived from the pet food industry and research on space foods. This technology provides a chewable soft-moist food bar with oral gratification and can serve as a vehicle for all the necessary nutrients except the offending one (Labuza 1980). There is also the possibility of using confectionery technology to produce a drier, chewable product with a lower moisture content such that adverse chemical reactions during distribution and storage are minimized. This is the concept used to make the imitation fruit leathers that are well accepted by young children.

Another approach would be to mix the medical food components with a nonreactive texturizer (starch, gum, protein, and water, etc.), and then extrude at high pressure and temperature to produce a dry expanded product with a crisp texture (Dziezak 1989). The low moisture achieved would minimize adverse chemical reactions during storage. Alternatively, components could be mixed with a structuring agent and water, frozen into bars, freeze dried, and then compressed, a technology currently utilized by the U.S. Army Natick Labs for both the military and space feeding programs.

Another technology that could be utilized is the wet gel texturization method used to convert fish paste (kamaboko) into imitation shrimp, crab, and lobster meat, (surimi process). The problem is that the product would be subject to microbial growth because of the high moisture content and so would need to be refrigerated or frozen. Given the current distribution channel used for medical foods, this approach may be precluded in some areas of the world. Possibly, controlled atmosphere-active packaging technology could be used to increase the shelf life of these refrigerated or frozen products (Labuza and Breene 1989). The U.S. Army Natick Labs have developed a controlled atmosphere bread product with a 9-month shelf life that was used in Operation Desert Storm. This technology could be used as a delivery system for medical food products.

With respect to sterilization, the classical methods have been batch retort heating of containers, and higher temperature short time (HTST) batch heating to reduce quality loss. HTST processing of liquids followed by aseptic filling of the liquid medical food product into a multilaminate juice box (mentioned earlier) may increase the interest in consumption of these products by children who have metabolic errors. One could also envision ohmic heating (direct discharge of electrical currents in the liquid food) being used to make soups containing food particulates for medical food use or utilizing ultra high pressure technology (Farr 1990) to pasteurize whole fresh foods for the purpose of killing offending bacteria so that compromised patients can eat normal foods without the chance of bacterial infection. The use of gamma irradiation has also been used to sterilize fresh food products for feeding to those who have loss all ability to fend off bacteria and must live in an enclosed sterile environment (personal communication 1991, John Verderveen, FDA). In addition, the use of high energy pulsed white light, a technique pioneered by a division of Maxwell Labs of San Diego, CA (Dunn et al. 1989) and the use of microwave sterilization (Mudgett 1989; Decareau 1986) may be used so that heat induced reactions of the dietary components are minimized. Palaniappa et al. (1990) have reviewed the effects of various electromagnetic frequency treatments on microbial death and nutrient/quality changes in sterilized food products.

Another area to pursue is that of the improvement of the flavor of medical food products. Anecdotal evidence suggests that a product with a bitter off-flavor result in poor acceptance, a primary reason for people not adhering to their dietary regime. Solutions for flavor improvement would be the development of biologically available flavorless amino acid derivatives or dipeptides, microencapsulation of the amino acids in slow-dissolving encapsulating agents (such as is done with drugs), and the addition of aroma agents to stimulate either a desire to consume or to turn off the rejection syndrome, so-called **aroma therapy** (Labuza 1984; Schiffman and Coney 1984; Levine and Labuza 1990).

Biotechnology is someday likely to have a major impact on the development of medical food products. Harlander and Labuza (1986) have reviewed the potential applications of biotechnology to the food industry in general. Potential applications with respect to medical food products include: (a) utilizing genetic manipulation to delete undesirable biochemical components during the growth of the plant or animal, such as deleting the gene for gluten in flour (Sharp et al. 1984), (b) adding genes for synthesis of specific therapeutic compounds to convert normal foods into specialized medical foods, (c) utilizing enzyme bioreactor systems to convert normal food products into needed medical foods (e.g., conversion of vegetables oils into MCT oil), or (d) using dairy starter cultures that would ferment milk into palatable products, but at the same time remove offending nutrients like phenylalanine or cholesterol or produce desirable ones like carnitine or taurine. Little thought has been given to the application of biotechnology to medical food products. It is interesting that a texturized protein product from a *Fusarium* mold, "**mycoprotein**," made by a biotechnology process in the world's largest airlift fermentor located in Billingham, U.K., is now a major protein source for meat pies.

THE AGING POPULATION AS A MEDICAL FOOD NEEDS GROUP

In recent years food scientists and nutritionists have increased their efforts to expand their knowledge and to focus on products that will help to reduce or control diet-related chronic diseases such as heart disease, atherosclerosis, hypertension, cancer, and osteoporosis, as well as the creation of specialty products for weight control. Beginning in the 1970s the direction of the food industry was aimed towards fat, cholesterol, and sodium reduction. New technologies and ingredients have fostered a further interest in "functional foods" for the health conscious and elderly population.

In the past, the elderly have received very little attention as a target population for specialized nutritional or functional food products. This will change in the next few decades as the elderly continue to increase as a percentage of the total population. (Dychtwald 1986). Many physiological functions slow down with age such as basal metabolic rate, heart output, lung capacity, and nerve conduction (Shock 1962). In spite of these changes, eating habits and often the amount of food eaten do not change significantly, contributing to obesity and other chronic disease. Social and psychological factors can also alter lifestyles and ultimately dietary habits. For example, economic pressures often increase as a result of lower income during retirement.

Because many of the elderly live alone, eating habits may change and there may be a lack of motivation in meal preparation. Many elderly people also suffer from anxiety and are depressed, which can seriously influence their dietary intake, wants, and needs (Roe 1983). However the elderly are becoming more motivated to maintain good health, in part because of the debilitating aspects of aging and the associated medical costs. Today, nearly 20% of the U.S. population is "**retired**," and this percentage is increasing as a result of improved medical care, a better food supply, and a better understanding of diet over the past decade. Consideration of the relationship of diet to disease is essential, including the relation of: (a) saturated fats and atherosclerosis, (b) sodium and hypertension, (c) fiber and cancer, (d) fish oils and the immune system, (e) polyunsaturated fats and cancer, and (f) calcium and osteoporosis among others. Health claims for foods with these ingredients will be strictly controlled by the FDA by the new regulations which go into effect in 1993 (Labuza 1987). Of additional concern is that the increased taste and odor threshold of the elderly must be taken into account in producing appealing and tasteful foods for this group, an arena in which aroma therapy might be of help (Schiffman and Coney 1984). Recent work on the consequences of the consumption of Maillard reaction products as well as oxidized lipids on aging also suggests that significant work on controlling these deteriorative reactions in any food during distribution is critical (Baynes 1989). With the projected trend of an ever-increasing older population allied with an increasing awareness of the needs of the maturing American, the food industry has a role in meeting the nutritional needs and wants of this population group by some type of functional foods. Two examples of what industry can do will suffice.

In the 1980s, major breakthroughs took place with respect to the development of foods to contribute to health and reduce chronic disease. As data became available showing the high incidence of osteoporosis in elderly women, food technologists looked for ways to add calcium to the diet (Gain 1990). The elderly tend not to drink milk and may shun other dairy products like cheese, which are high in calories from fat and contain saturated fatty acids as the fat source.

Thus, orange juice was chosen to be fortified with calcium, a difficult step because many calcium sources are insoluble in acid solutions. More recent is the introduction of a low-fat, yogurt-orange juice product. At this time, not enough data are available to assess whether this approach will reduce the incidence of osteoporosis. The second major change was the increased use of fiber, especially oat bran, in many different kinds of foods to give them a health function. The driving force was the building relationship between fiber intake and reduced coronary heart disease and colon cancer (Klurfeld 1987; Anderson 1989). Most of the products with added fiber were cereals and cereal baked goods, although oat bran was added to other products such as liquid drink mixes and salad dressings. Because fiber is not a nutrient, there is much less support for its use from some of the nutrition community, and, as with calcium in orange juice, the long-term benefits are not yet known.

FRONTIERS

A broad range of food products and medical foods (functional foods) have been produced by various food technologies for use in the nutrition and health care arena. By understanding the basic principles of formulation, food processing, and clinical nutrition, these modalities can be utilized to prevent the wastage of body mass; help to prevent, control, and/or alleviate acute, genetic, and chronic disease conditions; and provide proper nutrition to most population groups. These points are well made in the National Academy of Sciences report "**Designing Foods**," which discusses the future trends and developments in the creation of nutritionally based products for the normal population, the report from FASEB on medical food products (Anon. 1992), and the FASEB, Life Sciences Research Office report on scientific guidelines for review of enteral food products for medical purposes (Talbott 1990).

In the future, genetic engineering of the ingredients available to the food technologist will certainly contribute to significant modifications to engineer food products consistent with disease states to achieve optimal health and well-being. Better clinical evidence relating specific dietary components to health and disease will certainly be needed. Development will have to comply with the necessary regulations and conform to valid medical claims about the usefulness of specific foods. Food and pharmaceutical companies will need to invest in the area where the population at risk are few, but the moral and ethical obligations of society are high. There may be the need for the industry to apply its technologies through some consortium so that those with specific medical disorders can benefit from modern food technologies. Finally, governments and the insurance industry must recognize that medical foods are critical in sustaining life and maintaining the health and well-being of those with specific and rare medical problems. Because of the low numbers of people with a certain disease, there may be little incentive for the industry to manufacture products on such a low scale. If the cost of these foods of "***sufficient medical merit***" could be covered by medical insurance and used only under medical supervision, this would stimulate both development and manufacture by the food and pharmaceutical industry.

References

Alexander, J.W., and Gottschlish, M.M. 1990. Nutritional immunomodulation in burn patients. *Crit. Care Med.* 18:S149.

Alverdy, J.C. 1990. Effects of glutamine-supplemented diets on immunology of the gut. *JPEN* 14:149.

Alverdy, J.C.; Chi, M.S.; and Sheldon, G.F. 1985. The effect of parenteral nutrition on gastrointestinal immunity: The importance of enteral stimulation. *Ann. Surg.* 202:681.

American Dietetic Association. 1986. Congressional Briefing on the Cost Effectiveness of Nutrition Support: Speaker and Presentation Summaries. Available from American Dietetic Association, Chicago, IL.

Anderson, J. 1989. Recent advance in carbohydrate nutrition and metabolism in diabetes mellitus. *J. of American College of Nutrition* 8(5):61S.

Anderton, A. 1985. Growth of bacteria in enteral feeding solution. *J. Med. Microbio.* 20:63.

Anon. 1988. FDA 1988/89 Compliance Policy Program 7321.002—Medical Foods—Import and Domestic.

Anon. 1992. Federated Societies for Experimental Biology Symposium Development of Medical Foods for Rare Diseases. Life Sciences Research Office, Bethesda, MD.

Babayan, V.K. 1987. Medical chain triglycerides and structured lipids. *Lipids* 22:417.

Bach, A.C., and Babayan, V.K. 1982. Medium-chain triglycerides—an update. *Am. J. Clin. Nutr.* 36:950.

Baynes, J. 1989. *The Maillard Reaction in Aging, Diabetes, and Nutrition*. New York: A.R. Liss Press, Inc.

Bell, S.J.; Pasulka, P.S.; and Blackburn, G.L. 1989. Enteral formulas. In *Dietitian's Handbook of Enteral and Parenteral Nutrition*, A. Skipper, ed., p. 279. Rockville, MD: Aspen Publishers Inc.

Belli, D.C.; Leny, E.; Darling, P.; Leroy, C.; Lepage, G.; Giguere, R.; and Roy, C.C. 1987. Taurine improves the adsorption of a fat meal in patients with cystic fibrosis. *Pediatrics* 80:517.

Bichel, H.; Gerrard, J.; and Hickmaus, E. 1954. The influence of phenylalanine intake on the chemistry and behavior of a phenylketaoic child. *Acta Pediatricia* 43:64.

Birkhahn, R.H., and Border, J.R. 1981. Ultimate or supplemental energy sources. *J. Parenteral and Enteral Nutrition* 5:24–31.

Bistrian, B.R.; Blackburn, G.L.; Hallowell, E.; and Heddle, R. 1974. Protein status of general surgical patients. *JAMA* 230:858.

Bistrian, B.R.; Blackburn, G.L.; Vitale, J.; Cochran, D.; and Naylor, J. 1976. Prevalence of malnutrition in general medical patients. *JAMA* 235:1567.

Blackburn, G.L.; Moldawer, L.L.; Usui, S.; Bothe, A.; O'Keefe, S.J.D.; and Bistrian, B.R. 1979. Branched chain amino acid administration and metabolism during starvation, injury, and infection. *Surgery* 86:307.

Bohles, H.; Segerer, J.; and Fekl, W. 1984. Improved N-retention during L-carnitine-supplemented total parenteral nutrition. *Journal of Parenteral and Enteral Nutrition* 8:9.

Borum, P. 1983. Carnitine. *Ann. Rev. Nutr.* 3:233.

Brennan, M.F.; Cerra, F.; Daly, J.M.; Fischev, J.E.; Moldawer, L.L.; Smith, R.J.; Vinars, E.; Wannemacher, R.; and Young, V.R. 1986. Report of a research workshop: Branched-chain amino acids in stress and injury. *JPEN* 10:446.

Cerra, F.B.; Mazuski, J.; Teasley, K.; Nuwer, N.; Lysne, J.; Shronts, E.; and Konstantinides, F. 1983. Nitrogen retention in critically ill patients is proportional to the branched chain amino acid load. *Crit. Care Med.* 11:775.

Cheney, M.C. 1989. Medical foods: The international scene. *Food Drug Cosmetic Law Journal* 44:517.

Decareau, R.V. 1986. Microwave processing throughout the world. *Food Technol.* 40:99–105.

Deitch, E.A.; Winterton, J.; Li, M.; and Berg, R. 1987. The gut as a portal entry for bacteria: Role of protein malnutrition. *Ann. Surg.* 205:681.

Dunn, J.E.; Clark, R.W.; Asmus, J.F.; Pearlman, J.S.; Boyer, K.; Painchaud, F.; and Hofmann, G.A. 1989. U.S. Patent 4,871,559. *Methods for Preservation of Foodstuffs.* San Diego: Maxwell Labs.

Dychtwald, K. 1986. *Age Wave.* Los Angeles: J.P. Tarcher, Inc.

Dziezak, J. 1989. Single and twin extruders in food processing. *Food Technol.* 43(4):163–174.

Echenique, M.M.; Bistrian, R.B.; Moldawer, L.L.; Palombo, J.D.; Miller, M.M.; and Blackburn, G.L. 1984. Improvement in amino acid use in critically ill patients with parenteral formulas enriched with branched chain amino acids. *Surg. Gynec. Obst.* 159:233.

Fanslow, W.C.; Kulkaini, A.D.; Van Buren, C.T.; et al. 1988. Effect of nucleotide restriction and supplementation on resistance to experimental murine candidiasis. *JPEN* 12(1):49.

FAO/WHO Expert Group. 1965. Protein Requirements. FAO Nutri Meet. Rep Sev. No. 37. *Food and Agric. Org. Rome.*

FAO/WHO Food Standards Programme (1991) Codex Alimentarius Commission 19th Session. Report of the 17th Session of the Codex Committee on Nutrition and Foods for Special Dietary Uses, Bonn-Bad Godesberg, Germany, February 18–22, 1991 Codex Alimentarius.

Farr, D. 1990. High pressure technology in the food industry. *Trends in Food Science and Technology*, July:14–16.

Fisher, K.; Talbot, J.; and Carr, C. 1977. *A Review of Foods for Medical Purposes*. Life Science Research Office of the Federation of America Societies for Experimental Biology. *FDA Contract No. 223-75-2090*.

Food and Drug Administration. 1987. *Food Labeling; Public Messages on Food Labels and Labeling*. Food and Drug Admin. Federal Register 52:28843.

———. 1989. *Compliance Program Guidance Manual*. Chapter 21. Program No. 7321.002 (1989–91). Available from Food and Drug Administration, Washington, DC.

———. 1990. *Food Labeling; Mandatory Status of Nutrition Labeling*. Federal Register 55:29487–29517.

Freund, H.; Atamlan, S.; Holroyde, J.; and Fischer, J.G. 1979. Plasma amino acids as predictors of the severity and outcome of sepsis. *Ann. Surg*. 190:571.

Fritz, I.B. 1959. Action of carnitine on long chain fatty acid oxidation by the liver. *Am. J. Physiol*. 197–297.

Gain, S. 1990. Will calcium supplementation preserve bone integrity? *Nutrition Reviews* 48(1):26.

Greenstein, J.P.; Birnbaum, S.M.; Winitz, M.; and Otey, M.C. 1957. Quantitative nutritional studies with water soluble, chemically defined diets. I. Growth, reproduction, and lactation in rats. *Arch. Biophy*. 72:396.

Grimble, G.K.; Rees, R.G.; Keohane, P.P.; Cartwright, T.; Desreumaux, M.; and Silk, D.B.A. 1987. "Effect of peptide chain length on absorption of egg protein hydrolysate in normal human jejunium." *Gastroenterology* 92:135.

Harlander, S.K., and Labuza, T.P. 1986. *Biotechnology and Food Processing*. NJ: Noyes Publishing Co.

Hattan, D.G., and Mackey, D. 1989. A review of medical foods: Enterally administered formulations used in the treatment of disease and disorders. *Food Drug Cosmetic Law Journal* 44:479.

Hayes, K.C. 1988. "Vitamin-Like" Molecular-Taurine. In *1988 Modern Nutrition in Health and Disease*, M.E. Shills and V.R. Young, eds. Philadelphia: Lea and Febiger.

Heimburger, D.C., and Weinsier, R.L. 1985. Guidelines for evaluating and categorizing enteral feeding formulas according to therapeutic equivalence. *JPEN* 9:61.

Hertzman, P.A.; Falk, H.; Kilbourne, S.P.; and Schulman, L. 1991. The eosinophilia myalgia syndrome. *J. Rheumatol*. 18:867.

Kinney, J.M. 1976. Energy requirements for parenteral nutrition. In *Total Parenteral Nutrition*, J.E. Fischer, ed., p. 135. Boston: Little, Brown & Co.

Klurfeld, D. 1987. The role of dietary fiber in gastrointestinal disease. *J. Am. Dietetic Assn*. 87(9):1173.

Koretz, R.L., and Meyer, J.H. 1980. Elemental diets: Facts and fantasies. *Gastroenterology* 78:393.

Krey, S.; Porcelli, K.; Lockett, G.; Karlberg, P.; and Shronts, E. 1989. Enteral nutrition. In *Nutrition Support Dietetics Core Curriculum*. Silver Springs, MD: ASPEN.

Kulkarni, A.D.; Fanslow, W.C.; Rudolph, F.B.; and Van Buren, C.T. 1986. Effect of dietary nucleotides on response to bacterial infections. *JPEN* 10:169.

Labuza, T.P. 1977. Defining research and development priorities for foods to match the needs of renal deficient consumers. *Food Prod. Devel.* 11(1):64–68.

———. 1978. The technology of intermediate moisture foods in the U.S. In *Food Microbiology and Technology*. Milano, Italy: Medicina Viva Press.

———. 1980. The effect of water activity on reaction kinetics of food deterioration. *Food Technol.* 34:36–41,59.

———. 1981. Food laws and regulation: The impact on food research. *Food Drug Cosmetic Law J.* 36(6):293–302.

———. 1984. From moo foods to mood foods: A historical perspective of 1984. *Food Technol.* 38(11):114–121.

———. 1985. Industry's role in diet and health. *Cereal Foods World* 30(12):827–830.

———. 1986. Future food products: What will technology be serving up to consumers in the years ahead. *J. Assoc. Food and Drug Officials* 50(2):67–84.

———. 1987. A perspective in health claims in food labeling. *Cereal Foods World* 32:276.

Labuza, T.P., and Breene, W. 1989. Active packaging concepts for CA/MA packaging of foods. *J. Food Proc. and Preserv.* 13(1):1–69.

Labuza, T.P., and Massaro, S. 1990. Browning and amino acid losses in model total parenteral nutrition solutions. *J. Food Science* 55:821.

Lacey, J.M., and Wilmore, D.W. 1990. Is glutamine a conditionally essential amino acid? *Nutr. Rev.* 48:297.

Levine, A.S., and Labuza, T.P. 1990. Food systems: The relationship between health and food science/technology. *Environmental Health Perspectives* 86:233–238.

Levine, A.; Labuza, T.P.; and Morley, . 1985. Food technology and medical health. *N. Engl. J. Med.* 312:628–634.

Lo, G.S. 1990. Manufacturer Representative of Protein Technologies. Personal Communincation.

Lo, G.S.; Goldberg, A.P.; Lim, A.; et al. 1986. Soy fiber improves lipid and carbohydrate metabolism in primary hyperlipidemia subjects. *Atherosclerosis* 622:239.

Matthews, D.M., and Adibi, S.A. 1976. Progress in gastroenterology—Peptide absorption. *Gastroenterology* 71:151.

Merrill, R., and Hutt, P.B. 1990. *Food and Drug Law Case and Materials* Mineola, NY: Foundation Press.

Mountford, M.K., and Cristol, R.E. 1989. The enteral formula market in the United States. *Food Drug Cosmetic Law Journal* 44:479.

Mudgett, R.E. 1989. *Microwave Food Processing*. Scientific Status Summary. Chicago: Institute of Technologists.

Paige, D.M., and Bayless, T.M. 1981. *Lactose Digestion*. Baltimore, MD: Johns Hopkins University Press.

Palaniappa, S.; Sastry, S.; and Richler, E.R. 1990. Effects of electricity on microorganisms: A review. *J. Food Proc. and Preserv.* 14:393–414.

Raubicheck, C.J. 1989. Labeling a medical food. *Food Drug Cosmetic Journal* 44:533.

Recommended Dietary Allowances. 1989. 10th Edition National Research Council, National Academy Press, Washington, DC.

Roe, D.A. 1983. *Geriatric Nutrition.* Englewood Cliffs, NJ: Prentice Hall, Inc.

Rombeau, J.L. 1990. A review of the effects of glutamine-enriched diets on experimentally induced enterocolitis. *JPEN* 14:1005.

Rombeau, J.L., and Caldwell, M.D. 1986. *Enteral and Tube Feeding.* Philadelphia: W.B. Saunders Company.

Scarbrough, F.E. 1989. Medical foods. An introduction. *Food Drug Cosmetic Law Journal* 44:463.

Schiffman, S.S., and Coney, E. 1984. Changes in taste and smell with age. Nutritional aspects. In *Nutrition and Gerontology,* J.M. Ordy, D. Hann, and Alfred Slater, eds., p. 43. New York: Raner Press.

Schmidl, M.K. 1983. Liquid Elemental Diet. U.S. Patent #4414238.

Schmidl, M.K., and Labuza, T.P. 1985. Low calorie formulations: Cutting calories and keeping quality. *Proc. Prep. Foods* 154(11):118–120.

———. 1990. The history, current status and future of nutritional products. In *Food Product Development: From Concept to Marketplace.* E. Graf and I. Saguy, eds. New York: Van Nostrand Rheinhold Publishers.

Schmidl, M.K.; Massaro, S.; and Labuza, T.P. 1988. Parenteral and enteral food systems. *Food Technol.* 42(7):77–87.

Serafino, J.M. 1989. Good manufacturing practices and quality control for medical foods. *Food Drug Cosmetic Law Journal* 44:523.

Sharp, W.R.; Evans, D.A.; and Annirato, P.V. 1984. Plant genetic engineering: designing crops to meet food industry specifications. *Food Technol.* 38(2):112–119.

Shaver, H.J.; Laper, J.A.; and Lutes, R.A. 1980. Nutritional status of nursing home patients. *Journal of Parenteral and Enteral Nutrition* 4:367.

Shils, M. 1975. *Defined Formulas for Medical Purposes.* Chicago: American Medical Association.

Shils, M.E. 1988. Enteral (tube) and parenteral nutrition support. In *Modern Nutrition in Health and Disease.* M.E. Shils and V.R. Young, eds., p. 1023. Philadelphia: Lea and Febiger.

Slaven, J.L. 1987. Dietary Fiber, Classification, Chemical Analyses and Food Sources. *J. Am. Dietetic Assn.* 87(9):1165.

Shock, N.W. 1962. The physiology of aging. *Sci. Am.* 206(10):100.

Souba, W.W.; Smith, R.J.; and Wilmore, D.W. 1985. Glutamine metabolism by the intestinal tract. *JPEN* 9:608.

Steinburg, E.P., and Anderson, G.F. 1989. Implication of Medicare's Prospective Payment System for Specialized Nutrition Services, *Nutrition in Clinical Practice,* February: p. 12–28.

Sturman, J.A., and Hayes, K.C. 1980. The biology of taurine in nutrition and development. *Adv. Nutr. Res.* 3:231.

Talbot, J.M. 1990. *Guidelines for the Scientific Review of Enteral Food Products for Special Medical Purposes.* Life Science Research Office of Federation of American Societies for Experimental Biology. FDA Contract No. 223-88-2124.

Testasecca, D. 1987. Effects of carnitine administration to multiple injury patients receiving total parenteral nutrition. *J. Clin. Pharm. Ther. Tox.* 25:56.

Thompson, W.R.; Stephens, R.V.; Randall, H.T.; and Bower, J.R. 1969. Use of the "Space Diet" in the management of a patient with extreme short bowel syndrome. *Am. J. Surg.* 117:449.

Twomey, P.L., and Patching, S.C. 1985. Cost effectiveness of nutrition support. *Journal of Parenteral and Enteral Nutrition* 9:3.

U.S. Congress. 1988. 100th Congress. Organ Drug Amendments of 1988. P.L. 100-290 Amendment of Section 526(a)(1) of the Federal Food, Drug and Cosmetic Act. 21 U.S.C. 360bb(ax1). Library of Congress, Washington, DC.

———. 1990. 101th Congress. Nutrition Labeling and Education Act of 1990. P.L. 101-535 Amendment of Section 403 of the Federal Food, Drug and Cosmetic Act 21 U.S.C. 343. Library of Congress, Washington, DC.

van Buren, C.T.; Kulkaini, A.D.; Fanslow, W.C.; and Rudolph, F.B. 1985. Dietary nucleotides, a requirement for helper/inducer T lymphocytes. *Transplantation* 40:694.

Winitz, M., Graff, J. Gallagher, N., Narken, A., and Seedman, D.A. 1965. Evaluation of chemical diets as nutrition for man-in-space. *Nature* 205:741–743.

Part III

Health Functionality of Food Components

Chapter 9

Dietary Fiber

Aliza Stark and
Zecharia Madar

DEFINITION OF DIETARY FIBER

The first step in deciphering the function of dietary fiber is understanding that it is not a uniform substance. Dietary fiber is a mixture of many complex organic substances, each having unique physical and chemical properties. Although a single definition has yet to be agreed upon, dietary fiber is commonly defined as "plant polysaccharides and lignin which are resistant to hydrolysis by the digestive enzymes of man" (Trowell et al. 1976). Plant cell wall material containing cellulose, hemicellulose, pectic substances, and lignin are the major components of dietary fiber (Selvendran, Stevens, and Du Pont 1987). In addition, mucilages, gums, algal polysaccharides, and synthetic polysaccharides are also considered dietary fiber. Except for lignin, dietary fibers are carbohydrate in nature, and it has been suggested that the measurement of nonstarch polysaccharides (NSP) is more accurate than determination of "dietary fiber" (Englyst et al. 1987). On the other hand, it has been proposed that starch that is not digested in the small intestine and reaches the colon (resistant starch) is chemically and physically similar to other nondigestable polysaccharides and should therefore be included in the definition of dietary fiber (Asp 1990).

Several laboratories have developed methodologies to quantitate the dietary fiber content of foods. Early nutrient tables reported dietary fiber as "crude fiber," using a method where much of the fiber was lost during extraction. The estimates of crude fiber do not include most hemicelluloses and soluble fibers and may be one-fifth of the true dietary fiber value (Dreher 1987a). Over the years other methodologies such as the detergent methods of Goering and van Soest (1970), the Southgate method (Southgate 1969), and various enzymatic methods (Hellendoorn, Noordhuff, and Slagman 1975; Asp and Johansson 1981; Furda 1981) have been proposed. The Association of Official Analytical Chemists (AOAC) has agreed on a composite of these methods as a standard for measuring total dietary fiber based on enzymatic degradation and gravimetric determination of the residue (Prosky et al. 1985 Prosky and DeVries, 1992). Major drawbacks of this method are its inability to quantify soluble fibers or identify sugar components. Despite disagreements on the definition and the method of analysis, there is no doubt that the intake of dietary fiber has important metabolic and physiological effects.

CHARACTERIZATION OF DIETARY FIBER

Physicochemical Properties of Dietary Fiber

In an attempt to clarify the concept of dietary fiber, many scientists have opted to define individual fiber sources by their physical properties. One widely used classification is that of water soluble or gel-forming viscous fibers and water insoluble fibers. This distinction is convenient because many of the physiological effects of fiber seem to be based on this property. Soluble fibers are highly fermentable and are associated with carbohydrate and lipid metabolism, while insoluble fibers contribute to fecal bulk and reduced transit times (Madar and Odes 1990a). Particle size, water holding capacity, viscosity, cation exchange capability, and binding potential are specific for every fiber source (Dreher 1987b). These properties may change when a food undergoes cooking and digestion. The chemical structure of the fiber is a major factor in determining its physical properties. Cellulose, for example, contains tightly packed linear polysaccharides that are insoluble in water and resistant to hydration and swelling. In contrast, pectin is a charged viscous polysaccharide that is readily soluble in water and has a high ion binding potential. Binding properties vary among fiber components; lignin and hemicellulose adsorb bile salts, while cellulose alone has a very low capacity for bile salt adsorption (Story 1986). In plant tissues a mixture of polysaccharides and lignin is produced, which makes it difficult to accurately predict their physiological effects in the body. This explains why the effects of oat bran are so different from those of wheat bran, and why the physiological effects of dietary fiber are not always related to the physicochemical properties of the individual polysaccharides they contain.

Eastwood and Morris (1992) describe dietary fiber as a "water-laden sponge" moving through the intestine. The structure and surface activity contributed by the water insoluble fibers combined with the gel-forming viscous properties of the water soluble fiber network provide the fiber matrix with the ability to carry out such activities as cation exchange and gel filtration.

The Nutritional Value of Dietary Fiber

For many years, dietary fiber was not considered to have a significant nutritional value. We now know that many fibers are fermented in the large intestine to produce hydrogen, methane, carbon dioxide, and short-chain fatty acids. The short-chain fatty acids are rapidly absorbed from the gastrointestinal tract and contribute to the energy balance of the body (Cummings 1991). Because not all fibers are degraded equally by bacteria, an average of 2 kcal/g fiber has been suggested by the British Nutrition Foundation Task Force (Anon. 1990). Actual values range from 0 to 3 kcal/g for nonfermentable and highly fermentable fibers, respectively (Livesey 1990).

METABOLIC AND PHYSIOLOGICAL EFFECTS OF DIETARY FIBER

Gastrointestinal Tract

The small intestine is the site of the digestion and the absorption of large quantities of food. Dietary fiber has a significant influence on the rate and effectiveness of nutrient absorption. When plant tissues are eaten, much of the cell structure remains intact. Due to the fiber content, a barrier effect occurs and hydrolytic enzymes such as amylase are not able to associate with their substrates (Wursch, Del Vedovo, and Koellreutter 1986). Dietary fiber is also able to bind and adsorb water, enzymes, cations, and inorganic micronutrients, making them unavailable for digestion and absorption. Many components of dietary fiber have been shown to bind bile salts (Story 1986), which may interfere with intestinal lipid absorption. The rate of gastric emptying and intestinal transit time is also affected by fiber in the gastrointestinal tract. Soluble fiber tends to slow mouth-to-cecum transit times (Jenkins et al. 1978; Brown et al. 1988) by decreasing the rate of gastric emptying and increasing intestinal transit times. Increases in the viscosity of intestinal contents reduce the rate of nutrient transport and decrease access of nutrients to mucosal surface (Johnson 1990). Thus, peristaltic mixing is limited, enzyme-substrate contact and micelle formation are reduced, and absorption is slowed. The intake of fiber also modifies secretory enzyme activity. Conflicting data exist concerning the effects of fiber on pancreatic enzyme activities. Reports indicate that some dietary fibers reduce *in vitro* enzyme activities (Isaksson, Lundquist, and Ihse 1982; Dutta and Hlasko 1985), which may be compensated for by an increased secretion (Isaksson et al. 1983). While other studies indicate that no changes occur in enzyme levels after ingestion of pectin or guar gum (Calvert et al. 1985), other studies report an increased activity in the presence of pectin (Dunaif and Schneeman 1981).

Significant changes in intestinal morphology following long-term fiber intake have been observed (Vahouny and Cassidy 1986). Changes in the structure of intestinal villi occur, and other factors, such as cell migration rate, labeling index, and goblet cell number are also affected. These adaptation mechanisms also influence nutrient absorption from the small intestine.

Undigested food particles, fiber, and mucus arrive in the colon, where they are exposed to many species of bacteria. Soluble fibers are largely fermented, while insoluble fibers pass through the colon virtually unchanged. Insoluble fibers such as wheat bran and cellulose increase stool weight and frequency, soften feces, increase fecal bulk, and reduce intestinal transit time (Stevens et al. 1988; Wrick et al. 1983; Muller-Lissner 1988). Viscous undigestible fibers such as ispaghula and psyllium act similarly to insoluble fibers in the colon (Tomlin and Read 1988; Stevens et al. 1988). Degradable fibers have a minimal effect on stool volume, but highly fermentable fibers such as guar, as they lose their viscosity, reduce transit time in the colon (Tomlin and Read 1988). Nonfermentable fibers contribute to fecal bulking by immobilizing water in its matrix. Wheat bran undergoes minimal degradation in the colon, and its fecal bulking abilities can be predicted by its water-holding properties (Eastwood et al. 1983). Particle size is also critical in determining bulking ability. Finely ground bran has significantly less effect on fecal volume and transit time than coarse bran (Wrick et al. 1983). Fecal output and decreased transit time are also dependent on the variety of the fiber eaten, i.e., fiber from fruit and vegetable fiber are less effective than wheat bran (Cummings et al. 1978). It is important to note that there is no linear quantitative relationship between fiber intake and stool output and transit time. These parameters are determined by such factors as fiber variety and particle size. Accurate predictions of physiological effects are difficult to assess.

Carbohydrate Metabolism

Although insoluble dietary fiber has little or no effect on carbohydrate metabolism, water soluble fiber and foods rich in viscous fiber have been found to reduce postprandial glucose excursions and improve insulin profiles. Guar gum (a storage polysaccharide of the cluster bean) has been widely tested in both healthy and diabetic subjects. Repeatedly, guar gum has lowered glucose and insulin responses to glucose drinks or mixed meals (Jenkins et al. 1977; Ellis et al. 1988; LeBlanc et al. 1991). Long-term studies using diabetic subjects have also confirmed the beneficial effects of guar gum (Jenkins et al. 1980; Ebling et al. 1988). Other soluble fiber sources such as pectin, soy polysaccharides, and fenugreek are also attributed with hypoglycemic properties (Jenkins et al. 1978; Madar et al. 1988a, b; Lo et al. 1986). These fiber sources were administered as powders or incorporated into foods to increase palatability and compliance. Although isolated dietary fibers are often used in experiments, many studies have focused on the incorporation of high fiber foods into the daily diet.

The rate of glucose absorption and uptake in the body is dependent on the composition and structure of the food eaten. Jenkins et al. (1981) proposed the concept of a glycemic index (GI), which compares carbohydrate-containing foods on the basis of postprandial glucose profiles in relation to a standard of glucose or white bread. Although every component of a food (protein, fat, water, etc.) will influence the rate of carbohydrate absorption, fiber content is a major determinant in the postprandial glucose response (Nishimune et al. 1991; Wolever 1990). Studies using low GI diets as a basis for improving glucose tolerance show modest but significant improvements in glycemic control in both healthy (Jenkins et al. 1987) and diabetic (Brand et al. 1991) subjects.

Numerous mechanisms by which fiber affects carbohydrate metabolism have been proposed. Increased viscosity of the gastrointestinal contents is thought to be a major factor affecting the rate of glucose absorption. Highly viscous chyme slows gastric emptying and reduces the rate of intestinal absorption of glucose (Jenkins et al. 1978; Meyer et al. 1988; Di Lorenzo et al. 1988). Topping et al. (1988) showed that, in rats, a diet rich in highly viscous methylcellulose significantly lowered blood glucose levels when compared to methylcellulose of low viscosity. This was accounted for by the increased transit time from the stomach to the intestine, which delayed absorption of nutrients, and the unavailability of starch and sugars trapped in the methylcellulose matrix. High viscosity in the small intestine reduces the rate of diffusion of enzymes and digested sugars at the level of the intestinal mucosa (Johnson 1990). A combination of slowing of gastric emptying and hindered intestinal absorption appears to be responsible for the reductions in postprandial glucose concentrations. It has been suggested (Vinik and Jenkins 1988) that dietary fiber causes changes in the levels of the gut hormones that determine gastrointestinal motility, nutrient absorption, and insulin secretion. Gastric inhibitory polypeptide (GIP), glucagon, and somatostatin may all play a role in mediating the hypoglycemic effects of certain dietary fibers.

Using the euglycemic clamp technique, Fukagawa et al. (1990) reported an increase in peripheral insulin sensitivity when healthy subjects were fed a high carbohydrate, high fiber diet. In insulin dependent diabetic subjects, a high carbohydrate, high fiber diet decreased the basal insulin requirements and increased the carbohydrate disposal per unit insulin (Anderson et al. 1991). Although acute effects of fiber intake on insulin sensitivity can be attributed to a lower rate of glucose absorption, adequate explanations for the cause of long-term indirect effects of fiber intake have yet to be determined.

The prescription of high fiber diets and isolated fiber supplements is very common in the treatment of diabetic patients and is thought to aid in stabilizing blood glucose levels, increasing insulin sensitivity, and preventing long-term diabetic complications. Despite the general agreement on the benefits of the intake of fiber in the management of diabetes, contradictory reports have been published (Hollenbeck, Coulston, and Reaven 1986; Beattie et al. 1988). In addition, in some cases, undesirable side effects such as diarrhea, flatulence, and bloating occur because of an increased fiber intake. Care must be taken in choosing high fiber foods or fiber supplements that have been shown to induce desired responses.

Lipid Metabolism

The effects of dietary fiber on lipid metabolism have been widely investigated. In most studies, water soluble fibers have been shown to have hypocholesterolemic properties, while insoluble fibers have little or no effect on cholesterol metabolism (Kritchevsky 1988). Many studies have been criticized because appropriate control diets were not used. It has been proposed (Swain et al. 1990) that decreased cholesterol levels attributed to fiber supplementation were probably due to changes in the consumption of saturated fat during the experimental period rather than the intake of fiber. Furthermore, there has been much debate on the mechanisms by which soluble dietary fibers affect lipid metabolism. Despite these issues, the overall evidence indicates that soluble fibers lower plasma cholesterol levels. In the most recent studies summarized in Table 9-1, great care was taken to ensure that control groups received either an appropriate placebo or an insoluble dietary fiber supplement. Results indicate that a variety of different soluble fibers, including guar, psyllium, pectin, and oat bran, have hypocholesterolemic properties.

The mechanisms by which dietary fiber affects lipid metabolism are not well understood. The most obvious explanation is that physicochemical changes in the gastrointestinal contents (i.e., increased viscosity) interfere with micelle formation and lipid absorption. It has also been suggested that increased sterol excretion is responsible for the cholesterol-lowering effects of fiber. Certain varieties of dietary fibers can bind bile salts and neutral sterols (Vahouny and Cassidy 1985; Story 1986) and thus enhance their removal from the body. The bile salts excreted in the feces are removed from enterohepatic circulation and smaller amounts of bile salts are available for lipid absorption. In addition, serum and liver cholesterol are used for bile acid synthesis to replenish depleted pools. Studies in rats have consistently shown that fibers that reduce plasma cholesterol increase neutral sterol or bile acid excretion (Kritchevsky 1988; Illman and Topping 1985; Arjmandi et al. 1992). In clinical trials at

Table 9-1. Effects of Dietary Fiber on Plasma Cholesterol Levels—Human Studies

Fiber Variety	Dose g/day	Time Period (weeks)	Plasma Total Cholesterol Baseline (mmol/liter)	Final	Change (%)	Reference
Guar gum	12	8	7.4	6.5	−12	Kristen et al. 1989
Wheat bran	15	3	4.39	4.55	NS	
Rice bran	15	3	4.39	4.46	NS	Sanders and Reddy 1992
Rice bran	30	3	4.39	4.38	NS	
Psyllium	10	8	6.39	5.45	−15	Anderson et al. 1988
Cellulose placebo	10	8	6.45	6.22	−4.0	
Guar gum	10	4	6.44	5.82	−9.7	
Psyllium 6.3						Haskell et al. 1992
Pectin 3.9	15	4	6.44	5.92	−8.3	
Guar gum 3.3						
Locust bean 1.5						
Psyllium	12	6	6.51	5.96	−8.4	Anderson et al. 1992
Wheat bran	20	6	6.56	6.56	0	
Wheat bran	6.8 g fiber/ 1,000 kcal	2	5.36	5.11	−4.7	Kashtan et al. 1992
Oat bran		2	5.88	5.24	−11	
Oat bran	87	6	4.80	4.44	−8.0	Swain et al. 1990
Low fiber wheat control	93	6	4.80	4.46	−7.0	
Guar gum	15	4	6.23	5.58	−10	Superko et al. 1988

NS = Not Significant.

least one study, using beans as a fiber source, did not support this hypothesis (Anderson et al. 1984).

The products of bacterial fermentation of dietary fiber may also play a role in lipid metabolism. Short-chain fatty acids (SCFA) are produced in relatively large quantities in the colon from bacterial breakdown of dietary fiber. These SCFA are rapidly absorbed via the hepatic portal vein (Cummings 1991), and it has been proposed that propionate produced by fermentation inhibits hepatic cholesterol synthesis. This theory is based on *in vitro* experiments using isolated hepatocytes (Chen, Anderson, and Jennings 1984; Wright, Anderson, and Bridges 1990) and feeding experiments with dietary propionate (Chen, Anderson, and Jennings 1984; Illman et al. 1988). Much controversy has arisen concerning the mechanisms by which propionate lowers plasma cholesterol. Illman et al. (1988) suggest that the inhibition of cholesterol synthesis in isolated hepatocytes has been achieved at concentrations that are of little physiological importance and redistribution of cholesterol from the plasma to the liver occurs as a result of dietary propionate feeding. In addition, increased hepatic cholesterol synthesis has been observed in rats that were fed pectin (Nishina and Freedland 1990) and oat bran (Illman and Topping 1985). The role of the SCFA in lipid metabolism and the mechanisms by which dietary fibers lower plasma cholesterol levels remain to be elucidated.

Gastrointestinal Disorders

The accumulated evidence suggests that dietary fiber intake is directly related to normal functioning of the gastrointestinal tract. In addition, dietary fiber is commonly used as a therapeutic agent to relieve many gastrointestinal problems. Constipation is a common problem treated with different varieties of dietary fiber. Because the volume and texture of stools are influenced by fiber intake, many individuals suffering from constipation have found that dietary fiber treatment will relieve their symptoms. One of the most commonly used fibers in the treatment of constipation is wheat bran. Although the benefits of wheat bran treatment have been documented, overall, less than 50% of patients treated with wheat bran show clinical improvement (Muller-Lissner 1988; Klein 1988). Other fiber sources, such as psyllium, have laxative properties and reportedly can reduce the symptoms of constipation (Borgia et al. 1983; Marlett et al. 1987). Despite the widespread use of psyllium in the treatment of constipation, side effects of flatulence and abdominal pain may occur, and psyllium itself has been reported to cause constipation (Madar and Odes 1990b). A combination of psyllium, celandin, and aloe vera fibers reportedly increased the frequency of bowel movements, softened stools, and decreased laxative dependence in constipated subjects (Odes and Madar 1991). Studies using less common sources of fiber have also been shown to aid in the treatment of constipation. Spent grain, a waste product of the beer industry, reduced the symptoms of 79% of patients in a pilot study (Odes et al. 1986).

Irritable bowel syndrome (IBS) is a general term for a wide range of gastrointestinal disorders. Both diarrhea and constipation may be symptoms of IBS along with abdominal pain or bloating. Because no adequate operational definition has been agreed upon, IBS patients are commonly a heterogeneous group of subjects with many different symptoms. Other difficulties in testing the efficacy of fiber treatments in these patients arise when no truly objective measurement of improvement is available and a strong placebo effect is observed. Despite the problems in designing clinical studies, many have been carried out, but few provide convincing evidence for the beneficial effects of fiber. Wheat bran is often prescribed for treatment in IBS patients and in some cases, particularly in constipated individuals, this treatment provides symptomatic relief (Cann, Reed, and Holdsworth 1984). In some individuals, wheat bran exacerbates symptoms. Ispaghula was found to have some beneficial effects in treating constipated IBS patients, but had no effect on patients suffering from diarrhea (Prior and Whorwell 1987).

Diverticulosis is a degenerative disease characterized by herniations of the mucosal lining projecting through the muscle wall of the colon at a point where a nutrient artery penetrates the muscularis. It is commonly associated with a low fiber diet and high intracolonic pressure. Diverticulosis is often treated with wheat bran, which has been shown to increase stool weight (Muller-Lissner 1988) and decrease colonic pressure (Findlay et al. 1974). In a retrospective study (Leahy et al. 1985) subjects with colonic diverticulosis on a high fiber diet experienced significantly fewer symptoms and complications associated with diverticulosis when compared to a low fiber control group. A long-term study in rats (Fisher et al. 1985) also indicated that a fiber-rich diet prevented the development of diverticuli. Insufficient evidence is available to accurately determine the beneficial role of dietary fiber in diverticular disease. There is a need to further investigate both the etiology and treatment of this disorder.

Colorectal Cancer

Many epidemiological, clinical, and experimental studies have been undertaken in an attempt to clarify the connection between dietary fiber and colon cancer. Considerable evidence has been accumulated in epidemiological studies that supports a connection between reduced risk of colon cancer and a diet rich in fiber (reviewed in: Trock, Lanza, and Greenwald 1990). But it should be noted that several studies found little or no connection between fiber intake and cancer development (Potter and McMichael 1986; Miller et al. 1983).

Studies using animal models have produced conflicting results that are difficult to interpret. Fiber has been associated with a decreased risk of cancer, an increased risk of cancer, and in some cases, no association has been established. Different results arise from the variability in rodent species, the choice and the dose of carcinogen, the experimental diet (% fiber and % fat), the length of the experiment, and inappropriate control groups. The use of unacceptable statistical analysis further complicates interpretation of the results (Klurfeld 1990). It is also important to note that the human diet consists of foods containing fiber, while the experiments with animals are carried out entirely with fiber isolates.

Insoluble dietary fibers are known to prevent the development of colon cancer, while soluble fibers are often associated with tumor enhancement. In a recent review by Jacobs (1990) evaluating eight studies with animals having chemically induced colon cancer, the addition of pectin enhanced tumor development in half of the studies. Only in one study, did pectin have a protective effect. Pectin and guar have also been shown to increase cell proliferation in the colon (Lupton, Coder, and Jacobs 1988). Psyllium is the only soluble fiber that appears to inhibit tumor formation (Roberts-Anderson, Mehta, and Wilson 1987). This may be due to its low fermentability. Most experiments use pharmaceutical doses of fiber; it is possible that smaller doses may produce different results.

Insoluble fiber sources such as wheat bran and cellulose are generally considered to protect against colon cancer development. However, many conflicting data appear in the literature. Alberts et al. (1990) has shown that wheat bran included in the daily diet decreases cell turnover rates in the intestine and thus lowers the risk of colon cancer development. Heitman et al. (1989) showed a protective effect of cellulose (31–57% decreased incidence of adenocarcinomas). But reports of wheat bran stimulating epithelial cell growth

(Jacobs and Lupton 1986) and cellulose increasing tumorigenicity (Klurfeld et al. 1986) also have been published.

There are many proposed mechanisms by which dietary fiber may affect the development of colon cancer, and probably no single mechanism is exclusively responsible. Dietary fiber increases fecal bulk and therefore dilutes its contents (Gazzaniga and Lupton 1987), which decreases interactions between the intestinal mucosa and carcinogens or tumor-promoting agents. Furthermore, dietary fiber decreases intestinal transit times, thus limiting the time mucosal cells are exposed to carcinogens and tumor promoters (Wrick et al. 1983). Some types of dietary fiber adsorb mutagenic agents (Roberton et al. 1991), which would lead to their excretion in the feces and decrease their contact with colonic mucosal cells.

Colonic bacteria may also play a role in cancer development. Among their activities, bacteria metabolize several compounds present in the colon to carcinogens and they also produce secondary bile acids, which act as promoters of colon tumors (Reddy 1981). Several studies have shown that the intake of dietary fiber influences the activity of bacterial enzymes, which may play a role in promoting (fermentable fiber) or inhibiting (nonfermentable fiber) carcinogenesis (Reddy et al. 1992; Mallett, Wise, and Rowland 1984). Changes in colonic pH due to bacterial fermentation of fiber have also been considered a factor in cancer development. Based on epidemiologic data, acidic stools have been negatively associated with colon cancer (Walker, Walker, and Walker 1986; Bruce 1987), but this hypothesis has not been supported by experimental results (Lupton, Coder, and Jacobs 1988). Changes in pH influence enzyme activities and may affect the chemical availability of some carcinogens and bile salts.

Much attention has been given to the production of SCFA during the fermentation of fiber and colonic epithelial cell proliferation. It has been shown that infusions of SCFA in the rat colon increase cell proliferation (Sakata 1987; Kripke et al. 1989). On the other hand, *in vitro* experiments show that butyrate is a strong inhibitor of cell growth, blocking the cell cycle at mitosis and promoting cell differentiation (Kruh, Defer, and Tichonicky 1991). It remains to be seen how butyrate, produced from fiber fermentation, affects colon cancer development.

In experiments with animals, dietary fiber has been shown to both enhance and inhibit tumorigenesis. Much of the conflicting results can be attributed to experimental design. The questions remaining are how relevant the data are to human subjects and what dietary recommendations should be made to promote good health.

At the present time, long-term intervention studies are being carried out in human subjects. These studies should provide more conclusive data concerning the role of dietary fiber in colon carcinogenesis. Although this type of research is the most difficult to carry out successfully, it provides the most reliable and applicable information for formulating dietary recommendations for the population at large.

The association between high dietary fiber intake and good health is valid in many cases such as bowel regularity or modulation of blood glucose levels. In other cases, such as colon cancer, the association is tentative at best. Based on the knowledge available, generalizations concerning the preventative and therapeutic nature of dietary fiber should be made with caution, and consideration of the fiber source and its physiological effects should be noted.

References

Alberts, D.S.; Einspahr, J.; Rees-McGee, S.; Ramanujam, P.; Buller, M.K.; Clark, L.; Ritenbaugh, C.; Atwood, J.; Pethigal, P.; Earnest, D.; Villar, H.; Phelps, J.; Lipkin, M.; Wargovich, M.; and Meyskens, F.L. 1990. Effects of dietary wheat bran fiber on rectal epithelial cell proliferation in patients with resection for colorectal cancers. *J. Natl. Cancer Inst.* 82:1280–1285.

Anderson, J.W.; Story, L.; Sieling, B.; Chen, W.L.; Petro, M.S.; and Story, J. 1984. Hypocholesterolemic effects of oat-bran or bean intake for hypercholesterolemic men. *Am. J. Clin. Nutr.* 40:1146–1155.

Anderson, J.W.; Zettwoch, N.; Feldman, T.; Tietyen-Clark, J. Oeltgen, P.; and Bishop, C.W. 1988. Cholesterol-lowering effects of psyllium hydrophillic mucilloid for hypercholesterolemic men. *Arch. Intern. Med.* 148:292–296.

Anderson, J.W.; Zeigler, J.A.; Deakins, D.A.; Floore, T.L.; Dillon, D.W.; Wood, C.L.; Oeltgen, P.R.; and Whitley, R.J. 1991. Metabolic effects of high carbohydrate, high-fiber diets for insulin-dependent diabetic individuals. *Am. J. Clin. Nutr.* 54:936–943.

Anderson, J.W.; Riddell-Mason, S.; Gustafson, N.J.; Smith, S.F.; and Mackey, M. 1992. Cholesterol-lowering effects of psyllium-enriched cereal as an adjunct to a prudent diet in the treatment of mild to moderate hypercholesterolemia. *Am. J. Clin. Nutr.* 56:93–98.

Anon. 1990. Energy values of complex carbohydrates. In *Complex Carbohydrates in Foods: The Report of the British Nutrition Foundation's Task Force*, pp. 56–66. London: Chapman and Hall.

Arjmandi, B.H.; Ahn, J.; Nathani, S.; and Reeves, R.D. 1992. Dietary soluble fiber and cholesterol affect serum cholesterol concentration, hepatic portal venous short-chain fatty acid concentrations and fecal sterol excretion in rats. *J. Nutr.* 122:246–253.

Asp, N.B. 1990. Delimitation problems in definition and analysis of dietary fiber. In *New Developments in Dietary Fiber-Physiological, Physicochemical, and Analytical Aspects*, I. Furda and C.J. Brine, eds., pp. 227–236. New York: Plenum Press.

Asp, N.B., and Johansson, C.G. 1981. Techniques for measuring dietary fiber; principal aims of methods and a comparison of results obtained by different techniques. In *The analysis of dietary fiber in food*, W.P.T. James and O. Theander, eds., pp. 173–190. New York: Marcel Dekker Inc.

Beattie, V.A.; Edwards, C.A.; Hosker, J.P.; Cullen, D.R.; Ward, J.D.; and Read, N.W. 1988. Does adding fibre to a low energy, high carbohydrate, low fat diet confer any benefit to the management of newly diagnosed overweight type II diabetics? *Br. Med. J.* 296:1147–1149.

Borgia, M.; Sepe, N.; Brancato, V.; Costa, G.; Simone, P.; Borgia, R.; and Lugli, R. 1983. Treatment of chronic constipation by a bulk-forming laxative. *J. Int. Med. Res.* 11:124–127.

Brand, J.C.; Colagiuri, C.; Crossman, S.; Allen, A.; and Truswell, A.S. 1991. Low glycemic index carbohydrate foods improve glucose control in non-insulin dependent diabetes mellitus (NIDDM). *Diabetes Care.* 14:95–101.

Brown, N.J.; Worlding, J.; Rumsey, P.D.E.; and Read, N.W. 1988. The effect of guar gum on the distribution of radiolabelled meal in the gastrointestinal tract of the rat. *Br. J. Nutr.* 59:223–231.

Bruce, W.R. 1987. Recent hypotheses for the origin of colorectal cancer. *Cancer Res.* 47:4237–4242.

Calvert, R.; Schneeman, B.O.; Satchithanandam, S.; Cassidy, M.M.; and Vahouney, G. 1985. Dietary fiber and intestinal adaptation: Effects on intestinal and pancreatic digestive enzyme activities. *Am. J. Clin. Nutr.* 41:1249–1256.

Cann, P.A.; Read, N.W.; and Holdsworth, C.D. 1984. What is the benefit of coarse wheat bran in patients with irritable bowel syndrome? *Gut* 24:168–173.

Chen, W.L.; Anderson, J.W.; and Jennings, D. 1984. Propionate may mediate the hypocholesterolemic effects of certain soluble plant fibers in cholesterol-fed rats. *Proc. Soc. Exp. Biol. Med.* 175:215–218.

Cummings, J.H. 1991. Production and metabolism of short-chain fatty acids in humans. In *Short-Chain Fatty Acids: Metabolism and Clinical Importance*, pp. 11–16. Report of the Tenth Ross Conference on Medical Research, 2–4 December 1990, at Sanibel Island, Florida.

Cummings, J.H.; Southgate, D.A.T.; Branch, W.; Houston, H.; Jenkins, D.J.A.; and James, W.P.T. 1978. Colonic response to dietary fiber from carrot, cabbage, apple, bran and guar gum. *Lancet* 1:5–9.

Di Lorenzo, C.; Williams, C.M.; Hajnal, F.; and Venezuela, J.E. 1988. Pectin delays gastric emptying and increases satiety in obese subjects. *Gastroenterology* 95:1211–1215.

Dreher, M.L. 1987a. Dietary fiber methodology. In *Handbook of Dietary Fiber-An Applied Approach*, pp. 53–114. New York: Marcel Dekker Inc.

———. 1987b. Physicochemical and functional properties of dietary fiber as related to bowel function and food use. In *Handbook of Dietary Fiber-An Applied Approach*, pp. 137–182. New York: Marcel Dekker Inc.

Dunaif, G., and Schneeman, B.O. 1981. The effect of dietary fiber on human pancreatic enzyme activity in vitro. *Am. J. Clin. Nutr.* 34:1034–1035.

Dutta, S.K., and Hlasko, J. 1985. Dietary fiber in pancreatic disease: Effect of high fiber diet on fat malabsorption in pancreatic insufficiency and in vitro study of the interaction of dietary fiber with pancreatic enzymes. *Am. J. Clin. Nutr.* 41:517–525.

Eastwood, M.A., and Morris, E.R. 1992. Physical properties of dietary fiber that influence physiological function: A model for polymers along the gastrointestinal tract. *Am. J. Clin. Nutr.* 55:436–442.

Eastwood, M.A.; Robertson, J.A.; Brydon, W.G.; and MacDonald, D. 1983. Measurement of water holding properties of fibre and their faecal bulking ability in man. *Br. J. Nutr.* 50:539–547.

Ebling, P.; Hannele, Y.J.; Aro, A.; Helve, E.; Sinisalo, M.; and Koivisto, V.A. 1988. Glucose and lipid metabolism and insulin sensitivity in type I diabetes: the effect of guar gum. *Am. J. Clin. Nutr.* 48:98–103.

Ellis, P.R.; Burley, V.J.; Leeds, A.R.; and Peterson, D.B. 1988. A guar-enriched wholemeal bread reduces postprandial glucose and insulin responses. *J. Hum. Nutr. Diet* 1:77–84.

Englyst, H.N.; Trowell, H.; Southgate, D.A.T.; and Cummings, J.H. 1987. Dietary fiber and resistant starch. *Am. J. Clin. Nutr.* 46:873–874.

Findlay, J.M.; Smith, A.N.; Mitchell, W.D.; Anderson, A.J.B.; and Eastwood, M.A. 1974. Effects of unprocessed bran on colon function in normal subjects and in diverticular disease. *Lancet* 2:146–149.

Fisher, N.; Berry, C.S.; Fearn, T.; Gregory, J.A.; and Hardy, J. 1985. Cereal dietary fiber consumption and diverticular disease: A lifespan study in rats. *Am. J. Clin. Nutr.* 42:788–804.

Fukagawa, N.K.; Anderson, J.W.; Hageman, G.; Young, V.R.; and Minaker, K.L. 1990. High-carbohydrate, high-fiber diets increase peripheral insulin sensitivity in healthy young and old adults. *Am. J. Clin. Nutr.* 52:524–528.

Furda, I. 1981. Simultaneous analysis of soluble and insoluble dietary fiber. In *The Analysis of Dietary Fiber in Food*, W.P.T. James and O. Theander, eds., pp. 163–172. New York: Marcel Dekker Inc.

Gazzaniga, J.M., and Lupton, J.R. 1987. Dilution effect of dietary fiber sources: An in vivo study in the rat. *Nutr. Res.* 7:1261–1268.

Goering, H.K., and van Soest, P.J. 1970. Forage fiber analysis. *USDA Agricultural Handbook No. 379.* pp. 1–20. Washington, D.C.: U.S. Government Printing Office.

Haskell, W.L.; Spiller, G.A.; Jensen, C.D.; Ellis, B.K.; and Gates, J.E. 1992. Role of water-soluble dietary fiber in the management of elevated plasma cholesterol in healthy subjects. *Am. J. Cardiol.* 69:433–439.

Heitman, D.W.; Ord, V.A.; Hunter, K.E.; and Cameron, I.L. 1989. Effect of dietary cellulose on cell proliferation and progression of 1,2-dimethylhydrazine-induced colon carcinogenesis in rats. *Cancer Res.* 49:5581–5585.

Hellendoorn, E.W.; Noordhoff, M.G.; and Slagman, J. 1975. Enzymatic determination of the indigestable residue (dietary fibre) content of human food. *J. Sci. Food Agric.* 26:1461–1468.

Hollenbeck, C.B.; Coulston, A.M.; and Reaven, G.M. 1986. To what extent does increased dietary fiber improve glucose and lipid metabolism in patients with noninsulin-dependent diabetes mellitus (NIDDM)? *Am. J. Clin. Nutr.* 43:16–24.

Illman, R.J., and Topping, D.L. 1985. Effects of dietary oat bran on faecal steroid excretion, plasma volatile fatty acid and lipid synthesis in rats. *Nutr. Res.* 5:839–846.

Illman, R.J.; Topping, D.L.; McIntosh, G.H.; Trimble, R.P. §orer, G.B.; Taylor, M.N., and Cheng, B. 1988. Hypocholesterolemic effects of dietary propionate: Studies in whole animals and perfused rat liver. *Ann. Nutr. Metab.* 32:97–107.

Isaksson, G.; Lundquist, I.; and Ihse, I. 1982. Effect of dietary fibre on pancreatic enzyme activity in vitro. *Gastroenterology* 82:918–924.

Isaksson, G.; Lilja, P.; Lunquist, I.; and Ihse, I. 1983. Influence of dietary fiber on exocrine pancreatic function in the rat. *Digestion* 27:57–62.

Jacobs, L.R. 1990. Influence of soluble fibers on experimental colon carcinogenesis. In *Dietary Fiber: Chemistry, Physiology and Health Effects*, D. Kritchevsky, C. Bonfield, and J.W. Anderson, eds., pp. 402–415. New York:Plenum Press.

Jacobs, L.R., and Lupton, J.R. 1986. Relationship between colonic luminal pH, cell proliferation, and colon carcinogenesis in 1,2-dimethylhydrazine treated rats fed high fiber diets. *Cancer Res.* 46:1727–1734.

Jenkins, D.J.A.; Leeds, A.R.; Gassull, M.A.; Cochet, B.; and Alberti, K.G.M.M. 1977. Decrease in postprandial insulin and glucose concentrations by guar and pectin. *Ann. Int. Med.* 86:20–23.

Jenkins, D.J.A.; Wolever, T.M.S.; Leeds, A.R.; Gassull, M.A.; Haisman, P.; Dilawari, J.; Goff, D.V.; Metz, G.L., and Alberti, K.G.M.M. 1978. Dietary fibres, fibre analogues and glucose tolerance: importance of viscosity. *Br. Med. J.* 1:1392–1394.

Jenkins, D.J.A.; Wolever, T.M.S.; Taylor, R.H.; Reynolds, D.; Neneham, R.; and Hockaday, T.D.R. 1980. Diabetic glucose control, lipids, and trace elements on long-term guar. *Br. Med. J.* 280:1353–1354.

Jenkins, D.J.A.; Thomas, D.M.; Wolever, T.M.S.; Taylor, R.H.; Barker, H.; Fielden, H.; Baldwin, J.M.; Bowling, A.C.; Newman, H.C.; Jenkins, A.L.; and Goff, D.V. 1981. Glycemic index of foods: A physiological basis for carbohydrate exchange. *Am. J. Clin. Nutr.* 34:362–366.

Jenkins, D.J.A.; Wolever, T.M.S.; Collier, G.R.; Ocana, A.; Rao, A.V.; Buckley, G.; Lam, Y.; Mayer, A.; and Thompson, L.U. 1987. Metabolic effects of a low-glycemic-index diet. *Am. J. Clin. Nutr.* 46:968–975.

Johnson, I.T. 1990. The biological effects of dietary fibre in the small intestine. In *Dietary fibre: Chemical and Biological Aspects*, D.A.T. Southgate, K. Waldron, I.T. Johnson, and G.R. Fenwick, eds., pp. 151–163. Cambridge: The Royal Society of Chemistry.

Kashtan, H.; Stern, H.S.; Jenkins, D.J.A.; Jenkins, A.L.; Hay, K.; Marcon, N.; Minkin, S.; and Bruce, W.R. 1992. Wheat bran and oat bran supplements' effects on blood lipids and lipoproteins. *Am. J. Clin. Nutr.* 55:976–980.

Klein, K. 1988. Controlled treatment trials in the irritable bowel syndrome: A critique. *Gastroenterology* 95:232–241.

Klurfeld, D.M. 1990. Insoluble dietary fiber and experimental colon cancer: Are we asking the proper questions? In *Dietary Fiber: Chemistry, Physiology and Health Ef-*

fects, D. Kritchevsky, C. Bonfield, and J.W. Anderson, eds., pp. 402–415. New York: Plenum Press.

Klurfeld, D.M.; Weber, M.M.; Buck, C.L.; and Kritchevsky, D. 1986. Dose-response of colonic carcinogenesis to different amounts and types of cellulose. *Fed. Proc.* 45:1076.

Kripke, S.A.; Fox, A.D.; Berman, J.M.; Settle, R.G.; and Rombeau, J.L. 1989. Stimulation of intestinal mucosal growth with intracolonic infusion of short-chain fatty acids. *JPEN*. 13:109–116.

Kristen, R.; Domning, B.; Nelson, K.; Nemeth, K.; Oremek, G.; Hubner-Steiner, U.; and Speck, U. 1989. Influence of guar on serum lipids in patients with hyperlipidaemia. *Eur. J. Clin. Pharmacol.* 37:117–120.

Kritchevsky, D. 1988. Dietary fiber. *Ann. Rev. Nutr.* 8:301–328.

Kruh, J.; Defer, N.; and Tichonicky, L. 1991. Molecular and cellular effects of sodium butyrate. In *Short-Chain Fatty Acids: Metabolism and Clinical Importance,* pp. 45–50. Report of the Tenth Ross Conference on Medical Research, 2–4 December 1990, at Sanibel Island, Florida.

Leahy, A.L.; Ellis, R.M.; and Quill, D.S. 1985. High fibre diet in symptomatic diverticular disease of the colon. *Ann. R. Doll. Surg. Engl.* 67:173–174.

LeBlanc, J.; Nadeau, A.; Mercier, I.; McKay, C.; and Samson, P. 1991. Effect of guar gum on insulinogenic and thermogenic response to glucose. *Nutr. Res.* 11:133–139.

Livesey, G. 1990. Energy values of unavailable carbohydrates and diets: An inquiry and analysis. *Am. J. Clin. Nutr.* 51:617–637.

Lo, G.S.; Goldberg, A.P.; Lim, A.; Grundhauser, J.J.; Anderson, L.; and Schonfeld, G. 1986. Soy fiber improves lipid and carbohydrate metabolism in primary hyperlipidemic subjects. *Atherosclerosis* 62:239–248.

Lupton, J.R.; Coder, D.M.; and Jacobs, L.R. 1988. Long-term effects of fermentable fibers on rat colonic pH and epithelial cell cycle. *J. Nutr.* 118:840–845.

Madar, Z.; Arieli, B.; Trostler, N.; and Norynberg, C. 1988a. Effect of consuming soybean dietary fiber on fasting and postprandial glucose and insulin levels in type II diabetes. *J. Clin. Biochem. Nutr.* 4:165–173.

Madar, Z.; Abel, R.; Samish, S.; and Arad, J. 1988b. Glucose-lowering effect of fenugreek in non-insulin dependent diabetics. *Eur. J. Clin. Nutr.* 43:51–54.

Madar, Z., and Odes, H.S. 1990a. Dietary fiber in metabolic diseases. In *Dietary Fiber Research,* R. Paoletti, ed., pp. 1–65. Basel:Krager.

———. The therapeutic role of psyllium in health and disease. In *Dietary Fiber Research,* R. Paoletti, ed., pp. 90–96. Basel:Krager.

Mallet, A.K.; Wise, A.; and Rowland, I.R. 1984. Hydrocolloid food additives and rat caecal microbial enzyme activities. *Food Chem. Toxicol.* 22:415–418.

Marlett, J.A.; Li, B.U.K.; Patrow, C.J.; and Bass, P. 1987. Comparative laxation of psyllium with and without senna in an ambulatory constipated population. *Am. J. Gastroenterol.* 82:333–337.

Meyer, J.H.; Gu, Y.G.; Jehn, D.; and Taylor, I.L. 1988. Intragastric vs intraintestinal viscous polymers and glucose tolerance after liquid meals of glucose. *Am. J. Clin. Nutr.* 48:260–266.

Miller, A.B.; Howe, G.R.; Jain, M.; Craib, K.J.P.; and Harrison, L. 1983. Food items and food groups as risk factors in a case-control study of diet and colo-rectal cancer. *Int. J. Cancer* 32:155–161.

Muller-Lissner, S.A. 1988. Effect of wheat bran on weight of stool and gastrointestinal transit time: A meta analysis. *Br. Med. J.* 296:615–617.

Nishimune, T.; Yakushiji, T.; Sumimoto, T.; Taguchi, S.; Konishi, Y.; Nakahara, S.; Ichikawa, T.; and Kunita, N. 1991. Glycemic response and fiber content of some foods. *Am. J. Clin. Nutr.* 54:414–419.

Nishina, P.M., and Freedland, R.A. 1990. The effects of dietary fiber feeding on cholesterol metabolism in rats. *J. Nutr.* 120:800–805.

Odes, H.S.; Madar, Z.; Trop, M.; Namir, S.; Gross, J.; and Cohen, T. 1986. Pilot study of the efficacy of spent grain dietary fiber in the treatment of constipation. *Isr. J. Med. Sci.* 22:12–15.

Odes, H.S., and Madar, Z. 1991. A double-blind trial of celandin, aloevera and psyllium laxative preparation in adult patients with constipation. *Digestion* 49:65–71.

Potter, J.D., and McMichael, A.J. 1986. Diet and cancer of the colon and rectum: A case-control study. *J. Natl. Cancer Inst.* 76:557–569.

Prior, A., and Whorwell, P.J. 1987. Double blind study of ispaghula in the irritable bowel syndrome. *Gut* 28:1510–1513.

Prosky, L.; Asp, N.; Furda, J.; DeVries, J.; Schweizer, T.; and Harland, B. 1985. Determination of total dietary fiber in foods and food products: Collaborative study. *J. Assoc. Off. Anal. Chem.* 68:677–679.

Prosky, L. and DeVries, J. 1992. Controlling Dietary Fiber in Food Products. Van Nostrand Reinhold.

Reddy, B.S. 1981. Diet and bile acids. *Cancer Res.* 41:3766–3768.

Reddy, B.S.; Engle, A.; Simi, B.; and Goldman, M. 1992. Effect of dietary fiber on colonic bacterial enzymes and bile acids in relation to colon cancer. *Gastroenterology* 102:1475–1482.

Roberton, A.M.; Ferguson, L.R.; Hollands, H.J.; and Harris, P.J. 1991. Adsorption of a hydrophobic mutagen to dietary fiber preparations. *Mutation Res.* 262:195–202.

Roberts-Anderson, J.; Mehta, T.; and Wilson, R.B. 1987. Reduction of DMH-induced colon tumors in rats fed psyllium husk or cellulose. *Nutr. Cancer* 10:129–136.

Sakata, T. 1987. Stimulatory effect of short-chain fatty acids on epithelial cell proliferation in the rat intestine: A possible explanation for trophic effects of fermentable fibre gut microbes and luminal trophic factors. *Br. J. Nutr.* 58:95–103.

Sanders, T.A.B., and Reddy, S. 1992. The influence of rice bran on plasma lipids and lipoproteins in human volunteers. *Eur. J. Clin. Nutr.* 46:167–172.

Selvendran, R.R.; Stevens, B.J.H.; and Du Pont, M.S. 1987. Dietary fiber: Chemistry, analysis, and properties. *Adv. Fd. Res.* 31:117–208.

Southgate, D.A.T. 1969. Determination of carbohydrate in foods II. Unavailable carbohydrates. *J. Sci. Food Agric.* 20:331–335.

Stevens, J.; Van Soest, P.J.; Robertson, J.B.; and Levitsky, D.A. 1988. Comparison of the effects of psyllium and wheat bran on gastrointestinal transit time and stool characteristics. *J. Am. Diet. Assoc.* 88:323–326.

Story, J.A. 1986. Modification of steroid excretion in response to dietary fiber. In *Dietary Fiber Basic and Clinical Aspects*, G.V. Vahouny and D. Kritchevsky, eds., pp. 253–264. New York: Plenum.

Superko, H.R.; Haskell, W.L.; Sawrey-Kubicek, L.; and Farquhar, J.W. 1988. Effects of solid and liquid guar gum on plasma cholesterol and triglyceride concentrations in moderate hypercholesterolemia. *Am. J. Cardiol.* 62:51–55.

Swain, J.F.; Rouse, I.L.; Curley, C.B.; and Sacks, F.M. 1990. Comparison of the effects of oat bran and low-fiber wheat on serum lipoprotein levels and blood pressure. *N. Engl. J. Med.* 332:147–152.

Tomlin, J., and Read, N.W. 1988. The relation between bacterial degradation of viscous polysaccharides and stool output in human beings. *Br. J. Nutr.* 60:467–475.

Topping, D.L.; Oakenfull, D.; Trimble, R.P.; and Illman, R.J. 1988. A viscous fibre (methylcellulose) lowers blood glucose and plasma triacylglycerols and increases liver glycogen independently of volatile acid production in the rat. *Br. J. Nutr.* 59:21–30.

Trock, B.; Lanza, E.; and Greenwald, P. 1990. Dietary fiber, vegetables, and colon cancer: Critical review and meta-analyses of the epidemiological evidence. *J. Natl. Cancer Inst.* 82:650–661.

Trowell, H.; Southgate, D.A.T.; Wolever, T.M.S.; Leeds, A.R. αssull, M.A.; and Jenkins, D.A. 1976. Dietary fibre redefined. *Lancet* 1:967.

Vahouny, G.V., and Cassidy, M.M. 1985. Dietary fibers and absorption of nutrients. *Proc. Soc. Exp. Biol. Med.* 180:432–446.

———. 1986. Dietary fiber and intestinal adaptation. In *Dietary Fiber Basic and Clinical Aspects*, G.V. Vahouny and D. Kritchevsky, eds., pp. 253–264. New York:Plenum.

Vinik, A.I., and Jenkins, D.J.A. 1988. Dietary fiber in management of diabetes. *Diabetes Care* 11:160–173.

Walker, A.R.P.; Walker, B.F.; and Walker, A.J., 1986. Fecal pH, dietary fibre intake and proneness to colon cancer in four South African populations. *Br. J. Cancer.* 53:489–495.

Wolever, T.M.S. 1990. Relationship between dietary fiber content and composition in foods and the glycemic index. *Am. J. Clin. Nutr.* 51:72–75.

Wrick, K.L.; Robertson, J.B.; Van Soest, P.J.; Lewis, B.A.; Rivers, J.M.; Roe, D.A.; and Hackler, L.R. 1983. The influence of dietary fiber source on human intestinal transit and stool output. *J. Nutr.* 113:1464–1479.

Wright, R.S.; Anderson, J.W.; and Bridges, S.R. 1990. Propionate inhibits hepatocyte lipid synthesis. *Proc. Soc. Exp. Biol. Med.* 195:26–29.

Wursch, P.; Del Vedovo, S.; and Koellreutter, B. 1986. Cell structure and starch nature as key determinants of the digestion rate of starch in legume. *Am. J. Clin. Nutr.* 43:25–29.

Chapter 10

Special Physiological Functions of Newly Developed Mono- and Oligosaccharides

Tsuneyuki Oku

INTRODUCTION

In the last few years, new types of mono- and oligosaccharides are being developed actively, mainly in Japan, as bulk sucrose substitutes with beneficial effects. The functional properties of these saccharides to be used in processed foods should be as close as possible to those of sucrose in providing bulk properties and a good sweetness. However, these saccharides differ from sucrose with regard to several physiological properties as is discussed later. Several types of mono- and oligosaccharides are already used as sucrose substitutes in processed foods such as soft drinks, candies, caramels, chocolates, cookies, cakes, breads, canned fruits, soft creams, jams and jellies, and pudding.

Although most of the mono- and oligosaccharides are constituents of natural plant foods, they are currently mass produced from sucrose, lactose, glucose, or starch derivatives, using specific bacterial enzymes. Sugar alcohols, except for erythritol, are produced by hydrogenation of mono- or disaccharides such as glucose, maltose, lactose, and partly hydrolyzed starch derivatives. Generally, the larger the molecular weight of the bulk sucrose substitute, the weaker its intensity of sweetness. The sweetness of highly purified oligosaccharides is usually less than half of sucrose, usually less than one-third, and the sweetened taste is mild.

STRUCTURAL CHARACTERISTICS OF FUNCTIONAL MONO- AND OLIGOSACCHARIDES

The newly developed mono- and oligosaccharides with special physiological functions are listed in Table 10-1. These saccharides are manufactured by chemical and/or enzymatic modification of mono- and oligosaccharides such as sucrose, lactose, maltose, glucose, and starch derivatives, which are the natural constituents of common foods. These newly developed oligosaccharides are classified according to their structure in two main groups, sucrose and lactose derivatives and other derivatives. Fig. 10-1 shows the structural characteristics of the main mono- and oligosaccharides having special physiological functions.

Neosugar is a mixture of fructo-oligosaccharides in which two or three molecules of fructose are bound to the fructose residue of sucrose by a β-2.1 linkage, and is produced enzymatically from sucrose. The main functional components of the soybean oligosaccharide are raffinose and stachyose in which one or two molecules of galactose are bound to the glucose residue of sucrose by a α-1.6 linkage. This oligosaccharide consists of 30% raffinose and stachyose, 50% sucrose, and 20% monosaccharides and other. Lactosucrose (galactosyl-sucrose) has a structure similar to that of raffinose, but the linkage of the galactose moiety to glucose is a β-1.4 linkage, not a α-1.6 linkage. Coupling sugar is a mixture of glucosyl-sucrose and maltosyl-sucrose in which one or two molecules of glucose are bound to the glucose residue of sucrose by a α-1.4 linkage. Palatinose is a structural isomer of sucrose consisting of glucose and fructose residues bound by a α-1.6 linkage.

Galacto-oligosaccharide is a polymer in which one to four molecules of galactose are bound to the galactose residue of lactose. There are two types of galacto-oligosaccharides. One contains a galactose

Table 10-1. Mono- and Oligosaccharides with Beneficial Effects for Health

	Sweetness % of sucrose
1. Oligosaccharides	
1) Neosugar: A mixture of fructo-oligosaccharides such as kestose ($G_{\alpha1}{-}_{2}F_{1}{-}_{2\beta}F$), nystose ($G_{\alpha1}{-}_{2}F_{1}{-}_{2\beta}F_{1}{-}_{2\beta}F$), and fructofuranosyl nystose ($G_{\alpha1}{-}_{2}F_{1}{-}_{2\beta}F_{1}{-}_{2\beta}F_{1}{-}_{2\beta}F$)	30–60
2) Galacto-oligosaccharides: 2 types basic bond: ($-_{4}Gal_{\beta1}{-}_{4}Gal_{\beta1}{-}_{4}G$) ($-_{6}Gal_{\beta1}{-}_{6}Gal_{\beta1}{-}_{4}G$)	20–40
3) Xylo-oligosaccharide ($Xyl_{\beta1}{-}_{4}Xyl_{\beta1}{-}_{4}Xyl_{\beta1}-$)	ca. 50
4) Isomalto-oligosaccharide: A mixture of isomaltotriose ($G_{\alpha1}{-}_{6}G_{\alpha1}{-}_{6}G$), panose ($G_{\alpha1}{-}_{4}G_{\alpha1}{-}_{6}G$), isomaltose ($G_{\alpha1}{-}_{6}G$), maltose ($G_{\alpha1}{-}_{4}G$), and glucose (G)	ca. 50
5) Soybean oligosaccharide: A mixture of raffinose ($Gal_{\alpha1}{-}_{6}G_{\alpha1}{-}_{2}F$), stachyose ($Gal_{\alpha1}{-}_{6}Gal_{\alpha1}{-}_{6}G_{\alpha1}{-}_{2}F$), sucrose ($G_{\alpha1}{-}_{2}F$), and monosaccharides (G, F)	ca. 70
6) Lactosucrose ($Gal_{\beta1}{-}_{4}G_{\alpha1}{-}_{2}F$)	35–60
7) Lactulose ($Gal_{\beta1}{-}_{4}F$)	60–70
8) Coupling sugar: A mixture of glucosyl-sucrose ($G_{\alpha1}{-}_{4}G_{\alpha1}{-}_{2}F$) and maltosyl-sucrose ($G_{\alpha1}{-}_{4}G_{\alpha1}{-}_{4}G_{\alpha1}{-}_{2}F$)	50–60
9) Palatinose ($G_{\alpha1}{-}_{6}F$)	37–45
2. Disaccharide alcohols	
1) Maltitol ($G_{\alpha1}{-}_{4}Sol$)	80–95
2) Lactitol ($Gal_{\beta1}{-}_{4}Sol$)	30–40
3) Palatinit: An equivalent mixture of isomaltitol ($G_{\alpha1}{-}_{6}Sol$) and glucopyranosyl-α 1,6-mannitol ($G_{\alpha1}{-}_{6}Man$)	30–40
3. Monosaccharides	
1) Erythritol (tetrose alcohol)	75–85
2) Sorbitol (hexose alcohol)	60–70
3) Mannitol (hexose alcohol)	ca. 50
4) Sorbose (hexose)	55–65

linkage of β-1.4 and the other contains a β-1.6 linkage. They are produced independently by different manufacturing processes. Lactulose is a lactose derivative in which the glucose residue of lactose is replaced by fructose. Other saccharides are xylo-oligosaccharide containing pentose, and isomalto-oligosaccharide, which consists of isomaltotriose and panose.

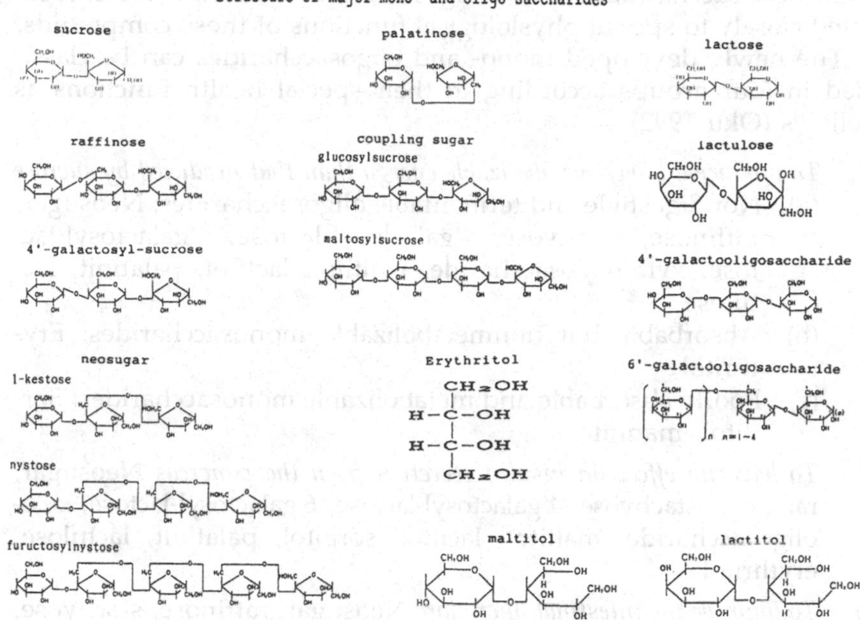

FIGURE 10-1. The Structure of Major Mono- and Oligosaccharides.

THE CLASSIFICATION OF MONO- AND OLIGOSACCHARIDES WITH SPECIAL PHYSIOLOGICAL FUNCTIONS

New types of mono- or oligosaccharides with beneficial effects for health have some unique properties for digestion, absorption, fermentation, and metabolism.

Some oligosaccharides with special physiological functions are not hydrolyzed or poorly hydrolyzed by the digestive enzymes present in the small intestine. Some other oligosaccharides are digested completely and absorbed from the small intestine. In addition, there are two types of monosaccharides with special physiological functions: those that are poorly absorbable and readily metabolizable, and those that are absorbable but nonmetabolizable.

Mono- and oligosaccharides that are not absorbed and/or not digested in the small intestine reach the large intestine where they become available for degradation by the colonic bacteria. The fact that these saccharides are non-digestible, but are fermentable, is related closely to special physiological functions of these compounds.

The newly developed mono- and oligosaccharides can be classified in four groups according to their special health functions as follows (Oku 1992):

1. *To produce a lower metabolizable energy than that produced by sucrose*
   (a) Nondigestible and fermentable oligosaccharides: Neosugar, raffinose, stachyose, 4′galactosyl-lactose, 6′galactosyl-lactose, xylo-oligosaccharide, maltitol, lactitol, palatinit, lactulose
   (b) Absorbable but nonmetabolizable monosaccharides: Erythritol
   (c) Poorly absorbable and metabolizable monosaccharides: Sorbitol, mannitol
2. *To have no effect on insulin secretion from the pancreas* Neosugar, raffinose, stachyose, 4′galactosyl-lactose, 6′galactosyl-lactose, xylo-oligosaccharide, maltitol, lactitol, sorbitol, palatinit, lactulose, erythritol
3. *To improve the intestinal microflora* Neosugar, raffinose, stachyose, 4′galactosyl-lactose, 6′galactosyl-lactose, xylo-oligosaccharide, lactulose
4. *To prevent dental caries*
   (a) Digestible oligosaccharides (normal energy): Coupling sugar, palatinose
   (b) Reduced-energy mono- and oligosaccharides: Neosugar, raffinose, stachyose, 4′galactosyl-lactose 6′galactosyl-lactose, xylo-oligosaccharide, maltitol, lactitol, sorbitol, palatinit, lactulose, erythritol

THE METABOLISM OF A NONDIGESTIBLE OLIGOSACCHARIDE, NEOSUGAR

Newly developed sucrose substitutes, except for coupling sugar and palatinose, are not or partly hydrolyzed by the digestive enzymes. It was formerly considered by human nutritionists that food components that are not digested should be excreted mostly in the feces, and that they do not serve as energy sources in the body. However, recent studies on the utilization and the physiological functions of new types of mono- and oligosaccharides and dietary fibers show that the nondigestible and/or the nonabsorbable saccharides are metabolized by the intestinal bacteria, and these properties are closely related to their special physiological functions. Neosugar is a typical nondigestible oligosaccharide which is metabolized through a unique pathway.

1-kestose (GF_2) and Nystose (GF_3), components of Neosugar, were found not to be hydrolyzed by the digestive enzymes. Also, ^{14}C-Neosugar administered intravenously was not degraded by any other organ's enzymes and was readily excreted into the urine without any decomposition (Oku, Tokunaga, and Hosoya 1984). These results suggested that the ingested Neosugar would not be metabolized to CO_2 and would not have any physiological functions. However, contrary to this expectation, when ^{14}C-Neosugar was administered orally to conventional rats, about 60% of the total radioactivity administered was exhaled as $^{14}CO_2$ within 24 hours, similar to the results obtained with ^{14}C-sucrose (Tokunaga, Oku, and Hosoya 1989; Oku 1992). There was a delay of about 3 hours in $^{14}CO_2$ release in animals that were administered with ^{14}C-Neosugar as compared to animals fed with ^{14}C-sucrose. Furthermore, the conversion of Neosugar to $^{14}CO_2$ after oral administration was strongly inhibited in germ-free and antibiotic-treated rats.

These results show that the Neosugar molecules that are not digested and are not absorbed by the small intestine reach the large intestine where they are fermented completely by the intestinal bacteria. The short-chain fatty acids thus produced are absorbed readily, are further metabolized to CO_2, and provide energy to the host (Tokunaga, Oku, and Hosoya 1989). When Neosugar was incubated anaerobically with the cecal content of conventional rats, short-chain fatty acids (SCFA) such as acetic acid, propionic acid, butyric acid, and valeric acid, and CO_2 were produced readily (Tokunaga, Oku, and Hosoya 1989). Moreover, ^{14}C-Neosugar injected directly to the cecum of conventional rats was metabolized to $^{14}CO_2$ and intact Neosugar was not detected in the feces of these rats (Tokunaga, Oku, and Hosoya 1986). Similar results were obtained from experiments with human subjects and using ^{14}C-Neosugar (Hosoya, Dhorranintra, and Hidaka 1988).

The conclusion drawn from the experiments with Neosugar is shown in Fig. 10-2. The figure demonstrates that a fermentable oligosaccharide that is not digested and/or not absorbed is utilized by the intestinal bacteria, with the production of a metabolizable energy. As a result, the composition of the intestinal microflora changes; i.e., the percentage of beneficial bacteria such as *Bifidobacterium* and

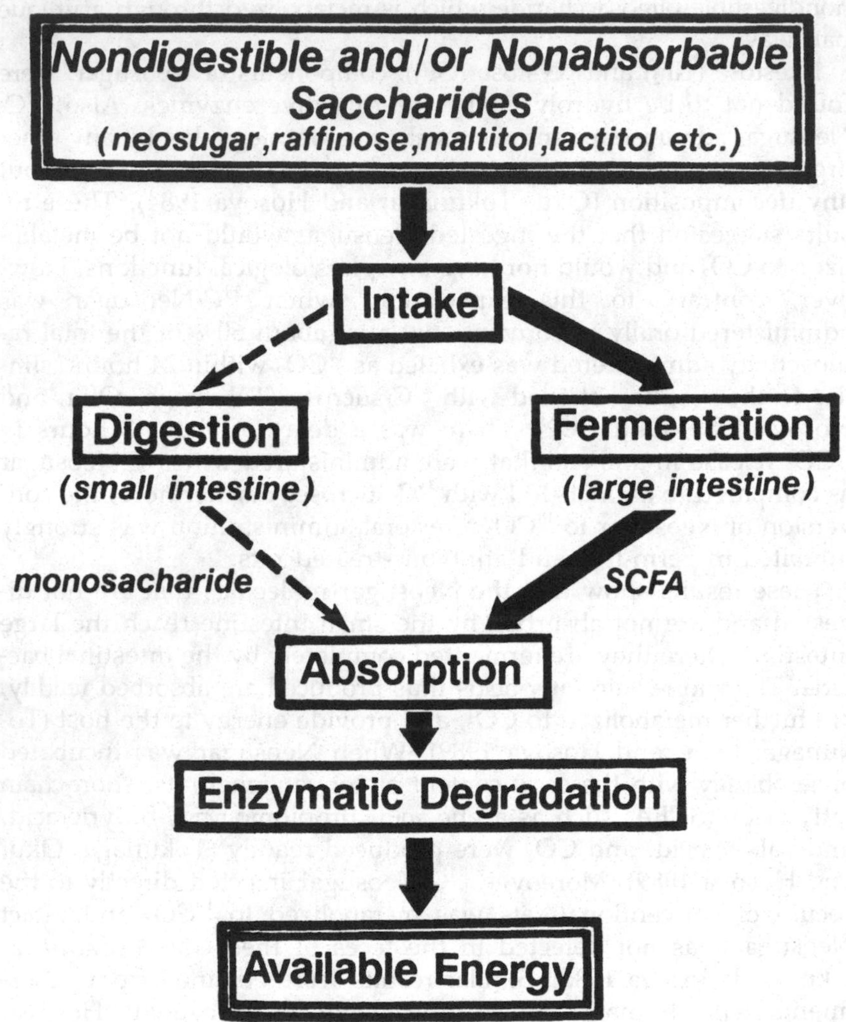

FIGURE 10-2. The Pathway of Energy Production of Nondigestible and/or Nonabsorbable Saccharides. (SCFA = Short Chain Fatty Acids.)

Lactobacillus increases and the percentage of harmful microbes such as *Clostridium* decreases. Some special physiological functions are caused via the improvement of the intestinal microflora. Furthermore, this figure suggests that the metabolizable energy of the nondigestible and fermentable carbohydrates cannot be evaluated by the currently used methods for common foods listed in the standard food table.

The nondigestible oligosaccharides, such as raffinose, stachyose, 4′galactosyl-lactose, 6′galactosyl-lactose, xylo-oligosaccharide, 4′galactosyl-sucrose, and lactulose, that escape digestion and absorption in the small intestine, reach the large intestine and are fermented by the intestinal bacteria in a similar manner to that of Neosugar.

SPECIAL PHYSIOLOGICAL FUNCTIONS OF MONO- AND OLIGOSACCHARIDES

1. Mono- and Oligosaccharides Have Low Metabolizable Energy

The Metabolism of Erythritol and the Available Energy Produced

In order to prevent obesity and decrease body weight, the intake of energy should be controlled. Reduced-energy mono- and oligosaccharides are developed for this purpose and are used to replace sucrose in foods.

Erythritol is a typical reduced-energy monosaccharide which is absorbable but nonmetabolizable. It is a tetrose alcohol (Fig. 10-1) and has about 75% of the sweetness of sucrose. It is synthesized during the cultivation of *Aureobasidium* sp., a yeast mutant, on glucose. When ^{14}C-erythritol was administered orally to conventional rats, more than 90% of the erythritol ingested was excreted into the urine without any degradation within 24 hours, and about 6% was excreted as expired $^{14}CO_2$ (Noda and Oku 1992). The fecal excretion was less than 1%. However, when ^{14}C-glycerol, which is an absorbable and metabolizable triose alcohol, was administered orally to conventional rats, about 60% of the total radioactivity was excreted as expired $^{14}CO_2$, and about 2% was excreted into the urine as metabolites within 24 hours. Furthermore, in balance studies using animals (Oku and Noda 1990) and human subjects (Oku 1990)

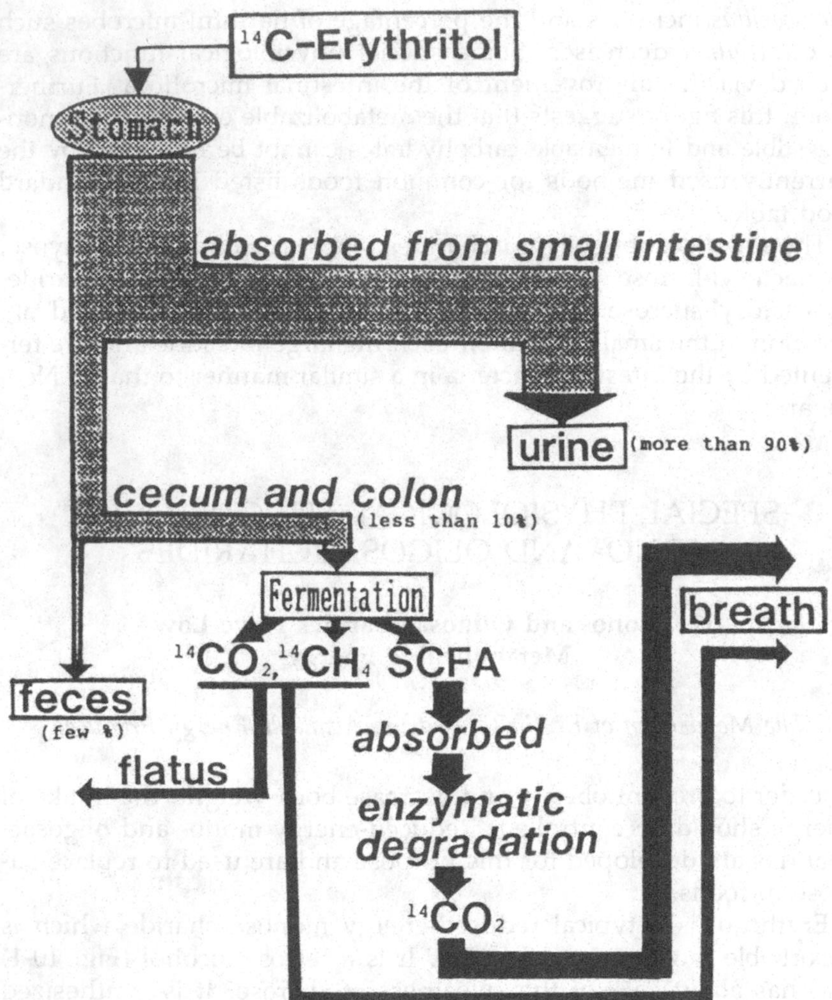

FIGURE 10-3. A Summary of the Metabolism and Disposition of ^{14}C-Erythritol in Rats.

more than 90% of the ingested erythritol was excreted into the urine in its native form within 24 or 48 hours.

In conclusion, more than 90% of the erythritol administered orally is absorbed readily by the small intestine and excreted quickly into the urine without any degradation. The remaining erythritol, less than 10%, is fermented in the large intestine (Fig. 10-3). Therefore, the available energy of erythritol is about 0.2 to 0.3 kcal/g (Oku 1992). This value comprises only 10% of the value of sucrose, and is the lowest value of the reduced-energy bulk sweeteners at present.

The Evaluation of the Available Energy Produced by Nondigestible Oligosaccharides

Some reduced-energy oligosaccharides are not digested or scarcely digested in the small intestine, and have bulk properties similar to sucrose. Mono- and oligosaccharides escaping digestion and absorption in the small intestine reach the large intestine and are fermented by the intestinal bacteria (McNeil 1984). In this fermentation, short-chain fatty acids, such as acetic acid, propionic acid, and butyric acid, CO_2, CH_4, and H_2 are synthesized as end-products by the intestinal bacteria. The short-chain fatty acids are absorbed from the large intestine and further metabolized by the host to produce energy. Therefore, determination of the relationship between the substrate and the products resulting from the fermentation process will allow us to estimate the metabolizable energy.

Table 10-2 shows several fermentation equations proposed previously (Hungate 1966; Miller and Wolin 1979; Smith and Bryant 1979; Livesey and Elia 1988). From these equations we can measure the stoichiometric relation between the substrate and the products of the fermentations. However, the molecular ratio of each product to substrate is different depending on the individual equation.

When the combustion energy of each short-chain fatty acid (Table 10-3) is taken into consideration in the calculation of these equations (Table 10-2), the total energy included in the short-chain fatty acids produced by the intestinal bacteria can be calculated (Oku 1991). When each value of the combustion energy was substituted in the above equations, the total energy obtained was between 2.51 and 2.86 kcal/g as shown in Table 10-2. The average value was 2.71 kcal/g. These values are thus similar, in spite of the fermentation equations being very different. The result also indicates that the potential energy decreases from 4 to 2.71 kcal/g if the carbohydrate is fermented completely by the intestinal bacteria.

On the other hand, the potential energy of short-chain fatty acids produced in the lower intestine is believed to be utilized ineffectively and only partially by the host. For example, the combustion energy of acetic acid is 3.48 kcal/g, but the available energy used

Table 10-2. Proposed equations for colonic fermentation of carbohydrates and estimation of combustion energy of short-chain fatty acids produced.

Equation	Reference
58 $C_6H_{12}O_6$———62 acetate + 22 propionate + 16 butyrate + 60.5 CO_2 + 33.5 CH_4 + 27 H_2O 2.78 kcal/g	(Hungate 1966)
34.5 $C_6H_{12}O_6$———48 acetate + 11 propionate + 5 butyrate + 34.25 CO_2 + 23.75 CH_4 + 10.5 H_2O 2.70 kcal/g	(Miller and Wolin 1979)
58 $C_6H_{12}O_6$ + 36 H_2O———60 acetate + 24 propionate + 16 butyrate + 92 CO_2 + 256 [H] 2.86 kcal/g	(Livesey and Elia 1988)
37.73 $C_6H_{12}O_6$———34.5 acetate + 9.7 propionate + 8.6 butyrate + 38.2 CO_2 + 18.8 CH_4 + 6.13 $C_6H_{10}O_3$ 2.51 kcal/g	(Smith and Bryant 1979)

in the standard food table is 2.40 kcal/g. If the difference between 2.40 and 3.48 kcal/g is based on the utilization efficiency of acetic acid, the apparent utilization efficiency is calculated as 69% (2.40/3.48 × 100 = 69). There are no such data for propionic acid and butyric acid. Therefore, this value will be used here as the apparent utilization efficiency of short-chain fatty acids, although the catabolic pathways of propionic acid and butyric acid are clearly different from that of acetic acid. The total available energy of short-chain fatty acids produced from 1 g of substrate is obtained by the multiplication of 69% by 2.71 kcal/g = 1.87 kcal/g. This value is less than 50% of the value for sucrose (4 kcal/g) and thus indicates that when a carbohydrate is fermented completely by the intestinal bacteria, the available energy is about 2 kcal/g.

Table 10-3. Combustion Energy of Short-Chain Fatty Acids (SCFA)

SCFA	Combustion energy	
	kcal/g	kcal/mol
Acetic acid	3.48	209.0
Propionic acid	5.0	370.4
Butyric acid	6.0	528.7

In order to estimate the available energy of carbohydrates that escapes digestion and absorption in the small intestine, three factors should be determined: the net digestion and absorption ratio subtracting the urinary excretion ratio of intact material, the fecal excretion ratio, and the fermentation ratio. As mentioned above, the gross available energy of oligosaccharides such as Neosugar, 4'galactosyl-lactose, raffinose, maltitol, lactitol, and lactulose can be estimated as about 2 kcal/g.

2. Mono- and Oligosaccharides do not Affect Insulin Secretion from the Pancreas

Ingested Neosugar is not digested in the small intestine and does not produce monosaccharides. Therefore, it appears that Neosugar does not affect insulin secretion from the pancreas. Thus, the oral administration of Neosugar to human subjects affected neither blood glucose nor insulin levels (Yamada et al. 1990). Similar results were obtained with test substances such as xylo-oligosaccharide, galacto-oligosaccharides, galactosyl-lactose, maltitol, lactitol, palatinit, and erythritol. Accordingly, these mono- and oligosaccharides can be used as bulk sweeteners with pleasant taste for patients having diabetes mellitus. However, it should be noted that these oligosaccharides contribute about 2 kcal/g to the host as energy sources.

On the other hand, the intake of large amounts of these non-digestible and/or partly absorbable mono- and oligosaccharides causes overt diarrhea, although minor symptoms such as abdominal distention and flatulence often arise in animals and humans. This diarrhea is due to osmogenic retention of fluid in both the small and the large intestine, but this is recovered within a few days, because the intestinal bacteria can utilize those saccharides at an increased rate. The tolerance to the ingestion of these sucrose substitutes, expressed as the relationship between dose and symptoms, shows considerable individual variation. The maximal dose that does not cause diarrhea in adult Japanese is shown in Table 10-4 (personal communication). These values show the maximum dose level of sucrose substitutes per kg of body weight per one ingestion. Obviously, if these sucrose substitutes are ingested two or three times per day, the maximum permissible dose per day will be higher.

3. Sucrose Substitutes Prevent Dental Caries

There are three essential conditions for the development of dental caries: the ingestion of a fermentable sugar, the infection by dental

Table 10-4. Maximum Permissible Dose Not Causing Diarrhea in Adult Japanese

Saccharides	Maximum Permissible Dose (g/kg body weight/day) Male	Female
Neosugar	0.3	0.4
4'galactosyl-sucrose	0.6	0.6
4'galacto-oligosaccharide	0.28	0.28
6'galactosyl-lactose	0.3	0.3
Xylo-oligosaccharide	0.12	0.12
Maltitol	0.3	0.3
Palatinit	0.3	0.3
Erythritol	0.66	0.8
Sorbitol	0.17	0.24

caries-inducing bacteria, and the eruption of the teeth. If one of these conditions is not met, dental caries will not occur. Accumulation of *Streptococcus mutans* in the teeth is facilitated by the biosynthesis of extracellular insoluble glucans from sucrose that increase the cellular adhesive properties and serve as a matrix for plaque formation. Dietary sucrose is believed to play a major role in dental caries formation. Therefore, the idea of developing sucrose substitutes to prevent dental caries is based on the prevention of sucrose ingestion and the subsequent formation of insoluble glucans.

There are two types of noncariogenic sucrose substitutes. Sugar substitutes of the first type are digested completely in the small intestine, while those of the second type are not digested or poorly absorbed. Examples of the former type are coupling sugar and palatinose. Sugar substitutes of the second type are reduced-energy mono- and oligosaccharides such as erythritol, sorbitol, Neosugar, maltitol, lactitol, and palatinit. Insoluble glucans and short-chain fatty acids are not produced from palatinose and coupling sugar because these carbohydrates are not metabolized by *S. mutans*. The low pH (less than pH 5.5) obtained by this fermentation favors demineralization of the enamel of the tooth. However, the reduction in pH is not induced by the incubation of *S. mutans* in the presence of palatinose (Ooshima et al. 1983). Also the production of insoluble glucan is suppressed when *S. mutans* is incubated with coupling sugar (Ikeda et al. 1978).

The α-1.4 linkage between the glucose residues in the coupling sugar is first hydrolyzed by the enzyme maltase, and then the sucrose unit obtained is further hydrolyzed by the enzyme sucrase (Yamada et al. 1981). The α-1.6 linkage of palatinose is hydrolyzed slowly by the enzyme isomaltase (Goda and Hosoya 1983). Therefore, the ingestion of coupling sugar or palatinose increases the blood glucose level and stimulates insulin secretion, while the available energy is similar to that of sucrose. These digestible noncariogenic sucrose substitutes are convenient sweeteners for growing children who require much energy.

4. Oligosaccharides Improve the Intestinal Microflora

Most of the oligosaccharides that are nondigestible or partly digestible are fermented easily by the intestinal bacteria in the large intestine, and are not excreted into the feces in their native form. The intestinal bacteria are present mainly in the large intestine. Therefore, in order to improve the intestinal microflora by the ingestion of oligosaccharides, these compounds must escape the digestion and the absorption processes in the small intestine and reach the large intestine. Oligosaccharides such as Neosugar, soybean oligosaccharide containing raffinose and stachyose (Benno et al. 1987), xylo-oligosaccharide (Okazaki, Fujikawa, and Matsumoto 1990), 4'galactosyl-lactose (Ohtsuka et al. 1989), 6'galactosyl-lactose (Tanaka et al. 1983), 4'galactosyl-sucrose (Fujita et al. 1991), and lactulose (Sahota, Bramley, and Menzies 1982) have these properties. In contrast to these sugars, sucrose, maltose, and glucose, which are digested and/or absorbed in the small intestine, have no effect on the intestinal microflora, because they do not reach the large intestine.

The intestinal bacteria metabolize oligosaccharides readily and produce large amounts of short-chain fatty acids. As a result, the pH in the lumen of the large intestine decreases to an acidic pH. At the same time, the total number of the intestinal microbes increases and enhances the fecal volume. The beneficial bacteria such as *Bifidobacterium* sp. and *Lactobacillus* sp. are resistant to the acidic environment, whereas the harmful bacteria such as *Clostridium* sp. are sensitive to the acidic conditions. Therefore, the proliferation of *Bifidobacterium* sp. and other useful bacteria is stimulated and that of the harmful bacteria is suppressed (Hidaka et al. 1984). The increase in the growth of *Bifidobacterium* sp. is accompanied by the production of nitrogen derivatives such as ammonia, indole, phenol, and skatole, and by the elimination of carcinogenic substances during the fermentation. Furthermore, constipation and the conditions of the feces are improved. In order to improve the intestinal microflora, lactulose, a derivative of lactose, was first developed as a medicine. Also Neosugar has been added to livestock diets to increase their efficiency through the improvement of the intestinal microflora (Hidaka et al. 1984).

CONCLUSIONS

The development of pleasant-tasting sucrose substitutes that are entirely safe and have beneficial effects for health would enable us to enjoy sweetened food, while at the same time these materials would prevent chronic diseases such as obesity, heart disease, diabetes mellitus, colonic cancer, and dental caries. In the near future, many such sucrose substitutes with beneficial effects will be developed.

In Japan, processed foods containing these sucrose substitutes and other food components having special physiological functions are recognized by law as foods for specified health uses. These foods can be labeled as having beneficial effects for health only after the approval by the Ministry of Health and Welfare. These foods will be available for sale in the very near future. (Thirteen foods are permitted as foods for specified health uses as of September 30, 1993)

References

Benno, Y.; Endo, K.; Shiragami, N.; Sayama, K.; and Mitsuoka, T. 1987. Effect of raffinose intake on human fecal microflora. *Bifidobact. Microflora* 6:59–63.

Fujita, K.; Hara, K.; Sakai, S.; Miyake, T.; Yamashita, M.; and Tsunetomi, Y. 1991. Effect of 4G-β-D-galactosylsucrose (lactosucrose) on intestinal flora and its digestibility in humans. *Denpun Kagaku* 38:249–55 (in Japanese).

Goda, T., and Hosoya, N. 1983. Hydrolysis of palatinose by rat intestinal sucrase-isomaltase complex. *J. Jpn. Soc. Nutr. Food Sci.* 36:168–73 (in Japanese).

Hidaka, H.; Hara, T.; Eida, T.; Okada, A.; Shimada, K.; and Mitsuoka, T. 1984. Effect of fructooligosaccharides on human intestinal flora. In *Intestinal Microflora and Dietary Factors*, Tomotari Mitsuoka, ed., pp. 39–64. Tokyo: Center for Academic Publications (in Japanese).

Hosoya, N.; Dhorranintra, B.; and Hidaka, H. 1988. Utilization of [U-14C] fructooligosaccharides in man as energy resource. *J. Clin. Biochem. Nutr.* 5:67–74.

Hungate, R.E. 1966. *The Rumen and its Microbes*. New York: Academic Press.

Ikeda, T.; Shiota, T.; McGhee, J.R.; Otaka, S.; Michalek, S.M.; Ochiai, K.; Hirasawa, M.; and Sugimoto, K. 1978. Virulence of *Streptococcus mutans*: comparison of the effects of a coupling sugar and sucrose on certain metabolic activities and cariogenicity. *Infect. Immun.* 19:477–80.

Koizumi, N.; Fujii, M.; Ninomiya, R.; Inoue, Y.; Kagawa, T.; and Tsukamoto, T. 1983. Study on transitory laxative effects of sorbitol and maltitol. Chemosphere 12:45–53.

Livesey, G., and Elia, M. 1988. Estimation of energy expenditure, net carbohydrate utilization, and net fat oxidation and synthesis by indirect calorimetry: Evaluation of errors with special reference to the detailed composition of fuels. *Am. J. Clin. Nutr.* 47:608–28.

McNeil, N.I. 1984. The contribution of the large intestine to energy supplies in man. *Am. J. Clin. Nutr.* 39:338–42.

Miller, T.L., and Wolin, M.J. 1979. Fermentations by saccharitic intestinal bacteria. *Am. J. Clin. Nutr.* 32:164–72.

Noda, K., and Oku, T. 1992. Metabolism and disposition of erythritol after oral administration to rats. *J. Nutr.* 122:1266–72.

Official materials to apply as food for specified health uses in Japanese law.

Ohtsuka, K.; Benno, Y.; Endo, K.; Ueda, H.; Ozawa, O.; Uchida, T.; and Mitsuoka, T. 1989. Effect of 4'-galactosyllactose intake on human fecal microflora. *Bifidus* 2:143–49.

Okazaki, M.; Fujikawa, S.; and Matsumoto, N. 1990. Effect of xylooligosaccharide on the growth of *Bifidobacteria*. *Bifidobact. Microflora* 9:77–86.

Oku, T. 1990. Erythritol balance study and estimation of metabolizable energy of erythritol. In *Caloric Evaluation of Carbohydrates*, Norimasa Hosoya, ed., pp. 65–75. Tokyo: Research Foundation for Sugar Metabolism.

———. 1991. Caloric evaluation of reduced-energy bulking sweeteners. In *Obesity: Dietary Factors and Control*, Dale R. Romsos, Jean Himms-Hagen, and Masashige Suzuki, eds., pp. 169–80. Tokyo: Japan Scientific Societies Press/Basel: Karger.

———. 1992. Dietary fiber and new sugars. In *Natural Resources and Human Health*, Sigeaki Baba, Olayiwola Akerele, and Yuji Kawaguchi, eds., pp. 159–70. Amsterdam: Elsevier.

Oku, T., and Noda, K. 1990. Influence of chronic ingestion of newly developed sweetener, erythritol, on growth and gastrointestinal function of the rats. *Nutr. Res.* 10:989–96.

Oku, T.; Tokunaga, T.; and Hosoya, N. 1984. Nondigestibility of new sweetener, "Neosugar," in the rat. *J. Nutr.* 114:1574–81.

Oku, T.; and Okazaki, M. 1993. Estimation of maximum non-effective level on transitory laxative of new sweetener erythritol in normal human subjects. (in preparation)

Ooshima, T.; Izumitani, A.; Sobue, S.; Okahashi, N.; and Hamada, S. 1983. Noncariogenicity of the disaccharide palatinose in experimental dental caries of rats. *Infect. Immun.* 39:43–49.

Sahota, S.S.; Bramley, P.M.; and Menzies, I.S. 1982. The fermentation of lactulose by colonic bacteria. *J. Gen. Microbiol.* 120:319–25.

Smith, C.H.J., and Bryant, M.P. 1979. Introduction to metabolic activities of intestinal bacteria. *Am. J. Clin. Nutr.* 32:149–57.

Tanaka, R.; Takayama, H.; Morotomi, M.; Kuroshima, T.; Ueyama, S.; Matsumoto, K.; Kuroda, A.; and Mutai, M. 1983. Effect of administration of TOS and *Bifidobacterium breve* 4006 on the human fecal flora. *Bifidobact. Microflora* 2:17–24.

Tokunaga, T.; Oku, T.; and Hosoya, N. 1986. Influence of chronic intake of new sweetener fructooligosaccharide (Neosugar) on growth and gastrointestinal function of the rat. *J. Nutr. Sci. Vitaminol.* 32:111–21.

———. 1989. Utilization and excretion of a new sweetener, fructooligosaccharide (Neosugar), in rats. *J. Nutr.* 119:553–59.

Yamada, K.; Goda, T.; Hosoya, N.; and Moriuchi, S. 1981. Hydrolysis of glucosylsucrose and maltosyl-sucrose by rat intestinal disaccharidases. *J. Jpn. Soc. Nutr. Food Sci.* 34:133–37 (in Japanese).

Yamada, K.; Hidaka, H.; Inaoka, H.; Iwamoto, Y.; and Kuzuya, K. 1990. Plasma fructosemic and glycosemic responses to fructooligosaccharides in rats and healthy human subjects. *Shoka To Kyushu* 13:88–91 (in Japanese).

Chapter 11

Sugar Alcohols

Kauko K. Mäkinen

CHEMICAL DEFINITIONS

The term "sugar alcohol" refers in chemical colloquialism to reduction products of "sugars" and indicates that all oxygen atoms in a simple sugar alcohol molecule are present in the form of hydroxyl groups. The sugar alcohols are "polyols." This term refers to chemical compounds containing three or more hydroxyl groups, and is synonymous with another customary term, polyhydric alcohol. The polyols can be divided into acyclic polyols (alditols or glycitols, which are true sugar alcohols), and the cyclic polyols. An example of the former is sorbitol, while a common cyclic polyol is *myo*-inositol. The most important sugar alcohols that will be discussed in this chapter are based on five- or six-carbon skeletons, and are therefore called pentitols and hexitols. The alditols can be regarded as derivatives of hydrocarbons. For example, xylitol is a pentahydroxypentane, or *xylo*-pentane-1,2,3,4,5 pentol. All of the theoretically possible four-carbon sugar alcohols, tetritols (threitol and erythritol), are known, as are all stereoisomerically possible pentitols and hexitols. All above polyols are based on a single monosaccharide skeleton. Among disaccharide sugar alcohols maltitol, lactitol, and palatinit have received attention in nutrition. Complex, long-chain polyhydroxy alcohols have been produced for various food uses by chemical reduction of oligosaccharides and dextrins. Such procedures normally lead to the formation of mixtures of polyols of various lengths.

As a chemical group, simple, pure sugar alcohols are crystalline substances varying in taste from faintly sweet (galactitol and lactitol) to very sweet (erythritol and xylitol). This discussion will be predominantly restricted to pentitols and hexitols among the simple alditols which can be best considered as reduction products of monosaccharides, and to a few disaccharide sugar alcohols. For example, although glycerol can be regarded as a simple sugar alcohol (methanol and ethylene glycol would theoretically be the lowest homologues) and an important dietary ingredient by virtue of its presence in triglycerides, space does not allow its treatise. For the same reason several interesting synthetic derivatives of sugar alcohols must be excluded. From several naturally occurring sugar alcohols (such as rhamnitol; corresponds to 6-deoxy-L-mannose) there is too little information at this stage. Table 11-1 lists several natural, dietary sugar alcohols and a few synthetic products that already have a place in human nutrition or in certain medical uses, or which have a potential to have one in the future. Although this author prefers the term alditol for simple straight-chain polyols derived from aldoses, the term sugar alcohol will nevertheless be employed in this chapter.

HISTORIC AND EVOLUTIONARY ASPECTS

Humans and other animal species have consumed certain small amounts of sugar alcohols during the entire evolution. Although there understandably cannot be any direct reference in any ancient scripts to the consumption of sugar alcohols (although there are plenty of historical references to "sugar" and other sweet items), it is likely that various fruits, fronds, leaves, and other plant parts consumed by man and other animals have contained certain concentration levels of known sugar alcohols. The term "manna" is used for a food that was miraculously supplied to the Israelites in their journey through the wilderness. Manna was also used by Arabic nomads (Bodenheimer 1947; Reicke and Rost 1950). Manna also means the sweetish dried exudate of a European ash (especially *Fraxinus ornus*) that contains D-mannitol, and a similar product excreted by two closely related species of scale insects (superfamily Coccoidea). These insects are *Trabutina mannipara* and *Majacoccus serpentinus*, feeding on tamarisk (*Tamarix mannifera*). It is important to emphasize the habitual consumption of such sweet exudates by man since prehistoric times, and the fact that D-mannitol, for example, is frequently found in high concentrations in exudates of plants (Stacey 1974). Some seaweeds were reported to contain up to 35% mannitol (in dry weight) (Lohmar 1962; Stacey 1974).

For the understanding of the role of dietary sugar alcohols as one category of functional foods, it is necessary to refer to the expediency of the evolution of these carbohydrates. During the evolution of the biosphere, carbohydrates were probably produced by the polymerization of formaldehyde. The strongly reducing conditions were most likely inimical to the existence of free aldoses and ketoses (Horecker 1969). The latter would become stabilized by formation of glycosidic bonds with other molecules (such as purines, pyrimidines, and other monosaccharides). The remaining aldoses and ketoses would eventually undergo reduction to corresponding polyhydric alcohols. Therefore, evolution embraced certain sugar alcohols at the very first stages of the existence of life.

COMMON SUGAR ALCOHOL PROPERTIES

Virtually all sugar alcohols share the same type of carbon skeleton with other natural, dietary carbohydrates, and the sugar alcohols can even be assayed as "sugars" in chemical total sugar analyses. All sugar alcohols can be converted chemically or enzymatically to the corresponding aldoses and ketoses, which in turn are reducible to the sugar alcohol form. What, then, are the particular chemical features that make sugar alcohols a special group of carbohydrates? Some of the common denominators of sugar alcohols that make them biologically unique are as follows:

(a) *The absence of reducing carbonyl groups*—This fact makes sugar alcohols chemically somewhat less reactive than the corresponding aldoses and ketoses. The sugar alcohols thus "avoid" certain chemical reactions that take place at a high rate with several aldoses and ketoses. The relative chemical inertness is also reflected in the fact that—in the human oral cavity—the sugar alcohols are less reactive and do not normally participate in extensive acid formation in dental plaque.

Table 11-1. Examples of sugar alcohols and their properties.[1]

Trivial and Chemical Names (Sweetness in Relation to Sucrose)	Occurrence	Uses and Selected Properties
Erythritol, 1,2,3,4-butanetetrol, "*meso*-erythritol," tetrahydroxybutane (a tetritol) (up to 2.0)[2]	Algae, grasses, lichens fungi, certain bacteria, human urine, fermented molasses, etc.	Absorbed from the GI[3] tract of white rat and chicken; is not a precursor of liver glycogen. Has been used (along with its tetranitrate) as a vasodilator. Dental and physiologic significance still inadequately known.
D-Arabitol, 1,2,3,4,5-pentanepentol, D-arabinitol, D-lyxitol (a pentitol) (about 0.6)	Lichens, fungi, yeasts, mushrooms, plants, human body fluids (especially cerebrospinal fluid) and tissues (brain)	Poorly metabolized by mammals (low-caloric use); excreted to a remarkable extent in the urine. Not tested in long-term dental trials. L-arabitol (found in nature) may be sweeter.
Ribitol, adonitol (a pentitol) (about 0.8)	Lichens, fungi, yeasts, higher plants, human body fluids, honeydews of wax scale insects (*Ceroplastes*); as a constituent in certain lipopolysaccharides and riboflavin.	A precursor of liver glycogen in rat and guinea pig; enters the pentose phosphate pathway. Less antiketogenic than xylitol; more slowly metabolized than xylitol.
Xylitol, *xylo*-pentane- 1,2,3,4,5-pentol (a pentitol) (1.0)	Certain microorganisms, lower and higher plants (at much lower levels than sorbitol and D-mannitol), animal tissues.	Slowly absorbed from the GI tract in man; a precursor of liver glycogen. Used in oral and intravenous nutrition and as sugar substitute for diabetes and in anticaries preparations.

Sorbitol, D-glucitol (a hexitol) (0.5–0.6)	Algae, fungi, many higher plants, animal tissues.	Absorbed more slowly than glucose; used as a sugar substitute for diabetes. A precursor of liver glycogen. Extensively used in food manufacturing and other industries. In medicine as a cathartic substance and in confectionaries as a low-cariogenic bulk sweetener. Fermented by certain cariogenic bacteria.
D-Mannitol, manna sugar (a hexitol) (0.45–0.6)	Bacteria, algae, grasses, etc., including higher plants.	Absorbed more slowly than sorbitol and xylitol; a precursor of liver glycogen. The laxative effect is higher than with sorbitol and xylitol. Much of the ingested mannitol appears unchanged in the urine. Dental effects almost similar to those of sorbitol. A natural diuretic, diagnostic aid (renal function); in other medical tests.
Galactitol, dulcitol (a hexitol) (0.4)	Algae, fungi, higher plants.	Acts as a precursor of liver glycogen. Poorly absorbed. Because of low sweetness, has not been subjected to extensive dental studies.

Table 11-1. Continued

Trivial and Chemical Names (Sweetness in Relation to Sucrose)	Occurrence	Uses and Selected Properties
Lactitol, β-D-galactopyranosyl- 1,4-sorbitol (0.35–0.5)	Occurrence in nature is uncertain.	Slowly hydrolyzed in the GI tract by duodenal glycosidases to galactose and sorbitol. The carbon skeleton is eventually fully fermented in the large intestine. May be classified as dentally harmless; long-term human studies are lacking.
Maltitol, α-D-glucopyranosyl- 1,4-sorbitol (0.6–0.75)	Occurrence in nature is uncertain.	Slowly hydrolyzed in the intestines to glucose and sorbitol. Most likely dentally harmless. Used as a sweetener in foods (may yield ca. 2 kcal/g).
Isomaltitol, α-D-glucopyranosyl- 1,6-sorbitol; isomalt (0.45)	Occurrence in nature is uncertain.	Rats have shown reduced caloric utilization. Not subjected to long-term dental studies; may behave like lactitol and maltitol.
Palatinit, equimolar mixture of α-D-glucopyranosyl-1, 6-sorbitol, and α-D-glucopyranosyl-1, 6-D-mannitol (about 0.5)	Occurrence in nature is uncertain.	Close to 50% caloric utilization has been reported. Metabolism and dental effects most likely comparable to those of lactitol and maltitol.

Maltotriitol, a trisaccharide consisting of D-glucose and sorbitol (α-1,4- and α-1,6- bonds) (0.7)	Not found in nature.	Forms glucose and sorbitol by intestinal enzymes. Slowly fermented by dental plaque; inhibits fermentation of cariogenic sugars.
Maltotetraitol, a tetrasaccharide consisting of D-glucose and sorbitol (0.7)	Not found in nature.	Behaves almost like maltotriitol (both are weak inhibitors of digestive amylases).
Lycasin, a hydrogenated starch syrup; contains sorbitol, maltitol, trimeric and higher homologues (maltotetraitol, -pentaitol, -hexaitol, etc.) (0.5)	An industrial product.	Composition may vary. The pancreatic amylase forms oligosaccharides that are further hydrolyzed by duodenal maltase. The hydrolysis of smallest sorbitol- containing fragments proceeds slowly. A bulk sweetener. Cariogenicity almost equal to that of sorbitol.
Malbit, a hydrogenated starch syrup containing 86–90% maltitol, some sorbitol, and higher homologues (0.9)	An industrial product.	Reduced calorie content. Absorption, metabolism, and dental effects almost similar to those of maltitol.

[1]Only sugar alcohols that occur in nature or that have been used in human nutrition, or that may have potential of becoming "nutraceuticals," have been included. The theoretical number of tetritols, pentitols, etc., is higher.

[2]Sweetness is not a constant parameter, but depends on the chemical and physical nature of the food ingested, and on the concentration of the carbohydrate involved. The values given represent those frequently reported in the literature.

[3]Gastrointestinal tract.

(b) *The reducing power*—In spite of the relative inertness and the absence of reducing *carbonyl* groups, the sugar alcohols do actively participate in metabolic reactions where their "extra" hydrogens can be deposited on other metabolites such as coenzymes (NADP and NAD, for example) and other acceptors to form chemically reduced products of metabolism.

(c) *Complex formation*—By virtue of their polyoxy nature, many sugar alcohols form interesting although chemically weak complexes with several polyvalent cations. For various physiologic and nutritional purposes the complexes with Ca^{2+}, Fe^{3+}, Cu^{2+} and possibly several trace elements in general, are important.

(d) *Hydrophilicity*—The presence of the maximum possible number of hydroxyl groups in a carbohydrate structure makes virtually all sugar alcohols very hydrophilic (although the solubility of galactitol and D-mannitol in water is lower). At least some lower homologues can compete with water molecules for the hydration layer of proteins, other biomolecules, and also metal cations (without true complex formation). The consequences of this can be seen in the fact that in aqueous solutions—and this may be valid in biological fluids as well—the sugar alcohols indirectly strengthen hydrophobic interactions between proteins, stabilizing them against thermal and other denaturation or damaging processes (Lewin 1974; Mäkinen 1985). In chemical parlance also glucose and sucrose are polyhydric alcohols. Such compounds can act as catalysts in the hydrolysis of esters and other compounds. This reaction may be attributed to the nucleophilic reactivity of those carbohydrates, as a result of the anion formed by ionization of a hydroxy group. Therefore, the sugar alcohols also form such highly basic alkoxide ions (with pK_a values in the range of 12 to 13; such anions have an exceptionally strong solvation energy).

There are several physiologic corollaries of the above qualities of sugar alcohols, one of them being their "osmotically active" nature. This quality not only affects the function of sugar alcohols as food, but also in numerous intracellular events and medical applications (as in the intravenous use of D-mannitol in the lowering of intracranial pressure, and in renal function studies that remain outside of our present scope). The use of D-mannitol as a diuretic should be specifically mentioned. Because of their polyol nature, some sugar alcohols (D-mannitol, for example) with the right configuration can act as free radical scavengers in biological and experimental systems. Under certain chemical conditions sugar alcohols can function as chaotropic agents. Chaotropic agents break up organized water structures in aqueous systems. Such agents have the capacity to affect reactions that obtain their energy from the release of structured water. At least in theory these substances could be used to disperse suspensions of dental plaque and salivary deposits. This property of sugar alcohols may indirectly be associated with their dentally advantageous effects.

OCCURRENCE IN NATURE

Regarding sugar alcohols as "functional foods" presumes their presence in relatively high quantities in the human diet or that their specific effects on the body are elicited even by low concentration levels. No sugar alcohol constitutes a major source of calories in human nutrition. The common sugar alcohol properties indicated above provide the chemical background for the specific effects sugar alcohols may exert on man's physiology even if relatively small quantities are consumed, or if larger quantities are consumed during a limited period of time. The scientific literature is replete with chemical data on the presence of sugar alcohols in food items, or in plant and animal material in general (Carr and Krantz 1945; Touster and Shaw 1962; Lohmar 1962; Wasshüttl, Riederer and Baucher, 1973). The most ubiquitous sugar alcohols are sorbitol and D-mannitol, the concentrations of which in some plants and plant exudates may be very high. For example, the red seaweed *Bostrychia scorpioides* was reported to contain as much as 13.6% sorbitol (Haas and Hill 1932). The berries of *Sorbus* species (mountain ash) may contain as much as 10% (Strain 1934, 1937). Already more than 100 years ago, French researchers indicated that fruits of the family Rosaceae (pears, apples, cherries, prunes, peaches, apricots, etc.) contain appreciable amounts of sorbitol (cited by Lohmar 1962). D-Mannitol is also widespread among plants, but unlike sorbitol, it is found more frequently and at higher levels in exudates of plants. For example, the exudates of the olive and plane trees are rich in mannitol; 80 to 90% of the plane tree exudate has been reported to be pure mannitol (cited by Lohmar 1962). So high are the levels of mannitol in the sap of *Fraxinus rotundifolis* (flowering or manna ash) that for a time D-mannitol was obtained commercially from this source. However, certain marine algae have offered the greatest potential natural source of D-mannitol, because it is present in virtually all brown seaweeds

(Lohmar 1962). D-Mannitol seems to be a product of photosynthesis in the fronds which may contain—at certain times of the year and at certain depths—over 20% of D-mannitol (Stacey 1974). Xylitol is also a ubiquitous sugar alcohol, but it is present in plant material in much lower quantities than the above hexitols. The highest reported levels are near one gram per 100 g dry weight in certain greengages (a plum variety), while somewhat lower levels are present in raspberries, strawberries, bananas, etc. (Washüttl, Riederer and Baucher, 1973). However, it is still somewhat uncertain whether the 935 mg/100 g concentration reported twenty years ago for greengages is correct, because earlier researchers treated the plant extracts with yeast to remove fermentable sugars; the yeast cells may have converted some xylose to xylitol. Xylitol is also present in fungi, several microorganisms, and in the human body (in which the liver alone can synthesize 5 to 15 g of xylitol daily).

In summary, D-mannitol seems to be the most common of the alditols in fungi, green plants, and plant exudates; sorbitol is present in numerous berries, fungi, algae, and generally in higher plants; and arabinitol is common in fungi and lichens. Alditols in general have been claimed to be one of the principal soluble carbohydrate components of many fungi and lichens (Stacey 1974). Free ribitol is present at least in *Adonis* (pheasant's eye of the Ranunculaceae, or buttercap family), and *Bupleurum* (thoroughwax), galactitol in the Celastraceae (staff-tree) family (Stacey 1974), red seaweed, yeasts, and mannas of higher plant life, etc. Madagascar manna has been claimed to be almost pure galactitol (Bouchardat 1872). Volemitol (D-*glycero*-D-*manno*-heptitol) is present, for example, in *Pelvetia canaliculata* (an alga; Fucaceae family; Lindberg and Paju 1954), while perseitol (D-*glycero*-D-*galacto*-heptitol) and an octitol (D-*erythro*-D-*galacto*-octitol) are constituents of the avocado pear (Stacey 1974). The remaining alditols that have been found in angiosperms are restricted to a few genera or even a single species, for example allitol in *Itea* spp. (Virginia willow) (Stacey 1974). Most of the alditols mentioned above also occur in microorganisms. However, our current knowledge is far from complete. A thorough analytical study of sugar alcohols in plants and plant-derived foods would reveal more widespread occurrence of sugar alcohols, the presence of which in plants depends on seasonal variations. It is important to realize—in view of the spirit of the functional food concept—that alditols are frequently found to be major products of photosynthesis (Stacey 1974), the basis of most life forms.

METABOLISM

The monosaccharide sugar alcohols are more slowly absorbed from the small intestine than are sucrose and glucose; no specific transport system has been detected for the pentitols and hexitols. The absorption, therefore, seems to take place by means of passive diffusion along a concentration gradient. The absorbed portion varies to a certain extent from sugar alcohol to sugar alcohol. From a larger xylitol dose perhaps some 30% is absorbed, whereas it is likely that smaller single doses—such as those present in chewing gum—may be more completely absorbed. Field experience also suggests that xylitol mixed in a regular diet is utilized to a higher extent than from single liquid doses, which may cause transient osmotic diarrhea and are removed from the body in feces. Field experience also indicates that the absorption of sorbitol is somewhat slower than that of xylitol; perhaps only about 20% of ingested sorbitol is absorbed. The oral tolerance of D-mannitol is normally slightly lower than that of sorbitol.

The disaccharide sugar alcohols maltitol, isomaltitol, and lactitol are not absorbed as such from the small intestine. They are slowly hydrolyzed by glycosidases present in the intestinal mucosa, liberating the building blocks of these disaccharides, which are then absorbed according to physiologic laws. Consequently, maltitol yields glucose and sorbitol, palatinit yields glucose, and D-mannitol and sorbitol, while lactitol forms galactose and sorbitol. The liberated glucose joins the normal dietary glucose pool and is rapidly absorbed in healthy individuals, whereas the liberated monosaccharide sugar alcohols are slowly absorbed. The unabsorbed portions of xylitol, sorbitol, and D-mannitol subsequently enter the more distal sections of the intestine where they are utilized by the commensal intestinal flora. The major end-products of this fermentation are short-chain carboxylic acids (acetic acid, propionic acid, butyric acid, etc.) and also gaseous products (H_2, CO_2, perhaps some methane, etc.). The above acids are absorbed to a high extent for further metabolism in the body along existing pathways. After ingestion of large single doses, some sugar alcohols may be present in feces.

The absorbed carbon skeletons of xylitol, sorbitol, and D-mannitol (and of other monosaccharide sugar alcohols such as galactitol which is also slowly absorbed) are transported by the portal vein into the liver where they are converted into glucose and glycogen by means of existing metabolic pathways. Xylitol can also enter other organs, and even the erythrocytes possess the necessary enzymes for the metabolism of xylitol. Such tissues contain the so-called glucuronate-xylulose cycle (glucuronic acid cycle), which helps convert xylitol into glucose (Bässler 1978). The portion of sorbitol that enters the hepatocytes in the liver is dehydrogenated by a dehydrogenase. The product of this reaction is fructose, which enters the normal fructose metabolism of the liver cells. Consequently, a portion of the carbons present in sugar alcohols are eventually found in glucose, glycogen, triglycerides, fatty acids, and other metabolites, or they are "circulated" in the tricarboxylic acid cycle whereby complete oxidation to CO_2 and water takes place. All known dietary sugar alcohols and obviously the synthetic polyols shown in Table 11-1 follow the above general absorption and metabolic mechanisms: slow absorption in the duodenum and—in the case of disaccharides and longer molecules—possible initial and slow enzymatic hydrolysis to constituent monosaccharides, followed by an active bacterial metabolism of the unabsorbed sugar alcohol portions later during the digestive process. The absorption and metabolism of the most important sugar alcohols have been summarized (Förster 1978; Bässler 1978; Mäkinen 1978; United States Food and Drug Administration 1978; Allison 1979a,b; Wang and van Eys 1981; Senti 1986; Dills 1989).

OSMOTIC EFFECTS

All slowly absorbed carbohydrates can cause diarrhea if they are consumed as larger single doses. This is a normal, physiologic, osmotic response to the presence of slowly absorbed food ingredients in the gut lumen; virtually all foods consumed in disproportionate levels, and especially lactose (in milk) can lead to osmotic diarrhea and flatulence in several individuals. The severity of such symptoms depends on the individual's size, age, other type of diet consumed, and the structure of the food taken. Generally single xylitol doses of about 20 g are unlikely to cause gastrointestinal side effects. However, humans become adapted to higher consumption levels (Förster, Quadbeck and Gottstein, 1982). The oral tolerance of man toward the other sugar alcohols is somewhat smaller, sorbitol and D-mannitol may elicit similar effects at somewhat lower consumption levels. It is important to emphasize that the development of osmotic diarrhea after consumption of sugar alcohols is not a diseased state, and that persons consuming higher amounts of sugar alcohols, such as diabetic subjects, normally learn to adjust their consumption levels. The quantities of sugar alcohols needed for dental purposes, such as 5 to 10 g of xylitol daily, normally cause no side effects, even in small children (Åkerblom et al. 1982). Folk medicine and some modern physicians have regarded the osmotic effect of sorbitol and D-mannitol as positive; the emptying of the gastrointestinal tract (catharsis) in today's medical practice still partly leans on sorbitol. Related to this is the earlier use of manna as a general demulcent.

CALORIE VALUE

The calorie value ascribed to sugar alcohols should in theory be the same [approximately 4 kcal/g (17.6 kJ/g)] as determined for dietary carbohydrates in general. However, the incomplete absorption of sugar alcohols, especially from larger single doses, suggests that perhaps a lower calorie value should be given to these substances. Values of 2.0–2.4 kcal/g, for example, have later been proposed for sugar alcohols. However, it is possible that small single doses, such as those derived from chewing gums, will undergo more complete absorption. Furthermore, scientists proposing overall lower caloric values for sugar alcohols have sometimes ignored the absorption of the alditol-derived carbon skeletons from the large intestine after gut bacteria have fermented all or a certain portion of the remaining sugar alcohols to absorbable carboxylic acids. The amount of sugar alcohols ingested for dental purposes is so small that at least those consumers should not be concerned by the scientific debate on the true calorie content of sugar alcohols.

THE POSITION OF SUGAR ALCOHOLS IN THE DIABETIC DIET

Xylitol and sorbitol, and to a lesser extent some other polyols, have been used as sweeteners in the diabetic diet. While sorbitol's and lactitol's lower sweetness makes them organoleptically less suitable than xylitol for this purpose, the problem can be partly abolished by adding intense sweeteners. Xylitol's high sweetness and better tolerance have made it a dominant carbohydrate sweetener in the diabetic diet in [the former] Soviet Union, and xylitol is currently used for that purpose in other countries as well. The metabolic basis for this use of sugar alcohols is that their initial metabolic steps, including their membrane transport, are insulin-independent. The slow absorption and noninvolvement of insulin at these metabolic phases cause a delay in the need of insulin, which first becomes evident after the carbon skeleton of those sugar alcohols has been converted, mostly in the liver, into glucose, the utilization of which inevitably requires insulin. Enteral (orally consumed) xylitol causes no or only a small increase in blood glucose levels in healthy and even in diabetic subjects, with no or only a small elevation in plasma insulin levels (Mellinghoff 1961; Bässler et al. 1962; Yamagata et al. 1965; Scheinin and Mäkinen 1975; Förster 1978). Moderate doses (about 40 g) of sorbitol also do not form as high blood glucose levels, and as high insulin requirements, as does the consumption of comparable amounts of sucrose, by insulin-dependent subjects (Steinke et al. 1961). Oral loading tests with maltitol have caused some increase in blood glucose. The results obtained with disaccharide sugar alcohols are generally encouraging, however, and indicate that such compounds can be used, along with xylitol and sorbitol, as sweeteners in the diabetic diet. Space does not allow a more thorough treatise of this topic, but it is justifiable to advocate the use of certain sugar alcohols in the diabetic diet and in the so-called glycemic control of a patient.

It is understandable that most diabetic subjects would like to consume regular sweetened items; sweetness is a normal hedonistic chemodeterminant of many human foods. Artificial sweeteners do not satisfy energy needs. Blind reliance on the use of intense sweeteners frequently ignores the fact that in several food items the manufacturers must replace the "doomed" sugar-like bulking agents with other carbohydrates, fats, and proteins in order to make the food palatable. This may in some cases cause overweight and diabetic subjects to gain additional calories from artificially sweetened food. However, intense sweeteners are ideal in various beverages that should not normally be sweetened with sugar alcohols at all.

SUGAR ALCOHOLS AS CARIES-REDUCING SALIVA STIMULANTS

The "reducing power" stored in the sugar alcohol molecules constitutes an important chemodeterminant from the point of view of bacterial metabolism in the dental plaque. It is not possible to review in this context the chemical and microbiologic features of cariogenic oral flora and caries-associated dental plaque; other texts have reviewed the relevant features (Loesche 1986). The initiation and progression of dental caries is associated with the fermentation of sucrose and certain other dietary carbohydrates into lactic acid (although acetic acid, propionic acid, and other acids may also play a role). The acidity of plaque and the production of insoluble dextran-like extracellular polysaccharides in plaque by cariogenic organisms (especially so-called *Streptococcus mutans*) contribute to dental caries. A particular key point is the increased adhesiveness of dental plaque and of the cells of mutans streptococci and other organisms onto saliva-coated hydroxyapatite surfaces (tooth enamel). The above-mentioned excessive reducing power of certain dietary sugar alcohols is manifested in the fact that they do not give rise to as many moles of lactic acid in dental plaque (if any) as do sucrose, glucose, and fructose (Loesche 1982). This is so because fermentation normally presumes that the substrate and the end-products (acids) must be stoichiometrically balanced in the number of carbon, oxygen, and hydrogen atoms. The regular dietary carbohydrates (sucrose, glucose, and fructose) have an elemental composition equaling $(CH_2O)_n$. The sugar alcohols have two additional hydrogen atoms, which must be present in some of the end-products. The general sugar alcohol structure $(CH_2O)_n \cdot 2H$ may thus give rise to some lactic acid, acetic acid, and formic acid (which will subsequently decompose enzymatically and nonenzymatically into CO_2 and water, a reduced product). The amounts of various acids formed will depend on whether hexitols or pentitols are involved. The fermentation of sorbitol and D-mannitol yields under normal oral conditions one mole each of lactic acid, formic acid, and ethanol, whereas one mole of xylitol should give one mole each of acetic acid, formic acid, and ethanol. Consequently, by virtue of the extra hydrogen atoms present in the sugar alcohol molecules, they yield on a molar basis smaller amounts of acids (and in some cases no lactic acid), whereas most plaque organisms are capable of rapidly fermenting dietary $(CH_2O)_n$ skeletons. The proposed non- and anticariogenicity of xylitol may also be based on the ability of xylitol to interact with salivary calcium (Mäkinen 1985, 1989).

The sugar alcohols that have been most frequently studied as potential caries-limiting sugar substitutes are sorbitol and xylitol. The most relevant vehicles of sugar alcohols are chewing gums. Lozenges are related chewable products because their mastication stimulates salivation. Clinical studies have shown that sorbitol-containing sweets reduce dental caries (Bánóczy et al. 1981), but chewing gum programs have not yet provided the same kind of caries prevention as xylitol gum programs have generally yielded (Mäkinen 1989). By 1992 reports from about ten controlled clinical caries trials in humans had been published (reviewed in Mäkinen and Isokangas 1988; Mäkinen 1992). The percentage reduction of the incidence of dental caries caused by xylitol compared with controls (who used no gum or sugar gum) has varied between 30% and at least 85%, with several reports indicating the demonstration of "healing" of caries lesions. Healing of a caries lesion is tantamount to its natural, physiologic remineralization (rehardening), and should not be confused with wound healing, which involves various vascular reactions and participation of inflammatory mediators. The source of the calcium and phosphate in remineralization in the oral cavity is predominantly saliva, which is supersaturated with regard to calcium phosphate. The underlying microbiologic, physicochemical, and bioinorganic features of the noncariogenicity and anticariogenicity of xylitol have been outlined elsewhere (Mäkinen 1989). The dietary sweet pentitols should in theory exert advantageous dental effects because they are not readily (or not at all) metabolized by cariogenic oral streptococci. Sorbitol and D-mannitol, on the other hand, are normal substrates of mutans streptococci, facilitating their growth, although the formation of acids and extracellular polysaccharides from these hexitols is very small. However, there are several reports on the adaptation of the human oral flora to sorbitol (cited by Mäkinen 1992). Although the low sweetness of sorbitol impairs its dietary uses, the low production costs of sorbitol will most likely keep it available as a bulking agent in odontologically relevant products. Under conditions where dental caries is extensive, it may actually be possible to also achieve significant caries reduction by means of a sorbitol gum, or any other gum that does not contain easily fermentable carbohydrates.

Because the prerequisite of the use of any sugar alcohol as a sugar alternative is the sweetness of the substance, relatively few sugar alcohols will come into use as dietary sweeteners. However, because the missing sweetness can be compensated for with aspartame, acesulfame K, and other intense sweeteners (or by adding xylitol), sugar alcohols such as lactitol, maltitol, palatinit, and complex hydrogenation products of dextrins have found use in candies, desserts, and chewing gums.

Consumption of a xylitol diet may also affect the gingival and periodontal health. Xylitol consumption induced the growth of dental plaque that was less inflammatory than that grown in the presence of sucrose (Tenuovo, Mielityinen and Paunio 1981). "Xylitol plaque" seems to be less irritant than "sucrose plaque" (reviewed in Mäkinen and Isokangas 1988).

ANIMAL FEEDING

Mixtures of sugar alcohols present in the residue after most xylitol has been removed (in the manufacturing of xylitol) were shown to be a good supplement in piglets' creep feed. They lowered the mortality of piglets and the incidence of diarrhea, and improved feed conversion efficiency (Näsi and Alaviuhkola 1980, 1981). Similar polyol mixtures (containing xylitol, arabinitol, mannitol, sorbitol, rhamnitol, galactitol, some reducing sugars and other sugar alcohols) increased the udder health in dairy cows (Korhonen, Rintamaki and Antila, 1977). A part of the sugar alcohols remain unfermented in the rumen (there are differences between sugar alcohols), but even the unabsorbed portion is eventually absorbed and metabolized. Similar sugar alcohol mixtures have been used to treat and prevent ketosis in cattle (Poutiainen, Tuori and Sirvio, 1976; Korhonen, Rintamaki and Antila, 1977). The sugar alcohols, being additional sources of glucose, contribute to the protection against disturbances in metabolism taking place in the liver. Dietary xylitol was associated with elevated activity levels of salivary peroxidase in monkeys (Mäkinen et al. 1978). Space does not allow a more comprehensive treatment of the role of sugar alcohols in animal feeding. Today, several commercial applications exist that have been proposed to improve the health of domestic animals.

OTHER APPLICATIONS

The increased use of xylitol for dental purposes and in infusion therapy, the application of D-mannitol for numerous medical purposes, and the use of sorbitol as a major bulking agent have perhaps masked several interesting research avenues proposed during the past decades. Therefore, reference to older literature is deliberately made as follows. There is evidence that the consumption of certain sugar alcohols increases the absorption of certain vitamins and other nutrients in pharmaceuticals. When the sugar alcohols were included in the diet of rats, the requirements for several, if not all, of the B group vitamins decreased (Rofe et al. 1982). This may result from the stimulation of enteral vitamin synthesis. Consequently, sorbitol and xylitol feeding was shown to have a thiamine-sparing effect (Rofe et al. 1982). Xylitol has also been tested as a therapeutic agent in glucose 6-phosphate dehydrogenase deficiency (van Eys et al. 1974), in the preservation of red blood cells (Quadflieg and Brand 1978; Strauss 1981), and in the amelioration of drug-induced hemolysis (Ukab et al. 1981). Both xylitol and sorbitol have been tested as potential antiulcer agents (Danon and Assouline 1979; Assouline and Danon 1981). The antiulcer activity of hypertonic solutions of xylitol and sorbitol was also noted by other researchers (Danon and Assouline 1979); the polyol-effect may presume prostaglandin mediation. Intraduodenal introduction of xylitol in man inhibited the basal gastric secretion and had a stimulating effect on the secretion of enzymes by the pancreas (Vainshtein, Pivikova and Maksudova, 1973). However, there are species differences; in dogs xylitol (and ethyl alcohol) stimulated the secretion of acid gastric juice while sorbitol was inert in this respect (Sysoev Kremer and Shlygim, 1980). Some researchers have indicated that xylitol is a potential nutrient that should be investigated as a substitute for glucose to yield proportionally higher energy metabolism in host cells than in tumor cells (Sato, Wang and van Eys 1981). Earlier studies investigated that role of xylitol in the treatment of diabetes (Bässler et al. 1962; the literature is quite replete with this topic, involving both enteral and parenteral administration of sugar alcohols), bile duct and liver disorders, ketonemia (Touissant, Roggenkamp and Bässler, 1967), and prevention of adrenocorticol suppression during steroid therapy. Xylitol may also selectively affect (via increased NADH levels?) the so-called mixed function oxidase system in rats (Smith 1982). Xylitol, and to a lesser extent sorbitol, significantly increased the concentrations of citrate and calcium in bone (Knuuttila, Svanberg and Hamalainen, 1989). It is interesting to note that a single oral administration of xylitol (0.75 to 1.5 g/kg) increased the auditory threshold values of patients with Meniere's disease (Palchun et al. 1982, 1983). Xylitol may also inhibit several food-spoilage organisms (Mäkinen and Söderling 1981).

When evaluating the above studies, a reader unfamiliar with the chemical properties of these substances may be surprised by the multifaceted nature of the conditions mentioned. The sugar alcohols are not panaceas. Many of the listed effects become understandable because of the ability of xylitol (and some other sugar alcohols) to produce abundant NADH and NADPH in hepatocytes and certain other cells when xylitol is given intravenously, and in the gastrointestinal tract when given enterally. Larger sugar alcohol levels thus affect coenzyme levels, which in turn regulate numerous metabolic reactions and hormonal effects. The reduced forms of those coenzymes form because xylitol activates the metabolic pattern in both the pentose phosphate shunt and the glucuronic acid cycle. The consequence is decreased redox potential accompanied by other reactions. Such reactions indirectly lead also to the fact that intravenously administered xylitol mobilizes fat depots in humans; reduction in weight can also be seen after enteral consumption of larger quantities of xylitol in man and experimental animals. However, such consumption levels are largely unrealistic for everyday use by average consumers.

There is a very large body of evidence for the benefits of the use of sugar alcohols such as xylitol in parenteral nutrition. Especially German and Japanese clinicians and scientists have performed commendable work in this field, and the above scattered references do not do justice to all the achievements. A considerable effort has also been put forward by researchers in the [former] Soviet Union. The fact that the current parenteral and infusion-based use of xylitol is safe also lends support for its special functions and safety as a functional food. Although this review excluded cyclic sugar alcohols, it should be mentioned that some of them may display increased importance as functional foods. For example, *myo*-inositol possesses several interesting physiologic properties, one of which should be mentioned: its possible beneficial effect on the "free-radical disturbances" caused by high glucose concentrations in embryonic dysmorphogenesis associated with diabetic pregnancy (Eriksson et al. 1991).

In conclusion, although sugar alcohols may not serve specifically in the biosynthesis or breakdown of cellular elements, some sugar alcohols are frequently found to be major products of photosynthesis, act as active metabolites in a variety of plants, and occur as intermediates in the carbohydrate metabolism of man and other mammals. It is possible that certain sugar alcohols are useful in the transformation of hydrogen to coenzymes. The ubiquitous nature of several dietary sugar alcohols indicates that they are normal carbohydrate constituents of man's diet, and can serve special purposes in certain clinical situations.

References

Åkerblom, H.K.; Koivukangas, T.; Puukka, R.; and Mononen, M. 1982. The tolerance of increasing amounts of dietary xylitol in children. *Int. J. Vit. Nutr. Res. (Suppl)* 22:53–66.

Allison, R.G. 1979a. *Dietary Sugars in Health and Disease III. Sorbitol.* Prepared for FDA under Contract Number FDA 223-75-2090. Bethesda, MD: Life Sciences Research Office.

Allison, R.G. 1979b. *Dietary Sugars in Health and Disease IV. Mannitol.* Prepared for FDA under Contract Number FDA 223-75-2090. Bethesda, MD: Life Sciences Research Office.

Assouline, G., and Danon, A. 1981. Hyperosmotic xylitol, prostaglandins and gastric mucosal barrier. *Prostaglandin Med.* 7:63–70.

Bánóczy, J.; Hadas, E.; Esztáry, I.; Marosi, I.; and Nemes, J. 1981. Three-year results with sorbitol in clinical longitudinal experiments. *J. Int. Ass. Dent. Child.* 12:59–63.

Bässler, K.H. 1978. Biochemistry of xylitol, in *Xylitol*, J.N. Council ed., pp. 35–41. London: Applied Science Publishers.

Bässler, K.H.; Prellwitz, W.; Unbehaun, V.; and Lang, K. 1962. Xylitstoffwechsel beim Menschen: Zur Frage der Eignung von Xylit als Zucker-Ersatz beim Diabetiker. *Klin. Wschr.* 40:791–793.

Bodenheimer, F.S. 1947. The manna of Sinai. *Bibl. Archaeol.* 10:2–6.

Bouchardat, G. 1872. Sur la dulcite et les sucres en général. *Ann. Chim. Phys.* 27:68–109.

Carr, C.J., and Krantz, J.C., Jr. 1945. Metabolism of the sugar alcohols and their derivatives. *Advan. Carbohydr. Chem.* 1:175–192.

Danon, A., and Assouline, G. 1979. Antiulcer activity of hypertonic solutions in the rat: Possible role of prostaglandins. *Eur. J. Pharmacol.* 58:425–431.

Dills, W.L. 1989. Sugar alcohols as bulk sweeteners. *Ann. Rev. Nutr.* 9:161–186.

Eriksson, U.J.; Borg, L.A.H.; Forsberg, H.; and Styrud, J. 1991. Diabetic embryopathy. Studies with animal and *in vitro* models. *Diabetes* 40 (Suppl. 2):94–98.

Förster, H. 1978. Tolerance in the human. Adults and children. In *Xylitol*, J.N. Council, ed., pp. 43–66. London: Applied Science Publishers.

Förster, H.; Quadbeck, R.; and Gottstein, U. 1982. Metabolic tolerance to high doses of oral xylitol in human volunteers not previously adapted to xylitol. *Int. J. Vit. Nutr. Res. (Suppl)* 22:67–88.

Haas, P., and Hill, T.G. 1932. The occurrence of sugar alcohols in marine algae. II. Sorbitol. *Biochem. J.* 26:987–990.

Horecker, B.L. 1969. The role of pentitols and other polyols in evolutionary development. In *International Symposium on Metabolism, Physiology and Clinical Use of Pentoses and Pentitols*, B.L. Horecker, K. Lang, and Y. Takagi, eds., pp. 5–25. Berlin: Springer.

Knuuttila, M.; Svanberg, M.; and Hämäläinen, M. 1989. Alterations in rat bone composition related to polyol supplementation of the diet. *Bone & Mineral* 6:25–31.

Korhonen, H.; Rintamäki, O.; and Antila, M. 1977. A polyol mixture of molasses treated beet pulp in the silage based diet of dairy cows. II. The effect on the lactoperoxidase and thiocyanate content of milk and the udder health. *J. Sci. Agric. Soc. Finl.* 49:330–345.

Lewin, S. 1974. *Displacement of Water and its Control of Biochemical Reactions*, pp. 1–344. London and New York: Academic Press.

Lindberg, B., and Paju, J. 1954. Low-molecular carbohydrates in algae IV. *Acta Chem. Scand.* 8:817–820.

Loesche, W.J. 1982. *Dental Caries: A Treatable Infection*, p. 271. Ann Arbor, Michigan: The University of Michigan Dental Publications.

———. 1986. Role of *Streptococcus mutans* in human dental decay. *Microbiol. Rev.* 50:353–380.

Lohmar, R.L. 1962. The polyols. In *The Carbohydrates, Chemistry, Biochemistry, Physiology*, W. Pigman, ed., pp. 241–298. New York: Academic Press.

Mäkinen. 1978. *Biochemical Principles of the Use of Xylitol in Medicine and Nutrition with Special Consideration of Dental Aspects*, pp. 7–21. Basel and Stuttgart: Birkhäuser Verlag.

Mäkinen, K.K. 1985. New biochemical aspects of sweeteners. *Internat. Dent. J.* 35:23–35.

———. 1989. Latest dental studies on xylitol and mechanism of action of xylitol in caries limitation. In *Progress in Sweeteners*, T.H. Grenby, ed., Chapter 13. London and New York: Elsevier Applied Science.

———. 1992. Dietary prevention of dental caries by xylitol—clinical effectiveness and safety. *J. Appl. Nutr.* 44:16–28.

Mäkinen, K.K., and Isokangas, P. 1988. Relationship between carbohydrate sweeteners and oral diseases. *Prog. Food. Nutr. Sci.* 12:73–109.

Mäkinen, K.K., and Söderling, E. 1980. A quantitative study of mannitol, sorbitol, and xylose in wild berries and commercial fruits. *J. Food Sci.* 45:367–371,374.

———. 1981. Effect of xylitol on some food-spoilage microorganisms. *J. Food Sci.* 46:950–951.

Mäkinen, K.K.; Bowen, W.H.; Dalgard, D.; and Fitzgerald, G. 1978. Effect of peroral administration of xylitol on exocrine secretions of monkeys. *J. Nutr.* 108:779–789.

Mellinghoff, G.H. 1961. Über die Verwendbarkeit des Xylit als Ersatzzucker bei Diabetikern. *Klin. Wschr.* 39:447.

Mielityinen, H.; Tenovuo, J.; Söderling, E.; and Paunio, K. 1983. Effect of xylitol and sucrose plaque on release of lysosomal enzymes from bones and macrophages *in vitro*. *Acta Odont. Scand.* 41:173–180.

Näsi, M., and Alaviuhkola, T. 1980. Polyol mixture supplementation in the diet of breeding sows and piglets. *J. Sci. Agric. Soc. Finl.* 52:50–58.

———. 1981. Polyol mixture supplementation as a sweetener and/or feed additive in the diet of piglets. *J. Sci. Agric. Soc. Finl.* 53:57–63.

Palchun, V.T.; Aslamazova, V.I.; Buyanovskaya, O.A.; and Polyakova, T.S. 1982. Employment of xylite for intralabyrinthine hydropsy detection. *Vestn. Otorinolaringol.* 4 (Jul–Aug):35–38 (in Russian).

———. 1983. Diagnostic informativity of the drugs used to reveal intralabyrinthine hydrops according to the data of audiologic and biochemical studies. *Zh. Ushn. Nos. Gorl. Bolezn.* 43(2):27–31 (in Russian).

Poutiainen, E.; Tuori, M.; and Sirviö, I. 1976. The fermentation of polyalcohols by rumen microbes *in vitro*. *Proc. Nutr. Soc.* 35:140A–141A.

Quadflieg, K.-H., and Brand, K. 1978. Carbon and hydrogen metabolism of xylitol and various sugars in human erythrocytes. *Hoppe-Seyler's Z. Physiol. Chem.* 359:29–36.

Reicke, B., and Rost, L. 1950. *Biblisch-Historiches Handwörterbuch*, pp. 1141–1143. Göttingen: Zweiter Band (H-O), Vandenhoeck and Ruprecht.

Rofe, A.M.; Krishnan, R.; Bais, R.; Edwards, J.B.; and Conyers, R.A.J. 1982. A mechanism for the thiamin-sparing action of dietary xylitol in the rat. *Aust. J. Exp. Biol. Med. Sci.* 60:101–111.

Sato, J.; Wang, Y.-M.; and van Eys, J. 1981. Metabolism of xylitol and glucose in rats bearing hepatocellular carcinomas. *Cancer Res.* 41:3192–3199.

Scheinin, A., and Mäkinen, K.K. 1975. Turku sugar studies I-XXI. *Acta Odont. Scand.* 33 (Suppl. 70):1–350.

Senti, F.R. 1986. *Health Aspects of Sugar Alcohols and Lactose*. Prepared for FDA under Contract Number FDA 223-83-2020 (Task Order #7). Bethesda, MD: Life Sciences Research Office.

Smith, J.T. 1982. Effect of xylitol feeding on the mixed acid oxidase system. *Nutr. Reports Int.* 26:347–353.

Stacey, B.E. 1974. Plant polyols. In *Plant Carbohydrate Biochemistry*, J.B. Pridham, ed., pp. 47–59. London and New York: Academic Press.

Steinke, J.; Wood, F.C.; Domenge, L; Marble, A.; and Renold, A.E. 1961. Evaluation of sorbitol in the diet of diabetic children at camp. *Diabetes* 10:218.

Strain, H.H. 1934. *d*-Sorbitol: A new source, method of isolation, properties and derivatives. *J. Am. Chem. Soc.* 56:1756–1759.

———. 1937. Sources of *d*-sorbitol. *J. Am. Chem. Soc.* 59:2264–2266.

Strauss, D. 1981. The two-step preservation of red blood cells. A contribution to improve their viability and therapeutic efficiency. *Acta Biol. Med. Germ.* 40:721–725.

Sysoev, Y.A.; Kremer, Y.N.; and Shlygin, G.K. 1980. Secretory function of the stomach and the energy component used in the parenteral mixtures nutrition. *Vopr. Pitan.* 1:45–49 (in Russian).

Tenovuo, J.; Mielityinen, H.; and Paunio, K. 1981. The effect of dental plaque grown in the presence of xylitol or sucrose on bone resorption *in vitro*. *Pharmacol. Therap. Dent.* 6:35–43.

Touissant, W.; Roggenkamp, K.; and Bässler, K.H. 1967. Behandlung der Ketonämie im Kindesalter mit Ksylit. *Z. Kinderheilk.* 98:146–154.

Touster, O., and Shaw, D.R.D. 1962. Biochemistry of the acyclic polyols. *Physiol. Rev.* 42:181–225.

Ukab, W.A.; Sato, J.; Wang, Y.-M.; and van Eys, J. 1981. Xylitol mediated amelioration of acetylphenylhydrazine-induced hemolysis in rabbits. *Metabolism* 30:1053–1059.

United States Food and Drug Administration. 1978. *Dietary Sugars in Health and Disease II. Xylitol*. Prepared for FDA under Contract Number FDA 223-75-2090. Bethesda, MD: Life Science Research Office.

Vainshtein, S.G.; Pivikova, M.I.; and Maksudova, D. 1973. Xylitol action on the gastric secretion and the external secretory function of the pancreas in patients with duodenal ulcer. *Vopr*. No. 1:14–17 (in Russian).

van Eys, J.; Wang, Y.-M.; Chan, S.; Voravarn, S.; Tanphaichitr, S.; and King, S.M. 1974. Xylitol as a therapeutic agent in glucose 6-phosphate dehydrogenase deficiency. In *Sugars in Nutrition*, H.L. Sipple and K.W. McNutt, eds., pp. 613–631. New York: Academic Press.

Wang, Y.-M., and van Eys, J. 1981. Nutritional significance of fructose and sugar alcohols. *Ann. Rev. Nutr*. 1:437–475.

Washüttl, J.; Riederer, P.; and Bancher, E. 1973. A qualitative and quantitative study of sugar alcohols in several goods. *J. Food Sci*. 38:1262–1263.

Yamagata, S.; Goto, Y.; Ohneda, A.; Anzai, M.; Kawashima, S.; Chiba, M.; Maruhama, Y.; and Yamauchi, Y. 1965. Clinical effects of xylitol on carbohydrate and lipid metabolism in diabetes. *Lancet* ii:918.

Chapter 12

Amino Acids, Peptides, and Proteins

Wayne E. Marshall

INTRODUCTION

The ingestion of amino acids, peptides, and proteins from a variety of food sources is essential for maintaining health. Humans, as do other animals, use protein chiefly for its amino acid content. Proteins are converted to large and small peptides and individual amino acids by gastric and duodenal proteases (Castro 1991). Large peptides are hydrolyzed to small peptides, usually di- and tripeptides, by intestinal peptidases. Amino acids, dipeptides, and tripeptides leave the intestine and enter the hepatic portal system, but the peptides leave the liver and enter the peripheral blood as amino acids. The blood transports the amino acids to individual cells where they are placed in a cytosolic "pool." They are utilized from the "pool" to synthesize proteins essential for growth and maintenance of healthy tissue. The continual hydrolysis and synthesis of proteins is central to their utilization. However, certain proteins and their hydrolytic products also carry out important functions unrelated to their primary metabolic mission.

Recent advances in biomedical research have helped reveal some of the complex relationships between nutrition and disease. This research has suggested that food proteins, peptides, and amino acids may be useful in the treatment of a number of pathological conditions arising from disease or injury.

The recognition of a health restoration role of dietary proteins, apart from their traditional health maintenance role, has become a popular research and marketing endeavor in Japan and, to a lesser extent, in Western Europe and the United States. In Japan and Western Europe, these food proteins are but one class of health-enhancing ingredients collectively called "functional foods." In the United States they are sometimes referred to as "nutraceuticals." In both cases, they are defined as substances that provide medical or health benefits, including the prevention and/or treatment of certain diseases. Although Western Europe and the United States have contributed considerable knowledge to recognizing specific proteins as functional foods and elucidating their mechanism of action, the Japanese have been quick to utilize this knowledge to begin marketing these products.

This chapter will focus on the consideration of amino acids, peptides, and proteins from dietary sources as functional foods. The purpose of the review is to highlight several of the most active research areas in terms of what is being done and how the knowledge can be used for therapeutic purposes.

AMINO ACIDS

Amino Acids as Therapeutic Agents

Nutritionally, only the essential amino acids (His, Ile, Leu, Lys, Met, Phe, Thr, Trp, Val) are required from exogenous sources. The others can be synthesized *in vivo*. Inadequate levels of essential amino acids result in depression of food intake and retardation of growth. These consequences may be seen among the world's poor where protein deficient diets are common.

Although the beneficial effects of amino acid supplementation for individuals with protein deficient diets are well known, little if any evidence conclusively demonstrates the physiological benefits of amino acid supplementation of the diets of healthy individuals. In a typical U.S. diet, about 100 g of protein are consumed daily. If we assume that this diet consists, on average, of an intake of "moderate

Table 12-1. Daily Dietary Intake[a] and Requirements[b] and Manufacturer's Suggested Dietary Supplementation[c] of Amino Acids in the United States

Amino Acid	Amount Consumed (g/day)	Requirements for Essential Amino Acids (g/day)	Doses Suggested on Product Labels (g/day)
Alanine	5.8	—	not listed
Arginine	5.4	—	0.5
Asparagine	7.4	—	not listed
Aspartic acid	3.2	—	not listed
Cysteine + Methionine	3.6	0.91	1.0–2.0
Glutamine	9.6	—	>0.5
Glutamic acid	8.0	—	0.5–3.0
Glycine	7.6	—	0.5–4.0
Histidine	1.6	0.56–0.84	3.0
Isoleucine	5.2	0.7	0.5
Leucine	7.4	0.98	0.5
Lysine	5.0	0.84	0.5–4.5
Phenylalanine + Tyrosine	6.8	0.98	1.0–3.5
Proline	5.6	—	not listed
Serine	7.2	—	not listed
Threonine	4.0	0.49	0.5
Tryptophan	0.8	0.25	0.5–2.0
Valine	4.8	0.7	0.5

[a]Mean daily dietary intake of 100 g of a protein of "moderate quality." Soy glycine was used to approximate the amino acid content of a typical U.S. mixed diet.
[b]Dietary requirements for a 70-kg person calculated from estimated requirements for essential amino acids for adults (World Health Organization 1985).
[c]Summary compiled from label information of a random sampling of amino acid dietary supplement products sold over the counter (Life Sciences Research Office 1992).

quality" protein, then the individual has ingested more essential amino acids than the minimum required (Table 12-1). Table 12-1 also presents information on suggested dosages of particular amino acids as dietary supplements to demonstrate the fallacy of consuming amino acid dietary supplements. In many cases, the suggested supplemental amounts plus the required levels of amino acids are below the actual dietary levels.

The literature contains numerous examples of investigations into the possible beneficial effects of free amino acids taken as dietary supplements. Table 12-2 lists examples of such investigations, but is not intended to be inclusive of all abnormal physiological states that have been investigated. Examination of this literature, which includes anecdotal reports and uncontrolled trials, reveals conflicting results concerning the efficacy of free amino acids used in the therapy of certain disorders. Additionally, the therapeutic response within a particular disorder appears heterogeneous. A limited number of free amino acid preparations have proven useful for the treatment of particular disorders and are available as prescription drugs (see, for example, the Physicians Desk Reference [1993]). Many more of these products are available as dietary supplements, where scientific evidence of their effectiveness may be lacking. There is some concern about the safety of amino acid supplements for the general public. An expert panel investigating the safety of amino acids used as dietary supplements has concluded that some amino acids may pose a health threat when ingested in high doses over prolonged periods (Life Sciences Research Office 1992). Consequently, caution should be exercised and individuals should be under medical supervision while undergoing treatment with amino acid preparations.

Examples of Amino Acids Used as Therapeutic Agents

Treatment of Disease and Injury

Under conditions of physiological stress, such as injury or disease, the body may increase its demand for certain amino acids to promote healing or maintain metabolic function. Recent research has determined that high levels of the hydrophobic, branched-chain amino acids leucine, isoleucine, and valine may aid healing in multiple trauma (Brennan et al. 1986) and in burn patients (Alexander and Gottschlish 1990). The physiological mechanism for this beneficial effect is not completely understood. During abnormal metabolic states, secondary to the body's reaction to stress-inducing trauma, normal utilization of fatty acids and glucose is impaired. Under these conditions, catabolism of the branched-chain amino acids by body tissue compensates for the diminished stores of fatty acids and glucose, especially in muscle, and helps preserve normal metabolic processes.

Glutamine appears to be important in maintaining the integrity of the gastrointestinal tract in both healthy and diseased states. In

Table 12-2. Examples of Investigations of Proposed Therapeutic Effects of Some Orally Administered Amino Acids

Amino Acids	Effect Investigated	References
Arginine	Treatment of hypertension	Kato et al. (1991)
Aspartic acid and asparagine	Treatment of drug addiction	Koyuncuoglu (1983)
		Kruse (1961)
	Management of chronic fatigue	Eriksson (1985)
	Treatment of cirrhosis	
Cysteine and cystine	Treatment of acetaminophen poisoning	Mofenson, Caraccio, and Greensher (1992); Prescott and Critchley (1983)
Glutamic acid	Relief of mental retardation and epilepsy	Mudge (1985); Reynolds (1982)
Glutamine	Treatment of cystinuria	Jaeger et al. (1986); Miyagi, Nakada, and Ohshiro (1979)
	Treatment of alcoholism	Korein (1979)
Histidine	Treatment of rheumatoid arthritis	Pinals et al. (1977)
Leucine	Treatment of Duchenne muscular dystrophy	Mendell et al. (1984)
Lysine	Treatment and prevention of herpes simplex lesions (cold scores)	McCune et al. (1984); Simon van Melle, and Ramelet (1985)
Methionine	Improvement of inflammatory liver disease	Council on Pharm. and Chem. (1947) Prescott and Critchley (1983); Vale, Meredith, and Goulding (1981)
	Treatment of acetaminophen poisoning	
Phenylalanine	Treatment of pain	Balagot et al. (1983); Walsh et al. (1986)
	Prevention or treatment of depression	Young (1987)
	Treatment of hyperactivity	Zametkin, Karoum, and Rapoport (1987)
	Treatment of attention deficit disorder	Wood, Reimherr, and Wender (1985)
	Mood changes and arousal	Ryan-Harshman, Leiter, and Anderson (1987)

Table 12-2. Continued

Amino Acids	Effect Investigated	References
Threonine	Modification of amyotrophic lateral sclerosis	Patten and Klein (1988)
Tyrosine	Treatment of Parkinson's disease	Cotzias, Papavasiliou, and Mena (1973)
	Attention deficit disorder	Nemzer et al. (1986)
	Treatment of narcolepsy	Mouret et al. (1988)
	Treatment of hypertension	Benedict, Anderson, and Sole (1983)
	Treatment of depression	Gelenberg et al. (1990)
Tryptophan	Sleep aid	Schneider-Helmert and Spinweber (1986)
	Affective disorder	Chouinard, Young, and Annable (1985); van Praag and Lemus (1986); Young (1986)
	Treatment of pain	King (1980); Seltzer et al. (1982)

Modified from Life Sciences Research Office (1992).

animal studies, Rombeau (1990) has shown that glutamine-supplemented diets have a tropic effect on the gastrointestinal tract, thus improving intestinal integrity. Diets low in glutamine can result in degenerative changes in the intestinal mucosa with consequent malabsorption of foods (Souba, Smith, and Wilmore 1985). This becomes important in chronically ill patients where maintenance of gastrointestinal integrity not only prevents malabsorption and intestinal diarrhea, but also prevents enteric bacteria from entering the blood. In these individuals, administration of additional glutamine, either in the diet or therapeutically, may be useful.

Influence on Central Nervous System (CNS) Function

The amino acids aspartic acid (aspartate), glutamic acid (glutamate), phenylalanine, tyrosine, and tryptophan either directly or indirectly influence CNS function. Phenylalanine, tyrosine, and tryptophan are transported across the blood-brain barrier and are converted by neural tissue to the neurotransmitters 5-hydroxytryptamine (serotonin) (Trp), dopamine, norepinephrine, and epinephrine (Phe, Tyr). These neurotransmitters modulate several physiological and psychological processes including an individual's mental state or mood. This aspect of their function is described in detail in Chapter 7.

Meal composition affects blood levels and, hence, brain levels of phenylalanine, tyrosine, and tryptophan. Protein-free, carbohydrate-rich meals increase brain tryptophan, but not brain phenylalanine or tyrosine, due to the effects of insulin secretion (Fernstrom 1991). Conversely, a protein-rich meal will raise blood levels of all three amino acids, but only brain tyrosine will increase. The tyrosine content of many food proteins is greater than that of phenylalanine and tryptophan. Therefore, tyrosine has a competitive advantage for the active transport system, which regulates the concentration of specific amino acids in the brain capillaries (Fernstrom 1990). The fat content of a meal does not appear to affect the brain levels of the three precursors (Fernstrom 1991). Although diet clearly alters the brain concentrations of tryptophan, tyrosine, and phenylalanine, no carefully controlled, conclusive scientific evidence is available to demonstrate the influence of diet-induced neurotransmitter levels on brain function in healthy individuals (Fernstrom 1991). To produce a therapeutic effect, these amino acids must be taken in dosages far exceeding the usual dietary dose, and therefore they should be treated as drugs and not as functional foods.

Aspartate and glutamate are usually present in high concentrations in the CNS where they act as excitatory neurotransmitters, eliciting the depolarization of neural membranes (Cooper, Bloom, and Roth 1986). A balanced diet maintains adequate levels of these neurotransmitters because most food proteins contain abundant aspartate and glutamate. In addition, these amino acids are ubiquitous in prepared foods as aspartame and monosodium glutamate (MSG). In rare instances where inadequate concentrations exist in the CNS, response times to external and internal stimuli may be delayed. Dietary aspartate and glutamate cause little if any increase in plasma levels of these amino acids because they are mostly metabolized before entering the systemic circulation (Fernstrom 1991).

Because aspartate and glutamate are two of the more common amino acids ingested in the diet, concern has been raised over the prospect of too much of these amino acids being consumed. The prevailing evidence suggests no observed neurotoxicity or neuronal degeneration in either humans fed large oral doses (200 mg per kg body weight) of aspartame (Stegink, Filer, and Baker 1981) or monkeys fed 2,000 mg per kg body weight of aspartame combined with 1,000 mg per kg body weight of glutamate (Reynolds et al. 1980). In the latter study, high plasma levels of aspartate and glutamate were observed, but neuronal degeneration was not noted.

Although glutamate is apparently harmless when ingested in reasonable excess, a considerable amount of public opinion has been expressed in the United States regarding the presence of MSG in foods. Much of the public outcry was generated from media reports associating MSG to the symptom complex known as the "Chinese restaurant syndrome." Perhaps this and other factors have led to a decline in consumer acceptance of MSG, particularly in the United States and Europe.

PEPTIDES AND PROTEINS

Functional Peptides Derived from Milk

Some of the most extensively studied food-derived peptides with biological activity are the peptides from bovine milk proteins. These peptides have been reported to have opioid or morphine-like activity (casomorphins, lactorphins), immune system stimulating activity (immunopeptides), and the ability to bind minerals, thereby enhancing mineral solubilization and utilization (caseinophosphopeptides) (Table 12-3). Other food-derived peptides have been found in *in vitro* assays to have opioid activity. These include peptides isolated from bovine blood, which could be a component of red meat (Brantl et al. 1986) and peptides from the vegetable proteins soy, wheat, barley, and corn (Zioudrou, Streaty, and Klee 1979). These and most other bioactive peptides have been isolated from *in vitro* enzyme digests of the precursor proteins. Thus far, only two peptides, beta(β)-casomorphin-11 (Meisel and Frister 1989) and α_{s1}-caseinophosphopeptide-9 (Meisel and Frister 1988) have been characterized as *in vivo* digestion products.

Peptides with Opioid Activity

Of all the peptides with opioid activity assayed thus far, two casomorphins, β-casomorphin-5 and β-casomorphin-4-amide were the most potent (Meisel and Schlimme 1990). In contrast, the lactorphins had relatively weak activity. The physiological function of these

Table 12-3. Bioactive Peptides from Bovine Milk Proteins and Their Physiological Effects

Bioactive Peptide	Protein Precursor	Physiological Effect
Casomorphins	α-, β-casein	Increase intestinal water and electrolyte absorption; increase gastrointestinal transit time; anti-diarrhetic effect
α-, β-lactorphins	α-lactalbumin	Similar to casomorphins
Immunopeptides	α-, β-casein	Stimulation of immune system, especially macrophages and T lymphocytes
Caseinophosphopeptides	α-, β-casein	Mineral binding, especially calcium; enhancement of intestinal calcium absorption

peptides appears to be on the gastrointestinal tract. They enhance water and electrolyte absorption in both the large and small intestine and thus exert a potent antidiarrhetic effect (Daniel, Vohlwinkel, and Rehner 1990).

Peptides as Immunostimulants

Immunostimulating peptides have been identified from enzymatic digests of casein, particularly from the α_{s1}- and the β-casein fractions (Parker et al. 1984). These peptides enhance the phagocytic activity of macrophages from both mice and humans and enhance resistance against certain bacteria in mice. Their physiological mode of action is not known, but they may stimulate the proliferation and maturation of immune system cells. These cells appear to be particularly effective in the defense against a wide range of bacteria, especially enteric bacteria (Parker et al. 1984, Migliore-Samour and Jollès 1988).

Caseinophosphopeptides in Mineral Absorption

Of all the bioactive peptides produced by enzymatic activity both *in vitro* and *in vivo,* the caseinophosphopeptides (CPPs) are the best characterized and most intensely studied. These peptides, derived from α_{s1}-, α_{s2}- and β-caseins, are highly phosphorylated and contain consecutive serine phosphate and glutamyl residues in the sequence X-SerP-SerP-SerP-Glu-(Glu)-Y (Meisel, Frister, and Schlimme 1989). The highly anionic character of these peptides renders them resistant to further proteolytic attack and allows them to form soluble complexes with calcium. CPP can effectively prevent the formation of insoluble calcium phosphate complexes over a range of Ca:P ratios (Sato, Noguchi, and Naito 1986; Berrocal et al. 1989).

CPP and their influence on calcium bioavailability was recognized over forty years ago when tryptic digests of casein were found to enhance bone calcification in rachitic children in the absence of vitamin D (Mellander 1950). Subsequent studies have shown that the presence of CPP in the distal small intestine enhances the passive absorption of calcium across the luminal border. However, the role of CPP in directly enhancing active calcium transport is controversial (Kitts and Yuan 1992).

Studies have been carried out to assess the importance of CPP in the utilization of absorbed calcium for bone mineralization. Sato, Noguchi, and Naito (1986) reported that the presence of casein or CPP in a rat diet enhanced the deposition of radiolabeled calcium into femoral tissue. However, Yuan and Kitts (1991) and Yuan et al. (1991) failed to confirm an increase in calcium absorption with an increase in femoral deposition of radiolabeled calcium, bone calcium mineralization, and bone strength in animals fed casein or soy protein diets. As with the above-mentioned studies on the efficacy of CPP in active calcium absorption, the efficacy of CPP in calcium utilization also remains controversial. Many of the discrepancies in the literature may result from comparisons between *in situ* studies using ligated intestinal segments and whole animal tests. These studies are limited by lack of dietary balance tests, fractional absorption studies, bone composition, and bone functionality (strength) tests (Kitts and Yuan 1992). By using carefully controlled experiments and measuring a variety of different parameters, the role of CPP in calcium metabolism can be resolved.

Bioactive Peptides as Functional Foods

Peptides with biological activity could be produced in and from foods in several ways. The most common methods are (a) processing of foods using heat and/or acid/alkali conditions that hydrolyze proteins, (b) enzymatic hydrolysis of food proteins during the course of digesting a meal, and (c) microbial activity in fermented foods. Related to the last method, Hamel, Kielwein, and Teschemacher (1985) have found bioactive casomorphins in fermented milk products inoculated with caseolytic bacteria.

Although bioactive peptides probably do exist in a number of processed and fermented foods, their true effects on physiological functions in humans are unknown. In a healthy individual, eating a varied diet, the presence of bioactive peptides may help keep the nervous, immune, and digestive systems in a well-maintained state. The future potential value of bioactive peptides in the diet may be their ability to affect certain pathological conditions, although this has yet to be proven. For example, if proven to be effective, casomorphins could be administered in the treatment of diarrhea, immunopeptides for immunodeficiencies, and CPP for tooth and bone disease due to lack of normal mineral utilization. Indeed, casein-derived peptides, particularly CPP and β-casomorphins, have already found potential applications as dietary supplements (Brulé et al. 1982).

Antihypertensive Peptides

Another group of bioactive peptides that have been investigated recently are the antihypertensive peptides. Japan has been at the forefront of research in this area and a recent comprehensive review on the subject has appeared (Ariyoshi, 1993). This review documents the wide variety of food sources from which peptides with antihypertensive activity have been isolated. These sources include gelatin, bovine casein, bonito, sardine and tuna muscle, and from the major corn protein α-zein. Antihypertensive activity was also observed in the peptide fraction isolated from rice glutelin (Muramoto and Kawamura 1991) and rice prolamin (Saito et al. 1991). In all cases noted above, the bioactive peptides are not present in the foods *per se*, but arise from proteolytic digests of the food protein. In his review, Ariyoshi (1993) did note, however, reports of naturally occurring antihypertensive activity in hot-water extracts (heat stable peptides) of various soybean products, teas, shellfish, several fruits and buckwheat. All of these peptides appear to inhibit the conversion of angiotensin I to angiotensin II. Angiotensin II increases peripheral resistance by causing a generalized vasoconstriction of the arterioles, thus elevating blood pressure. The reports mentioned above are to be considered preliminary and need to be repeated and extended. Furthermore, the efficacy and safe conditions of use of these peptides in animals, much less in humans, remains to be proven. This is an important area in which bioactive peptides may be found to be useful for pharmaceutical purposes. There are many hypertensive people in the world and many take traditional antihypertensive drugs. If these antihypertensive peptides, derived from food proteins, prove effective and safe under expected conditions of use in humans, then they could provide a welcome addition to currently used therapies.

Vegetable Proteins and the Cholesterol Connection

Considerable scientific evidence from both animal and human studies has demonstrated that a diet rich in vegetable protein may lower blood cholesterol and possibly prevent the occurrence of atherosclerotic lesions. Of the vegetable proteins examined, soy protein produced the most pronounced beneficial effect (Carroll et al. 1978, Fumagalli, Paoletti, and Howard 1978). Therefore, much of the literature in this area describes the use of soy protein, and more specifically, soy protein isolate. Meeker and Kesten (1940, 1941) were the first to clearly demonstrate a difference in atherogenic potential between animal and vegetable proteins in the diet. Rabbits fed casein as the protein source developed severe atherosclerotic lesions in the aorta, while rabbits fed soy protein isolate showed few if any lesions. In an update of these studies, Carroll, Huff, and Roberts (1977) conducted long-term, rabbit feeding experiments and obtained results similar to those of Meeker and Kesten. Even after 11 months, no atherosclerotic lesions were found in rabbits fed soy protein isolate as the sole protein source. A particularly notable study by Huff, Hamilton, and Carroll (1977) demonstrated the sparing effect of a soy protein diet in rabbits. They showed that soy protein could reduce the high plasma cholesterol levels produced by casein when combined with this protein (Table 12-4). These results lend credence to the view that a balance of both animal and vegetable proteins constitutes a healthy diet.

While experiments with animals, particularly rabbits, show interesting trends, studies must be conducted with humans to show the utility of such dietary approaches. Hodges et al. (1967) were among the first to demonstrate a clear effect of vegetable protein on lowering serum cholesterol levels in hypercholesterolemic men. Serum

Table 12-4. Effect of Casein and Isolated Soy Protein Mixtures on Plasma Cholesterol Levels in Rabbits

Dietary Protein		Plasma Cholesterol (mg/100 ml)		
Casein (%)	Isolated Soy Protein (%)	Initial	14 days	28 days
100	0	52	167	239
75	25	76	168	138
50	50	52	98	70
0	100	66	78	67

Modified from Huff, Hamilton, and Carroll (1977).

cholesterol decreased over 40% by the end of the study. In addition, the decrease was seen at relatively high fat intakes (45% of calories) as well as at low fat intakes (15% of calories). A number of other studies have shown not only a decrease in plasma cholesterol, but also a decrease in plasma triglycerides when hypercholesterolemic male and female subjects were placed on a soy protein diet (Mercer et al. 1987). Additional research has observed the cholesterol lowering potential of soy protein in hypercholesterolemic outpatients (Goldberg et al. 1982), hypercholesterolemic children (Widhalm 1986) and normocholesterolemic women (Carroll et al. 1978).

Despite numerous studies attesting to the beneficial effect of dietary soy protein on blood lipid management, the mechanism by which it lowers blood lipid concentrations is uncertain, particularly in humans. One current area of investigation centers around the amino acid composition of soy protein. Jackson, Morrisett, and Gotto (1976) induced hypercholesterolemia in rabbits by employing a mixture of amino acids identical in composition to soy protein in their food. Recently, Lyapkov (1992) measured changes in plasma amino acid levels in humans initially fed a diet high in animal protein and then switched to a diet where the sole protein component was soy protein. Plasma cholesterol and arginine concentrations were lowered, the lysine/arginine ratio was significantly lower, and significant decreases were seen in levels of plasma lysine, isoleucine, and phenylalanine. The hypothesis to explain these results was that the reduced plasma concentrations of lysine and arginine limited the synthesis of apoprotein E, a protein rich in arginine. Apoprotein E participates in the binding and transportation of cholesterol in both high density lipoproteins and very low density lipoproteins, and a diminution of this protein would lead to a decrease in plasma cholesterol. Other mechanisms center around observed differences between animal and soy protein and include interactions between protein and minerals, rates of protein digestibility, and hormonal responses to dietary proteins (Forsythe, Green, and Anderson 1986). In regard to hormonal responses, soy protein was thought to stimulate glucagon production, thereby inhibiting both cholesterol and triglyceride synthesis (Sugano et al. 1982).

Over the past 25 years, there have been data to indicate that the inclusion of soy protein in the human diet may provide cholesterol and triglyceride lowering benefits to certain individuals who are hypercholesterolemic and/or hypertriglyceridemic. Although its beneficial nature is well documented, soy protein has received attention as a protein supplement or as an animal protein replacement primarily in Oriental diets. In the 1970s, soy-based foods marketed in the United States were considered to have unacceptable flavor and texture. With the advent of second generation soy products, most of the objectionable flavor and texture have been removed, but negative impressions are hard to overcome. Some of that stigma still remains. However, with U.S. consumers more health conscious in the 1990s, perhaps vegetable proteins and their food products will take their place as potential disease-preventative, functional foods for the 1990s.

CONCLUSIONS

This chapter has attempted to present a broad overview of the use of amino acids, peptides, and proteins as functional foods. Current scientific evidence has clearly indicated that only a handful of amino acids, peptides, and proteins show promise as functional foods. These dietary components do not, in general, appear to exert pronounced effects in healthy individuals and therefore will find little value as enhancers of normal physiological function. They may be important adjuncts in the treatment of certain pathological conditions. Because of this, their main use will be in pharmaceutical applications, possibly as functional foods.

Soy protein isolates, for example, are considered GRAS (Generally Regarded As Safe) by the U.S. FDA because they have been included in foods for many years. However, for amino acids, and particularly the bioactive peptides, to be touted as curative, considerable time and effort will have to be made to demonstrate the efficacy and safety of these materials. A "Nutraceutical Initiative" has recently been proposed to set up an alternate system of government review and approval of functional foods apart from the normal U.S. FDA process (Anon. 1992). Proponents of this approach hope that such a modified review will shorten the time taken for approval and provide incentive to U.S. food processors to undertake development of functional foods.

References

Alexander, J.W., and Gottschlish, M.M. 1990. Nutritional immunomodulation in burn patients. *Crit. Care Med.* 18:S149–153.

Anon. 1992. Highlights of "The nutraceutical initiative: a proposal for economic and regulatory reform." *Food Technol.* 4:77–79.

Ariyoshi, Y. 1993. Angiotensin-converting enzyme inhibitors derived from food proteins. *Trends Food Sci. Technol.* 4:139–144.

Balagot, R.C.; Ehrenpreis, S.; Greenberg, J.; and Hyodo, M. 1983. D-phenylalanine in human chronic pain. In *Degradation of Endogenous Opioids: Its Relevance in Human Pathology and Therapy. Vol. 5,* S. Ehrenpreis and F. Sicuteri, eds., pp. 207–215. New York: Raven Press.

Benedict, C.R.; Anderson, G.H.; and Sole, M.J. 1983. The influence of oral tyrosine and tryptophan feeding on plasma catecholamines in man. *Am. J. Clin. Nutr.* 38:429–435.

Berrocal, R.; Chanton, S.; Juillerat, M.A.; Pavillard, B.; Sherz, J.-C.; and Jost, R. 1989. Tryptic phosphopeptides from whole casein. II. Physicochemical properties related to the solubilization of calcium. *J. Dairy Res.* 56:335–341.

Brantl, V.; Gramsch, C.; Lottspeich, F.; Mertz, R.; Jaeger, K.-H.; and Herz, A. 1986. Novel opioid peptides derived from hemoglobin: Hemorphins. *Eur. J. Pharmacol.* 125:309–310.

Brennan, M.F.; Cerra, F.; Daly, J.M.; Fischev, J.E.; Moldawer, L.L.; Smith, R.J.; Vinars, E.; Wannemacher, R.; and Young, V.R. 1986. Report of a research workshop: Branched-chain amino acids in stress and injury. *J. Parenteral Enteral Nutr.* 10:446–452.

Brulé, G.; Roger, L.; Fauquant, J.; and Piot, M. 1982. Phosphopeptides from casein-based material. U.S. Patent 4,358,465.

Carroll, K.K.; Giovannetti, P.M.; Huff, M.W.; Moase, O.; Roberts, D.C.K.; and Wolfe, B.M. 1978. Hypercholesterolemic effect of substituting soybean protein for animal protein in the diet of healthy young women. *Am. J. Clin. Nutr.* 31:1312–1321.

Carroll, K.K.; Huff, M.W.; and Roberts, D.C.K. 1977. Dietary protein, hypercholesterolemia, and atherosclerosis. In *Atherosclerosis IV*, G. Schettler, ed., pp. 445–448. New York: Springer Verlag.

Castro, G.A. 1991. Digestion and absorption. In *Gastrointestinal Physiology*. 4th ed., L.R. Johnson, ed., pp. 108–130. St. Louis: Mosby Year Book.

Chouinard, G.; Young, S.N.; and Annable, L. 1985. A controlled clinical trial of L-tryptophan in acute mania. *Biol. Psychiatry* 20:546–557.

Cooper, J.R.; Bloom, F.E.; and Roth R.H. 1986. *The Biochemical Basis of Neuropharmacology*. London: Oxford University Press.

Cotzias, C.G.; Papavasiliou, P.S.; and Mena, I. 1973. L-m-tyrosine and Parkinsonism. *J. Am. Med. Assoc.* 223:83.

Council on Pharmacy and Chemistry. 1947. The status of methionine in the prevention and treatment of liver injury. *J. Am. Med. Assoc.* 133:107.

Daniel, H.; Vohwinkel, M.; and Rehner, G. 1990. Effect of casein and β-casomorphins on gastrointestinal motility in rats. *J. Nutr.* 120(3):252–257.

Eriksson, L.S. 1985. Administration of aspartate to patients with liver cirrhosis. *Clin. Nutr.* 4(Suppl.):88–96.

Fernstrom, J.D. 1990. Aromatic amino acids and monoamine synthesis in the central nervous system: Influence of the diet. *J. Nutr. Biochem.* 1(10):508–517.

———. 1991. The influence of dietary protein and amino acids on brain function. *Trends Food Sci. Technol.* 8:201–204.

Forsythe, W.A.; Green, M.S.; and Anderson, J.B.B. 1986. Dietary protein effects on cholesterol and lipoprotein concentrations: A review. *J. Am. Coll. Nutr.* 5:533–549.

Fumagalli, R.; Paoletti, R.; and Howard, A.N. 1978. Hypercholesterolemic effect of soya. *Life Sci.* 22(11):947–952.

Gelenberg, A.J.; Wojcik, J.D.; Falk, W.E.; Baldessarini, R.J.; Zeisel, S.H. ; Schoenfeld, D.; and Mok, G.S. 1990. Tyrosine for depression: A double-blind trial. *J. Affective Disord.* 19:125–132.

Goldberg, A.P.; Lim, A.; Kolar, J.B.; Grundhauser, J.J.; Steinke, F.H.; and Schonfeld, G. 1982. Soybean protein independently lowers plasma cholesterol levels in primary hypercholesterolemia. *Atherosclerosis* 43(2–3):355–368.

Hamel, V.; Kielwein, G.; and Teschemacher, H. 1985. β-casomorphin immunoreactive material in cows' milk incubated with various bacteria species. *J. Dairy Res.* 52:139–148.

Hodges, R.E.; Krehl, W.A.; Stone, D.B.; and Lopez, A. 1967. Dietary carbohydrates and low cholesterol diets: Effects on serum lipids of man. *Am. J. Clin. Nutr.* 20(2):198–208.

Huff, M.W.; Hamilton, R.M.G.; and Carroll, K.K. 1977. Plasma cholesterol levels in rabbits fed low fat, cholesterol-free, semipurified diets: Effects of dietary proteins, protein hydrolysates and amino acid mixtures. *Atherosclerosis* 8:187–195.

Jackson, R.L.; Morrisett, J.D.; and Gotto, A.M., Jr. 1976. Lipoprotein structure and metabolism. *Physiol. Rev.* 56(2):259–316.

Jaeger, P.; Portmann, L.; Saunders, A.; Rosenberg, L.E.; and Thier, S.O. 1986. Anticystinuric effects of glutamine and of dietary sodium restriction. *N. Engl. J. Med.* 315:1120–1123.

Kato, R.; Nakagi, T.; Saruta, Y.; Suzuki, H.; and Hishikawa, K. 1991. Antihypertensives containing L-arginine or its salts. Japanese Patent 4,282,313.

King, R.B. 1980. Pain and tryptophan. *J. Neurosurg.* 53:44–52.

Kitts, D.D., and Yuan, Y.V. 1992. Caseinophosphopeptides and calcium bioavailability. *Trends Food Sci. Technol.* 2:31–35.

Korein, J. 1979. Oral L-glutamine well tolerated. *N. Engl. J. Med.* 301:1066.

Koyuncuoglu, H. 1983. The treatment with L-aspartic acid of persons addicted to opiates. *Bull. Narc.* 35:11–15.

Kruse, C.A. 1961. Treatment of fatigue with aspartic acid salts. *Northwest Med.* 60:597–603.

Life Sciences Research Office. 1992. Safety of amino acids used as dietary supplements. S.A. Anderson and D.J. Raiten, eds., Bethesda: Federation of American Societies for Experimental Biology.

Lyapkov, B.G. 1992. Clinical and experimental approaches to studying the hypocholesterolemic effects of soybean proteins. In *New Protein Foods in Human Health: Nutrition, Prevention, and Therapy,* F.H. Steinke, D.H. Waggle, and M.N. Volgarev, eds., pp. 179–183. Boca Raton: CRC Press.

McCune, M.A.; Perry, H.O.; Muller, S.A.; and O'Fallon, W.M. 1984. Treatment of recurrent herpes simplex infections with L-lysine monohydrochloride. *Cutis* 34:366–373.

Meeker, D.R., and Kesten, H.D. 1940. Experimental atherosclerosis and high protein diets. *Proc. Soc. Exp. Biol. Med.* 45:543–545.

———. 1941. Effect of high protein diets on experimental atherosclerosis in rabbits. *Arch. Path.* 31:147–162.

Meisel, H., and Frister, H. 1988. Chemical characterization of a caseinophosphopeptide from *in vivo* digests from a casein diet. *Biol. Chem. Hoppe-Seyler* 369:1275–1279.

———. 1989. Chemical characterization of bioactive peptides from *in vivo* digests of casein. *J. Dairy Res.* 56:343–349.

Meisel, H.; Frister, H.; and Schlimme, E. 1989. Biologically active peptides in milk proteins. *Z. Ernährungswiss.* 28:267–278.

Meisel, H., and Schlimme, E. 1990. Milk proteins: Precursors of bioactive peptides. *Trends Food Sci. Technol.* 8:41–43.

Mellander, O. 1950. The physiological importance of the casein phosphopeptide calcium salts. II. Peroral calcium dosage in infants. Some aspects of the pathogenesis of rickets. *Acta Soc. Med. Ups.* 55:247–255.

Mendell, J.R.; Griggs, R.C.; Moxley, R.T.; III; Fenichel, G.M.; Brooke, M.H.; Miller, J.P.; Province, M.A.; Dodson, W.E.; and the CIDD Group. 1984. Clinical investigation in Duchenne muscular dystrophy. IV. Double-blind controlled trial of leucine. *Muscle Nerve* 7:535–541.

Mercer, N.H.; Carroll, K.K.; Giovannetti, P.M.; Steinke, F.H.; and Wolfe, B.M. 1987. Effects of human plasma lipids of substituting soybean protein isolate for milk protein in the diet. *Nutr. Rep. Int.* 35:279–287.

Migliore-Samour, D., and P. Jollès. 1988. Casein, a prohormone with an immunomodulating role for the newborn? *Experientia* 44:188–193.

Miyagi, K.; Nakada, F.; and Ohshiro, S. 1979. Effect of glutamine on cystine excretion in a patient with cystinuria. *N. Engl. J. Med.* 301:196–198.

Mofenson, H.C.; Caraccio, T.R.; and Greensher, J. 1992. Acute poisonings. In *Conn's Current Therapy,* R.E. Rakel, ed., pp. 1099–1139. Philadelphia: W.B. Saunders Company.

Mouret, J.; Lemoine, P.; Sanchez, P.; Robelin, N.; Taillard, J.; and Canini, F. 1988. Treatment of narcolepsy with L-tyrosine. *Lancet* 2:1458–1459.

Mudge, G.H. 1985. Agents affecting volume and composition of body fluids. In *The Pharmacological Basis of Therapeutics*, A.G. Gilman, L.S. Goodman, T.W. Rall, and F. Murad, eds., New York: MacMillan Publishing Company.

Muramoto, M., and Kawamura, Y. 1991. Properties of rice proteins and angiotensin-converting enzyme-inhibiting peptides from proteins. *Shokuhin Kogyo* 34(22):18–26.

Nemzer, E.D.; Arnold, L.E.; Votolato, N.A.; and McConnell, H. 1986. Amino acid supplementation as therapy for attention deficit disorder. *J. Am. Acad. Child Psychiatry* 25:509–513.

Parker, F.; Migliore-Samour, D.; Floc'h, F.; Zerial, A.; Werner, G.H.; Jollès, J.; Casaretto, M.; Zahn, H.; and Jollès, P. 1984. Immunostimulating hexapeptide from human casein: amino acid sequence, synthesis and biological properties. *Eur. J. Biochem.* 145:677–682.

Patten, B.M., and Klein, L.M. 1988. L-threonine and the modification of ALS. *Neurology* 38(Suppl. 1):354–355.

Physicians' Desk Reference. 1993. 47th ed. Montvale: Medical Economics Data.

Pinals, R.S.; Harris, E.D.; Burnett, J.B.; and Gerber, D.A. 1977. Treatment of rheumatoid arthritis with L-histidine: A randomized, placebo-controlled, double-blind trial. *J. Rheumatol.* 4:414–419.

Prescott, L.F., and Critchley, J.A.J.H. 1983. The treatment of acetaminophen poisoning. *Ann. Rev. Pharmacol. Toxicol.* 23:87–101.

Reynolds, J.E.F., ed. 1982. *Martindale: The Extra Pharmacopoeia*. 28th ed. London: Pharmaceutical Press.

Reynolds, W.A.; Stegink, L.D.; Filer, L.J., Jr.; and Renn, E. 1980. Aspartame administration to the infant monkey: Hypothalmic morphology and plasma amino acid levels. *Anat. Rec.* 198(1):73–85.

Rombeau, J.L. 1990. A review of the effects of glutamine-enriched diets on experimentally induced enterocolitis. *J. Parenteral Enteral Nutr.* 14:100S–105S.

Ryan-Harshman, M.; Leiter, L.A.; and Anderson, G.H. 1987. Phenylalanine and aspartame fail to alter feeding behavior, mood and arousal in men. *Physiol. Behav.* 39:247–253.

Saito, Y.; Kawato, S.; Abe, Y.; and Imayasu, S. 1991. Oral angiotensin-converting enzyme inhibitors for hypertension control. Japanese Patent 4,279,529.

Sato, R.; Noguchi, T.; and Naito, H. 1986. Casein phosphopeptide (CPP) enhances calcium adsorption from the ligated segment of rat small intestine. *J. Nutr. Sci. Vitaminol.* 32:67–76.

Schneider-Helmert, D., and Spinweber, C.L. 1986. Evaluation of L-tryptophan for treatment of insomnia: A review. *Psychopharmacology* 89:1–7.

Seltzer, S.; Stoch, R.; Marcus, R.; and Jackson, E. 1982. Alteration of human pain thresholds by nutritional manipulation and L-tryptophan supplementation. *Pain* 13:385–393.

Simon, C.A.; van Melle, G.D.; and Ramelet, A.A. 1985. Failure of lysine in frequently recurrent herpes simplex infection. *Arch. Dermatol.* 121:167–168.

Souba, W.W.; Smith, R.J.; and Wilmore, D.W. 1985. Glutamine metabolism by the intestinal tract. *J. Parenteral Enteral Nutr.* 9:608.

Stegink, L.D.; Filer, L.J., Jr.; and Baker, G.L. 1981. Plasma and erythrocyte concentrations of free amino acids in adult humans administered abuse doses of aspartame. *J. Toxicol. Environ. Health* 7(2):291–305.

Sugano, M.; Ishiwaki, N.; Nagata, Y.; and Imaizumi, K. 1982. Effects of arginine and lysine addition to casein and soya-bean protein on serum lipids, apolipoproteins, insulin and glucagon in rats. *Br. J. Nutr.* 48(2):211–221.

Vale, J.A.; Meredith, T.J.; and Goulding, R. 1981. Treatment of acetaminophen poisoning. *Arch. Intern. Med.* 141:394–396.

van Praag, H.M., and Lemus, C. 1986. Monoamine precursors in the treatment of psychiatric disorders. In *Nutrition and the Brain.* Vol. 7, R.J. Wurtman and J.J. Wurtman, eds., pp. 89–138. New York: Raven Press.

Walsh, N.E.; Ramamurthy, S.; Schoenfeld, L.; and Hoffman, J. 1986. Analgesic effectiveness of D-phenylalanine in chronic pain patients. *Arch. Phys. Med. Rehabil.* 67:436–439.

Widhalm, K. 1986. Effects of diet on serum cholesterol lipids and lipoproteins in hypercholesterolemic children. In *Nutritional Effects of Cholesterol Metabolism,* A.C. Beyen, ed., pp. 133–140. Transmondial Voorhuzen.

Wood, D.R.; Reimherr, F.W.; and Wender, P.H. 1985. Treatment of attention deficit disorder with DL-phenylalanine. *Psychiatry Res.* 16:21–26.

World Health Organization. 1985. *Energy and protein requirements.* Report of a Joint FAO/UN Expert Consultation. Technical Report Series No. 724. Geneva: World Health Organization.

Young, S.N. 1986. The clinical psychopharmacology of tryptophan. In *Nutrition and the Brain.* Vol. 7, R.J. Wurtman and J.J. Wurtman, eds., pp. 49–88. New York: Raven Press.

———. 1987. Facts and myths related to the use and regulation of phenylalanine and other amino acids. In *Dietary Phenylalanine and Brain Function,* R.J. Wurtman and E. Ritter-Walker, eds., pp. 341–347. Boston: Birkäuser.

Yuan, Y.V., and Kitts, D.D. 1991. Confirmation of calcium absorption and femoral utilization in spontaneously hypertensive rats fed casein phosphopeptide diets. *Nutr. Res.* 11(11):1257–1272.

Yuan, Y.V.; Kitts, D.D.; Nagasawa, T.; and Nakai, S. 1991. Paracellular calcium absorption, femur mineralization and biomechanics in rats fed selected dietary proteins. *Food Chem.* 39:125–137.

Zametkin, A.J.; Karoum, F.; and Rapoport, J.L. 1987. Treatment of hyperactive children with D-phenylalanine. *Am. J. Psychiatry* 144:792–794.

Zioudrou, C.; Streaty, R.A.; and Klee, W.A. 1979. Opioid peptides derived from food proteins. The exorphins. *J. Biol. Chem.* 254(7):2446–2449.

Chapter 13

Vitamins for Optimal Health

Harish Padh

INTRODUCTION

Recognition of micronutrients including vitamins was a natural consequence of the availability of purified macronutrients in diet. When purified protein, fat, and carbohydrate became available, it soon was apparent that they were not sufficient to sustain life and growth. In 1906, J. G. Hopkins of England observed that a small amount of vital substances, expected to be amines, were required in the diet. The term *vitamines* for "vital amines" was coined by Funk in 1911. The last letter "e" from vitamine was soon dropped when it was realized that all vital factors were not amines. Vitamins are classified according to their solubility as water soluble or fat soluble. Individual members are also designated by alphabet letters *A*, B_1, B_2, *C*, etc., a trend started by McCollum and Davis at the University of Wisconsin and Osborne and Mendel at Yale. The early history of vitamins indeed makes very fascinating reading (McCollum 1957; Clemetson 1988). It is now recommended that chemical names for the vitamins (e.g., ascorbic acid) should be used; however, designated alphabet letters continue to be used (e.g., vitamin C).

In this chapter chemistry, physiology, and deficiency symptoms of individual vitamins will be briefly discussed. In depth treatment of these and other aspects of vitamins can be found elsewhere (Goodman 1984; Diplock 1985; Clemetson 1988; Shils and Young 1988; NRC 1989b; Tver and Russell 1989; Hunt and Groff 1990; Robinson et al. 1990; Machlin 1991). Because some of the vitamins function as antioxidants, a discussion of antioxidant roles of vitamins will follow. Vitamin requirements, their roles in optimizing health, connection to chronic diseases, and effects of megadose intake will then be discussed.

FAT SOLUBLE VITAMINS

Vitamins A, D, E, and K are well-characterized fat soluble vitamins (Diplock 1985). Each vitamin is actually a group of substances having vitamin activity. Proper digestion, absorption, and utilization of fat soluble vitamins require proper lipid digestion, absorption, and liver functions. Conditions leading to obstruction in bile secretion or metabolism of chylomicrons are likely to affect availability of these vitamins. Some of the fat soluble vitamins like vitamins A and D need specific transport proteins for their circulation in blood. Fat soluble vitamins are stored in the body; thus, prolonged hypervitaminosis can result in serious clinical conditions and even death.

Vitamin A: Retinol

History

McCollum gave the name *fat-soluble A* to an ether extract of egg yolk and butterfat that was found to be required for normal growth. Steenbock at the University of Wisconsin recognized that the yellow pigments in leaves, carotenes, had vitamin A activity as well. The body's ability to convert carotenoids to vitamin A was recognized, and carotenoids are known as precursors or provitamin A.

Chemistry

Retinol (see Fig. 13-1), and its ester, aldehyde, and acid derivatives have vitamin A activity. It is found only in the animal kingdom.

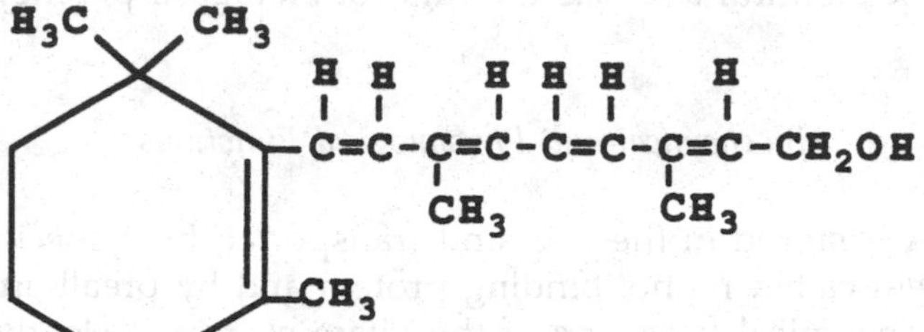

Vitamin A, Retinol

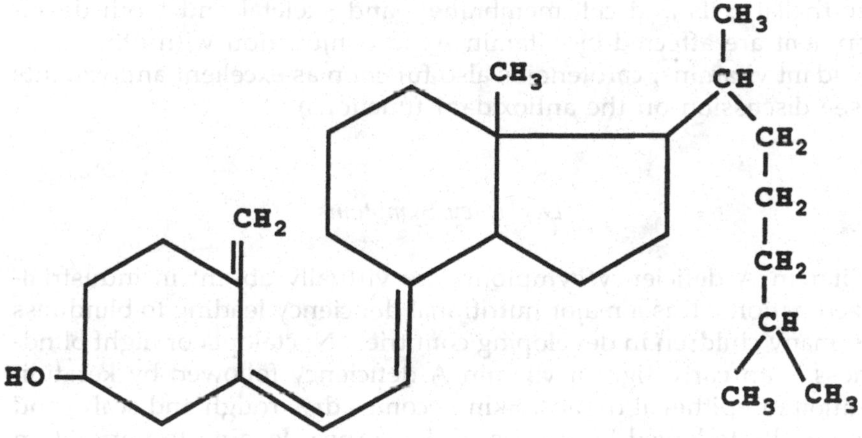

Vitamin D, Cholecalciferol

FIGURE 13-1. Vitamins A and D.

However, plants synthesize carotenoids, which can be converted to vitamin A in the intestinal mucosa. Cryptoxanthin and α-, β-, and γ- carotenes have vitamin A activity, and among them β-carotene is the most plentiful and has the highest biological potency.

Biochemical and Physiological Functions

Vitamin A is stored in the liver and transported by a specific transport protein called retinol binding protein and by prealbumin. The overall biochemical functions of the vitamin are not clearly understood. The best known function is in the aid of dark vision. Vitamin A in a retinaldehyde form is the prosthetic group of the retinal photosensitive pigments rhodopsin and iodopsin. Light initiates configurational change in retinaldehyde, resulting in its detachment from the protein and initiating a nerve impulse that is transmitted to the brain via the optic nerve.

Vitamin A is also required for normal differentiation and proliferation of cells, particularly epithelial cells. In addition, integrity of epithelial cells and cell membranes, and skeletal and tooth development are affected by vitamin A. In conjunction with other antioxidant vitamins, carotenoids also function as excellent antioxidants (see discussion on the antioxidant functions).

Deficiency Symptoms

Vitamin A deficiency symptoms are virtually absent in industrialized nations. It is a major nutritional deficiency leading to blindness in many children in developing countries. Nyctalopia or night blindness is an early sign of vitamin A deficiency followed by keratinization of epithelial tissues. Skin becomes dry, rough and scaly, and is usually followed by xerosis of the cornea, leading to perforation and loss of vision.

Hypervitaminosis

Prolonged excessive intake leads to vitamin A toxicity. Anorexia, desquamation of the skin, bone pain and fragility, headaches, irritability, and hypercalcemia are some of the manifestations of excessive intake of vitamin A. Excessive carotene does not produce such adverse effects.

Food Sources

Milk and eggs and their products contain vitamin A. However, we derive most of vitamin A from plant carotenes present in green stem and leafy vegetables (spinach, broccoli, asparagus) and yellow fruits and vegetables (carrots, squash, peaches, mangos). Vitamin A activity is fairly stable during normal cooking.

Optimal Health

Because of their effect on cellular differentiation, retinol and the carotenoids, especially β-carotene, are considered important in cancer prevention. Some epidemiologic studies have supported such a link (NRC 1989b). However, further studies are required in this direction. Availability of nontoxic derivatives of retinol will be helpful.

Vitamin D: Calciferols

History

The benefits of treating rickets with cod liver oil was known in the Middle Ages. In 1924 Hess and Steenbock independently identified antirackitic factor, which was subsequently isolated in crystalline form and was called calciferol.

Chemistry

Vitamin D_2 (calciferol) and D_3 (cholecalciferol, see Fig. 13-1) are the chief antirackitic factors. Calciferol is formed by the action of ultraviolet light on ergosterol found in plants. Vitamin D_3 is formed from 7-dehydrocholesterol in skin by exposure to the sunshine and is found only in animal cells. These derivatives are fairly stable to heat, alkalies, and oxidation.

Biochemical and Physiological Functions

Vitamin D synthesized in the skin enters blood circulation and binds to a specific transport protein. Excess amounts are mainly stored in the liver as vitamin D. Vitamin D is physiologically inactive. Its conversion in the liver to 25-hydroxycholecalciferol and further conversion in the kidney to 1,25-dihydroxycholecalciferol or calcitriol generates an active form of vitamin D. In reality vitamin D is a prohormone and 1,25-dihydroxy vitamin D_3 is a hormone. The generation of the active form of vitamin D is controlled by the level of parathyroid hormone in serum. Calcitriol promotes absorption of phosphorus and is also responsible for the synthesis of calcium transporting proteins in the intestinal mucosa. The active dihydroxy form of vitamin D promotes mobilization of calcium and phosphorous from bone.

Deficiency Symptoms

Infantile rickets patients exhibit improper formation of bones, bowing of legs, pot belly, poor muscle tone, restlessness, and irritability. Patients have higher levels of alkaline phosphatase. Osteomalacia, tetany, and poor dental health are sometimes found in adults.

Hypervitaminosis

Excessive intake for a prolonged time leads to toxicity symptoms including nausea, vomiting, polyuria, thirst, and diarrhea. Subsequently calcification of softer tissues like blood vessels, stomach, heart, kidney follows.

Food Sources

Exposure to sunlight, which converts a cholesterol derivative to vitamin D, and fortified foods, especially milk and its products, eggs, liver, and fish are common sources of vitamin D. Milk with its contents of calcium and phosphorous is ideal for vitamin D fortification.

Vitamin E: Tocopherols

History

A fat soluble factor found necessary for reproduction was named antisterility factor. The Greek words *tokos* meaning "birth" and *pheros* meaning "to carry" led to the name tocopherol.

Chemistry

Tocopherols (Fig. 13-2) and tocotrienols constitute a group of compounds having vitamin E activity. Among them α-tocopherol is the most active. Although stable in acidic and high temperature conditions, it is readily oxidized by metal ions or rancid fats.

Biochemical and Physiological Functions

It is present in the body in free form and stored in muscle, liver, and fat tissues. No specific transport protein is identified. Its functions are poorly understood. It maintains integrity of cell membranes by preventing oxidation of polyunsaturated fatty acids. It also prevents oxidation of vitamin A and therefore has a sparing effect on vitamin A. Being an antioxidant, it also has a sparing effect on another antioxidant nutrient, vitamin C. In its functions, vitamin E shares a role with selenium. It is not clear how the antioxidant function of vitamin E is related to its deficiency symptoms. The requirement of vitamin E depends on an intake of polyunsaturated fatty acids.

Deficiency Symptoms

Severe deficiency leads to increased fragility and hemolysis of red blood cells and impaired sensation and neuromuscular functions. Reproductive failure, necrosis of the liver, and muscular dystrophy are seen with variable frequency.

Hypervitaminosis

Adverse effects of a large intake of vitamin E are relatively rare. Impaired blood coagulation and elevation of serum lipids are sometimes observed.

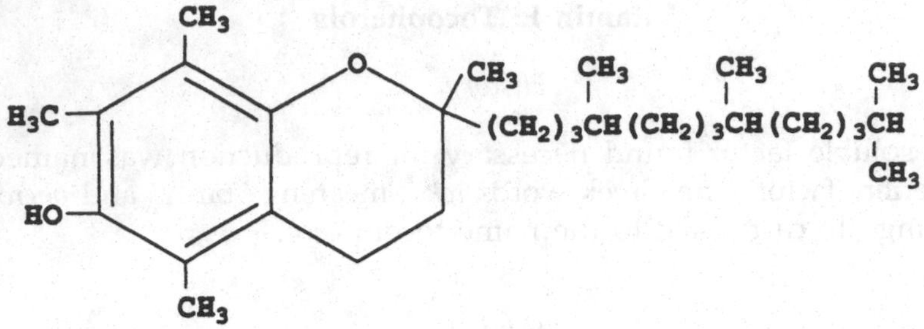

Vitamin E, α-Tocopherol

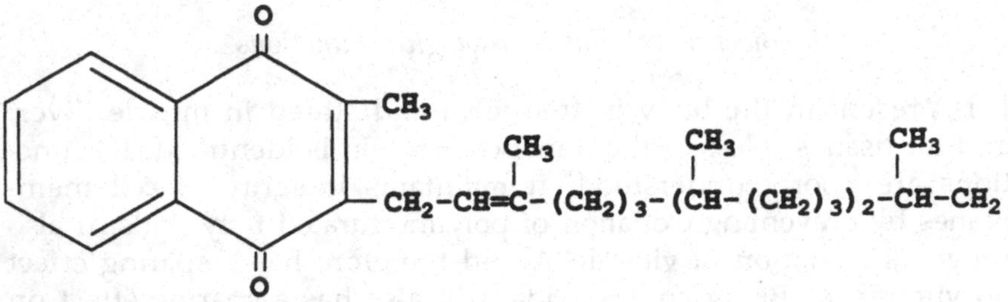

Vitamin K

FIGURE 13-2. Vitamins E and K.

Food Sources

Vegetable oils, nuts, whole grains, and green vegetables are among the chief dietary sources. It is present in very low amounts in foods of animal origin.

Vitamin K: Quinones

History

Dr. Dam of Copenhagen in 1935 found the "Koagulations vitamin" which promoted blood clotting and prevented fetal hemorrhage, hence, the name vitamin K.

Chemistry

A number of quinone derivatives have vitamin K (Fig. 13-2) activity. Among them are phylloquinone (vitamin K_1) from plants, menaquinone (vitamin K_2) from bacteria and putrefied fishmeal, and a synthetic potent derivative menadione. Vitamin K is resistant to heat, but susceptible to extreme pH, oxidation, and light.

Biochemical and Physiological Functions

Newborn infants have a poor supply of this vitamin and their intestinal flora is not yet established for vitamin K synthesis. Vitamin K is required as a cofactor for post-translation modification of prothrombin. Selective glutamic acid residues in prothrombin (and few other proteins) are converted to γ-carboxyglutamic acid. Only after this modification is prothrombin in a functional form. The effect of hypervitaminosis is not known.

Deficiency Symptoms

Vitamin K deficiency leads to increased hemorrhage due to lower prothrombin levels, especially in premature or anoxic infants. Oral therapy with antibiotics and sulfa drugs may interfere with intestinal bacterial synthesis of vitamin K. Conditions leading to inadequate lipid digestion, absorption, or storage may result in vitamin K deficiency. Similarly, therapy with the anticoagulant *Dicumarol* may lead to antagonism of vitamin K action.

Food Sources

About half of the requirement is met from plant foods and the other half by bacterial synthesis in the intestine. Green leafy vegetables, including cabbage, provide a good source of vitamin K.

WATER SOLUBLE VITAMINS

Overt deficiency of water soluble vitamins is uncommon in western nations except in susceptible groups like alcoholics. Unlike fat soluble vitamins, water soluble vitamins, with the exception of vitamin B_{12}, are not significantly stored in the body. It is, however, erroneously assumed that excesses of water soluble vitamins in megadoses will be excreted and therefore safe. There are several reports that megadoses of water soluble vitamins could cause harmful side effects in some individuals (Alhadeff et al. 1984).

Vitamin C: Ascorbic Acid

History

In early times, long sea voyages deprived sailors of fresh fruits and vegetables, often resulting in death from scurvy. The beneficial effects of citrus fruits in the treatment of scurvy was recognized long before the antiscorbutic factor was identified in 1932 as ascorbic acid, independently in the laboratories of Szent-Györgyi and King. The work of Aschoff and Koch in 1919 followed by that of Hojer in 1924 and Wolbach and Howe in 1926 defined the link between a scorbutic diet and the defect in collagen synthesis in connective tissues and its reversal by a diet containing fresh fruit juices.

Chemistry

Ascorbic acid (Fig. 13-3) and its oxidized form, L-dehydroascorbic acid, possess vitamin C activity. Further oxidation of dehydroascorbic acid leads to complete loss of vitamin activity. Ascorbic acid is highly soluble in water and is easily destroyed by heat, light, traces of copper or iron, and oxidative enzymes.

Biochemical and Physiological Functions

Ascorbic acid is an essential nutrient only for humans and a few other species who lack L-gulono-γ-lactone oxidase (EC 1.1.3.8), the last enzyme in the ascorbic acid biosynthesis from glucose. Ever since the discovery of vitamin C, scientists have been intrigued as to how ascorbic acid deficiency can lead to such diverse symptoms as exhibited in scurvy. The historic link between collagen synthesis and ascorbic acid dominated our thinking about the biochemical functions of ascorbic acid. Collagen synthesis is an elaborate process of

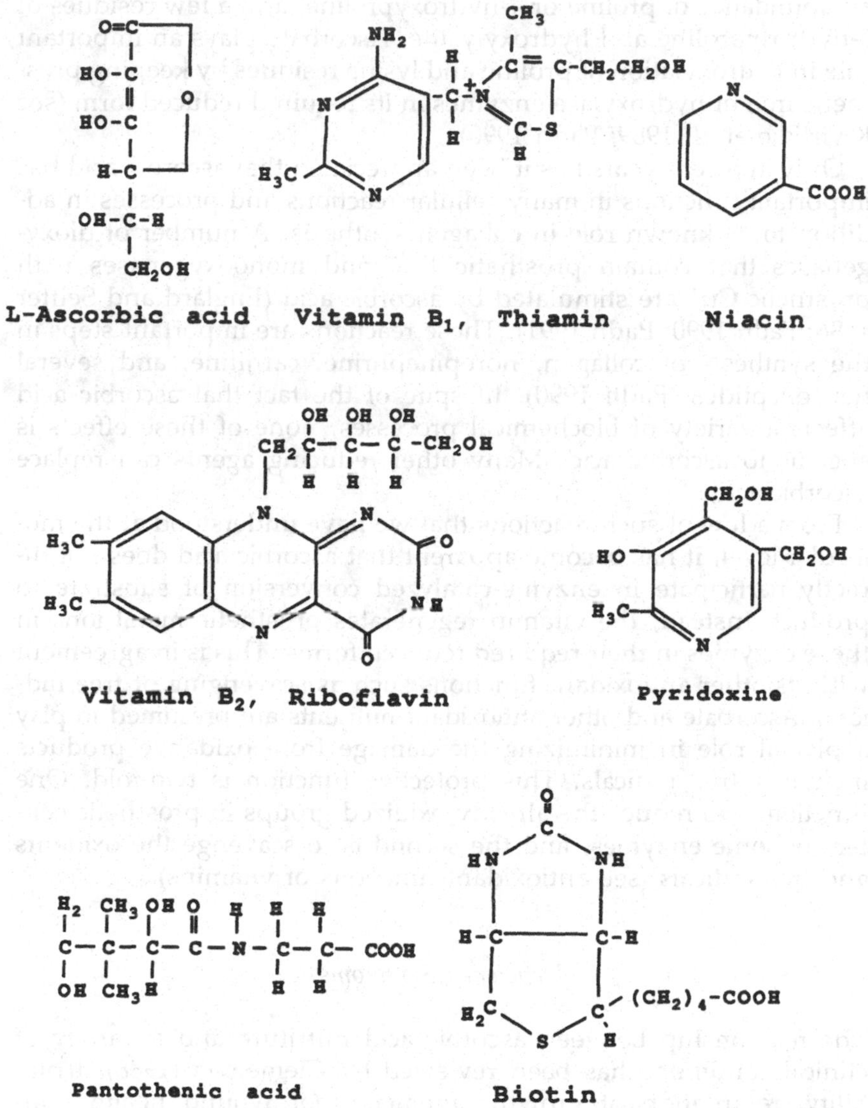

L-Ascorbic acid

Vitamin B_1, Thiamin

Niacin

Vitamin B_2, Riboflavin

Pyridoxine

Pantothenic acid

Biotin

FIGURE 13-3. Some Water Soluble Vitamins.

protein synthesis, post-translational modifications, protein secretion, and extracellular matrix formation. Collagen is a unique animal protein as up to one-third of its amino acid residues are glycine with an abundance of proline or 4-hydroxyproline, and a few residues of 3-hydroxyproline and hydroxylysine. Ascorbate plays an important role in hydroxylation of proline and lysine residues by keeping prosthetic iron of hydroxylase enzymes in its required reduced form (see Kivirikko et al. 1989; Padh 1990).

Only in recent years has it been appreciated that ascorbic acid has important functions in many cellular reactions and processes in addition to its known role in collagen synthesis. A number of dioxygenases that contain prosthetic Fe^{2+} and monooxygenases with prosthetic Cu^{+} are stimulated by ascorbic acid (Englard and Seifter 1986; Padh 1990; Padh 1991). These reactions are important steps in the synthesis of collagen, norepinephrine, carnitine, and several neuropeptides (Padh 1990). In spite of the fact that ascorbic acid affects a variety of biochemical processes, none of these effects is specific to ascorbic acid. Many other reducing agents can replace ascorbic acid.

From a few of such reactions that we have understood at the molecular level, it has become apparent that ascorbic acid does not directly participate in enzyme-catalyzed conversion of substrate to product. Instead, the vitamin regenerates prosthetic metal ions in these enzymes in their required reduced forms. This is in agreement with its other antioxidant functions such as scavenging of free radicals. Ascorbate and other antioxidant nutrients are presumed to play a pivotal role in minimizing the damage from oxidative products including free radicals. This protective function is two-fold. One function is to reduce the already oxidized groups in prosthetic centers of some enzymes, and the second is to scavenge the oxidants and free radicals (see antioxidant functions of vitamins).

Deficiency Symptoms

The relationship between ascorbic acid nutriture and a variety of clinical conditions has been reviewed by Clemetson (1988). Irritability, retardation of growth, anemia, poor wound healing, increased tendency to bleed, and susceptibility to infections are signs of ascorbate deficiency. Petechiae, bleeding gums, weak cartilages, and tenderness in the legs are some of the hallmark symptoms of scurvy.

Hypervitaminosis

With Linus Pauling's strong advocation (1970a,b), thousands of people are ingesting amounts of ascorbic acid up to 200 times RDA without significant toxicity. However, occasionally complications like kidney stones, gastrointestinal disturbances, interference with copper and iron metabolism, and a conditioning to megadoses may occur.

Food Sources

Vegetables and fruits provide most of the vitamin C in a diet. Meat, fish, eggs, and milk provide insignificant amounts of ascorbic acid. Ascorbic acid is a labile vitamin; thus, it is likely to incur some loss during cooking. The presence of metal ions like copper or iron significantly enhances destruction during cooking. Because of the increased use of ascorbic acid as a preservative in foods and drinks, and the increased availability of fresh fruits and vegetables throughout the year, consumption of vitamin C has increased during the past several decades.

Vitamin B_1: Thiamin

History

In 1926 pure thiamin was obtained from rice bran. In 1936 Dr. R.R. Williams published the structure and synthesis of this anti-beriberi factor. The presence of sulfur led to the name *thiamin* (Fig. 13-3).

Chemistry

It is stable in heat and in acidic conditions, but labile in alkaline conditions.

Biochemical and Physiological Functions

The active form of this vitamin is thiamin pyrophosphate, a required coenzyme for a number of important metabolic steps like oxidative decarboxylation of pyruvate to acetyl CoA and α-ketoglutaric acid to succinic acid. These two steps in energy metabolism are highly complex and require other vitamins like niacin and pantothenic acid. Thiamin is expected to have an additional role in the proper functioning of the nervous system.

Deficiency Symptoms

Variable early symptoms of beriberi include fatigue, irritability, lethargy, depression, weight loss, gastrointestinal disturbances, weak limbs leading to peripheral neuritis, and frequently cardiovascular complications. Infantile beriberi is occasionally observed in poor populations.

Food Sources

Wheat germ, whole grains, and enriched flour products are excellent sources. Meat products also provide a fair amount of this vitamin.

Vitamin B_2: Riboflavin

History

Riboflavin was the first coenzyme vitamin to be recognized. Its name is derived from its components: ribose and the fluorescent pigment of a group "flavins" (Fig. 13-3).

Chemistry

Stable to heat, acids, and oxidizing agents. Labile to light and alkaline conditions.

Biochemical and Physiological Functions

Riboflavin is a constituent of FMN and FAD, which are prosthetic groups for many hydrogen transfer enzymes involved in the metabolisms of energy, amino acids, and purine.

Deficiency Symptoms

Dermatitis around nose, cheilosis, altered coloration and texture of lips and tongue, fatigued and sensitive eyes, and bloodshot and vascularized eyes are common deficiency symptoms.

Food Sources

Dairy products, grains, meat, fish, and eggs provide a good supply of this vitamin. Fruits and vegetables provide little riboflavin. Exposure to light leads to significant loss of this nutrient.

Niacin

History

Pellegra characterized by rough skin has been described in the early eighteenth century. Goldberg in the United States is credited with the characterization of pellagra preventing factor. This was followed by identification of the factor as nicotinic acid. The vitamin was named *niacin* to distinguish it from nicotine from tobacco.

Chemistry

The term *niacin* includes nicotinic acid (Fig. 13-3) and niacinamide. It is a fairly stable vitamin.

Biochemical and Physiological Functions

The amino acid tryptophan is a precursor of niacin, and therefore tryptophan-rich and niacin-poor foods like milk and eggs also prevent pellagra. Niacin is a constituent of coenzymes NAD and NADP, which are hydrogen acceptors for numerous reactions in the metabolism of carbohydrates, lipids, and amino acids. Therefore, requirements for niacin vary with the metabolic state of an individual.

Deficiency Symptoms

It usually takes several months to develop a niacin deficiency. Pellegra is rare in the United States, but is not uncommon in countries where corn is a staple diet. The gastrointestinal tract, skin, and nervous system are affected in pellagra. Variable symptoms include loss of weight, poor health, nausea, soreness in the mouth, anemia, symmetric dermatitis, confusion, irritability, and dementia.

Food sources

A diet adequate in protein will generally provide adequate niacin and also tryptophan. Meat, fish, poultry, and whole grains are good sources of niacin. Except for legumes and potatoes, fruits and vegetables provide insignificant amounts of this vitamin.

Vitamin B_6: Pyridoxine

History

In addition to riboflavin, vitamin B_2 preparations in early times were reported to contain another factor named pyridoxine or B_6. The factor was isolated in 1938 and its structure determined.

Chemistry

Vitamin B_6 is actually a group of compounds like pyridoxine (Fig. 13-3), pyridoxal and pyridoxamine. Vitamin B_6 is stable in heat and acid, but is labile in light and in alkaline conditions.

Biochemical and Physiological Functions

Coenzyme pyridoxal phosphate is an active form of the vitamin that is involved in numerous steps of amino acid metabolism like transamination, decarboxylation, and transulfuration. It also has a role in many other reactions including the conversion of tryptophan to niacin. Excessive intake of pyridoxine may lead to dependency on higher intake.

Deficiency Symptoms

Weakness, ataxia, vomiting, irritability, seizures, dermatitis around the mouth and eyes, and neuritic symptoms are observed in a vitamin B_6 deficiency. People taking isonicotinic acid hydrazide (INH) for the treatment of tuberculosis or steroid contraceptives are likely to experience vitamin B_6 deficiency. Isonicotinic acid hydrazide is an antagonist of vitamin B_6.

Food Sources

Meat, fish, poultry, potatoes, vegetables, and whole grains are good sources of vitamin B_6.

Pantothenic Acid

History

Although pantothenic acid was isolated in 1938, it drew interest only after 1950 when Lipmann showed that pantothenic acid was a part of coenzyme A, required for biological acetylation reactions.

Chemistry

Pantothenic acid (Fig. 13-3) in its salt form is quite stable during normal cooking procedures. It is labile in acidic and alkaline conditions.

Biochemical and Physiological Functions

Pantothenic acid is a part of coenzyme A and acyl carrier protein. Coenzyme A has a function in the transfer of the acetyl group in a number of reactions, e.g., synthesis of acetylcholine, cholesterol, and porphyrin, and oxidation of pyruvate and fatty acids. Acyl carrier protein is essential for fatty acid synthesis. Because of its central role in cellular metabolism, pantothenic acid in its coenzyme A form is found in most plant and animal tissues.

Deficiency Symptoms

Deficiency symptoms are observed when an antagonist is included with the deficient diet. Depression, loss of appetite, peripheral neuritis, and cramping pain are usually exhibited. Pantothenic acid deficiency is a likely cause of neuropathy found in alcoholics.

Recommended Allowances

No recommended allowances have been formulated for pantothenic acid. An intake of 4 to 7 mg per day is considered adequate for adults (NRC 1989a).

Food Sources

Pantothenic acid is widely distributed in foods of animal origin and in whole grains and beans. Fruits and vegetables are relatively poor in their content of pantothenic acid. Processing of foods, including grinding, results in significant loss of this vitamin.

Biotin

History

A biotin deficiency can be caused by a diet rich in raw egg white. Egg white contains avidin, a glycoprotein that binds biotin and makes it unavailable.

Chemistry

Biotin (Fig. 13-3), a sulphur-containing cyclic urea derivative, is fairly stable except in alkaline or oxidizing conditions.

Biochemical and Physiological Functions

Biotin is a coenzyme in many carboxylation, decarboxylation, and deamination reactions. In addition, it is required for fatty acid synthesis, purine synthesis, and the conversion of pyruvate to oxaloacetate.

Deficiency Symptoms

Feeding raw egg white in humans causes nausea, desquamation, lassitude, and anorexia with increased cholesterol and decreased hemoglobin in the blood.

Recommended Allowances

Provisional recommendation by the NRC (1989a) is 30–100 μg biotin per day for adults.

Food Sources

Organ meats, legumes, nuts, and egg yolk contain good amounts of biotin. Cereals and milk contain small amounts of biotin. The avidin present in egg white is a heat labile antagonist.

Folic Acid

History

The name folic acid is derived from a Latin word, *folium*, for "green leaves." It was isolated as a growth factor and as a factor for the treatment of megaloblastic anemia.

Chemistry

Folacin, a common name for folic acid (Fig. 13-4), pteroylglutamic acid, and other related compounds having the vitamin activity, is a complex molecule made up of pteridine, para aminobenzoic acid, and glutamic acid. There is a variable but considerable loss of folacin during cooking.

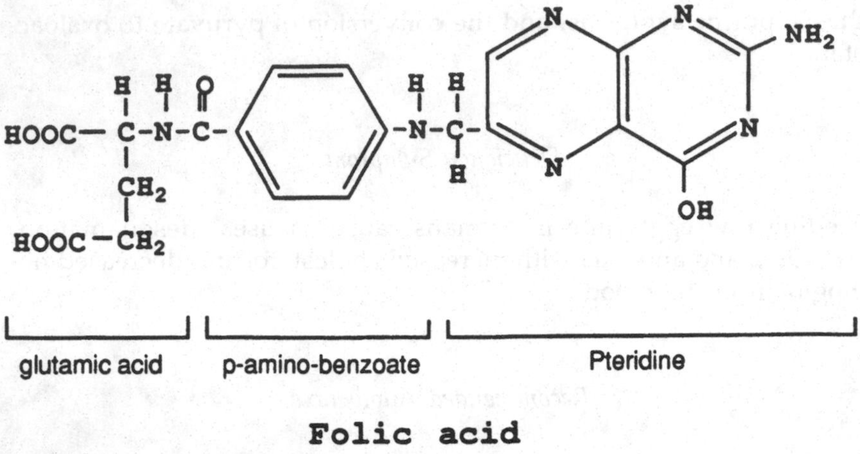

glutamic acid p-amino-benzoate Pteridine

Folic acid

FIGURE 13-4. Folic Acid.

Biochemical and Physiological Functions

Folacin is stored in the liver. Tetrahydrofolate is an active form of this vitamin. It is a coenzyme for numerous reactions involved in one-carbon transfers in the form of methyl, hydroxymethyl, or formyl groups. These reactions are involved in the synthesis of DNA, choline, methionine, serine, and glycine.

Deficiency Symptoms

Folacin deficiency leads to megaloblastic anemia in which proper maturation of red blood cells is hampered, resulting in fewer red blood cells and release of the large nucleated precursor cells. Folacin deficiency is sometimes observed during pregnancy and in the elderly.

Food Sources

Folacin is widely prevalent in many food groups. Organ meats, eggs, whole grains, and green vegetables are particularly rich in folacin.

Vitamin B_{12}: Cyanocobalamin

History

Pernicious anemia remained a fatal ailment until the 1920s. It was then revealed that people with this disease lacked *intrinsic factor* required for proper intestinal absorption of vitamin B_{12}.

Chemistry

Vitamin B_{12} is a complex organic molecule (Fig. 13-5) containing cobalt. Methylcobalamin, cyanocobalamin, and hydroxycobalamin are some of the forms of vitamin B_{12}. It is stable in heat, but sensitive to light and extreme pH. Loss during cooking is expected to be minimal.

Biochemical and Physiological Functions

The role of *intrinsic factor* is important in regulation of this vitamin. Liver is the prime storage site and may easily contain a few years supply of vitamin B_{12}. Vitamin B_{12} functions as a coenzyme for reactions involving one-carbon transfer, e.g., in the synthesis of methionine, choline, and nucleotides (for DNA replication).

Deficiency Symptoms

Vitamin B_{12} deficiency usually develops from the lack of *intrinsic factor* and not from the lack of vitamin B_{12} per se. Usual symptoms include megaloblastic anemia, dyspnea, anorexia, intestinal discomfort, weight loss, depression, and other neurological symptoms.

Food Sources

Liver, meats, fish, and poultry provide a good source of this vitamin. Dairy products are also good sources of this vitamin, especially for those who do not consume any meat. Plant products do not provide any vitamin B_{12}.

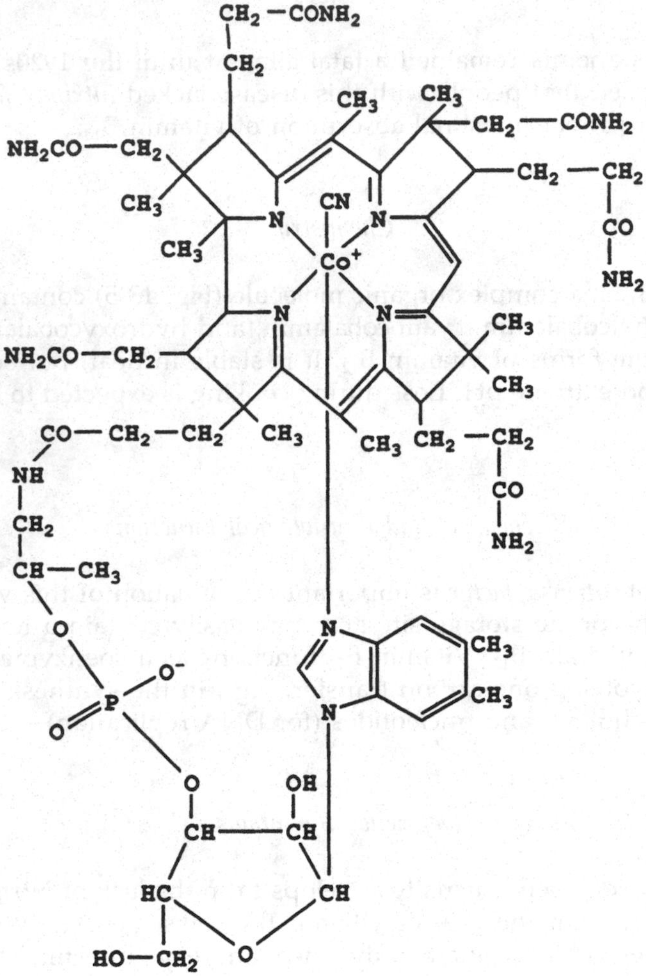

Vitamin B_{12}, cyanocobalamin

FIGURE 13-5. Vitamin B_{12}.

OTHER VITAMIN-LIKE FACTORS

Historically many other factors have been claimed as *vitamins*. In addition, the factors like choline, bioflavinoids, carnitine, lipoic acid, inositol, pyrroloquinoline quinone (PQQ), and para-aminobenzoic acid have important physiological functions, but their dietary need as vitamins has not been established (Shils and Young 1988; Hunt and Groff 1990; Cody 1991).

ANTIOXIDANT FUNCTIONS OF VITAMINS

Perhaps the most significant threat a cell continuously encounters is from oxidative molecules, including free radicals. Free radicals like hydroxy, hypochlorite, peroxy, alkoxy, superoxide, hydrogen peroxide, and singlet oxygen are generated by autoxidation, radiation, or from activities of some oxidases, dehydrogenases, and peroxidases. In addition, other sources of free radicals are tobacco smoke, hyperoxic air, solvents, pesticides, and certain pollutants, including ozone. Free radicals can be extremely damaging to biological systems (Halliwell and Gutteridge 1985). They are a threat to a variety of molecules, e.g., sulfhydryl proteins, enzymes containing transition metal ions, nucleic acids, and membrane lipids. In tissues the ratio of reduced to oxidized forms of ascorbic acid and glutathione is more than 30 to 300. Perhaps the most important necessity for a cell is to maintain its reducing state. To encounter oxidative forces, cells contain a number of antioxidant molecules. Among them are glutathione, ascorbic acid, tocopherols, carotenoids, superoxide dismutase, catalase, and selenium-dependent glutathione peroxidase (Ziegler 1985; Frei, England, and Ames 1989; Mascio, Murphy and Sies 1991). In healthy tissues, the antioxidants react and neutralize oxidative molecules including free radicals and also reduce molecules in cells that have already been inadvertently oxidized. It now appears that the major function of ascorbic acid is as an antioxidant, where it keeps molecules (specially enzyme bound copper and iron) in their required reduced forms. Its involvement in the synthesis of collagen, norepinephrine, or amidation of neuropeptides (Englard and Seifter 1986: Padh 1990) seems a part of its antioxidant function (Padh, 1991).

It is important to recall that evolutionary emergence of the ability to synthesize ascorbic acid in amphibians suggests that a greater need for ascorbate may have been linked with a step from an aquatic to a terrestrial mode of life, where the organisms faced higher oxygen tension and hot climate (Chatterjee 1978) A supportive argument is the fact that plants do not synthesize collagen or norepinephrine. Yet, plants in general and green leaves in particular are rich in ascorbate. We may recall that chlorophyll-containing cells carry out potent oxidative reactions known in biology.

The oxidative molecules, including free radicals, are suspected in the pathology of more than 70 medical conditions (Halliwell and Gutteridge 1990) ranging from atherosclerosis, cancer (Wittes 1985), and diabetes (Pecoraro and Chen 1987) to cataracts (Gerster 1989). Oxidative modification of proteins has been implicated in aging and many degenerative disorders (Johnson et al. 1986; Oliver et al. 1987; Carney et al. 1991). Although conclusive evidence for the role of oxidative molecules in pathology may be lacking, mounting evidence suggests that the battle between oxidative and reducing forces is altered in aging and during the pathogenesis of many degenerative disorders. In spite of the strong suggestive indications regarding the role of oxidative forces in the pathology of many disorders, this medical frontier has received scant attention in medical sciences and is almost completely ignored in basic sciences. The point is not about recommending supplements of antioxidant vitamins, but to acknowledge and appreciate their antioxidant functions in health and diseases. Specially, the interaction among antioxidant nutrients has not been fully appreciated. Next to pathogens, oxidative molecules may constitute the second most potent harmful force that a body continuously encounters. The study of this battle between oxidative and antioxidative forces offers a promising frontier in medicine for a number of disorders for which there is no effective cure or treatment at present.

Ascorbic acid, α-tocopherol, and β-carotene are excellent antioxidants and free radical scavenging nutrients protecting cells from damage by oxidants (for review see ref. Machlin and Bendich 1987; Anderson and Theron 1990; Padh 1991). These antioxidant vitamins can react with and scavenge many types of free radicals including singlet oxygen, superoxide, and hydroxy radicals. Ascorbate can regenerate the reduced form of α-tocopherol, perhaps accounting for observed sparing effects of these vitamins. A cell's major defense against free radicals and other oxidative damages include the antioxidant vitamins like ascorbate and α-tocopherol, enzymes like catalase and superoxide dismutase, and compounds like glutathione. However, in blood and other extracellular fluids, these vitamins are the major antioxidants. Ascorbate can also inactive and neutralize histamine (Chatterjee 1978; Aleo 1981; Clemetson 1988). Synergistic actions and sparing effects between these vitamins and with trace elements like selenium suggest that the optimum requirement of one nutrient will depend on the intake of the rest of the nutrients.

Our understanding of the biochemical functions of antioxidant vitamins does not permit us to correlate them with pathological symptoms found in acute deficiency symptoms, indication of a fundamental gap in our knowledge about their *in vivo* functions. Therefore, further studies of antioxidant activity of vitamins is warranted, especially as it relates to their deficiency symptoms. In that connection, it will be important to understand interactions among various antioxidant nutrients.

HOW MUCH OF VITAMINS DO WE REALLY NEED?

Recommended Dietary Allowances

Perhaps a simple question often asked is: How much vitamins in the diet can assure us that any lack of vitamins will not be the cause of any health problem that one might ever develop? Unfortunately, the science has not advanced enough to give a precise answer to this question. We have entrusted the National Research Council of the National Academy of Sciences, or its equivalents in other countries, to give us its best estimates, which is popularly known as the *Recommended Dietary Allowances* (RDA). The subcommittee of the Food and Nutrition Board of the National Research Council of the U.S.A. has recently published the Tenth Edition of the Recommended Dietary Allowances (NRC 1989a) A summary of their recommendations is given in Table 13-1.

Vitamin Requirements for Optimal Health

Historically vitamins have been viewed as trace nutrients that eliminated certain acute deficiency symptoms. This historically limited view has dominated our thinking about vitamins. In determining the RDA, considerations like the amount of a particular vitamin required for elimination of its acute deficiency symptoms are paramount. To that amount allowances are added to account for indi-

Table 13-1. Recommended Dietary Allowances Adapted from National Research Council (NRC 1989a)

Vitamin	Men	Women	Pregnant Women	Lactating with Women	Infants	Children to Age 11
Fat Soluble Vitamins:						
Vitamin A (retinol, μg)[a]	1,000	800	800	1,300	375	400–700
Vitamin D (cholecalciferol, μg)	5–10	5–10	10	10	7.5	10
Vitamin E (α-tocopherol, mg)	10	8	10	12	3–4	6–7
Vitamin K (μg)	65–80	55–65	65	65	5–10	15–30
Water Soluble Vitamins:						
Vitamin C (mg)[b]	60	60	70	90–95	30–35	40–45
Vitamin B_1 (Thiamin, mg)	1.5	1.1	1.5	1.6	0.3–0.4	0.8–1.2
Vitamin B_2 (Riboflavin, mg)	1.7	1.3	1.6	1.7–1.8	0.4–0.5	0.8–1.2
Niacin (mg)[c]	19	15	17	20	5–6	9–13
Vitamin B_6 (Pyridoxine, mg)	2.0	1.6	2.2	2.1	0.3–0.6	1.0–1.4
Vitamin B_{12} (μg)	2.0	2.0	2.2	2.6	0.3–0.5	0.7–1.4
Folic acid (μg)	200	180	400	260–280	25–35	50–100
Pantothenic acid	See text					
Biotin	See text					

[a]6 μg of β-carotene is equivalent of 1 μg of retinol.
[b]Smokers are now recommended a daily allowance of 100 mg/day, compared to 60 mg/day for nonsmoking adults.
[c]60 mg of dietary tryptophan is equivalent of 1 mg niacin.

vidual variations and for some margin of safety. In addition, turnover rate, body pool, and blood level of vitamins are considered by the subcommittee. After such considerations, the committee agrees on arbitrary amounts of vitamins that it recommends as dietary allowances for the normal healthy population. This approach has two significant limitations: (1) it is based on eliminating acute deficiency symptoms and not for optimal health, (2) it is difficult to define a normal healthy population.

Only recently have many people begun to ponder whether vitamins have other functions in optimizing health, especially on a long-term basis in fighting chronic complications like diabetes, cancer, atherosclerosis, cataracts, arthritis, or the process of aging (NRC 1989b). Recent understanding of the antioxidant functions of several vitamins is responsible for a major revision in our thinking in this field. We should ask what a particular vitamin does, rather than what disease it prevents. It is likely that the vitamins may do more than just prevent their acute deficiency symptoms. However, they are not to be taken as a "magic cure-all" medicine for diseases. Although no clear answer is available, the question remains whether the acute vitamin deficiency should be the only determinant for the RDA for vitamins (Hemila 1984). We should consider whether attaining blood and tissue saturation with a minimal intake of vitamins is a desirable physiological state.

In the tenth edition of the RDA, smokers are now recommended a daily allowance of 100 mg/day of vitamin C compared to 60 mg/day for nonsmoking adults (NRC 1989a). The RDA are formulated for "healthy" individuals in different age groups and physiological states such as pregnancy and lactation. So far the subcommittee has not set separate recommendations for any other subgroup (e.g., climatic or occupational). This I believe is the first time a subgroup of "smokers" is identified within a "healthy" adult population. This is a significant step, and hopefully is the precedent to consider requirements of various subgroups within a general population. The unprecedented identification of "smokers" for separate dietary recommendation is a recognition of the long-term effects of vitamins and not just acute deficiency symptoms. It would be prudent to identify more subgroups with altered vitamin requirements. Because the RDA is so widely respected as the official recommendation from the government and the science community, the subcommittee could also try to develop guidelines for nutrient requirements for people suffering from various chronic ailments. It is realized that the task will often be overwhelming and difficult. The first step could be to identify an illness with altered nutrient metabolism and requirements.

Groups Susceptible to Vitamin Deficiencies

It is generally believed that people in western countries have adequate (as defined by RDA) vitamin intakes. There are, however, susceptible groups within the general population who may have inadequate vitamin intakes. Such groups include smokers, dieters, people on medication including oral contraceptives, alcoholics, adolescents, the elderly, and people with diabetes and other chronic ailments (Gaby et al. 1991). The intake of vitamins A, D, or E, folate, pyridoxine, or ascorbic acid is frequently found inadequate in such groups. I will discuss two such subpopulations requiring special attention for their vitamin nutriture: the elderly and people with chronic ailments.

Vitamin Requirements for the Elderly

The subcommittee acknowledges that (a) it becomes increasingly difficult to define the term "healthy" with advancing age, and (b) the elderly have altered requirements for some nutrients (NRC 1989a). Various reports have shown that elderly persons, institutionalized or free living, have lower levels of vitamins in their blood or tissues (Goodwin, Goodwin, and Garry 1983; Cheng, Cohen, and Bhagavan 1985; Newton et al. 1985; Schneider et al. 1986; Kirsch and Bidlack 1987; VanderJagt Garry, and Bhagavan 1987; Gaby et al. 1991). The reasons for this are many, including lower intake, higher needs, poor utilization, physiological stress from the aging body, frequent medication, the likely presence of chronic illness in such a population, and a solitary lifestyle leading to lack of nutritious cooking habits. The elderly are likely to need special attention for their vitamin needs.

Vitamins and Chronic Disorders

It is becoming apparent that prolonged marginal insufficiency of antioxidant vitamins may have less obvious but more permanent effects on various organs. It is possible that such borderline deficiency may be an important factor in predisposing susceptible individuals to diabetes, atherosclerosis, or amyloidosis along with other chronic diseases (Anderson 1984; Pietrzik 1985; Wittes 1985; Pecoraro and Chen 1987; Tannenbaum and Wishnok 1987; Gerster 1989; Anderson and Theron 1990). Although the literature on such topics is vast and at times inconclusive or even controversial, it is becoming apparent that availability, utilization, or catabolism of antioxidant vitamins like ascorbic acid is altered in a number of chronic diseases (for reviews, see Clemetson 1988; NRC 1989b; Gaby et al. 1991). In many instances, vitamin supplementation has provided a favorable outcome. The antihistamine effect of ascorbate may have a significant role in such chronic conditions (Chatterjee 1978; Aleo 1981; Clemetson 1988). Long-term complications of many chronic ailments like diabetes, cataract formation, arthritis, joint lesions, periodontal diseases, cancer, and atherosclerosis are poorly understood. Understanding the effects of antioxidant nutrients and their metabolism during the progression of complications of such chronic diseases should be our priority. Any possible relief that moderate amounts (*not* megadoses!) of antioxidant nutrients can offer from such devastating complications should be seriously considered. It is important to remember that a great promise lies in retarding the development of such complications over a long period of time and not in any immediate relief from already developed complications.

Localized Tissue Deficiency

Another important concept requiring attention in chronic ailments is the question of local tissue deficiency of vitamins when the rest of the body has adequate supply. For example, periodontal gingivitis, cataracts, arthritis, or any condition involving locally inflamed tissue is likely to develop a local ascorbate insufficiency (Lunec and Blake 1985; Padh and Aleo 1987, 1989; Clemetson 1988; Gerster 1989). The chronic nature of such conditions may keep the involved tissues under vitamin insufficiency for a prolonged time, which may lead to undesirable consequences. In a very interesting finding Lunec and Blake (1985) observed that the ratio of dehydroascorbic acid to ascorbic acid increased remarkably in sera from patients with rheumatic arthritis. The authors always found more dehydroascorbic acid in the synovial fluid of these patients than the respective serum, indicating localized alteration in ascorbate metabolism. The high concentration of histamine (Chatterjee 1978; Aleo 1981) and activated complement (Padh and Aleo 1987, 1989) present in locally inflamed tissues may be instrumental in lowering the local concentration of ascorbic acid. Similarly, the local inflammation present in periodontal diseases may result in reduced collagen synthesis and lack of required support for involved teeth (Aleo 1981).

EXCESSIVE INTAKE OF VITAMINS

Fat soluble vitamins in general and vitamins A and D in particular have well-documented toxicity if ingested in excessive amounts for a length of time. Much larger intakes of water soluble vitamins and minerals have been advocated in public as well as professional press. At times these levels are up to 50 times RDA or even more. These recommendations are said to be for an "optimal" health or for prevention of infections like the common cold (Pauling 1970a,b). Beneficial effects of such large intakes of any nutrient have not been substantiated. The available data at present does not justify recommendation of larger intake of any nutrient. However, it is widely acknowledged that although vitamin C does not prevent or cure infections like the cold, it does significantly reduce the duration and severity of such infections. In this era of health consciousness, claims for the need and beneficial effects of vitamin supplements are often exaggerated. Until clear advantages of megadose intake of any nutrient is scientifically validated, the practice of self-medication with megadose intake of vitamins by many people is indeed disturbing and should be discouraged. It is erroneously assumed that water soluble vitamins in megadoses are safe. There are several reports that megadoses of water soluble vitamins could cause harmful side effects in some individuals (Alhadeff et al. 1984).

SUMMARY AND PROSPECTS

Most of the vitamins were discovered in the beginning of the twentieth century. In spite of intensive work, our knowledge of vitamins and their functions is still developing. We have a fairly good understanding of the metabolic steps that require most of the B complex vitamins. This group of vitamins function as coenzymes in a variety of reactions. Nevertheless that does not permit us to correlate their functions with their deficiency symptoms, indicating a gap in our understanding. This gap in our knowledge of vitamins is even wider for vitamins C, E, and A. Apart from the coenzyme functions of many vitamins, the antioxidant functions of vitamins C, E, A, and the carotenoids have recently been recognized. This important conceptual advance is primarily responsible for revitalization of the field of vitaminology. In addition, the role of oxidative molecules, including free radicals, in the pathogenesis of several chronic disorders has refocused attention on vitamin requirements. Future work will no doubt focus on interactions among oxidant molecules and antioxidant nutrients. Two other concepts receiving attention are (a) effects of prolonged suboptimal vitaminosis, and (b) localized vitamin deficiency in tissues. These two factors are very relevant to the pathogenesis of chronic disorders. We have historically looked at vitamins as micronutrients that eliminated certain acute deficiency symptoms. It is now time to ask if vitamins have roles beyond such elimination of acute deficiency symptoms in optimizing health and also in providing protection from many chronic ailments on a long-term basis.

References

Aleo, J.J. 1981. Diabetes and periodontal disease: Possible role of vitamin C deficiency. *J. Periodontol.* 52:251–4.

Alhadeff, L.C., Gualtieri, T.; and Lipton, M. 1984. Toxic effects of water soluble vitamins. *Nutrition Reviews* 42:33–40.

Anderson, R. 1984. The immunostimulatory, anti-inflammatory and anti-allergic properties of ascorbate. *Adv. Nutr. Res.* 6:19–45.

Anderson, R., and Theron, A.J. 1990. Physiological potential of ascorbate, β-carotene and α-tocopherol individually and in combination in the prevention of tissue-damage, carcinogenesis and immune dysfunction mediated by phagocyte-derived reactive oxidants. *World Rev. Nutr. Diet.* 62:27–58.

Carney, J.M.; Starke-Reed, P.E.; Oliver, C.N.; Landum, R.W.; Cheng, M.S.; Wu, J.F.; and Floyd, R.A. 1991. Reversal of age-related increase in brain protein oxidation, decrease in enzyme activity, and loss in temporal and spatial memory by chronic administration of the spin-trapping compound N-tert-butyl-α-phenylnitrone. *Proc. Natl. Acad. Sci. USA* 88:3633–36.

Chatterjee, I.B. 1978. Ascorbic acid metabolism. *World Rev. Nutr. Diet.* 30:69–87.

Cheng, L.; Cohen, M.; and Bhagavan, H.N. 1985. Vitamin C and the elderly. In *CRC Handbook of Nutrition in the Aged,* R.R. Watson. ed. Boca Raton, FL:CRC Press.

Clemetson, C.A.B. 1988. *Vitamin C.* vols. 1–3. Boca Raton: CRC Press.

Cody, M.M. 1991. Substances without vitamin status. In *Handbook of Vitamins.* 2d edition, L.J. Machlin, ed., pp. 565–582. New York: Dekker.

Diplock, A.T. 1985. *Fat Soluble Vitamins*. Lancaster, PA: Technomic Publishing.

Englard, S., and Seifter, S. 1986. The biochemical functions of ascorbic acid. *Ann. Rev. Nutr.* 6:365–406.

Frei, B.; England, L.; and Ames, B.N. 1989. Ascorbate is an outstanding antioxidant in human blood plasma. *Proc. Natl. Acad. Sci. (USA)* 86:6377–81.

Gaby, S.K., Bendich, A.; Singh, V.N.; and Machlin, L.J. 1991. *Vitamin Intake and Health:Scientific Review*. New York: Dekker.

Gerster, H. 1989. Antioxidant vitamins in cataract prevention. *Z. Ernahrungswissenschaft* 28:56–75.

Goodman, D.S. 1984. Vitamin A and retinoids in health and diseases. *N. Engl. J. Med.* 310:1023–31.

Goodwin, J.S.; Goodwin, J.M.; and Garry, P.J. 1983. Association between nutritional status and cognitive functioning in a healthy elderly population. *JAMA* 249:2917–21.

Halliwell, B., and Gutteridge, J.M.C. 1985. *Free-Radicals in Biology and Medicine*. Oxford: Clarendton.

———. 1990. Role of free radicals and catalytic metal ions in human disease: An overview. *Methods in Enzymology* 186:1–85.

Hemila, H.O. 1984. Nutrition need versus optimal intake. *Medical Hypothesis* 14: 135–9.

Hunt S.M., and Groff, J.L. 1990. *Advanced Nutrition and Human Metabolism*. St. Paul: West Publishing Company.

Johnson, J.E., Walford, R.; Harman, D.; and Miguel, J. 1986. *Free Radicals, Aging, and Degenerative Diseases*, pp. 1–588. New York: Alan R. Liss.

Kirsch, A., and Bidlack, W.R. 1987. Nutrition and the elderly: Vitamin status and efficacy of supplementation. *Nutrition* 3:305–14.

Kivirikko, K.I.; Myllyla, R.; and Pihlajaniemi, T. 1989. Protein hydroxylation: Prolyl 4-hydroxylase, an enzyme with four cosubstrates and a multifunctional subunit. *FASEB J.* 3:1609–17.

Lunec, J., and Blake, D.R. 1985. The determination of dehydroascorbic acid and ascorbic acid in the serum and synovial fluid of patients with rhematic arthritis. *Free Radical Res. Commun.* 1:31–39.

Machlin, L.J. 1991. *Handbook of Vitamins*. 2d ed. New York: Dekker.

Machlin, L., and Bendich, A. 1987. Free radical tissue damage: Protective role of antioxidant nutrients. *FASEB J.* 1:441–5.

Mascio, P.D.; Murphy, M.E.; and Sies, H. 1991. Antioxidant defence systems: Role of carotenoids, tocopherols, and thiols. *Am. J. Clin. Nutr.* 53:194S–200S.

McCollum, E.V. 1957. *A History of Nutrition*. Boston:Houghton Mifflin.

Newton, H.M.V.; Schorah, C.J.; Habibzadeh, N.; Morgan, D.B.; and Hullin, R.P. 1985. The cause and correlation of low blood vitamin C in elderly. *Am. J. Clin. Nutr.* 42:656–9.

NRC (National Research Council). 1989a. The Committee on Dietary Allowances. *Recommended Dietary Allowances*. 10th ed. Washington, D.C.:National Academic Press.

NRC (National Research Council). 1989b. *Diet and Health: Implications for Reducing Chronic Diseases Risk*. Washington, D.C.:National Academic Press.

Oliver, C.N.; Levine, R.L.; and Stadtman, E.R. 1987. A role of mixed function oxidation reactions in the accumulation of altered enzyme forms during aging. *J. Am. Geriatrics Soc.* 35:947–56.

Padh, H. 1990. Cellular functions of ascorbic acid. *Biochem. Cell Biol.* 68:1166–73.

———. 1991. Vitamin C: Newer insights into its biochemical functions. *Nutrition Reviews* 49:65–70.

Padh, H., and Aleo, J.J. 1987. Activation of serum complement leads to inhibition of ascorbic acid transport. *Proc. Soc. Exp. Biol. Med.* 185:153–7.

———. 1989. Ascorbic acid transport by 3T6 fibroblasts: Regulation by and purification of human serum complement factor. *J. Biol. Chem.* 264:6065–9.

Pauling, L. 1970a. *Vitamin C and Common Cold*. San Francisco:Freeman and Company.

———. 1970b. Evolution and the need for ascorbic acid. *Proc. Natl. Acad. Sci. USA* 67:1643–8.

Pecoraro, R.E., and Chen, M.S. 1987. Ascorbic acid metabolism in diabetes mellitus. *Ann. N.Y. Acad. Sci.* 498:248–58.

Pietrzik, K. 1985. Concept of borderline vitamin deficiencies. *Int. J. Vit. Nutr. Res.* (Suppl.) 27:61–73.

Robinson, C.H.; Lawler, M.R.; Chenoweth, W.L.; and Garwick, A.E. 1990. *Normal and Therapeutic Nutrition*. 17th ed. New York: Macmillan Publishing Company.

Schneider, E.L.; Vining, E.M.; Hadley, E.C.; and Farnham, S.A. 1986. Recommended dietary allowance for the health of the elderly. *N. Engl. J. Med.* 314:157–60.

Shils, M.E., and Young, V.R. 1988. *Modern Nutrition in Health and Diseases*. 7th ed. Philadelphia:Lea and Febiger.

Tannenbaum, S., and Wishnok, J.S. 1987. Inhibition of nitrosamine formation by ascorbic acid. *Ann. N.Y. Acad. Sci.* 498:354–63.

Tver, D.F., and Russell, P. 1989. *The Nutrition and Health Encyclopedia*. 2d ed. New York:Van Nostrand Reinhold.

VanderJagt, D.J.; Garry, P.J.; and Bhagavan, H.N. 1987. Ascorbic acid intake and plasma levels in healthy elderly people. *Am. J. Clin. Nutr.* 46:290–4.

Wittes, R.E. 1985. Vitamin C and cancer. *N. Engl. J. Med.* 312:178–9.

Ziegler, D.M. 1985. Role of reversible oxidation-reduction of enzyme thiols-disulfides in metabolic regulation. *Ann. Rev. Biochem.* 54:305–29.

Chapter 14

Lactic Acid Bacteria as Promoters of Human Health

Mary Ellen Sanders

INTRODUCTION

Great interest in the healthful effects of lactic acid bacteria exists in the research, commercial, and consuming communities. The designation "lactic acid bacteria" applies to a functional grouping of non-pathogenic, gram positive bacteria that have lactic acid as a primary metabolic end-product and are traditionally used in food fermentations. These bacteria include species of *Lactococcus, Lactobacillus, Streptococcus, Leuconostoc,* and *Pediococcus*. *Bifidobacterium* and *Enterococcus* species are also included in this group when addressing issues related to bacterial inoculants for health promotion. For some health-promoting applications, lactic acid bacteria of intestinal origin are specifically targeted. These include *Lactobacillus species of intestinal origin*, all species of *Bifidobacterium*, and *Enterococcus faecalis*. Research is sometimes conducted on the bacterial strains themselves, or sometimes on the food products that harbor these bacteria. Normally, the food products tested are milk products, including fermented milks such as yogurt and buttermilk, and unfermented milks with culture added. Recent reviews have been published covering this field in general (Alm 1991; Goldin 1989, 1990; Gorbach 1989, 1990; Bourlioux and Pochart 1988; de Simone et al. 1989; Fernandes, Shahani, and Amer 1987; Friend and Shahani 1984; Fuller 1991; Gurr 1984; Kim 1988; Kroger, Kurmann, and Rasic 1989; Lemonnier 1984; Rasic 1983; Renner 1991; Sellars 1989, 1991) or specifically focusing on lactose intolerance (Savaiano and Kotz 1988; Savaiano and Levitt 1987), gastrointestinal microbiology (Bianchi-Salvadori 1986; Brown 1977; Gorbach 1971, 1986; Savage 1977, 1983a, b; Tannock 1983, 1984, 1990), antimicrobial activities (Fernandes and Shahani 1989), anticarcinogenic properties (Fernandes and Shahani 1990), vaginal health (Redondo-Lopez, Cook, and Sobel 1990), Bifidobacteria (Bezkorovainy and Miller-Catchpole 1989; Hughes and Hoover 1991; Kurmann and Rasic 1991; Laroia and Martin 1990; Misra and Kuila 1991; Mitsuoka 1989, 1990; Modler, McKellar, and Yaguchi 1990; Poupard, Husain, and Norris 1973; Rasic and Kurmann 1983), *L. acidophilus* (Fonden 1989; Truslow 1986), and strain selection for dietary adjuncts (Gorbach and Goldin 1989; Klaenhammer 1982).

It becomes obvious when reviewing this field that a broad range of opinions exists among experts as to the general effectiveness of lactic cultures in promoting human health. This range of opinions comes from various levels of stringency in reviewing the published literature. In this review, emphasis is placed on human over animal studies, on *in vivo* over *in vitro* studies, and on statistical significance over trends. When considering the large number of publications available in this field, it is apparent that this research area has suffered from lack of coordinated efforts between the clinicians and the microbiologists, and that differences in strains, levels, model systems, and stringency of data interpretation lead to apparent inconsistencies in conclusions from published research.

Although research support is lacking for many claims of culture-induced promotion of human health, products containing lactic cultures are marketed as health-promoting worldwide. In many cases, the lactic cultures present in the foods provide many functional (acid production, flavor enhancement, textural improvements in a fermented product) and nutritional (ease of digestibility, improved availability of some nutrients) advantages. Some products also carry a complement of intestinal lactic bacteria that are added solely for their stated positive influence on human health. Although lactic cultures of intestinal origin have not been used widely as fermentative cultures, this trend appears to be reversing (Driessen and de Boer 1989), requiring these cultures to have functional properties beyond their stated ability to promote health. This review will examine some of the literature that supports or negates lactic cultures as promoters of human health.

HEALTH EFFECTS

Lactose Digestion

A large portion of the world's adult population is deficient in the intestinal lactose-digesting enzyme, lactase. Failure to digest lactose leads to its presence in the large intestine, providing a fermentation substrate for colonic flora. Symptoms include bloating, flatulence, gassiness, abdominal pain and diarrhea. Symptoms vary based on the amount of lactose consumed, the degree of lactase insufficiency, adaptation of the colonic flora, and the presence of alternative sources of lactase. The condition of lactose intolerance can be physiologically determined in a subject by providing a measured dose of lactose with subsequent monitoring of hydrogen (a biproduct of colonic fermentation of lactose) excreted in the breath. This objective measurement has provided a method for analysis of the effects of yogurt and lactic cultures on at least one symptom of lactose intolerance.

Recent human studies (summarized in Table 17-1) have provided convincing evidence that milk or milk products containing significant lactose loads can be consumed without symptoms by lactose intolerant individuals if high levels of certain starter cultures are present.

Yogurt repeatedly has been shown to significantly decrease breath hydrogen excretion in human clinical studies. This is often accompanied by a decrease in lactose intolerance symptoms. Results with *L. acidophilus* cultures are not as generally positive, although laboratory-dependent results have been reported. Unfermented acidophilus milk made with sonicated cells performed better than the regular product (McDonough et al. 1987). Sonication-induced permeabilization of the *L. acidophilus* cells likely improves the ability of intracellular bacterial lactase to reach lactose in the intestine. Strains that are naturally permeabilized by bile acids in the intestine (*L. bulgaricus* or *S. thermophilus*) can be effective without sonication. It has been shown that heat-treated yogurt is not as effective as unheated

Table 14-1. Summary of clinical experiments with human subjects on control of symptoms of lactose intolerance by lactic cultures. LI = intolerant; UYM = unfermented milk containing yogurt cultures UAM = unfermented *L. acidophilus* milk; + = positive result; − = negative result. All subjects consumed each of the meals. Breath hydrogen reduction was measured as changes in peak levels or cumulative hydrogen excretion and measurement was compared to milk or lactose dose.

Reference	No. LI Subjects	Products/Culture Tested	Significant Breath Hydrogen Reduction
Kim and Gilliland (1983)	12	UAM	+ UAM
Kolars et al. (1984)	10	yogurt	+ yogurt
Kolars et al. (1984)	10	yogurt lactulose lactose	+ yogurt
Savaiano et al. (1984)	9	yogurt buttermilk UAM heated yogurt	+ yogurt − buttermilk − UAM − milk − heated yogurt
McDonough et al. (1987)	14	yogurt heated yogurt UAM sonicated UAM	+ yogurt +/− heated yogurt
Wytock and DiPalma (1988)	8	3 yogurts	+ 2 yogurts − 1 yogurt
Lin, Savaiano, and Harlander (1991)	10	UYM UAM (3 strains tested) (10^7 or 10^8 bacteria/ml)	− (UYM 10^7) + (UYM 10^8) − (UAM 10^7) + (UAM 10^8 with strain LA-1 only)
Martini, Kukielka, and Savaiano (1991)	21	fermented milks with single strains; 7 yogurts	+ yogurts + ST yogurt + LB yogurt − LA yogurt − bifid yogurt
Martini et al. (1991)	22	yogurt yogurt + additional lactose	+ yogurt + yogurt and meal − yogurt and lactose

yogurt (McDonough et al. 1987; Savaiano et al. 1984), but it was unclear if this was due to a requirement for cellular viability or a heat-inactivation of lactase. Some decrease in symptoms was found when subjects consumed heated yogurt, even though breath hydrogen excretion was similar to patterns seen with milk consumption (Savaiano et al. 1984). McDonough et al. (1987), however, found heated yogurt to be somewhat effective at decreasing breath hydrogen excretion. Consumption of yogurt along with a meal still promoted lactose digestion, but yogurt did not protect against dietary lactose in excess of that present in yogurt. Levels of cells are important, with populations of about 10^8/ml of milk product needed for efficacy of yogurt cultures. Selection of strains with high levels of stable lactase may allow lower levels of culture to be more effective. However, lactase levels measured *in vitro* in dairy products may not provide a perfect correlation with *in vivo* lactose digestion: a variety of yogurts tested by Martini, Kukielka, and Savaiano (1991) showed a range of lactase levels, but were equivalent at digesting lactose *in vivo*. Milks fermented with single strains of *L. acidophilus* and *Bifidobacterium* did not reduce breath hydrogen as well as milks fermented with *L. bulgaricus* or *S. thermophilus*. Best performance was seen with milks fermented with both yogurt strains. Lower cell counts seen with the single strain fermented products (except for *Bifidobacterium*) perhaps accounted for the lower activity. Surveys of commercially available yogurts have shown that two of three (Wytock and DiPalma 1988) and seven of seven (Martini, Kukielka, and Savaiano 1991) decreased breath hydrogen excretion significantly, suggesting that most yogurts sufficiently protect against the lactose they carry. Formulation of a nonfermented culture-containing milk product, however, would require much higher levels and more appropriate strains than found in typical U.S. products containing 2 to 4×10^6/ml *L. acidophilus*.

The clinical evaluations in humans using objective criteria such as breath hydrogen have provided significant evidence that lactose intolerance symptoms can be avoided with the consumption of culture-containing dairy products. High levels of viable bacteria (5 to 10×10^7/ml) were necessary, and strains of *S. thermophilus* and *L. bulgaricus* were most efficacious in this application. Research suggests that these bacteria can be more effective than some more typical intestinal (such as *L. acidophilus* or *Bifidobacterium*) or fermentative (lactococci) bacteria in this role. Lactococci only digest phosphorylated lactose and bile-tolerant intestinal bacteria are not permeabilized in the intestine, shielding intracellular lactase from intestinal use. Since milk and yogurt products buffer stomach acid, ingested bacteria survive passage to the more distal portions of the gastrointestinal system. Clearly, fermented dairy products or culture-containing fluid milk products formulated with suitable populations of *S. thermophilus* and\or *L. bulgaricus* can be consumed by lactose intolerant people.

Diarrhea

The effect of lactic cultures on diarrhea in humans has been difficult to determine experimentally. This is largely due to the tremendous range of variables associated with diarrheal illnesses, including cause of diarrhea (viral, bacterial, antibiotic-induced), severity of illness, the sub-population affected, the site of infection (small versus large bowel), and specifics regarding the treatment used (which cultures at what levels). However, since 1980 there have been at least 13 clinical studies conducted on the effectiveness of lactic cultures in decreasing the symptoms or duration of diarrheal illnesses. These studies, along with other older studies, are summarized in Table 14-2.

Among these studies, a few were conducted on a sufficient number of subjects, with proper controls and statistical analysis of results, to consider their results convincing. Clements et al. (1983) examined neomycin-induced diarrhea in adults being treated for enterotoxigenic *Escherichia coli* infection. A granular preparation of 10^8 to 10^9 lactobacilli (Lactinex) was administered in milk four times per day after neomycin treatment commenced. One lot of Lactinex, administered to 20 patients showed a 50% lower attack rate (four cases) than that for 19 patients treated with a placebo (eight cases). The volume and number of diarrheal stools in the Lactinex group was also significantly less than in the placebo group. However, a second lot of Lactinex administered to 16 patients showed no protective effect. This lot-to-lot variability casts some doubt over the positive nature of these results. Isolauri et al. (1991) studied the ability of *Lactobacillus* GG to promote recovery of children from acute viral diarrhea. Diarrhea in children consuming either dried GG preparation or a milk fermented with GG was one day shorter than in children consuming pasteurized yogurt. Bellomo et al. (1980) studied the effect of *Enterococcus faecium* (SF68) on children suffering from diarrhea. The cause of diarrhea was not specified except to state that it appeared to be a sequela of respiratory infection or an-

Table 14-2. Summary of clinical experiments with human subjects on control of diarrhea by lactic cultures. LA = *Lactobacillus acidophilus*; LB = *Lactobacillus bulgaricus*; ST = *Streptococcus thermophilus*; ? = insufficient information to determine if results were truly positive. Number of subjects reported do not include control subjects.

Reference	Culture	Subject No./ Descript.	Diarrhea Cause	Results Reported	Comments	Overall Effect
Beck and Necheles (1961)	LA	59 adults	Various abdominal symptoms	+ ↓ symptoms	No controls	?
Pearce and Hamilton (1974)	Dried ST, LA and LB	94 children <3 yrs.	>10 mo. acute onset	–		–
Pozo-Olano et al. (1978)	Lactobacilli	26 healthy adults	Traveller's diarrhea	–	Low incidence of diarrhea	–
Bellomo et al. (1980)	Enterococcus faecium	53 children	Diarrhea of various causes	+ ↓ duration		+
Clements, Levine, and Black (1981)	Lactinex lactobacilli	23 healthy adults	ETEC	–		–
Clements et al. (1983)	Lactinex lactobacilli	23 healthy adults	ETEC	–		–

Clements et al. (1983)	Lactinex lactobacilli	36 adults with ETEC infection	Neomycine	+ ↓ symptoms	Lot-to-lot culture variability	+
Newcomer et al. (1983)	Sweet acidophilus-milk	61 adults	Irritable bowel syndrome	–		–
Colombel et al. (1987)	*B. longum*	10 healthy adults	Ery	+ ↓ symptoms	Small no. subjects	?
Gorbach, Chang, and Goldin (1987)	GG	5 adults with relapsing colitis	C. difficile colitis	+	No controls; small no. subjects	?
Hotta et al. (1987)	*B. breve B. bifidum* and/or *L. casei*	15 children	Intractable	+ ↓ duration	No controls	?
Salminen et al. (1988)	LA	11 adult female radiation therapy patients	Irradiation	+ ↓ incidence	Not blinded; subjective reporting	?
Siitonen et al. (1990)	GG	8 healthy adults	Ery	+ ↓ duration, symptoms	No statistics, small no. subjects	?
Oksanen et al. (1990)	GG	349 healthy adults	Traveller's diarrhea	+ ↓ incidence		+
Isolauri et al. (1991)	GG	47 children	Acute viral	+ ↓ duration		+

tibiotic treatment. Culture was administered in gelatin capsules containing 4×10^7 SF68 or 5×10^8 *Lactobacillus acidophilus*, 5×10^8 *Lactobacillus bulgaricus* and 4×10^9 *Lactococcus lactis*. SF68 and the multiculture blend were administered to 53 and 51 patients, respectively. After two days of treatment, 62% of the SF68-treated patients recovered as compared to 35% of the multi-culture-treated group. It is interesting that treatment with a multi-culture blend was chosen as a control. The positive effect of the SF68 may have been even stronger compared to a control where no viable cultures were used.

Oksanen et al. (1990) tested the effect of *Lactobacillus* GG on prevention of traveller's diarrhea. 756 people (10 to 80 years old) traveling from Finland to two destinations in southern Turkey were given either a placebo or freeze-dried GG in identical packets. The daily dose of GG was about 2×10^9 bacteria. A questionnaire recording incidence of diarrhea and related symptoms was filled out on the return trip by the participants. A high level of diarrhea incidence was found in the study (44% overall). Overall, of the 331 travelers with diarrhea, 178 of 383 subjects (46.5%) in the placebo group and 153 of 349 subjects (43.8%) in the GG group exhibited diarrhea. The differences were statistically significant. A greater protective effect was seen for patients traveling to a certain one of the two destinations and for those staying two weeks instead of one week. No protective effect was seen for travelers to the second destination. Although the protective effect was statistically significant, the positive effect was not dramatic, increasing the probability of staying well from about 60% to 80% for a one-week stay in one of the locations. These results were generated in a sufficient sample size so that they should be considered legitimate.

Many more studies claim positive results. Studies conducted by Colombel et al. (1987), Siitonen et al. (1990), and Gorbach, Chang, and Goldin (1987) were conducted with such a small number of subjects that the results are suggestive, but cannot be considered conclusive. Studies by Beck and Necheles (1961), Salminen et al. (1988), and Hotta et al. (1987) suffer from poor experimental design or data analysis. Suggestive positive results were obtained by Gorbach, Chang, and Goldin (1987) who studied treatment of five children with relapsing *Clostridium difficile* colitis. The patients had two to five relapses over a two- to ten-month period. Treatment for seven to ten days with 10^{10} viable GG bacteria resulted in immediate results, with only one relapse in the four-month to four-year follow up.

The uncontrolled study by Hotta et al. (1987) was interesting in that it tested the effect of *Bifidobacterium* or bifidus yogurt on pediatric intractable diarrhea (one to ten weeks in duration). Improvement was shown in all 15 patients within three to seven days of treatment.

Although some very positive effects of lactic cultures on the course of human diarrheal illness have been found, there is still not a preponderance of evidence to support a general beneficial effect. Positive effects are seen in certain subpopulations with certain causes of diarrhea. As continued clinical research focuses on the effects of high levels of appropriately selected strains on patients with known causes of diarrheal illnesses, it appears likely that the beneficial effects of lactic cultures in this regard will become clear.

Cholesterol Reduction

Mechanisms of a mediating effect of viable cultures on cholesterol levels have been proposed. Some bacteria harbor enzymes that are capable of modifying cholesterol. *Eubacterium* can anaerobically reduce cholesterol to coprostanol, while *Rhodococcus equi* can elicit cholesterol oxidation (Smith, Sullivan, and Goodman 1991). Although similar enzymatic capabilities have not been elucidated in lactic acid bacteria, some have been shown to metabolize bile. Bacterial enzymes playing a role in cholesterol assimilation *in vivo* would need to do so early in the small intestine. However, the sparse colonization of this area and its rapid emptying precludes much bacterial enzyme activity. Alternatively, physical association of cholesterol with the bacterial cells could enable the cells to act like a sponge, carrying the cholesterol out of the system. It is unlikely that bacteria harbor specific cholesterol-binding properties, and physiological conditions would favor solubility of such compounds through the small intestine. Bacterial activity in the large intestine would be insignificant, because cholesterol reaching this point is destined for excretion. Finally, lactic acid bacteria may produce a metabolite during growth that can affect cholesterol metabolism in humans. Research identifying such metabolic factors is inconclusive.

Although no evidence supports one mechanism of action, several studies have approached this issue empirically. These human, clinical studies have attempted to determine if a hypocholesterolemic effect is exerted by cultures or fermented dairy products; they are summarized in Table 14-3. There is no consensus in these few hu-

Table 14-3. Summary of human clinical studies on the effect of lactic cultures on serum cholesterol. TC = total cholesterol; T = triglycerides; HDL = high density lipids; LDL = low density lipids; ↓ = significant (at least 95% confidence) decrease; ↑ = significant (at least 95% confidence) increase; ♂ = in male subjects; ♀ = in female subjects; + = positive results; − = negative results. Number of subjects does not include control subjects.

Reference	Culture or Food Consumed	Subject No./Descript.	Analysis	Overall Results and Comments
Hepner et al. (1979)	Yogurt 4 and 12 weeks	9 + 11 ≥200 mg/dl serum cholesterol	TC, T	+ ↓TC 5%; viable cultures not necessary for effect
Rossouw et al. (1981)	yogurt 2L/day	11 boys, 16–18 yrs	TC, HDL, LDL	−
Thompson et al. (1982)	Sweet acidophilus milk, buttermilk, yogurt	12 + 11 + 13 healthy adults	TC, LDL, HDL	−
Bazzarre, Liu, Wu, and Yuhas (1983)	24 oz yogurt/day for 1 wk	21 healthy adults (18–30 yrs)	TC, HDL	+ ↓TC and ↑HDL in ♀; no effect in ♂
Jaspers, Massey, and Luedecke (1984)	Yogurt 681g/day, 14, 14, and 21 days	10 healthy adult males	TC, HDL, LDL	− Temporary effect ≤ 14 days in TC
Lin et al. (1989)	Lactinex (*L. acidophilus*, *L. bulgaricus*)	334 healthy adults	HDL, LDL, TC	−
Halpern et al. (1991)	450 g/day yogurt for 4 mo.	24 healthy adults (20–40 yrs)	TC, T, HDL, LDL	− No effect

man clinical studies to support a substantive positive effect for humans.

Several attempts have been made to determine the cholesterol-assimilating activity of different lactic cultures *in vitro* (Gilliland, Nelson, and Maxwell 1985; Gilliland and Walker 1990; Lin et al. 1989; Rasic et al. 1992; Danielson et al. 1989). Generally, results were generated by mixing culture and a water soluble cholesterol derivative in a growth medium containing bile and determining residual cholesterol after removal of cells by centrifugation. Although some strain differentiation was reported, no evidence was provided for specificity or tenacity of cholesterol binding under a variety of conditions. Binding due to physical association could be expected to be very different in a physiological environment than in a water based nutrient medium. The physiological significance of these results is not clear, although Gilliland, Nelson, and Maxwell (1985) did provide evidence that a strain selected *in vitro* for cholesterol assimilation performed better than a nonassimilating strain at inhibiting increases in serum cholesterol levels of pigs fed a high-cholesterol diet.

In evaluating this area of research, two questions emerge. Can consumption of lactic cultures result in a reduction of serum cholesterol levels in humans, and if so, will the level of reduction provide a substantial health benefit to the consumer? The link between dietary cholesterol and a reduced risk of morbidity or mortality caused by cardiovascular disease is still a matter for debate (Ahrens 1985). Clearly, this issue should be resolved before significant effort is placed in selecting or conducting clinical studies to determine if lactic cultures or their metabolites can be efficacious in this regard.

Limitations of some published human clinical studies include: failure to monitor or dictate total diet; choice of subjects with normal cholesterol levels instead of targeting populations who are recommended to control dietary cholesterol intake; and no discussion of substantive as opposed to statistical significance of results. *In vitro* and animal studies often do not adequately consider the physiological differences between these models and the target human population where an effect must be exerted.

Cancer Suppression

Drasar and Hill (1972) advanced a theory that the ability of intestinal bacteria to alter man's intestinal chemical environment could enable them to be a significant factor in the evolution of carcinogenesis.

Toward this end, convincing evidence generated in human subjects shows that some lactic cultures can alter the activity of some fecal enzymes that are thought to play a role in the development of colon cancer. These enzymes include β-glucuronidase, azoreductase, and nitroreductase. How the cultures bring about this change is speculative. However, it is likely that lactic cultures able to survive in the intestine produce organic acids that lower the pH of the bowel and its contents. This could effectively change the environment leading to altered metabolic activity of other microbial residents. This effect is postulated to have a general enhancing effect on the well-being of the intestine by preventing the growth of putrefactive bacteria (Modler, McKellar, and Yaguchi 1990), although human research supporting this claim in a general sense is deficient. Whether or not some lactic acid bacteria also produce specific bacteriocin-like compounds with a bacteriocidal effect *in vivo* against other intestinal bacteria is not known. To date, research suggests that these proteinaceous compounds are produced by many different lactobacilli, but their spectrum of activity, at least *in vitro*, is primarily limited to closely related bacteria (Klaenhammer 1988). Controlled experiments conducted in humans contrasting the effect of a bacteriocin-producing strain with a nonproducing mutant would go a long way toward explaining the role of these compounds *in vivo*.

Six studies (Table 14-4) have examined the effect of consumption of certain lactic cultures on fecal enzymes. The most positive results have been generated on nitroreductase and β-glucuronidase levels, both being significantly reduced in four studies of feeding *Lactobacillus*. Evidence on azoreductase suggests that this enzyme is not significantly affected by *L. acidophilus* feeding. Other enzymes have been insufficiently studied. These studies have been supported by similar positive results in rats (Goldin and Gorbach 1977; Goldin and Gorbach 1980) and mice (McConnell and Tannock 1991). Although there appears to be a clear role that lactobacilli can play in decreasing the activity of some enzymes harbored by intestinal bacteria, it must be emphasized that the exact role of these enzymes in the development of cancer in humans is not known. There is a need for epidemiological studies that may correlate consumption of acidophilus-containing products with incidence of colon cancer.

One epidemiological study focused on the role of fermented milk product (yogurt, buttermilk, curds, and kefir) consumption on the incidence of breast cancer in the Netherlands (van't Veer et al. 1989). The authors postulated that alteration of intestinal bacterial enzyme activity could influence the formation or withdrawal of estrogenic

Table 14-4. Summary of the effects of oral consumption of lactic cultures on fecal enzyme activity in humans, + = statistically significant positive results; − = negative results; ? = results not definitive.

Reference	Bacteria and Daily Dose	Reduction of Fecal Enzyme Activity
Ayebo, Angelo, and Shahani (1980)	*L. acidophilus* DDS1 in milk (4×10^9) (12 subjects)	? β-glucuronidase ? β-glucosidase (no statistical analysis)
Goldin, et al. (1980)	*L. acidophilus* (4×10^{10}) (7 subjects)	+ nitroreductase − azoreductase + β-glucuronidase − steroid 7-α-dehydroxylase
Goldin and Gorbach (1984a)	*L. acidophilus* NCFM (10^{10}) (7 subjects)	+ nitroreductase − azoreductase + β-glucuronidase
Goldin and Gorbach (1984b)	*L. acidophilus* NCFM (10^9) *L. acidophilus* N-2 (10^9) (22 subjects)	+ nitroreductase + azoreductase + β-glucuronidase
Marteau et al. (1990)	Milk fermented with *L. acidophilus* (10^9), *B. bifidum* (10^{10}), and mesophilic cultures (10^{10}) (9 subjects)	+nitroreductase −azoreductase −β-glucuronidase +β-glucosidase
Goldin et al. (1992)	*Lactobacillus* GG frozen concentrate-10^{10} (8 subjects)	+ β-glucuronidase

compounds or could alter bile metabolism. A statistically lower rate of fermented dairy product consumption was seen in 133 breast cancer cases than in 289 population controls. Data were based on recall of diet 12 months prior to diagnosis. Information was collected by interview three to six months after cancer treatment. Controls were asked to recall their diet during the 12 months prior to the interview. Although significant results were found, this study assumes that dietary recall is an accurate means of recording the diet. Furthermore, breast cancer is a slowly developing cancer; thus, dietary practices many years prior to diagnosis are likely to be as important or more important than those 12 months prior to diagnosis. This study could be defended if fermented milk product consumption is a lifetime pattern. However, this should be established experimentally before making conclusions.

In addition to these human studies, much research on the anticancer effect of lactic cultures has been published using animal or *in vitro* models. This research includes *in vitro* studies on the inhibition of mutagen activity by *L. acidophilus*, *B. bifidum*, and other nonintestinal lactic cultures (Hosono, Kashina, and Kada 1986; McGroatry, Hawthorn, and Reid 1988; Zhang and Ohta 1991a, b; Zhang, Ohta, and Hosono 1990), survival of human bladder tumor cells *in vitro* (McGroatry, Hawthorn, and Reid 1988) and animal studies on tumor suppression or incidence by *L. casei* (Kato et al. 1981;), *Bifidobacterium* species (Kohwi et al. 1978; Koo and Rao 1991; Sekine et al. 1985), *L. bulgaricus* (Bogdanov et al. 1975; Shackelford et al. 1983), and *L. acidophilus* (Goldin and Gorbach 1980). Results generated in studies show a decrease of mutagenicity, improved animal survival, or a decrease in tumors by treatment with viable cells, cell components, cell extracts, or fermented milks.

Care must be taken in extrapolation of this type of result to a general statement of an anticancer effect. First, these experiments are not performed with humans, and as such, cannot be directly applied to humans. Second, differentiation is not often made between an antitumor activity and an adjuvant activity, especially in certain protected animal models. Third, specificity of action is not resolved. Many microbes could probably elicit a similar effect. Antitumor activities have been reported for cells or cell components of *Corynebacterium parvum*, *Streptococcus pyogenes*, *Vibrio anguillarum*, *Listeria monocytogenes*, *Bacillus Calmette-Guerin* (BCG), and *Brucella abortus* since the 1970s (as referenced in Fernandes and Shahani 1990; McGroatry, Hawthorn, and Reid 1988). However, this approach to cancer treatment has not gained acceptance by the medical community. Fourth, the method of testing frequently involves intravenous, intraperitoneal, or intralesional injection of lactic cultures, not oral consumption. Therefore, these results cannot be applied to a dietary component. The difficulty of conducting experiments on the effect of lactic cultures on humans has created a need to find legitimate animal or *in vitro* systems to develop a basis for experimentation and conclusion. Responsible interpretation should recognize the deficiencies of these models as predictors of human responses. Further research in this area is clearly necessary.

Immune System Stimulation

A recent active area of research is the investigation of the effect of lactic cultures on immune system stimulation. With the threatening foothold of immune system deficiency diseases in human populations, the thought that dietary modifications could improve immune system functioning is very attractive. Research on immune system stimulation has focused on: (a) response of cell cultures to exposure to lactic bacteria or their cellular components, (b) response of mice to intraperitoneal injection of lactic bacteria or their cellular components, (c) response of mice to oral administration of lactic cultures or fermented milks, and (d) human feeding studies. Response assays in this research are broad based, including detection of specific immune system components, ability of animals to defend against bacteria infection, and antitumor effects. This discussion will focus on *in vivo* experiments for the following reasons. Research using tissue culture models, germ-free animals, or animals injected intravenously with live bacteria or bacterial components has shown impressive effects, but its significance is unclear. It is difficult to determine the human physiological significance of results from *in vitro* assays. A colonized gastrointestinal system is known to stimulate the immune system through the GALT (gut-associated lymphatic tissue). Because germ-free animals would not benefit from this exposure, results may be exaggerated in this model. Experiments conducted with injection of bacteria may not focus adequately on the means of effect that would be relevant in a food product.

Perdigon et al. (1989) determined the levels of macrophages and polymorphonucleases (as indicators of nonspecific immune response) and B and T lymphocytes (as indicators of humoral immune response) in mice fed yogurt or 10^9/day *L. casei, L. bulgaricus, L. acidophilus*, or *S. thermophilus*. All lactic cultures increased cell-mediated immunity; *L. casei* increased humoral immunity. *S. thermophilus* and *L. casei* inhibited colonization of the mouse liver and spleen by *Salmonella typhimurium*. Yogurt was effective, but only against low levels of pathogens.

Perdigon et al. (1986b) evaluated the effect of peroral (p.o.) and intraperitoneal (i.p.) injection of *L. casei* and *L. bulgaricus* on macrophage activation (as indicated by lysosomal enzyme activity, phagocytosis, and carbon clearance capability) in mice. In general, i.p. administration resulted in greater macrophage activation, but the study demonstrated that macrophages were activated by both

L. casei and *L. bulgaricus* administered p.o. or i.p. Using the same assays, Perdigon et al. (1987) determined the effect of *S. thermophilus* and *L. acidophilus* on the macrophage and lymphocyte function of mice. Enzymatic activities of peritoneal macrophages showed some increases in mice fed *L. acidophilus* as compared to the control mice, depending on enzyme assayed and day of sampling. Results with *S. thermophilus* were less significant. Both *S. thermophilus* and *L. acidophilus* cleared colloidal carbon significantly better than controls. These results show a positive contribution of consumption of lactic cultures to immune system stimulation in mice.

Kitazawa et al. (1991) induced macrophage stimulation by a slime-forming *Lactococcus lactis* ssp. *cremoris* strain. Numbers of peritoneal macrophages and cytotoxicity of the macrophages against Sarcoma-180 tumor cells was the functional assay used in the study. Numbers of peritoneal macrophages were stimulated initially, but returned to control levels by seven days after i.p. injection of the strain. *Lactococcus lactis* ssp. *cremoris* also substantially augmented the cytotoxic activity of peritoneal macrophages against tumor cells.

Perdigon et al. (1986a) elicited stimulated levels of macrophage and lymphocyte activity in mice fed unfermented milk containing *L. acidophilus* ATCC 4356 and *L. casei* CRL 431. The authors advocate the use of mixed cultures to elicit the strongest stimulation of host immune response. A similar conclusion was obtained from a study by the same research group in which *L. casei* and *L. acidophilus* were fed to mice infected with a lethal dose of *S. typhimurium* (Perdigon et al. 1990). Pretreatment of mice by feeding a mixture of the two lactic bacteria resulted in 100% survival after 21 days. Treatment with either culture alone slowed the rate of death, but after 21 days, survival matched that of the control group (20%). Liver and spleen colonization with *Salmonella* was inhibited in the group fed fermented milk. However, mice fed individually fermented milks showed a more rapid colonization of liver and spleen by *Salmonella* (two to three days for the mice fed *L. casei* or *L. acidophilus* fermented milks and seven days for the control group). When mice were fed the milk fermented with both lactic cultures (although the rate of colonization was faster by one day), the *Salmonella* were completely cleared by day seven, whereas in the control group approximately 10^5 *Salmonella* remained per organ. Significantly higher levels of circulating antibodies were found in mice fed the mixed cultured milk than individual or control mice. This study shows a significant effect from combined culture feeding.

A study in humans found an interesting result (Halpern et al. 1991). A rise in lymphocyte γ-interferon production was seen in young human adults consuming two cups of yogurt daily. The purpose of this paper was to determine if high levels of yogurt consumption (450 g/day for four months) showed a detrimental health effect on 24 young (20–40 years) adults as determined by extensive blood analysis. One control group (N = 24) consumed heat-treated yogurt; a second control group (N = 20) consumed no yogurt. The experiment lasted four months. Analysis included blood count (platelets, white and red blood cell, hamoglobin, and differential) and blood chemistry (urea nitrogen, creatinine, sodium, potassium, chloride, CO_2, anion gap, calcium, phosphorus, uric acid, ionized calcium, iron, total protein, albumin, globulin, cholesterol, alkaline phosphatase, GGT, and AST). In addition, γ interferon was determined in sera samples and in peripheral blood lymphocytes at some sampling times. No changes were seen in blood count. Blood chemistry showed higher ionized calcium in subjects consuming heated or unheated yogurt. Interferon was not detected in the sera of any subjects at any time. At the last sampling period, (120 days) subjects were also tested for lymphocyte γ-interferon production. Interferon production was markedly higher in lymphocytes obtained from subjects who consumed live active culture yogurt (20.9 IU/ml) than in the heated yogurt group (1.7 IU/ml) or control group (5.2 IU/ml). This observation suggests that yogurt consumption may be associated with an improved immune defense. Although the mechanism for this effect is unknown, binding of lactic acid bacteria to human peripheral lymphocytes may mediate the effect.

Yamazaki et al. (1985) followed the development of the immune responses in mice during monoassociation with *Bifidobacterium*. Low levels (10^2–10^4 colony forming units (cfu) per organ) were found in the liver, kidneys, and mesenteric lymph nodes through week two after bifidobacteria feeding, but not after four weeks. The humoral antibody response did not appear responsible for this activity, whereas cell-mediated immunity correlated well with the cessation of translocation.

The intestinal mucosa provides an effective barrier against the penetration of microbes to internal organs. Although the nature of this barrier is not well understood, translocation is more common in animals raised germ free. The majority of infectious diseases encountered by man are contracted from the large surface areas of the mucosal membranes. This may explain the large concentration of immunocompetent cells associated with these tissues. This has prompted de Simone et al. (1989) to offer the hypothesis that the exposure of the intestinal lining to the bacteria contained in fermented foods may be able to induce a local immunomodulation. In a review article, Tannock (1984) provides a similar speculation with regard to the normal intestinal flora as a whole. He suggests that flora antigens prime the immunological tissues of the host so that a degree of nonspecific resistance toward infection is produced, although direct evidence in humans is not cited. Any protective effect, however, can be overcome by the entry of large numbers of the pathogen. The critical question remains as to the extent of stimulation that could be expected when a person with a fully colonized immune system consumes lactic cultures.

The effect of addition of large numbers of immune system stimulating microbes to the intestinal system is not known. Immune stimulation is not necessarily a good occurrence. A primed immune system may be advantageous, but a fully active immune system may result in autoimmune disease. The importance of this effect on certain subpopulations may be a key to the success of this research. The elderly and those with certain diseases exist in an immunosuppressed state, and may benefit from system stimulation. There is a need for focused research that will provide a mechanism for action and determine the specificity of the effect.

Constipation

A few studies have attempted to address the effect of lactic cultures on constipation. Graf (1983) studied 25 adult outpatients with moderately severe constipation. Patients were given either milk fermented by mesophilic lactococci or acidophilus for four weeks each. Patients recorded abdomen and bowel behavior (cramps, flatulence, borborygmus, character of the stool). Patients also timed their bathroom stays, allowing calculation of mean time of passing stools. No statistical differences were seen in evaluation of any of the numerical results. This study shows no difference between acidophilus and lactococci fermented milk with regard to alleviating constipation. Unfortunately, no baseline information was provided, so that no comment can be made on the effect of fermented milks as compared to unfermented milk on constipation.

Constipation affects a majority of hospitalized geriatric patients, many of whom result to medication for symptom relief. To address this issue, Alm et al. (1983) studied 42 hospitalized geriatric patients

(36 women and 6 men aged 50–95) with regard to the effect of acidophilus milk on constipation. All patients had used several types of laxatives regularly for a long time. Most had moderate to severe mobility limitations. These subjects were divided into two groups. The experimental period for Group 1 was 56 days and for Group 2, 41 days, with intermittent workout periods of 36 and 40 days, respectively. Fermented acidophilus milk was given *ad libitum* at breakfast during the experimental period. The experimental protocol was crossover, so that patients served as their own control. The results showed that consumption of 200–300 ml acidophilus milk per day reduced the need for laxatives in constipated, hospitalized elderly patients. The acidophilus milk preparation was more effective than a buttermilk preparation. Unfortunately, no unfermented milk controls were conducted and statistical analysis of the data were not conducted. Although defecation rates were calculated, the results are difficult to interpret because different laxatives were administered as needed during the study. The one consistency in the results, however, was that the need for laxatives was reduced. This study shows a marginal effect of fermented acidophilus milk consumption on constipation.

Seki et al. (1978, in Japanese) studied the effect of *Bifidobacterium* on constipation in aged people. Eighteen constipated seniors were given milk cultured with bifidus or lactic cultures to determine the effect on constipation. A ten-day prefeeding period was followed by a 10-day lactic milk feeding period. Finally, the bifidus milk was fed for 10 days. Stool frequencies during these three periods were 4.8, 5.7, and 7.1, respectively. Although the bifidus milk feeding resulted in the most frequent stool elimination, the lack of crossover in the experimental design does not distinguish between an effect of the bifidus milk, or a postfeeding effect of the lactic milk.

From these few references, it is difficult to assess the effect of lactic cultures on constipation. However, a clinical study targeting this health effect would be relatively inexpensive to conduct, and could readily be done on humans.

Vaginitis

Hilton et al. (1992) tested the effect of ingested yogurt on preventing candidal vaginitis. Candidal vaginitis is a common cause of gynecologic infection, especially in pregnant women, and women receiving antibiotic or corticosteroid therapy. This study assessed the effect of daily ingestion of *L. acidophilus* yogurt on vulvovaginal candidal infections in women with recurrent candidal vaginitis. Patients for the study were not on antibiotic therapy and did not regularly ingest more than 16 ounces of yogurt weekly before the study. Patients did not alter dietary or sexual practices during the study. Patients served as their own controls and consumed eight ounces of yogurt daily for six months and no yogurt for six months. No effort was made to make this study blind to the patients or interviewer, but all laboratory samples were blinded. Columbo plain yogurt containing 10^8 *L. acidophilus*/ml was used. Moderate H_2O_2 production was seen by the yogurt *L. acidophilus*. Thirteen women completed the study (33 began the study). The mean number of infections per six months was 2.5 in the control arm and 0.38 in the yogurt arm ($P = 0.001$). Eight additional women refused to enter the control arm of the study because they experienced significant relief during the yogurt consumption arm of the study. These women had a chronic and prolonged history of vaginitis prior to the study. Authors acknowledge the need for a follow-up study using yogurt with killed culture as a control. This study is one of the few that provide statistically valid results with human subjects. The relationship between rectal and vaginal colonization is discussed, and suggests a mechanism for vaginal influence of consumed microbes.

CONSIDERATIONS FOR STRAIN SELECTION

When selecting strains for commercial application and for efficacy in health effects, the following parameters should be considered:

- Ability to commercially prepare and store culture to maintain viability and healthful attributes
- Maintenance of stated level of viability in food product through product pull date
- Accurate strain identification, including genus and species (increasingly this may require DNA homology confirmation), and record of strain origin
- Verification of fermentative and flavor profiles for the conditions of use
- Clinical evidence for stated health effect for particular strain being sold
- Laboratory verification of important parameters for physiological functioning, for example, high lactase levels, intestinal survival,

or *in vivo* macrophage stimulation, depending on targeted populations.

CONCLUSIONS

The state of research on healthful attributes of lactic cultures is still in its infancy. Although an overwhelming number of publications have been presented, only a few seem to contribute convincingly to our knowledge of health effects in humans. Research on the effect of lactic cultures on lactose intolerance symptoms in humans defines parameters for a positive effect for many people. Levels of 5 to 10×10^7/ml of *L. bulgaricus* or *S. thermophilus* in milk products with an average lactose load could be expected to be efficacious at preventing lactose intolerance symptoms. Feeding certain strains of *L. acidophilus* can be expected to decrease fecal enzymes that may be involved in colon cancer. Certain strains of lactic cultures appear to have an anti-diarrheal effect, although the strain levels, diarrheal causes, and affected subpopulations need to be more clearly defined. Some indications exist that lactic cultures may increase bowel regularity. The lack of direct human studies in the areas of immunostimulation and anticancer effects suggests that it is premature to make these claims. Negative results from cholesterol studies precludes any claim for anticholesterol effects.

Commercial interest in exploiting proposed healthful attributes of lactic cultures appears to be advancing faster than the definitive research that proves many of these claims. Responsible promotion of these products must consider attributes of the specific strain being used, efficacious levels, applicability toward specific subpopulations, and legitimate research support for healthful attributes. Some products are promoted today using testimonial information, uncontrolled studies, research not specifically conducted on the strain being sold, and general statements with no research support. In the long run, this will serve to hurt the overall market for these products. The public, unable to distinguish the reputable from the disreputable products, will lose faith in the entire product line. This is unfortunate, as recent research advances suggest that legitimate healthful attributes can be attributed to specific lactic strains, and these products could contribute to the overall health and well-being of consumers.

References

Ahrens, E.H., Jr. 1985. The diet-heart question in 1985: Has it really been settled? *Lancet* ii:1085–1087.

Alm, L. 1991. The therapeutic effects of various cultures—an overview. In *Therapeutic Properties of Fermented Milks,* R.K. Robinson, ed., pp. 45–64. New York: Elsevier Applied Science.

Alm, L., Humble, D.; Ryd-Kjellen, E.; and Setterberg, G. 1983. The effect of acidophilus milk in the treatment of constipation in hospitalized geriatric patients. In *Nutrition and the Intestinal Flora,* Bo Hallgren, ed., pp. 131–138. Stockholm, Sweden: Symposia of the Swedish Nutrition Foundation XV, Almqvist & Wiksell International.

Ayebo, A.D.; Angelo, I.A.; and Shahani, K.M. 1980. Effect of ingesting *Lactobacillus acidophilus* (sic) milk upon fecal flora and enzyme activity in humans. *Milchwissenschaft* 35:730–73.

Bazzarre, T.L.; Liu Wu, S.; and Yuhas, J.A. 1983. Total and HDL-cholesterol concentrations following yogurt and calcium supplementation. *Nutr. Rep. Int.* 28:1225–1232.

Beck, C., and Necheles, H. 1961. Beneficial effects of administration of *Lactobacillus acidophilus* (sic) in diarrheal and other intestinal disorders. *Amer. J. Gastroenterol.* 35:522–530.

Bellomo, G.; Mangiagle, A.; Nicastro, L.; and Frigerio, G. 1980. A controlled double-blind study of SF68 strain as a new biological preparation for the treatment of diarrhea in pediatrics. *Curr. Therap. Res.* 28:927–936.

Bezkorovainy, A., and Miller-Catchpole, R. 1989. *Biochemistry and Physiology of Bifidobacteria.* Boca Raton, Florida: CRC Press.

Bianchi-Salvadori, B. 1986. Intestinal microflora: The role of yogurt in the equilibrium of the gut ecosystem. *Int. J. Immunotherapy* (Suppl.) 11:9–18.

Bogdanov, E.G.; Dalev, P.G.; Gurevich, A.I.; Kolosov, M.N.; Mal'kova, V.P.; Plemyannikova, L.A.; and Sorokina, I.B. 1975. Antitumor glycopeptides from *Lactobacillus bulgaricus* cell wall. *FEBS Lett.* 57:259–261.

Bourlioux, R., and Pochart, P. 1988. Nutritional and health properties of yogurt. *Wld. Rev. Nutr. Diet.* 56:217–258.

Brown, J.P. 1977. Role of gut bacterial flora in nutrition and health: A review of recent advances in bacteriological techniques, metabolism, and factors affecting flora composition. *CRC Crit. Rev. Food Sci. Nutr.* 8:229–336.

Clements, M.L.; Levine, M.M.; Ristaino, P.A.; Daya, V.E.; and Hughes, T.P. 1983. Exogenous lactobacilli fed to man—their fate and ability to prevent diarrheal disease. *Prog. Fd. Nutr. Sci.* 7:29–37.

Clements, M.L.; Levine, M.M.; and Black, R.E. 1981. *Lactobacillus* prophylaxis for diarrhea due to enterotoxigenic *Escherichia coli. Antimicrob. Agents Chemother.* 20:104–108.

Colombel, J.F.; Cortot, A.; Neut, C.; and Romond, C. 1987. Yogurt with *Bifidobacterium longum* reduces erythromycin-induced gastrointestinal effects. *Lancet* ii:43.

Danielson, A.D.; Peo, E.R., Jr.; Shahani, K.M.; Lewis, A.J.; Whalen, P.J.; and Amer, M.A. 1989. Anticholesteremic property of *Lactobacillus acidophilus* yogurt fed to mature boars. *J. Anim. Sci.* 67:966–974.

de Simone, C.; Bianchi Salvadori, B.; Jirillo, E.; Baldinessi, L.; Bitonti, F.; and Vesely, R. 1989. Health properties of yogurt. In *Yogurt: Nutritional and Health Properties,* R.C. Chandan, ed., pp. 201–213. McLean, VA: National Yogurt Association.

Drasar, B.S., and Hill, J.F. 1972. Intestinal bacteria and cancer. *Amer. J. Clin. Nutr.* 25:1399–1404.

Driessen, F.M., and de Boer, R. 1989. Fermented milks with selected intestinal bacteria: A healthy trend in new products. *Neth. Milk Dairy J.* 43:367–382.

Fernandes, C.F., and Shahani, K.M. 1989. Modulation of antibiosis by lactobacilli and yogurt and its healthful and beneficial significance. In *Yogurt: Nutritional and Health Properties,* R.C. Chandan, ed., pp. 145–160. McLean, VA: National Yogurt Association.

———. 1990. Anticarcinogenic and immunological properties of dietary lactobacilli. *J. Food Protect.* 53:704–710.

Fernandes, C.F.; Shahani, K.M.; and Amer, M.A. 1987. Therapeutic role of dietary lactobacilli and lactobacillic fermented dairy products. *FEMS Microbiol. Rev.* 46:343–356.

Fonden, R. 1989. *Lactobacillus acidophilus.* In *Les Laits Fermentes. Actualite de la Recherche,* pp. 35–40. Montrouge, France: John Libbey Eurotext.

Friend, B.A., and Shahani, K.M. 1984. Nutritional and therapeutic aspects of lactobacilli. *J. App. Nutr.* 36:125–153.

Fuller, R. 1991. Probiotics in human medicine. *Gut* 32:439–442.

Gilliland, S.E.; Nelson, C.R.; and Maxwell, C. 1985. Assimilation of cholesterol by *Lactobacillus acidophilus. Appl. Environ. Microbiol.* 49:377–381.

Gilliland, S.E., and Walker, D.K. 1990. Factors to consider when selecting a culture of *Lactobacillus acidophilus* as a dietary adjunct to produce a hypocholesterolemic effect in humans. *J. Dairy Sci.* 73:905–911.

Goldin, B.R. 1989. Lactic acid bacteria: Implications for health. In *Les Laits Fermentes. Actualite de la Recherche,* pp. 95–104. Montrouge, France: John Libbey Eurotext.

———. 1990. Intestinal microflora: Metabolism of drugs and carcinogens. *Ann. Medicine* 22:43–48.

Goldin, B.R.; Swenson, L.; Dwyer, J.; Sexton, M.; and Gorbach, S.L. 1980. Effect of diet and *Lactobacillus acidophilus* supplements on human fecal bacterial enzymes. *J. Natl. Canc. Instit.* 64:255–261.

Goldin, B.R.; Gorbach, S.L.; Saxelin, M.; Barakat, S.; Gualtiere, L.; and Salminen, S. 1992. Survival of *Lactobacillus* species (Strain GG) in human gastrointestinal tract. *Digestive Diseases and Sci.* 37:121–128.

Goldin, B., and Gorbach, S.L. 1977. Alterations in fecal microflora enzymes related to diet, age, *Lactobacillus* (sic) supplements and dimethylhydrazine. *Cancer* 40:2421–2426.

———. 1980. Effect of *Lactobacillus acidophilus* dietary supplements on 1,2-dimethylhydrazine in dihydrochloride-induced intestinal cancer in rats. *J. Natl. Canc. Instit.* 64:263–265.

———. 1984a. The Effect of oral administration on *Lactobacillus* and antibiotics on intestinal bacterial activity and chemical induction of large bowel tumors. *Dev. Indus. Microbiol.* 25:139–150.

———. 1984b. The effect of milk and *Lactobacillus* (sic) feeding on human intestinal bacterial enzyme activity. *Amer. J. Clin. Nutr.* 39:756–761.

Gorbach, S.L. 1971. Progress in gastroenterology. Intestinal microflora. *Gastroenterol.* 60:1110–1129.

———. 1986. Function of the normal human microflora. *Scand. J. Infect. Dis. (Suppl) 49:17–30.*

———. *1989. Metabolism of carcinogens and drugs by the intestinal flora: Effects of Lactobacillus.* In *Les Laits Fermentes. Actualite de la Recherche,* pp. 85–94. Montrouge, France: John Libbey Eurotext.

———. 1990. Lactic acid bacteria and human health. *Ann. Medicine* 22:37–41.

Gorbach, S.L.; Chang, T.-W.; and Goldin, B. 1987. Successful treatment of relapsing *Clostridium* difficile (sic) colitis with *Lactobacillus* (sic) GG. *Lancet* ii:1519.

Gorbach, S.L., and Goldin, B.R. 1989. *Lactobacillus* (sic) strains and methods of selection. US Patent Number 4,839,281.

Graf, W. 1983. Studies on the therapeutic properties of acidophilus milk. In *Nutrition and the Intestinal Flora,* Bo Hallgren, ed., pp. 119–121. Stockholm, Sweden: Symposia of the Swedish Nutrition Foundation XV, Almqvist & Wiksell International.

Gurr, M.I. 1984. Fermented milks, intestinal microflora and nutrition. *International Dairy Federation Bulletin* 179:54–59.

Halpern, G.M.; Vruwing, K.G.; van de Water, J.; Keen, C.L.; and Gershwin, M.E. 1991. Influence of long-term yoghurt consumption in young adults. *Int. J. Immunotherapy* VII:205–210.

Hepner, G.; Fried, R.; St. Jeor, S.; Fusetti, L.; and Morin, R. 1979. Hypocholesterolemic effect of yogurt and milk. *Amer. J. Clin. Nutr.* 32:19–24.

Hilton, E.; Isenberg, H.D.; Alperstein, P.; France, K.; and Borenstein, M.T. 1992. Ingestion of yogurt containing *Lactobacillus acidophilus* as prophylaxis for candidal vaginitis. *Annals of Internal Medicine* 116:353–357.

Hosono, A.; Kashina, T.; and Kada, T. 1986. Antimutagenic properties of lactic acid-cultured milk on chemical and fecal mutagens. *J. Dairy Sci.* 69:2237–2242.

Hotta, M.; Sato, Y.; Iwata, S.; Yamashita, N.; Sunakawa, K.; Oikawa, T.; Tanaka, R.; Watanabe, K.; Takayama, H.; Yajima, M.; Sekiguchi, S.; Arai, S.; Sakurai, T.; and Mutai, M. 1987. Clinical effects of *Bifidobacterium* preparations on pediatric intractable diarrhea. *Keio. J. Med.* 36:298–314.

Hughes, D.B., and Hoover, D.G. 1991. *Bifidobacteria*: Their potential for use in American dairy products. *Food Technol.* 45:74–83.

Isolauri, E.; Juntunen, M.; Rautanen, T.; Sillanaukee, P.; and Koivula, T. 1991. A human *Lactobacillus* strain (*Lactobacillus casei* sp. strain GG) promotes recovery from acute diarrhea in children. *Pediatrics* 88:90–97.

Jaspers, D.A.; Massey, L.K.; and Luedecke, L.O. 1984. Effect of consuming yogurts prepared with 3 culture strains on human serum lipoproteins. *J. Food Sci.* 49:1178–1181.

Kato, I.; Kobayashi, S.; Yokokura, T.; and Mutai, M. 1981. Antitumor activity of *Lactobacillus casei* in mice. *Gann* 72:517–523.

Kim, H.S. 1988. Characterization of *lactobacilli* and *bifidobacteria* as applied to dietary adjuncts. *Cult. Dairy Prod. J.* August:6–9.

Kim, H.S., and Gilliland, S.E. 1983. *Lactobacillus acidophilus* as a dietary adjunct for milk to aid lactose digestion in humans. *J. Dairy Sci.* 66:959–966.

Kitazowa, H.; Nomura, M.; Itoh, T.; and Yamaguchi, T. 1991. Functional alteration of macrophages by a slime-forming *Lactobacillus lactis* ssp. *cremoris*. *J. Dairy Sci.* 74:2082–2088.

Klaenhammer, T.R. 1982. Microbiological considerations in selection and preparation of *Lactobacillus* strains for use as dietary adjuncts. *J. Dairy. Sci.* 65:1339–1349.

———. 1988. Bacteriocins of lactic acid bacteria. *Biochimie* 70:337–349.

Kohwi, T.; Imai, K.; Tamura, A.; and Hashimoto, Y. 1978. Antitumor effect of *Bifidobacterium infantis* in mice. *Gann* 69:613–618.

Kolars, J.C.; Levitt, M.D.; Aouji, M.; and Savaiano, D.A. 1984. Yogurt—An autodigesting source of lactose. *N. Engl. J. Med.* 310:1–3.

Koo, M., and Rao, A.V. 1991. Long-term effect of *Bifidobacteria* and neosugar on precursor lesions of colonic cancer in CF_1 mice. *Nutrition and Cancer* 16:249–257.

Kroger, M.; Kurmann, J.A.; and Rasic, J.L. 1989. Fermented milks—Past, present, and future. *Food Technol.* 43:92–98.

Kurmann, J.A., and Rasic, J. Lj. 1991. The health potential of products containing *Bifidobacteria*. In *Therapeutic Properties of Fermented Milks*, R.K. Robinson, ed., pp. 117–157. New York: Elsevier Applied Science.

Laroia, S., and Martin, J.H. 1990. *Bifidobacteria* as possible dietary adjuncts in cultured dairy products—A review. *Cult. Dairy Prod. J.* 25:18–22.

Lemonnier, D. 1984. Fermented milks and health. *International Dairy Federation Bulletin* 179:60–68.

Lin, M.-Y.; Savaiano, D.; Harlander, S. 1991. Influence of nonfermented dairy products containing bacterial starter cultures on lactose maldigestion in humans. *J. Dairy Sci.* 74:87–95.

Lin, S.Y.; Ayres, J.W.; Winkler, W., Jr.; and Sandine, W.E. 1989. *Lactobacillus* effects on cholesterol: In vitro and in vivo results. *J. Dairy Sci.* 72:2885–2899.

Marteau, P.; Pochart, P.; Flourie, B.; Pellier, P.; Santos, L.; Desjeux, J.F.; and Rambaud, J.C. 1990. Effect of chronic ingestion of a fermented dairy product containing *Lactobacillus acidophilus* and *Bifidobacterium bifidum* on metabolic activities of the colonic flora in humans. *Amer. J. Clin. Nutr.* 52:685–688.

Martini, M.C.; Kukielka, D.; and Savaiano, D.A. 1991. Lactose digestion from yogurt: Influence of a meal and additional lactose. *Amer. J. Clinic. Nutr.* 53:1253–1258.

Martini, M.C.; Lerebours, E.C.; Lin, W.-J.; Harlander, S.K.; Berrada, N.M.; Antoine, J.M.; and Savaiano, D.A. 1991. Strains and species of lactic acid bacteria in fermented milks (yogurts): Effect on in vivo lactose digestion. *Amer. J. Clin. Nutr.* 54:1041–1046.

McConnell, M.A., and Tannock, G.W. 1991. Lactobacilli and azoreductase activity in the murine cecum. *Appl. Environ. Microbiol.* 57:3664–3665.

McDonough, F.E.; Wong, N.P.; Hitchins, A.; and Bodwell, C.E. 1987. Alleviation of lactose malabsorption from sweet acidophilus milk. *Amer. J. Clin. Nutr.* 42:345–346.

McGroatry, J.A.; Hawthorn, A.-A.; and Reid, G. 1988. Anti-tumor activity of *Lactobacilli* in vitro. *Microbios Lett.* 39:105–112.

Misra, A.K., and Kuila, R.K. 1991. The selection of *Bifidobacteria* for the manufacture of fermented milks. *Austr. J. Dairy Technol.* May:24–26.

Mitsuoka, T. 1989. *Bifidobacterium* microecology. In *Les Laits Fermentes. Actualite de la Recherche,* pp. 41–48. Montrouge, France: John Libbey Eurotext.

———. 1990. *Bifidobacteria* and their role in human health. *J. Indust. Microbiol.* 6:263–268.

Modler, G.W.; McKellar, R.C.; and Yaguchi, M. 1990. *Bifidobacteria* and bifidogenic factors. *Can. Inst. Food Sci. Technol. J.* 23:29–41.

Newcomer, A.D.; Park, H.S.; O'Brien, P.C.; and McGill, D.B. 1983. Response of patients with irritable bowel syndrome and lactase deficiency using unfermented acidophilus milk. *Amer. J. Clin. Nutr.* 38:257–263.

Oksanen, P.J.; Salminen, S.; Saxelin, M.; Hamalainen, P.; Ihantola-Vormisto, A.; Muurasniemi-Isoviita, L.; Nikkari, S.; Oksanen, T.; Porsti, I.; Salminen, E.; Siitonen, S.; Stuckey, H.; Toppila, A. ˉd Vapaatalo, H. 1990. Prevention of travellers' diarrhoea by *Lactobacillus* GG. *Ann. Medicine* 22:53–56.

Pearce, J.L., and Hamilton, J.R. 1974. Controlled trial of orally administered lactobacilli in acute infantile diarrhea. *J. Pediat.* 84:261–262.

Perdigon, G.; Nader de Macias, M.E.; Alvarez, S.; Medici, M.; Oliver, G.; and Pesce de Ruiz Holgado, A. 1986a. Effect of a mixture of *Lactobacillus casei* and *Lactobacillus acidophilus* administered orally on the immune system in mice. *J. Food. Protect.* 49:986–989.

Perdigon, G.; Nader de Macias, M.E.; Alvarez, S.; Oliver, G.; and Pesce de Ruiz Holgado, A.A. 1986b. Effect of perorally administered lactobacilli on macrophage activation in mice. *Infection and Immunity* 53:404–410.

Perdigon, G.; Nader de Macias, M.E.; Alvarez, S.; Oliver, G.; and Pesce de Ruiz Holgado, A.A. 1987. Enhancement of immune response in mice fed with *Streptococcus thermophilus* and *Lactobacillus acidophilus. J. Dairy Sci.* 70:919–926.

Perdigon, G.; Alvarez, S.; Nader de Macias, M.E.; and Medici, M. 1989. Effect of lactic acid bacteria orally administered and of yoghurt on the immune system. In *Les Laits Fermentes. Actualite de la Recherche,* pp. 77–84. Montrouge, France: John Libbey Eurotext.

Perdigon, G.; Nader de Macias, M.E.; Alvarez, S.; Oliver, G.; and Pesce de Ruiz Holgado, A.A. 1990. Prevention of gastrointestinal infection using immunobiological methods with milk fermented with *Lactobacillus casei* and *Lactobacillus acidophilus. J. Dairy Res.* 57:255–264.

Poupard, J.A.; Husain, I.; and Norris, R.F. 1973. Biology of the bifidobacteria. Bacteriol. Rev. 37:136–165.

Pozo-Olano, J.D.; Warram, J.G., Jr., Gomez, R.G.; and Cavazos, M.G. 1978. Effect of a *Lactobacilli* preparation on traveler's diarrhea: A randomized, double blind clinical trial. *Gastroenterology* 74:829–830.

Rasic, J.Lj. 1983. The role of dairy foods containing bifido- and acidophilus bacteria in nutrition and health? *N. Europ. Dairy J.* 4:2–10.

Rasic, J.Lj., and Kurmann, J.A. 1983. *Bifidobacteria and their role. Microbiological, Nutritional- Physiological, Medical and Technological Aspects and Bibliography.* Basel, Switzerland: Birkhäuser Verlag.

Rasic, J.Lj.; Vujicic, I.R.; Skrinjar, M.; and Vulic, M. 1992. Assimilation of cholesterol by some cultures of lactic acid bacteria and *Bifidobacteria*. *Biotechnol. Bull.* 14:39–44.

Redondo-Lopez, V.; Cook, R.L.; and Sobel, J.D. 1990. Emerging role of *Lactobacilli* in the control and maintenance of the vaginal bacterial microflora. *Rev. Infect. Diseases* 12:856–872.

Renner, E. 1991. Cultured dairy products in human nutrition. *International Dairy Federation Bulletin* 255:2–24.

Rossouw, J.E.; Burger, E.-M.; van der Vyver, P.; and Ferreira, J.J. 1981. The effect of skim milk, yogurt, and full cream milk on human serum lipids. *Amer. J. Clin. Nutr.* 34:351–356.

Salminen, E.; Elomaa, I.; Minkkinen, J.; Vapaatalo, H.; and Salminen, S. 1988. Preservation of intestinal integrity during radiotherapy using live *Lactobacillus acidophilus* cultures. *Clin. Radiol.* 39:435–437.

Savage, D.C. 1977. Microbial ecology of the gastrointestinal tract. *Ann. Rev. Microbiol.* 31:107–133.

———. 1983a. Associations of indigenous microorganisms with gastrointestinal epithelial surfaces. In *Human Intestinal Microflora in Health and Disease*, D.J. Hentges, ed., pp. 55–78, New York:Academic Press, Inc.

———. 1983b. Morphological diversity among members of the gastrointestinal microflora. *International Rev. Cytol.* 82:305–334.

Savaiano, D.A.; Abou El Anovar, A.; Smith, D.E.; and Levitt, M.D. 1984. Lactose malabsorption from yogurt, pasteurized yogurt, sweet acidophilus milk, and cultured milk in lactose-deficient individuals. *Amer. J. Clin. Nutr.* 40:1219–1223.

Savaiano, D.A., and Kotz, C. 1988. Recent advances in the management of lactose intolerance. *Comtemporary Nutrition.* 13:10.

Savaiano, D.A., and Levitt, M.D. 1987. Milk intolerance and microbe-containing dairy foods. *J. Dairy Sci.* 70:397–406.

Seki, M.; Igarashi, Fukuda, Y.; Simamura, S.; Kawashima, T.; and Ogasa, K. 1978. The effect of *Bifidobacterium* cultured milk on the "regularity" among an aged group. *Nutrition and Foodstuff.* 31:379–387. (in Japanese)

Sekine, K.; Toida, T.; Saito, M.; Kuboyama, M.; Kawashima, T.; and Hashimoto, Y. 1985. A new morphologically characterized cell wall preparation (whole peptidoglycan) from *Bifidobacterium infantis* with a higher efficacy on the regression of an established tumor in mice. *Cancer Res.* 45:1300–1307.

Sellars, R.L. 1989. Health properties of yogurt. In *Yogurt: Nutritional and Health Properties*, R.C. Chandan, ed., pp. 115–144. McLean, VA: National Yogurt Association.

———. 1991. Acidophilus products. In *Therapeutic Properties of Fermented Milks*, R.K. Robinson, ed., pp. 81–116. New York: Elsevier Applied Science.

Shackelford, L.A.; Rao, D.R.; Chawan, C.B.; and Pulusani, S.R. 1983. Effect of feeding fermented milk on the incidence of chemically induced colon tumors in rats. *Nutr. Cancer* 5:159–164.

Siitonen, S.; Vapaatalo, H.; Salminen, S.; Gordin, A.; Saxelin, M.; Wikberg, R.; and Kirkkola, A.-L. 1990. Effect of *Lactobacillus* GG yogurt in prevention of antibiotic associated diarrhea. *Ann. Medicine* (Helinski) 22:57–59.

Smith, M.; Sullivan, C.; and Goodman, N. 1991. Reactivity of milk cholesterol with bacterial cholesterol oxidases. *J. Agric. Food Chem.* 39:2158–2162.

Tannock, G.W. 1983. Effect of dietary and environmental stress on the gastrointestinal microbiota. In *Human Intestinal Microflora in Health and Disease*, D.J. Hentges, ed., pp. 517–539. New York: Academic Press, Inc.

———. 1984. Control of gastrointestinal pathogens by normal flora. In *Current Perspectives in Microbial Ecology*, M.J. Klug and C.A. Reddy, eds., pp. 374–382. Washington, D.C.: Amer. Soc. Microbiol.

———. 1990. The microecology of *Lactobacilli* inhabiting the gastrointestinal tract. *Adv. Microb. Ecol.* 11:147–171.

Thompson, L.U.; Jenkins D.J.A.; Amer, M.A.V.; Reichert, R.; Jenkins, A.; and Kamulsky, J. 1982. The effect of fermented and unfermented milks on serum cholesterol. *Amer. J. Clin. Nutr.* 36:1106–1111.

Truslow, W.W. 1986. Lactobacillus (sic) and fermented foods. *J. Holistic Med.* 8:36–46.

van't Veer, P.; Dekker, J.M.; Lamers, J.W.J.; Kok, F.J.; Schouten, E.G.; Brants, H.A.M.; Sturmans, F; and Hermus, R.J.J. 1989. Consumption of fermented milk products and breast cancer: A case-control study in the Netherlands. *Cancer Res.* 49:4020–4023.

Wytock, D.H., and J.A. DiPalma. 1988. All yogurts are not created equal. *Am. J. Clin. Nutr.* 47:454–457.

Yamazaki, S.; Machii, K.; Tsuyuki, S.; Momose, H.; Kawashima, T.; and Ueda, K. 1985. Immunological responses to monoassociated *Bifidobacterium longum* and their relation to prevention of bacterial invasion. *Immunology* 56:43–50.

Zhang, X.B., and Ohta, Y. 1991a. In vitro binding of mutagenic pyrolyzates to lactic acid bacterial cells in human gastric juice. *J. Dairy Sci.* 74:752–757.

———. 1991b. Binding of mutagens by fractions of the cell wall skeleton of lactic acid bacteria on mutagens. *J. Dairy Sci.* 74:1477–1481.

Zhang, X.B.; Ohta, Y.; and Hosono, A. 1990. Antimutagenicity and binding of lactic acid bacteria from a Chinese cheese to mutagenic pyrolyzates. *J. Dairy Sci.* 73:2702–2710.

Chapter 15

Nutrition of Macrominerals and Trace Elements

John J.B. Anderson and
Jonathan C. Allen

INTRODUCTION

This chapter will focus on two categories of minerals, macrominerals and trace elements. The first category consists of minerals that have clearly been demonstrated to be deficient in the diets of large segments of populations in the developed and developing nations. The second includes those mineral nutrients that may be potentially deficient in populations, but whose deficiencies are difficult to establish by traditional assessment methods. The two well-established problem minerals are calcium and iron, and the potential (or possible) problem minerals include zinc, magnesium, chromium, selenium, and iodine (Table 15-1). Except for iodine, with which specific foods are generally fortified in much of the world, each of these mineral nutrients will be reviewed in the context of overcoming deficiencies through functional foods.

This chapter is divided into four major sections: General Aspects of the Minerals, Epithelial Transport of the Minerals, Issues of the

Table 15-1. Prevalent Dietary Mineral and Trace Element Deficiencies

More Prevalent	Less Prevalent[a]
Calcium	Magnesium
Iron	Zinc
	Chromium
	Selenium
	Iodine

[a]Assessment of deficiencies of these nutrients less clear-cut.

Problem Nutrients, and Health Effects of Specific Mineral Deficiencies. A Conclusions section follows.

GENERAL ASPECTS OF THE MINERALS

The dietary intakes of mineral nutrients by populations throughout the world are generally documented by surveys using instruments that estimate daily intakes by 24-hour dietary recalls, records for one or more days, or food frequencies representing a week or more. In this chapter, data from surveys in the United States will be utilized to illustrate the magnitude of the deficiencies of the problem minerals. The actual mean amounts consumed will be compared with the Recommended Dietary Allowances (RDAs), the common standard used in the United States, and to the new Reference Daily Intakes (RDIs), recently formulated by the Food and Drug Administration, a U.S. governmental agency. Finally, this section will conclude with an overview of the homeostatic mechanisms regulating the entry and exit of the key mineral nutrients, and their transport, storage, and overall metabolism.

Dietary Intakes of Mineral Nutrients

The low or inadequate mean consumptions of calcium and iron by females become apparent by approximately age 11, i.e., during the prepubertal period in the United States, where the average age of menarche is 12.5 years. Both the U.S. Department of Agriculture (USDA) (1977–1978) and National Health and Nutrition Examination Survey (NHANES) (1971–1974) data support the early onset of

Table 15-2. Daily Intakes of Calcium and Iron by Adult Americans from National Surveys as Percentages of the RDAs (NHANES I, 1971–74; USDA 1977–78)

		NHANES, 1971		USDA, 1977	
Mineral	Age	Men	Women	Men	Women
Calcium, mg	19–64	~100%	~60%	101%	70%
	65 +	~90	~55	89	71
Iron, mg	19–64	~130	~70	157	73
	65 +	~120	~95	142	108

deficient intakes of these two minerals among females, but not males to any great extent (Table 15-2).

The mean deficits in calcium across the entire lifecycle by females amount to approximately 30–50% below the age-specific RDAs (100%). The deficits of iron exist in women only during the menstruating years, but they also range from 30–50% below the age-specific RDAs (1989). Deficiencies in calcium and iron are relatively small among men because of the greater numbers of servings of the animal food groups, i.e., meats, fish, poultry, and legumes for iron, and dairy products for calcium, consumed by males. Mean zinc intakes by both male and female Americans have also been found to be significantly lower than the RDAs (Moser-Veillon 1990). The assessment of minerals available in the American diet has been documented by the Food and Drug Administration's Total Diet Study, 1982–1986 (Pennington, Young, and Wilson 1989).

RDAs and RDIs

Values of the RDAs and RDIs of the minerals under consideration in this chapter are given in Table 15-3. The RDIs are emphasized because they will be used on all food labels effective 1994.

The U.S. Food and Drug Administration and the U.S. Department of Agriculture have issued new regulations for nutritional labeling of foods that become effective in 1994. These regulations replace the term U.S. RDA (United States Recommended Daily Allowance), which has been in use on labels since 1973, with the term RDI (Reference Daily Intake). Food labels will use the term "Daily Values" to encompass both the term RDI and Daily Reference Values (DRV), which are intakes for nutrients that relate to

Table 15-3. Adult RDAs (1989) and RDIs (1993)

		RDA[a]		RDI
Mineral	Age (yr).	Male	Female	Male and Female
Calcium, mg	19–24	1,200	1,200	1,000[b]
	25 +	800	800	
Phosphorus, mg	19–24	1,200	1,200	1,000
	25 +	800	800	
Magnesium, mg	19–24	400	280	400
	25 +	350	280	
Iron, mg	19–24	12	15	15[b]
	25 +	10	15	
Zinc, mg	11 +	15	12	15
Iodine, μg	11 +	150	150	150
Selenium, μg	19 +	70	55	NA[a]
Chromium, μg	19 +	50–200	50–200	NA[c]
Copper, μg	11 +	1.5–3.0	1.5–3.0	2

[a]The values listed for chromium and copper are estimated Safe and Adequate Daily Dietary Intakes (SADDIs) rather than RDAs.
[b]RDI required on food label. Others are optional.
[c]NA. Not available. Nutrient content cannot be used on label.

chronic diseases, including sodium and potassium. The procedure used for selecting the RDI values is the same as for the U.S. RDA, with modifications due to rounding. The values are based on the RDAs (Recommended Daily Allowances) published by the National Academy of Science (NAS). RDAs are based on the best available scientific evidence at their time of publication and provide values for 18 different population groups, based on age, sex, and physiological state (Table 15-3). The RDI formulation uses the "population coverage approach," selecting the requirement for the group with the highest recommended intake. The RDIs use data from the 9th edition (1980) of the NAS RDAs, whereas when the U.S. RDAs were formulated, they used the 1968 edition of NAS RDAs and were never updated as NAS recommendations changed. Calcium and iron are the only minerals that are mandatory for labeling. As noted above (Tables 15-1 and 15-2), potential nutritional deficiencies are most likely for calcium and iron. The other minerals listed in Table 15-3 may be consumed in low quantities relative to physiological requirements, but in general these intakes are not deficient. Voluntary use of these other nutrients via supplementation may be advisable in certain parts of the world.

There are slight discrepancies between the list of minerals for which RDAs have been formulated and those on the new list of Daily Values. An RDI has been calculated for copper, although the NAS found only enough data on requirements to list it with the Estimated Safe and Adequate Daily Dietary Intake (ESADDI) table. The FDA found compelling evidence that an RDI for copper should be available, but did not have the same conclusion about manganese, fluoride, chromium, and molybdenum. There are no Generally Recognized as Safe (GRAS) mineral sources for adding fluoride, chromium, or selenium to foods. Although RDA values are published for selenium, no RDI exists for use of it in labeling.

Overview of Homeostatic Mechanisms Governing the Minerals

This section will highlight the mechanisms regulating mineral transport across epithelia, transport in blood, storage, and overall metabolism.

Much of the difficulty in determining the true dietary needs for minerals results from the efficiency with which animals can adapt to varying levels of intake. The homeostatic mechanisms that maintain relatively constant internal concentrations vary among minerals, but they include modulating the rate of absorption from the gastrointestinal (GI) tract, increased excretion if excess mineral is absorbed, and storage of excess mineral in innocuous forms in various tissues.

Homeostatis

The percentage absorption of cations, such as calcium, zinc, copper, and iron, decreases as intake within the normal dietary range increases. Calcium absorption is regulated hormonally. When serum calcium decreases, activation of vitamin D by the kidney to 1,25-dihydroxycholecalciferol (calcitriol) increases synthesis of a calcium-binding protein (calbindin) in the intestinal epithelial cells. Calbindin promotes the transport of calcium from the apical to basolateral regions of the cells located primarily in the upper half of the small bowel, and it thus increases net calcium transport. Similarly, absorption of zinc, copper, and iron is regulated by intracellular binding proteins in the intestine, but by a different mechanism. High dietary zinc, or a high level of zinc available to the intestinal cells from the plasma, induces intestinal metallothionein in the cell until either plasma levels decrease significantly, or the cells mature and are sloughed into the intestinal lumen for excretion. The same type of inhibition of absorption, termed the mucosal block, regulates iron. High dietary iron leads to increases in cytoplasmic ferritin, which prevents the iron from leaving the intestinal cell before the aged cells are sloughed off.

Minerals that are absorbed at a more constant percentage of that ingested are eliminated by variable excretion routes, either into the urine by the kidney, or through pancreatic fluid or bile into the gut. Sodium and potassium, for example, are nearly completely absorbed but their urinary excretion maintains a nearly constant plasma concentration. Iodine is readily absorbed and then excreted in the urine with little regulation by the kidney of the amount excreted. The thyroid, which synthesizes the only physiologically active form(s) of iodine, must compete with the kidney, so that it absorbs iodine from the blood and stores the mineral in a protein-bound form. Similarly, if excess iron is absorbed, high concentrations can be stored in liver ferritin. Because iron is not readily excreted, reducing the body burden of this mineral can be difficult.

Transport Proteins

For most essential minerals, binding to proteins and other molecules both in cells and in plasma exceeds the free ionic concentration. Serum proteins have been identified that are major transport molecules for many minerals, binding significant amounts of calcium, zinc, iron, and chromium.

Storage

Storage of the minerals is highly specific for each nutrient. The skeleton is a major retrievable store for several minerals, particularly calcium, phosphorus, magnesium, zinc, and fluoride. Iron is stored predominantly as ferritin, but other forms such as hemosiderin can be molecular storage aggregates under conditions of iron overload. Gut cells also store iron as ferritin, but as far as we know this enterocyte pool is not retrievable. Zinc is stored to a limited extent in the liver, as well as in the skeleton. The thyroid gland serves as the sole iodine store. Other minerals tend to be found in highest concentrations in the liver, but most tissues contain these same minerals. Hepatic stores can be potentially mobilized for distribution to other tissues during periods of need, but the mechanisms controlling retrieval of minerals from the liver are not understood.

Excretion

Renal and/or biliary excretory mechanisms exist for all the minerals except iron, but only a few of these are reasonably well understood. Skin losses of minerals via sweating or exfoliation may be significant, but these routes are not regulated by homeostatic control systems.

Summary

The physiological processes governing the entry and exit of minerals are not only complex, they also are different for practically each mineral nutrient. Thus, the knowledge and understanding of how each mineral is regulated in the body become important prior to formulating foods with additional amounts of a specific mineral in order to avoid adverse effects, toxicities, and undesirable interactions of minerals, including deficiencies.

EPITHELIAL TRANSPORT: NEW FINDINGS AND MECHANISMS

Transport mechanisms operating in gastrointestinal (GI), renal, and mammary epithelial tissues will be reviewed in this section. Emphasis will be again placed on calcium and iron, but magnesium, zinc, and other trace elements will also be briefly noted.

Gastrointestinal Transport

The mechanisms governing GI transport of the minerals are complex and are not entirely understood. These mineral uptake mechanisms currently remain active areas of investigation. In general, greater amounts of mineral nutrients when administered at high loads in foods or supplements can be "pushed" across the gut, but protective mechanisms operate to prevent overloads (see below for comments about iron overload).

Substantial progress has been made in understanding the regulation and mechanisms of absorption of certain minerals in recent years. Intestinal calcium absorption has a saturable, transcellular component and a nonsaturable, paracellular component (Bronner 1992). The active, transcellular component involves entry from the lumen into the intestinal cells, down an electrochemical gradient, passage through the cell, and pumping of the calcium ions out of the cells at the basolateral membrane with a Ca-ATPase. The regulatory step involved in changing the calcium transport rate is not entirely clear. Because of the low (50–100 nmol) intracellular calcium ion activity and the much higher concentration (1–5 mmol) in both the intestinal lumen and the serosal interstitial fluid and plasma, entry into the epithelial cells has been proposed as rate limiting (Stein 1992). However, comparing the calcium transport process through the intestines of vitamin D-deficient and -replete animals, the most important change appears to be in the synthesis of the vitamin D-dependent calcium-binding protein, calbindin, which presumably acts as a shuttle for calcium movement across the cytosol of the cell (Fullmer 1992). Transcellular movement of calcium through the enterocytes by sequestering it in lysosome-like vesicles has recently been proposed (Nemere 1992). Changes in the calcium-containing vesicle populations have been shown to be vitamin D dependent and contain calbindin immunoreactivity. Vesicular transport implies that calcium entry into the cell may involve endocytosis of an intestinal membrane calcium-binding protein.

Paracellular calcium transport also increases in the vitamin D-treated rats, in both the mucosal-to-serosal and serosal-to-mucosal directions. In the duodenum, but not the jejunum and ileum, there is net absorption of calcium by the paracellular pathway. This entry is not necessarily by diffusion down an electrochemical gradient; osmotic movement by a solvent drag phenomenon has also been implicated as a mechanism for the paracellular calcium pathway (Karbach 1992). This pathway could account for nearly all calcium absorption under normal circumstances in the lower two-thirds of the small intestine. The molecular mechanisms responsible for changes in paracellular transport rates of calcium are not fully understood.

Proposed mechanisms for zinc transport in the intestine are similar to those for calcium. A cysteine-rich intestinal protein (CRIP) is present in intestinal cytosol and may function in a manner analogous to calbindin. The amount of zinc bound to CRIP changed in proportion to the relative zinc transport. Because metallothionein concentration is more variable than CRIP in the absorbing cells, zinc transport rates may be regulated by competition for binding between metallothionein and CRIP (Hempe and Cousins 1992).

Much less progress has been made recently in understanding the absorptive mechanisms of other minerals and trace elements at the cellular and molecular levels. Rather, research has been directed more toward understanding dietary interactions of zinc with other nutrients. For example, a high zinc intake competitively inhibits copper absorption and, in clinical cases, it has resulted in copper deficiency (Botash et al. 1992). Inositol hexaphosphate (phytate) is a common component of grains and legumes that binds minerals with high affinity and, therefore, it may inhibit their absorption. Soaking, fermentation, and cooking of phytate-containing foods can partially degrade phytate to lower order inositol phosphates, that should have less inhibition of iron and zinc absorption (Sandberg 1991) and, probably, also of calcium. In test meals loaded with different inositol phosphates, hexa- and pentaphosphates inhibited zinc absorption, whereas inositol tetraphosphate did not. Analyses of inositol phosphates in naturally fermented breads was less closely correlated with their effect on zinc absorption (Sandstrom and Sandberg 1992). Calcium and zinc interact in determining the solubility of the inositol phosphate complexes (Simpson and Wise 1990). The chelating agents that are likely to be present in chyme, histidine, and picolinate can remove some zinc from the calcium-zinc-phytate precipitates.

Components of other foods, such as milk, can also alter mineral bioavailability. Caseins are phosphorylated at clusters of serine residues, which contain hydroxyl groups, in relatively close proximity in the amino acid chain. The phosphopeptides from tryptic digests of casein (CPP) have been characterized for a number of years (Manson and Annan 1971). *In vivo* production of these peptides by trypsin and other proteolytic enzymes in the gut may promote calcium uptake by intestinal cells by preventing precipitation of insoluble salts (Naito, Kawakami, and Imamura 1972). A more recent study (Kopra, Scholz-Ahrens, and Barth 1992) found that addition of CPP to a whey protein diet increased plasma calcium in vitamin D-deficient rats, but no other effects on calcium absorption were evident. Similarly, no difference in percentage retention of calcium from hydrolyzed casein and hydrolyzed whey formulas fed to Rhesus monkeys as observed, although both of these proteins resulted in greater ^{47}Ca retention than hydrolyzed soy protein formulas (Rudloff and Lonnerdal 1992). Possibly interaction of calcium with some other component in milk, such as citrate, is responsible for the increase in the observed absorption in some studies with casein phosphopeptides, but not in others.

Gastric acid secretion can also affect calcium absorption by the small intestine of adult subjects. For example, hyposecretors of acid and so-called achlorhydrics absorb calcium carbonate (taken as a solid pill) poorly while fasted, but they have no trouble absorbing calcium from calcium carbonate when administered following a meal (Recker 1985). Furthermore, calcium from calcium citrate (liquid or solid) is absorbed efficiently by these hypochlorhydric and achlorhydric subjects while fasting (Hunt and Johnson 1983; Nicar and Pak 1985; Pak, Poindexter, and Finlayson 1989).

The vitamin D status can also have a significant effect on calcium metabolism and bone mass of individuals, especially those who are elderly. The skin of an elderly person is less efficient in synthesizing vitamin D in response to ultraviolet-B light, and typically the elderly do not get outside much for exposure to sunlight. Populations living at latitudes distant from the equator are at the greatest risk for low vitamin D status (Webb et al. 1990). In addition, the calcium *and* vitamin D intakes of older subjects generally are considerably less than the RDAs. Thus, the vitamin D requirements of the elderly in much of the Northern Hemisphere are not being met by a combination of biosynthesis and diet (Holick 1986). When vitamin D status is low, those who consume low to marginal amounts of calcium compared to the RDA may have significantly reduced calcium absorption efficiency, which usually results in low bone mass (Fujita 1992).

Renal Transport

The efficiency of renal tubular reabsorption of most mineral nutrients remains high under nearly all conditions, but even small decrements in efficiency, e.g., 0.5 to 1.0%, can result in large urinary losses. This concept is best illustrated by factors, such as sodium, acid (H+) load, and amino acids, that interfere with calcium reabsorption from the tubular lumen into proximal (or distal) epithelial cells. Although the mechanisms have not been fully established, much experimental and clinical evidence supports the adverse effects of high sodium intake (King, Jackson, and Ashe 1964) and acid load (Lemann 1990) on partially inhibiting calcium reabsorption and, hence, increasing urinary calcium excretion. The data supporting an adverse effect of amino acids on calcium reabsorption operating through glucoregulatory hormones or other hormone-like factors are less convincing (Schuette and Linkswiler 1982; and Anderson, Thomsen, and Christiansen 1987), but they suggest that excessive protein intakes from animal products can be calciuretic by mechanisms that are at least partially independent of sodium or acid loads. One group of investigators has even hypothesized that the typically high protein intakes from animal products may be responsible for the high fracture rates in western nations (Abelow, Holford, and Insogna 1992).

The mechanisms governing hypercalciuria are diverse. At the cellular level, numerous studies have shown a sodium-calcium exchanger to be located in various segments of renal tubular cells. In studies of rabbit kidney (Brunnette, Maillous, and Lajeunesse 1992a,b) distal tubule cells had a low affinity, saturable, Ca++-uptake system in addition to the high affinity Na-Ca exchanger. Calcium absorption in the proximal tubule cells was not Na-dependent. An alkaline pH stimulated Ca++ transport by membrane vesicles from distal, but not proximal, tubule cells (Brunnette, Maillous, and Lajeunesse 1992a,b). However, in rat proximal tubules, the Na-Ca exchanger played an important role in regulation of calcium concentration of the tubule cells (Dominguez, Juhaszova, and Feister 1992). Species differences and location within the nephron appear to determine the reliance of kidney cells on the different molecular mechanisms for reabsorption of calcium. The redundancy of these mechanisms attests to the importance of precisely regulating calcium excretion. Dietary effects that might slightly alter this balance could have long-term metabolic consequences. The sodium-calcium competition for renal tubular transport (reabsorption) from the tubular lumen means that, at high sodium intakes, calcium reabsorptive efficiency will be lowered.

The increased acid load, generated in part by high animal protein intakes, also appears to operate by an ion competitive mechanism, but in this case hydrogen ions are secreted by tubular cells, which interferes with the reabsorption of calcium ions. In addition, mechanisms other than the acid load are most likely involved in protein-induced hypercalciuria. The mechanism or mechanisms producing the protein-induced hypercalciuria are poorly understood at this time, but the overall effects of acute increases in protein intake are clearly established (Anderson, Thomsen, and Christiansen 1987). It is not clear from actual experimentation, however, whether chronic ingestion of large amounts of animal protein will produce hypercalciuria of 20–30 mg of additional calcium each day. Lifelong vegetarian diets, either partial or total, consumed by Seventh-day Adventists do not appear to result in greater bone mass compared to omnivorous diets eaten by similarly aged elderly women (Tylavsky and Anderson 1988; Hunt et al. 1989). The potential skeletal benefits of various types of vegetarian diets deserve further investigation.

Much less is known about renal transport mechanisms for other minerals. The Na-Ca exchange system appears to transport magnesium as well; high Mg^{++} concentrations are competitively inhibitory. Zinc uptake by rabbit proximal tubule cells was shown to be stimulated by cysteine and histidine (Gachot et al. 1991). However, feeding a high level of cysteine to rats increased urinary zinc excretions (Hsu and Smith 1983). Most trace minerals are largely bound to plasma proteins that are not filtered through the glomeruli and, thus, they remain in the plasma. However, the fact that renal dialysis patients can become deficient in certain minerals implies that the physical chemistry of binding is not the only phenomenon involved in regulating the renal reabsorption of trace elements.

Mammary Transport by Lactating Women

The mineral transport into milk by mammary cells involves several discrete pathways (Neville, Allen, and Watters 1983). Lactose and milk proteins are secreted by exocytosis of secretory vesicles. In most species, casein micelles complex the large quantities of calcium, phosphate, magnesium, and zinc that are found in milk, so that exocytosis also transports most of the calcium, phosphate, and probably most of the protein-bound trace elements, such as selenium, iodine, and zinc. A second pathway for mineral secretion is that responsible for eccrine secretion of electrolytes through the apical plasma membrane. Sodium, potassium, and chloride move between the milk and mammary cytosol through channels or via cotransporters. Active transport of calcium through this membrane is also possible. Fat is secreted by the mammary cells with a unique pathway that functions like exocytosis in reverse; lipid droplets free in the cytoplasm are released into the milk coated with an apical plasma membrane-derived lipid bilayer. Small quantities of iron, copper, and other trace elements are associated with fat. Some milk proteins, such as immunoglobulin and transferrin (Sanchez et al. 1992), enter milk from the interstitial space via vesicular transcytosis. This process may transport iron, zinc, and manganese that bind to transferrin into milk. Movement of substances between plasma and milk through a paracellular pathway may be important during the formation of colostrum at the beginning of lactation, and also during involution and mastitis. This pathway can dramatically change the content of electrolytes (Allen 1990) and other minerals in milk (Allen, Keller, and Neville 1991).

Summary

Efficient epithelial transport of minerals is critical for both absorption and conservation of minerals. The goal of a functional food is to provide a greater quantity of a specific nutrient that will have a high probability of either enhancing the functional utilization of the mineral or improving its retention in tissue stores or functional units of cells. Thus, it would be desirable to enhance the entry of the mineral via gastrointestinal mechanisms and to reduce the excretion of the mineral, as long as no adverse effects accompanied either of these steps.

ISSUES OF MINERAL NUTRIENTS IN FOODS AND SUPPLEMENTS

This section will review four specific issues of problem minerals: minerals on the "generally recognized as safe" (GRAS) list of the FDA; fortification of foods with minerals and potential mineral-mineral interactions; supplementation of minerals through pills or tablets and potential interactions; and the use of additives in foods, such as phosphates, that have potentially adverse effects on the retention of problem mineral nutrients.

Minerals on the GRAS List

The current FDA list includes the following minerals: calcium, iron, magnesium, sodium, phosphorus, potassium, zinc, copper, manganese, and a few others (Table 15-4). In addition, the FDA has

Table 15-4. Selected Mineral Acids and Salts on the GRAS List for Various Purposes.

Multiple Purpose GRAS Food Substances	
Hydrochloric acid	
Phosphoric acid	
Silica aerogel	
[CATION]	[ANIONS]
Aluminum	sulfate
Calcium	citrate, phosphate
Potassium	citrate, glutamate
Sodium	carboxymethylcellulose, caseinate, phosphate, tripolyphosphate, {acid} pyrophosphate
Sodium aluminum phosphate	
Anticaking Agents	
[CATION]	[ANIONS]
Aluminum calcium	silicate
Calcium	silicate
Magnesium	silicate
Sodium	aluminosilicate
Tricalcium silicate	
Chemical Preservatives	
[CATION]	[ANIONS]
Calcium	ascorbate, sorbate
Potassium	bisulfite, metabisulfite, sorbate
Sodium	ascorbate, bisulfate, metabisulfite, sorbate, sulfite
Sulfur dioxide	

stated in a separate document that calcium supplementation up to 2,500 mg per day is considered safe.

The Code of Federal Regulations provides a partial list of substances that are generally recognized as safe for their intended use. A number of nutritionally essential mineral compounds are listed under the various functional categories (Table 15-4). The permissible amounts are not published, but good manufacturing practice requires that the quantity added to foods does not exceed that which is required to accomplish the intended physical, nutritional, or technical effect in food. The minimum effective amount of a food additive should be used, in recognition that all minerals can be toxic at a high enough dose. Further, the additive must be of food grade, or of sufficient purity that there is no risk of toxicity from any impurities. The substances on the GRAS list have many functions; not all are nutrients. (See Table 15-5).

When the list was established in 1958, there was a provision to review the safety of any substances called into question, and eventually, all of the compounds on the list. Because of the long history of safe use of most of these substances in practice, few of the mineral compounds have actually been thoroughly reviewed for safety.

All of the additives on the list of GRAS substances have specific functions. Anticaking agents act by coating the food particles to reduce their contact surface, and preferentially absorbing moisture. This allows dry foods to flow freely during processing or mixing (Baranowski 1990).

Chemical preservatives include the cation salts of sorbates, sulfites, nitrites, phosphates, and benzoate derivatives, and sodium chloride (Davidson and Juneja, 1990). Sorbates inhibit fungi and certain bacteria. Sulfites are used on fruits, fruit juices, wines, meats, shrimp, and pickles to retard the growth of pathogenic and spoilage microorganisms. Sulfites act as an oxidizing agent and can destroy thiamin in foods that are good sources of this vitamin. Certain asthmatic individuals are sensitive to sulfites in food or sulfur dioxide in the atmosphere. This finding has led to a ban on the use of sulfites on raw fruits and vegetables and labeling of other sulfite containing products. Sodium or potassium nitrate is used in cured meats to prevent growth of *Clostridium botulinum*. The nitrate inhibits numerous bacterial enzymes. Ascorbate, erythrobate, or other iron chelators enhance the inhibitory effects and are now required so that the nitrite concentration may be reduced. Toxic effects of nitrite relate to production of methemoglobinemia, and possible formation of nitrosamines, which are potent carcinogens. Over 30 phosphate salts are used in food processing; an antimicrobial activity is one of several functions of these additives. Other roles are emulsification, dispersion, metal chelation, water binding, and leavening (Dziezak 1990). Toxicological concerns from older studies relate to inhibition of calcium and iron absorption. Salt inhibits growth of microorganisms by reducing water activity. Concentrations required are dependent on pH, temperature, and interaction with other antimicrobial agents and processes. The high salt content of the American diet derives from this functional use of salt. Efforts underway in the food industry and nutritional community to reduce average salt intake are triggered by hypertension in sodium-sensitive persons.

Nutritional supplementation of foods has closely followed progress in nutritional science during the last century. When geographic

Table 15-5. Safe Consumption Limits of Minerals on the GRAS List.

Mineral		Upper Limit of Safety[a]	Animals[e]
Calcium		2,500	10,000
Potassium		18,000[b]	20,000
Iron		5,100	3,000
Zinc		2,000[c]	1,000
Copper		25[d]	
Manganese		10	400
Aluminum			200
Magnesium			3,000
Phosphorus			15,000
Dietary Supplements			
[CATION]	[ANIONS]		
Calcium	carbonate, citrate, glycerophosphate, oxide, pantothenate, phosphate, pyrophosphate		
Copper	gluconate		
Ferric	phosphate, pyrophosphate		
Ferrous	gluconate, lactate, sulfate		
Iron reduced			
Magnesium	oxide, phosphate, sulfate		
Manganese	citrate, gluconate, glycerophosphate, sulfate, oxide		
Potassium	chloride, glycerophosphate		
Sodium	pantothenate, phosphate		
Zinc	chloride, gluconate, oxide, stearate, sulfate		
Sequestrants			
[CATION]	[ANIONS]		
Calcium	citrate, diacetate, hexametaphosphate, phosphate (monobasic)		
Potassium	citrate, phosphate (dibasic)		
Sodium	citrate, gluconate, hexametaphosphate, metaphosphate, phosphate (monobasic, dibasic), pyrophosphate, tripolyphosphate		
Nutrients			
[CATION]	[ANIONS]		
Calcium	citrate, phosphate, pyrophosphate		
Choline	bitartrate, chloride		
Manganese	hypophosphate		
Sodium	phosphate		
Zinc	chloride, gluconate, oxide, stearate, sulfate		

[a]Data from NAS, NRC 1989 for adults, mg.
[b]Acute hyperkalemia.
[c]Acute toxicity.
[d]Possible chronic hypocupremia; mg/day.
[e]Values measured in or extrapolated to swine from other species as mg/kg of diet. From Subcommittee on Mineral Toxicity in Animals, NRC, NAS 1980. Mineral Tolerance of Domestic Animals. National Academy Press, Washington, D.C. pp. 5–7.

or population-based nutrient deficiencies have been recognized, restoration of the nutrient to the food supply of the affected groups has soon followed (Augustin and Scarbrough 1990). The public awareness of the nutritional content of foods has increased the use of nutritional additives, following the widespread nutritional labeling of foods. In general, minerals added to foods for nutritional purposes have been limited to those for which U.S. RDA values have been established: calcium, magnesium, phosphorus, copper, fluoride, iodine, iron, and zinc. Addition of fluoride is limited by law to water supplies and bottled water. Numerous forms of calcium are available for supplementation. Most studies have found the calcium bioavailability from various salts to be virtually identical (Heaney et al. 1990). Magnesium supplementation for nutritional purposes is not common, although several magnesium salts are permitted for this use. The RDI for phosphorus is set at 1,000 mg, providing a 1:1 Ca:P ratio. The phosphates added to foods during processing for functional uses, other than for nutrition, normally contribute to excessive phosphorus intake. However, nutritional supplementation with sodium or potassium phosphates is permitted. Copper supplementation with copper sulfate or copper gluconate is permitted for nutritional purposes, but it is not common because dietary copper deficiency is not a prevalent problem. However, newer dietary supplements that provide a high level of zinc may contain additional copper to prevent a competitive inhibition of zinc on copper absorption.

For iodine, the RDI of 150 μg/d is contained in 2 g iodized salt, which contains 76 μg l/g NaCl in the form of potassium iodide or cuprous iodide. Iodized salt consumption has eliminated iodine deficiencies in the United States (historically this had been a severe problem in the central states in the early decades of this century). Iron deficiency anemia is still a potential problem for women of childbearing age. Iron-containing dietary supplements and iron fortification of grains and cereals with the approved compounds have reduced this risk. Bioavailability of these salts is lower than the iron naturally contained in many foods. The RDI for zinc is 15 mg/d. Concerns exist for the adequacy of zinc intake in populations. Nutritional fortification of cereals and dietary supplements with zinc is in practice. In addition to the zinc salts listed in Table 15-4, zinc methionine sulfate has been approved as a dietary supplement. Manganese salts may be used in dietary supplements or for the nutrification of foods. Because manganese deficiency is unknown in humans, no RDI is set for this element.

Sequestrants, which are anions of organic acids that chelate heavy metals, form inactive complexes when added to food as the sodium, potassium, or calcium salts. These anions form stable complexes with iron or copper, which prevent the minerals from catalyzing the oxidation of lipids and related compounds. Sequestrants prevent metal-induced discoloration, turbidity, and flavor changes in foods (Lindsay 1985).

Fortification of Foods with Minerals

The question that arises is which frequently consumed foods can serve as vehicles for fortification. For calcium, only a few foods have been utilized in commercial formulations in the United States, but clearly many more products could be fortified, even if in smaller amounts. Snack food items have largely been ignored as vehicles for calcium fortification by food corporations, even though many of these foods might be ideal sources for fortification and readily marketed as healthful items under new regulations.

Iodine fortification of a food vehicle represents the classical prototype. Practically all nations now fortify one or more foods with iodine to prevent iodine deficiency and goiter.

Iron fortification via wheat and corn flour has existed in the United States for approximately 50 years. Many other nations follow the same practice. In addition, formula milks for infants are fortified with iron in the United States.

Fortification with zinc has only been widely used in infant formulas and baby foods. Magnesium, chromium, and selenium have not been used as fortificants in the United States; only nutrient supplementation of these mineral has been employed.

Supplementation with Minerals

Mineral supplementation has been practice for almost half a century in the United States. The marketing of pills or tablets containing absorbable mineral salts as a method of improving nutritional status by the pharmaceutical industry has been effective in a large segment of the more affluent population because of claims of health benefits. Iron supplementation of menstruating women, for example, has been well accepted by the medical profession, as well as nutritionists and pharmacists, for preventing iron deficiency and the more severe iron deficiency anemia. Similarly, calcium supplements have been widely touted, although compliance by women has not been as good as for iron, since the NIH Consensus Conference on Osteoporosis (1984) recommended increased calcium consumption by American women.

A recent prospective epidemiologic investigation of over 45,000 men over 40 years of age found that high calcium intakes, including calcium from supplements, led to a decline in symptomatic kidney stones (Curhan et al. 1993). If verified by other reports, this finding, which implies that calcium actually aids in preventing the formation of renal stones, means that the use of calcium as a supplement or food fortificant is not contraindicated in stone-forming individuals. Thus, additional amounts of calcium in diets is safe up to approximately 2,500 mg per day.

The health benefits of additional amounts of magnesium are beginning to influence consumer use of supplements of this mineral. Selenium and chromium supplements are rarely taken because the level of consciousness of deficiencies of these trace elements is low, and the health benefits of additional intakes have not yet been well substantiated.

Phosphate Additives and Potential Adverse Effects on Minerals

Phosphates have been used as functional additives, rather than as nutrients, for a long time to maintain shelflife or specific qualities of foods. The potential problem of many of these salts containing phosphates is the gut interactions that may make several minerals less bioavailable and, therefore, absorbable.

Another potential problem may arise for calcium because of the development of a phosphate-induced hyperparathyroidism that could contribute to osteopenia and subsequent osteoporotic-related fractures later in life. In a month-long investigation of young adult women given a low-calcium, high-phosphate diet (1:4 ratio), serum parathyroid hormone concentrations were significantly elevated, which suggests that bone turnover was also increased (Calvo, Kumar, and Heath 1990). Such a consumption pattern on a regular basis could conceivably contribute to a reduction in bone mass and density and place an individual woman (or man) at increased risk of fracture at a later stage of life. If high phosphate diets are consumed in the presence of adequate amounts of calcium, e.g., RDA values, then the risk of low bone mass and subsequents fractures is presumed to be greatly reduced.

Oral phosphate administration has also been shown to enhance the synthesis of the hormonal form of vitamin D, namely, 1,25-dihydroxvitamin D (Portale et al. 1986).

Summary

Calcium, iron, zinc, and iodine have been the primary mineral fortificants in the United States and in many other nations. Zinc, as noted earlier, has only been added to baby foods and to a few breakfast cereals, but it is a likely candidate for future food fortifications. Magnesium may be another micronutrient that will be added to selected foods. Populations benefiting the most from these fortificants are the very young, especially infants, and the elderly, but other segments of the populations such as menstruating women (iron) and females from prepuberty and beyond (calcium) may also gain improved nutritional *and* health status. It appears that we are just at the forefront of mineral fortification of foods.

HEALTH EFFECTS OF SPECIFIC MINERAL DEFICIENCIES

This section will deal with both clearly established as well as a few questionable mineral deficiencies. Thus, the health consequences of deficiencies of calcium, magnesium, iron, zinc, chromium, and selenium will be reviewed.

Calcium—Osteoporosis, Osteomalacia, Hypertension, Colon Cancer

Dietary calcium deficiency has been linked epidemiologically to several chronic diseases, including osteoporosis, osteomalacia, hypertension, and colon cancer. Each of these calcium-disease relationships will be briefly reviewed.

The calcium-bone association has now been investigated for several decades. It is far better supported by laboratory and clinical evidence than either the relationship between calcium and hypertension or the one between calcium and colon cancer. Several natural history or retrospective studies have demonstrated positive linkages between adequate lifetime calcium intakes and greater bone mass measurements or reduced fracture rates (Matkovic et al. 1979; Hurxthal and Vose 1969; Halioua and Anderson 1989; Anderson, 1991, 1992; Weaver 1992; Sandler et al., 1985). Other lines of evidence in support of this positive relationship come from supplementation studies during childhood and adolescence (Johnston et al. 1992; Recker et al. 1992; Reid et al. 1993), the perimenopause (Elders, Netelenbos, and Lips 1991), the first postmenopausal decade (Dawson-Hughes et al. 1990), and late life (Chapuy et al. 1992). In the report by Chapuy et al. (1992) of almost 2,000 octogenarian women living in and around Lyon, France, a daily supplement of 1,200 mg of calcium and 800 IU of vitamin D resulted in a reduction of approximately 50% of all nonvertebral fractures. Thus, the emerging data from supplemental and other types of studies suggest that inadequate calcium intakes from traditional foods can be overcome through the consumption of supplemental calcium with the reasonable assurance of improved bone mass and a significant reduction in the risk of fracture, even when instituted late in life. The mechanism for the improved bone mass and reduced fracture risk lies in both the enhanced intestinal calcium absorption and the increased calcium uptake and retention by bone tissue.

The relationship between calcium and the maintenance of a lower and healthier blood pressure during adult life has only been established in recent years (McCarron et al. 1984), but the mechanism for the benefit of calcium remains elusive.

Colon cancer is characterized by an increased rate of proliferation of enterocytes in the crypts throughout the large intestine, but more frequently in the sigmoid colon than in other regions. Secondary and tertiary bile acids, as well as fatty acids, have been invoked as the causative carcinogenic agents of this cancer, but other etiologic factors, including low calcium consumption have also been implicated in this disease. Recent trials involving calcium supplements administered to subjects with familial polyposis coli (Lipkin et al. 1989), which have high rates of colonic cell turnover but a benign status, and in patients with resected adenomas (Wargovich et al. 1992) have supported the positive association between calcium and reduced cell turnover rates which would be consistent with a cancer-like condition. A recent small, short-term, double-blinded, randomized, placebo-controlled clinical trial, however, has not been so supportive of the beneficial effects of calcium on reducing the proliferation of colonic epithelia in high-risk subjects (Bostick et al. 1993).

Magnesium—Coronary Heart Disease

The role of magnesium in the function of heart muscle, which has been unfolding in recent years, is critical for various metabolic pathways, especially those that generate ATP (Dyckner 1980; Seelig 1980). ATPase enzymes in the heart have a high requirement for magnesium. Magnesium deficiency may compromise the optimal operation of all of these pathways and help contribute to compromised muscle function, especially in elderly subjects. Magnesium deficiency also affects brain function and manifests itself in anorexia, personality changes, and other effects. Skeletal muscle function is also compromised by magnesium deficits (Elin 1988). All of these functional declines have been established in human subjects, but it is almost impossible to produce a clear-cut magnesium deficiency via inadequate consumption of this mineral from foods. Because of low dietary intakes coupled with gastrointestinal dysfunction and/or medication usage, the elderly are more likely to become magnesium deficient than any other segment of the population (Sherwood et al. 1986).

Iron—Iron Deficiency Anemia

Iron deficiency anemia, rather than simple iron deficiency, is characterized by decrements in red blood cell function, as well as in depressed function of numerous other cellular activities because of inadequate oxygen delivery. Several indices of iron status are assessed clinically, and one of these, serum ferritin, has become a highly useful marker for body iron stores (Herbert 1992).

The mechanistic roles of iron excess in both CHD and cancer are poorly understood. Several epidemiologic lines of evidence suggest that high intakes of iron by males and postmenopausal women may contribute to CHD because of the speculated increase in iron-stimulated oxidation of LDL lipid components. Similar epidemiologic evidence on excess iron consumption has been gathered for cancer, at several sites, but speculations about possible mechanisms remain unsupported by experimental data.

Iron overload in genetically prone individuals results, not so much because of high dietary intakes, but primarily because of a highly efficient GI absorptive mechanism (Herbert 1992).

Zinc—Early Growth Failure, Compromised Immune Defense

In recent years, zinc deficiency has been recognized as a problem of public health significance in particular groups. Sandstead (1991) cited an evaluation of dietary intakes in young women in the United States that concluded that 20–30% would have been at risk for zinc deficiency had they become pregnant. The RDA of women for zinc increases from 12 mg/day to 15 mg/day during pregnancy and 19 and 16 mg/day during the first and second six months of lactation, but many women do not achieve these intakes. A number of studies have found a correlation between various indices of maternal zinc status, such as amniotic fluid, serum, and leukocyte zinc concentration and infant birth weight (Favier 1992; Sievers et al. 1992).

The zinc in human milk provides the only source of zinc for breast-fed infants. Because the zinc concentration of human milk decreases steeply over the course of lactation, the zinc intakes in breast-fed infants can decline by approximately 75% in a 17-week period, whereas the intakes of zinc by bottle-fed infants decline at a slower rate after the first few months; limitations of zinc have been speculated to be a possible cause of growth delays. Zinc deficiency inhibits the synthesis or activity of growth hormone, thyroid hormones, and androgens such as testosterone, and these effects have been proposed as a mechanism by which zinc deficiency impairs growth rate (Favier 1992). While the correction of lower-than-normal growth rates by zinc supplementation has been well documented, the overall clinical significance is less well defined (Hambidge 1991).

The classical severe zinc deficiency has only been reported for a few subjects living in Iran and Egypt (Favier 1992), but the characteristics of marginal or suboptimal zinc deficiency have been more difficult to pinpoint. The preschool children in Denver reported to have marginal zinc deficiency showed poor growth and limited attention spans (Hambidge 1991). A well-characterized syndrome has not been established or verified by other investigators. Part of the lack of support may be that the poorly nourished preschoolers in Denver were deficient in many nutrients, including high quality protein, besides zinc, which could have affected their growth and mental functional status.

Immune defense compromised by zinc deficiency has been similarly elusive, but in recent years several reports have suggested that suboptimal zinc intakes contribute to depressed cellular immunity, especially of cells of the humoral arm (Chandra 1991). Results of this active field of investigation hold promise, but conclusions are not yet warranted.

Chromium—Impaired Glucose Tolerance and Diabetes Mellitus (Type II)

Functional deficits resulting from chromium deficiency have not been well established, even though researchers have shown in well-designed crossover studies that chromium supplementation improves glucose tolerance (Anderson et al. 1983). The diets of most Americans are low in chromium, but whether this observation translates to functional deficits has not been established with certainty. Further investigations are needed to clarify the meaning of the low chromium intakes.

Selenium—Cancer

A deficiency of selenium, which functions as part of an antioxidant enzyme system, glutathione oxidase, has not been experimentally linked to cancer, but epidemiologic evidence suggests that inadequate selenium consumption may be associated with increased overall rates of cancers. Selenium deficiency in western regions of China has been found to be associated with Keshan disease, a cardiomyopathy found only in the People's Republic of China (Keshan Disease Research Group 1979a,b).

Most research attention on the nutritional effects of selenium in the last 20 years has focused on its role as an antioxidant in the enzyme glutathione peroxidase. This enzyme converts hydrogen peroxide or lipid peroxides to alcohols by coupling peroxide reduction to oxidation of glutathione. This antioxidant mechanism is a likely explanation for the protective effect of selenium against cancer and heart disease. Numerous epidemiological studies have shown an inverse relationship between selenium intakes and cancer risk. In animal studies, dietary selenium has been found to be protective against tumor formation in rodents fed various types of carcinogens (Levander 1986).

Similarly, both epidemiological and animal experiments suggest that low selenium intakes and low plasma selenium concentrations increase the risk for coronary heart disease. Oxidative damage to cholesterol is a probable factor in the development of atherosclerotic plaques that lead to ischemic heart disease. Selenium-dependent glutathione peroxidase may reduce cholesterol oxidation and retard the progression of this disease.

Recently, another selenium-containing enzyme of humans has been identified and associated with pathophysiology during selenium deficiency. In extrathyroidal cells, iodothyronine deiodinase, type I, (ID-I) converts thyroxine (T_4) to triiodothyronine (T_3), the thyroid hormone with the greater physiological activity. The gene for the enzyme has been cloned and a codon in the DNA sequence was identified that translates to selenocysteine at the active site of the enzyme (Berry and Larsen 1993). In selenium-deficient rats, T_4 increased and T_3 decreased in plasma and liver, a primary tissue site of ID-I. In the thyroid, both hormones were decreased. In rats deficient in both selenium and iodine, plasma, liver, and thyroid T_4 concentrations were decreased, but liver T_3 was slightly increased relative to control rats. In the combined deficiency, type II iodothyronine deiodinase (ID-II), which does not contain selenium, was elevated, possibly enabling the brain to maintain adequate concentrations of T_3 (Beckett et al. 1993). Corvilain et al. (1993) described two syndromes of the iodine deficiency known as endemic cretinism. Nervous cretinism in New Guinea and Latin America showed mental retardation and goiter, but relatively little or no hypothyroidism. Endemic cretins in central Africa had less goiter, but they were more hypothyroid. Increased severity of clinical symptoms (growth depression and mental retardation) correlated with the degree of hypothyroidism in these populations. The African cretins were subsequently found to be deficient in selenium, as well as iodine. The authors postulated that the selenium deficiency led to low thyroidal glutathione peroxidase activity that allowed the hydrogen peroxide associated with iodination of thyroid proteins to build up and destroy the thyroid cells. In peripheral tissues, low ID-I may spare thyroxine (T4) for conversion to T_3 by ID-II in brain. In nervous cretinism (adequate selenium), the thyroid is stimulated to enlarge, but still produces only small amounts of thyroxine, which are deiodinated appropriately in peripheral tissues. Deficiency of T_3

during brain development results in deaf-mutism and other forms of mental retardation.

Iodine deficiency in developed countries has been nearly eliminated by the addition of iodine to salt. As efforts are made to reduce the risk of iodine deficiency to the 1 billion people who are wholly dependent on food from iodine-deficient soils, attention should be paid to the possibility of selenium deficiency in these populations as well. Because selenium deficiency appears to protect against some of the symptoms of iodine deficiency, selenium supplementation should not be undertaken without correction of the iodine deficiency (Vanderpas et al. 1993).

Summary

The significant number of chronic diseases affected by mineral deficiencies and the potentially millions of people with morbidity emphasize the importance of meeting the physiological requirements of these diverse nutrients throughout life via foods and/or supplements. The disease load, and mortality as well, resulting from osteoporosis in the United States alone accounts for over $10 billion. Projections by experts who estimate that the elderly population (65 and over) will increase in the United States from 12.5% in the early 1990s to approximately 25% by 2030 would suggest that the dollar costs of osteoporosis will more than double over this same period. Functional foods providing modest increments in the amounts of the essential problem minerals should be able to make significant impacts in reducing the morbidity and mortality associated with the aforementioned mineral-related chronic diseases.

SUMMARY AND CONCLUSIONS

Although this review of the minerals has focused on considerations of the dietary mineral intakes and RDAs of Americans, the general principles introduced herein have broad application to any nation or population throughout the world. The development of functional foods, designer foods, nutraceuticals, or foods by any other name that are fortified with a mineral nutrient began only a few years ago, but future introductions of such foods by food manufacturers and processors should be exponential because of the widespread prevalence of micronutrient deficiencies, both in developed and less developed countries. The potential benefits of functional foods on the nutritional status of populations are difficult to predict, but increased use of functional foods may eliminate mineral deficiencies and lead to improved health and quality of life.

References

Abelow, B.J.; Holford, T.R.; and Insogna, K.L. 1992. Cross-cultural association between dietary animal protein and hip fracture.: A hypothesis. *J. Nutr.* 50:14–18.

Allen, J.C. 1990. Milk synthesis and secretion rates in cows with milk composition changed by oxytocin. *J. Dairy Sci.* 73:975–984.

Allen, J.C.; Keller, R.P.; and Neville, M.C. 1991. Studies in human lactation. Milk composition and daily excretion rates of macronutrients in the first year of lactation. *Am. J. Clin. Nutr.* 54:69–80.

Anderson, J.J.B. 1991. Nutritional biochemistry of calcium and phosphorus. *J. Nutr. Biochem.* 2:300–307.

———. 1992. The role of nutrition in the functioning of skeletal tissue. *Nutr. Rev.* 50:388–394.

Anderson, J.J.B.; Thomsen, K.; and Christiansen, C. 1987. High protein meals, insular hormones and urinary calcium excretion in human subjects. In *Osteoporosis 1987*, C. Christiansen, J.S. Johansen, and B.J. Riis, eds., pp. 240–245. Copenhagen: Osteopress ApS.

Anderson, R.A.; Polansky, M.M.; Bryden, N.A.; Roginski, E.E.; Mertz, W.; and Glinsmann, W. 1983. Chromium supplementation of human subjects: Effects on glucose, insulin and lipid variables. *Metabolism* 32:894–899.

Augustin, J., and Scarbrough, F.E. 1990. Nutritional additives. In *Food Additives*, A.L. Branen, P.M. Davidson, and S. Salminen, eds., pp. 33–81. New York: Marcel Dekker, Inc.

Baranowski, E.S. 1990. Miscellaneous food additives. In *Food Additives*, A.L. Branen, P.M. Davidson, and S. Salminen, eds., pp. 511–578. New York: Marcel Dekker, Inc.

Beckett, G.J.; Nicol, F.; Rae, P.W.H.; Beech, S.; Guo, Y.; and Arthur, J.R. 1993. Effects of combined iodine and selenium deficiency on thyroid hormone metabolism in rats. *Am. J. Clin. Nutr.* (Suppl) 57:240S–243S.

Berry, M.J., and Larsen, P.R. 1993. Molecular cloning of the selenocysteine-containing enzyme type I iodothyronine deiodinase. *Am. J. Clin. Nutr.* (Suppl) 57:249S–255S.

Bostick, R.M.; Potter, J.D.; Fosdick, L.; Grambsch, P.; Lampe, J.W. ϖod, J.R.; Louis, T.A.; Ganz, R.; and Grandits, G. 1993. Calcium and colorectal epithelial cell proliferation: A preliminary randomized, double-blinded, placebo-controlled clinical trial. *J. Natl. Cancer Inst.* 85:132–141.

Botash, A.S.; Nasca, J.; Dobowy, R.; Weinberger, H.L.; and Oliphant, M. 1992. Zinc-induced copper deficiency in an infant. *Amer. J. Dis. Child* 146:709–711.

Bronner, F. 1992. Current concepts of calcium absorption: An overview. *J. Nutr.* 122:641–643.

Brunnette, M.G.; Mailloux, J.; and Lajeunesse, D. 1992a. Calcium transport through the luminal membrane of the distal tubule. I. Interrelationship with sodium. *Kidney Int.* 41:281–288.

———. 1992b. Calcium transport through the luminal membrane of the distal tubule. II. Effect of pH, electrical potential and calcium channel inhibitors. *Kidney Int.* 41:289–96.

Calvo, M.S.; Kumar, R.; and Heath, H. 1990. Persistently elevated parathyroid hormone secretion and action in young women after four weeks of ingesting high phosphorus, low calcium diets. *J. Clin. Endocrinol. Metab.* 70:1334–1340.

Chandra, R.K. 1991. Trace elements and immune responses. In *Trace Elements in Nutrition of Children—II*, R.K. Chandra, ed., pp. 201–214. Nestle Nutrition Workshop Series, Vol. 23. New York: Raven Press.

Chapuy, M.C.; Arlot, M.E.; Duboeuf, F.; Arnaud, S.; Delmas, P.D.; and Meunier, P.J. 1992. Vitamin D3 and calcium to prevent hip fractures in elderly women. *N. Engl. J. Med.* 327:1637–1642.

Corvilain, B.; Contempre, B.; Longombe, A.O.; Goyens, P.; Gervey-Decoster, C.; Lamy, F.; Vanderpas, J.B.; and Dumont, J.E. 1993. Selenium and the thyroid: How the relationship was established. *Am. J. Clin. Nutr.* (Suppl) 57:244S–248S.

Curhan, G.C.; Willet, W.C.; Rimm, E.B.; and Stampfer, M.J. 1993. A prospective study of dietary calcium and other nutrients and the risk of symptomatic kidney stones. *N. Engl. J. Med.* 328:833–838.

Davidson, P.M., and Juneja, V.K. 1990. Antimicrobial agents. In *Food Additives* A.L. Branen, P.M. Davidson, and S. Salminen, eds., pp. 83–137. New York: Marcel Dekker, Inc.

Dawson-Hughes, B.; Dallal, G.E.; Krall, E.A.; Sadowski, L.; Sahyoun, N.; and Tannenbaum, S. 1990. A controlled trial of calcium supplementation on bone density in postmenopausal women. *N. Engl. J. Med.* 323:878–883.

Dominguez, J.H.; Juhaszova, M.; and Feister, H.A. 1992. The renal sodium- calcium exchanger. *J. Lab. Clin. Med.* 119:640–649.

Dyckner, T. 1980. Serum magnesium in acute myocordial infarction. *Acta Med. Scand.* 207:59–66.

Dziezak, J.D. 1990. Phosphates improve many foods. *Food Technol.* 44(4): 80–92.

Elders, P.J.M.; Netelenbos, C.; and Lips, P. 1991. Calcium supplementation reduces vertebral bone loss in perimenopausal women: A controlled trial in 248 women between 46 and 55 years of age. *J. Clin. Endocrinol. Metab.* 73:533–540.

Elin, R.J. 1988. Magnesium metabolism in health and disease. *Dis. Month* 34:161–219.

Favier, A.E. 1992. Hormonal effects of zinc on growth in children. *Biol. Trace Elem. Res.* 32:383–398.

Fujita, T. 1992. Vitamin D in the treatment of osteoporosis. *Proc. Soc. Exp. Biol. Med.* 199:394–399.

Fullmer, C.S. 1992. Intestinal calcium absorption: Calcium entry. *J. Nutr.* 122:644–650.

Gachot, T.; Tauc, M.; Morat, L.; and Poujeol, P. 1991. Zinc uptake by proximal cells isolated from rabbit kidney: Effects of cysteine and histidine. *Pflugers Arch.* 419:583–587.

Halioua, L., and Anderson, J.J.B. 1989. Lifetime calcium intake and physical activity habits: Independent and combined effects on the radial bone of healthy premenopausal Caucasian women. *Am. J. Clin. Nutr.* 49:534– 541.

Hambidge, K.M. 1991. Zinc in the nutrition of children. In *Trace Elements in Nutrition of Children—II.* Chandra, R.K., ed., pp. 65–77. Nestle Nutrition Workshop Series, Vol. 23. New York: Raven Press.

Heaney, R.P.; Smith, K.T.; Recker, R.R.; and Hinders, S.M. 1989. Meal effects on calcium absorption. *Am. J. Clin. Nutr.* 49:372–376.

Heaney, R.P.; Weaver, C.M.; Fitzsimmons, M.L.; and Recker, R.R. 1990. Calcium absorptive consistency. *J. Bone Miner. Res.* 5:1139–1142.

Hempe, J.M., and Cousins, R.J. 1992. Cysteine-rich intestinal protein and intestinal metallothionein: An inverse relationship as a conceptual model for zinc absorption in rats. *J. Nutr.* 122:89–95.

Herbert, V. 1992. Everyone should be tested for iron disorders. *J. Am. Diet. Assoc.* 92:1502–1509.

Holick, M.F. 1986. Vitamin D requirements for the elderly. *Clin. Nutr.* 5: 121–129.

Hsu, J.M., and Smith, J.C. 1983. Cysteine feeding affects urinary zinc excretion in normal and ethanol-treated rats. *J. Nutr.* 113:2171–2177.

Hunt, I.F.; Murphy, N.J.; Henderson, C.; Clark, V.A.; Jacobs, R.; Johnston, P.K.; and Coulson, A.H. 1989. Bone mineral content in postmenopausal women: Comparison of omnivores and vegetarians. *Am. J. Clin. Nutr.* 50:517–523.

Hunt, J.N., and Johnson, C. 1983. Relation between gastric secretion of acid and urinary excretion of calcium after oral supplements of calcium. *Digest. Dis. Sci.* 28:417–421.

Hurxthal, L.M., and Vose, G.P. 1969. The relationship of dietary calcium intake to radiographic bone density in normal and osteoporotic persons. *Calcif. Tissue Res.* 4:245–256.

Johnston, C.C., Jr.; Miller, J.Z.; Slemenda, C.W.; Reister, T.K.; Hui, S.; Christian, J.C.; and Peacock, M. 1992. Calcium supplementation and increases in bone mineral density in children. *N. Engl. J. Med.* 327:82–87.

Karbach, U. 1992. Paracellular calcium transport across the small intestine. *J. Nutr.* 122:672–677.

Keshan Disease Research Group. 1979a. Observations on effects of sodium selenite in prevention of Keshan disease. *Chin. Med. J.* 92:471–476.

———. 1979b. Epidemiological studies on the etiological relationships of selenium and Keshan disease. *Chin. Med. J.* 92:477–482.

King, J.S.; Jackson, R.; and Ashe, B. 1964. Relation of sodium intake to urinary calcium excretion. *Invest. Urol.* 1:555–560.

Kopra, N.; Scholz-Ahrens, K.E.; and Barth, C.A. 1992 Effect of casein phosphopeptides on utilization of calcium in vitamin-D replete and vitamin-D deficient rats. *Milchwissenschaft* 47:488–492.

Lemann, J., Jr. 1990. The urinary excretion of calcium, magnesium and phosphorus. In *Primer on the Metabolic Bone Diseases and Disorders of Mineral Metabolism,* M.J.

Favus, ed., pp. 36–39. American Society for Kelseyville, CA: Bone and Mineral Research.

Levander, O.E. 1986. Selenium. In *Trace Elements in Human and Animal Nutrition*, Volume 2, 5th ed. W. Mertz, ed., pp. 209–279. Orlando: Academic Press.

Lindsay, R.C. 1985. Food additives. In O.R. Fennema, *Food Chemistry*. 2d ed., O.R. Fennema, pp. 629–687. New York: Marcel Dekker.

Lipkin, M.; Friedman, E.; Winawer, S.J.; and Newmark, H. 1989. Colonic epithelial cell proliferation in responders and nonresponders to supplemental dietary calcium. *Cancer Res.* 49:248–254.

Manson, W., and Annan, W.D. 1971. The structure of a phosphopeptide derived from β-casein. *Arch. Biochem. Biophys.* 145:16–25.

Matkovic, V.; Kostial, K.; Simonovic, I.; Buzina, R.; Brodarec, A. and Nordin, B.E.C. 1979. Bone status and fracture rates in two regions of Yugoslavia. *Am. J. Clin. Nutr.* 32:540–549.

McCarron, D.A.; Morris, C.D.; Henry, H.J.; and Stanton, J.L. 1984. Blood pressure and nutrient intake in the United States. *Science* 224:1392–1398.

Moser-Veillon, P.B. 1990. Zinc: Consumption patterns and dietary recommendations. *J. Am. Diet. Assoc.* 90:1089–1093.

Naito, H.; Kawakami, A.; and Imamura, T. 1972. In vivo formation of phosphopeptide with calcium binding property in the small intestinal tract of the rat fed casein. *Agric. Biol. Chem.* 36:409.

National Academy of Sciences. 1989. *Recommended Dietary Allowances*. 10th ed. Subcommittee on the Recommended Dietary Allowances, National Research Council, Washington, D.C.: National Academy Press.

Nemere, I. 1992. Vesicular calcium transport in chick intestine. *J. Nutr.* 122:657–661.

Neville, M.C.; Allen, J.C.; and Watters, C.D. 1983. Mechanisms of milk secretion. In *Human Lactation: Physiology, Nutrition and Breast-Feeding*, M.C. Neville, ed., pp. 49–102. New York: Plenum.

NHANES I, National Center for Health Statistics. 1977. *Dietary Intake Findings, United States, 1971–1974*. Vital and Health Statistics. Series 112, No. 202. DHEW Pub. No. (HRA) 77-1647. Washington, D.C.: U.S. Government Printing Office.

Nicar, M.J., and Pak, C.Y.C. 1985. Calcium bioavailability from calcium carbonate and calcium citrate. *J. Clin. Endocrinol. Metab.* 61:391–393.

NIH (National Institutes of Health). 1984. Consensus Conference on Osteoporosis, 1984. *J. Am. Med. Assoc.* 252:799–802.

Pak, C.Y.C.; Poindexter, J.; and Finlayson, B. 1989. A model system for assessing physicochemical factors affecting calcium absorbability from the intestinal tract. *J. Bone Miner. Res.* 4:119–127.

Pennington, J.A.; Young, B.E.; and Wilson, D.B. 1989. Nutritional elements in U.S. diets: Results from the Total Diet Study, 1982 to 1986. *J. Am. Diet. Assoc.* 89:659–664.

Portale, A.A.; Halloran, B.P.; Murphy, M.M.; and Morris, R.C. 1986. Oral intake of phosphorus can determine the serum concentration of 1,25- dihydroxyvitamin D by determining its production rate in humans. *J. Clin. Invest.* 77:7–12.

Recker, R.R. 1985. Calcium absorption and achlorhydria. *N. Engl. J. Med.* 313:70–73.

Recker, R.R.; Davies, K.M.; Hinders, S.M.; Heaney, R.P.; Stegman, M.R.; and Kimmel, D.B. 1992. Bone gain in young adult women. *J. Am. Med. Assoc.* 268: 2403–2408.

Reference Daily Intakes (RDIs). 1993. Federal Register. 58 (No. 3):2227–2228.

Reid, I.R.; Ames, R.W.; Evans, M.C.; Gamble, G.D.; and Sharpe, S.J. 1993. Effect of calcium supplementation on bone loss in postmenopausal women. *N. Engl. J. Med.* 328:460–464.

Rudloff, S., and Lonnerdal, B. 1992. Calcium and zinc retention from protein hydrolysate formulas in suckling rhesus monkeys. *Am. J. Dis. Child.* 146:588–91.

Sanchez, L.; Lujan, L.; Oria, R.; Castillo, H.; Perez, D.; Ena, J.M.; and Calvo, M. 1992. Synthesis of lactoferrin and transferrin in the lactating mammary gland of sheep. *J. Dairy Sci.* 75:1257–1262.

Sandberg, A.F. 1991. The effect of food processing on phytate hydrolysis and availability of iron and zinc. In *Nutritional and Toxicological Consequences of Food Processing*, M. Friedman, ed., pp. 499–508. New York: Plenum.

Sandler, R.B.; Slemenda, C.W.; LaPorte, R.E.; Cauley, J.A.; Schram, M.M.; Banesi, M.L.; and Kriska, A.M. 1985. Postmenopausal bone density and milk consumption in childhood and adolescence. *Am. J. Clin. Nutr.* 42:270–274.

Sandstead, H.H. 1991. Zinc deficiency: A public health problem? *Am. J. Dis. Child.* 145:853–859.

Sandstrom, B., and Sandberg, A.S. 1992. Inhibitory effects of isolated inositol phosphates on zinc absorption in humans. *J. Trace Elem. Electrolytes Health Dis.* 6:99–103.

Schuette, S.A., and Linkswiler, H.M. 1982. Effects of Ca and P metabolism in humans by adding meat, meat plus milk, or purified proteins plus Ca and P to a low protein diet. *J. Nutr.* 112:338–349.

Seelig, M. 1980. *Magnesium Deficiency in the Pathogenesis of Disease: Early Roots of Cardiovascular, Skeletal, and Renal Abnormalities.* New York: Plenum Medical Books.

Sherwood, R.A.; Aryanayagam, P.; Rocks, B.F.; and Mankikar, G.D. 1986. Hypomagnesemia in the elderly. *Gerontology* 32:105–109.

Sievers, E.; Oldigs, H.-D.; Dorner, K.; and Schaub, J. 1992. Longitudinal zinc balances in breast-fed and formula-fed infants. *Acta Paediatr.* 81: 1–6.

Simpson, C.J., and Wise, A. 1990. Binding of zinc and calcium to inositol phosphates (phytate) in vitro. *Br. J. Nutr.* 64:225–232.

Stein, W. 1992. Facilitated diffusion of calcium across the rat intestinal cell. *J. Nutr.* 122:651–656.

Tylavsky, F.A., and Anderson, J.J.B. 1988. Dietary factors in bone health of elderly lactoovovegetarian and omnivorous women. *Am. J. Clin. Nutr.* 48:842–849.

USDA, Human Nutrition Information Service, H-6. 1983. *Food Consumption: Households in the United States, Seasons and Year, 1977–78*. Washington, D.C.: U.S. Government Printing Office.

Vanderpas, J.B.; Contempre, B.; Duale, N.L.; Deckx, H.; Bebe, N. ∠ngombe, A.O.; Thilly, C.H.; Diplock, A.T.; and Dumont, J.E. 1993. Selenium deficiency mitigates hypothyroxinemia in iodine-deficient subjects. *Am. J. Clin. Nutr.* (Suppl) 57:271S–275S.

Wargovich, M.J.; Isbell, G.; Shabot, M.; Winn, R.; Lanza, F.; Hochman, L.; Larson, E.; Lynch, P.; Rouben, L.; and Levin, B. 1992. Calcium supplementation decreases rectal epithelial cell proliferation in subjects with sporadic adenoma. *Gastroenterology* 103:92–97.

Weaver, C.M. 1992. Calcium bioavailability and its relation to osteoporosis. *Proc. Soc. Exp. Biol. Med.* 200:157–160.

Webb, A.R.; Pilbeam, C.; Hanafin, N.; and Holick, M.F. 1990. A one-year study to evaluate the roles of exposure to sunlight and diet on the circulating concentrations of 25-OH-D in an elderly population in Boston. *Am. J. Clin. Nutr.* 51:1075–1081.

Chapter 16

Fatty Acids

Artemis P. Simopoulos

INTRODUCTION

In the last 20 years epidemiologic studies, clinical investigations, and animal experiments have expanded our knowledge of the properties of dietary fatty acids in health and disease and in growth and development. The impetus for the research and expansion in our knowledge resulted from research advances in prostaglandin metabolism, membrane structure and function, the role of ω-3 (omega-3) fatty acids in gene expression, and the epidemiologic investigations that demonstrated a much lower rate of coronary heart disease in Eskimos (Bang and Dyerberg 1972; Dyerberg, Bang, and Hjorne 1975; Bang, Dyerberg, and Hjorne 1976; Dyerberg, Bang, and Stofferson 1978; Dyerberg and Bang 1979). These epidemiologic observations were followed by epidemiologic studies in Japan (Hirai et al. 1980; Hirai et al. 1984), the United States (Phillipson et al. 1985; Shekelle et al. 1985; Harris et al. 1988), and elsewhere (Kromhout, Bosschieter, and Coulander 1985); and by clinical investigations on the hypolipidemic (Harris 1989), antithrombotic (Weber et al. 1986a), and anti-inflammatory (Lee et al. 1985; Kremer and Robinson 1991) aspects of ω-3 fatty acids and the mechanisms involved (Simopoulos, Kifer, and Martin 1986; Galli and Simopoulos 1989; Simopoulos et al. 1991).

As a result of these investigations, major changes have taken place in our understanding of the contributions of fatty acids to health and disease. The focus is no longer on ω-6 fatty acids in lowering LDL (Low Density Lipoprotein) cholesterol (a desirable effect), but in lowering HDL (High Density Lipoprotein) (an undesirable effect); on the effects of the structural changes that occur in linoleic acid as a result of the hydrogenation process, such as the formation of trans fatty acids which raise LDL cholesterol and lower HDL cholesterol (Zock and Katan 1992) and raise lipoprotein (a) (Nestel et al. 1992); the ratio of ω-6/ω-3 fatty acids in the diet; the essentiality of ω-3 fatty acids; and their metabolic effects in the development, prevention, and treatment of chronic diseases (Simopoulos 1991). These biological and functional effects of ω-3 fatty acids exert profound beneficial metabolic changes in coronary heart disease, hypertension, non-insulin-dependent diabetes mellitus, inflammatory and autoimmune disorders, and possibly cancer.

NOMENCLATURE

Fats are solid at room temperature and oils are fluid. Fats and oils consist of fatty acids. Fatty acids are classified by (a) their length: short chain (less than 6 carbons), medium chain (6 to 10 carbons), and long chain (12 or more carbons); and (b) in terms of their degree of saturation: saturated fatty acids lack double bonds, monounsaturated fatty acids have a single double bond, and polyunsaturated fatty acids (PUFA) have more than one double bond. Polyunsaturated fatty acids are subdivided into two classes based on the location of the first double bond counting from the methyl end of the fatty acid molecule. ω-3 (or n-3) fatty acids have their first double bond between the third and fourth carbon atoms, and the ω-6 (or n-6) fatty acids have their first double bond between the sixth and seventh carbon atoms (Fig. 16-1).

This designation of numbering from the methyl end of the molecule is contrary to the Geneva system of numbering the carbon atoms from the functional group (in the case of fatty acids, from the carboxyl group). However, the characteristics of oils from a nutritional standpoint depend on what exists near the methyl end, not the carboxyl end of the fatty acid molecule. Dr. Ralph Holman was

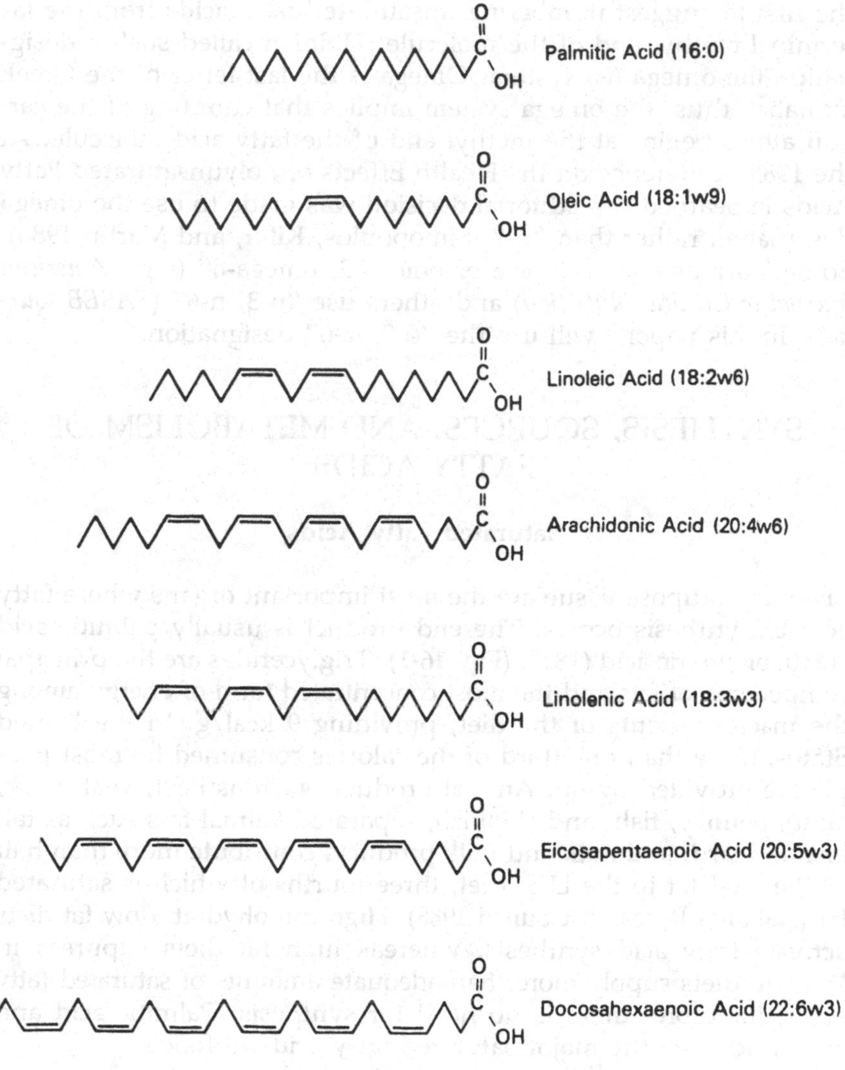

FIGURE 16-1. Structural formulas for saturates, ω-9, ω-6, and ω-3 fatty acids. The first number (before the colon) gives the number of carbon atoms in the molecule, and the second gives the number of double bonds. ω-9, ω-6 and ω-3 indicate the position of the first double bond in a given fatty acid molecule.

the first to suggest numbering unsaturated fatty acids from the far terminal methyl end of the molecule. Holman called such a designation the omega (ω) system. Omega is the last letter of the Greek alphabet; thus, the omega system implies that counting of the carbon atoms begins at the methyl end of the fatty acid molecule. At the 1985 Conference on the Health Effects of Polyunsaturated Fatty Acids in Seafoods an editorial decision was made to use the omega designation rather than "n-" (Simopoulos, Kifer, and Martin 1986). Some journals use "ω-3, ω-6 or omega-3, omega-6" (e.g., *American Journal of Clinical Nutrition*) and others use "n-3, n-6" (*FASEB Journal*). In this paper I will use the "ω-3, ω-6" designation.

SYNTHESIS, SOURCES, AND METABOLISM OF FATTY ACIDS

Saturated Fatty Acids

Liver and adipose tissue are the most important organs where fatty acid biosynthesis occurs. The end product is usually palmitic acid (16:0) or stearic acid (18:0) (Fig. 16-1). Triglycerides are the principal components of fats and the most concentrated form of energy among the macronutrients of the diet, providing 9 kcal/g. In the United States, more than one-third of the calories consumed by most people are provided by fat. Animal products such as beef, veal, pork, lamb, poultry, fish, and shellfish, separated animal fats such as tallow and lard, and milk and milk products contribute more than half of the total fat to the U.S. diet, three-fourths of which is saturated fat (National Research Council 1988). High carbohydrate/low fat diets activate fatty acid synthesis, whereas high fat diets suppress it. Western diets supply more than adequate amounts of saturated fatty acids; therefore, there is no need for synthesis. Palmitic acid and stearic acid are the major saturated fatty acids in foods.

Monounsaturated Fatty Acids

In order to function, cell membranes need unsaturated fatty acids to maintain fluidity. Through the process of desaturation a simple double bond is introduced between the ninth and tenth carbon atom by the enzyme delta (δ)-9 desaturase to form oleic acid (18:1ω9), the major monounsaturated fatty acid in the diet (Fig. 16-1). The enzyme is present in both plants and animals. The activity of δ-9 desaturase is markedly decreased by fasting and in patients with diabetes. Protein and insulin restore its activity (Brenner 1989).

Diets rich in cholesterol increase δ-9 desaturase activity, which results in an increased ratio of monounsaturates to saturates (16:1/16:0 or 18:1/18:0) and leads to an increase in membrane fluidity, thus offsetting the "hardening" effects of their enhanced cholesterol content. When monounsaturates are not available in the diet, δ-9 desaturase enters into action by increasing monounsaturates from saturates via the introduction of a single double bond into saturated fatty acids. 18:1ω9 is metabolized through a series of enzymatic steps involving desaturases and elongases to 18:2ω9, 20:2ω9, and 20:3ω9. The latter increases in plasma membrane phospholipids when there is an essential fatty acid (EFA) deficiency (Holman 1960).

Olive oil is the premier oil, consisting mostly of oleic acid. Other sources of oleic acid are peanut oil, canola oil, and safflower oil. Many of the studies have been performed with oleic acid, whereas others have used olive oil. The functional properties of oleic acid are not the same as those of olive oil. Olive oil contains oleic acid in various concentrations ranging from 65–82%, ω-6 fatty acids from 6–30%, and ω-3 fatty acids from 0.3–3%. Furthermore, olive oil contains squalene, which has anti-inflammatory properties (Budiarso 1990). Thus olive oil as a food has functional properties that go beyond the effects of oleic acid. Additional effects of olive oil include those on CH_2O metabolism (Trevisan et al. 1990).

Saturated and monounsaturated fatty acids can be readily synthesized from other nutrients in the diet such as glucose or amino acids, and thus are not essential dietary components. ω-3 and ω-6 fatty acids, on the other hand, cannot be synthesized and are therefore essential in the diet and must be obtained as such.

Essential Polyunsaturated Fatty Acids: ω-6 and ω-3

Plants can insert additional double bonds into oleic acid, which has one double bond, and form linoleic acid (LA) with two double bonds and alpha (α)-linolenic acid (LNA) with three double bonds (Fig. 16-1). These two families of essential fatty acids are not interconvertible.

Linolenic acid (18:3ω3), or α-linolenic acid (LNA), is found in the chloroplasts of green leafy vegetables, such as purslane and spinach, and in the seeds of flax, linseed, walnuts, etc. (Simopoulos 1987).

Linoleate series		Linolenate series	
C18:2w6	Linoleic acid	C18:3w3	Alpha-linolenic acid
Δ^6 desaturase		Δ^6 desaturase	
↓		↓	
C18:3w6	Gamma-linolenic acid	C18:4w3	
↓		↓	
C20:3w6	Dihomo-gamma-Linolenic Acid	C20:4w3	
Δ^5 desaturase		Δ^5 desaturase	
↓		↓	
C20:4w6	Arachidonic acid	C20:5w3	Eicosapentaenoic acid
↓		↓	
C22:4w6		C22:5w3	Docosapentaenoic acid
Δ^4 desaturase		Δ^4 desaturase	
↓		↓	
C22:5w6	Docosapentaenoic acid	C22:6w3	Docosahexaenoic acid

FIGURE 16-2. Essential fatty acid metabolism desaturation and elongation of ω-6 and ω-3.

Purslane, a wild plant and herb, is the richest source of 18:3ω3 of any green leafy vegetable examined to date (Simopoulos et al. 1992). Linoleic acid (LA) is found in large amounts in Western diets in corn oil, safflower oil, canola oil, and soybean oil (Adam 1989). It is very plentiful in nature and found in practically all plant seeds with the exception of palm, cocoa, and coconut. Eicosapentaenoic acid (EPA) and docosahexaenoic acid (DHA) are found in fish (Simopoulos, Kifer, and Martin 1986). In the U.S. food supply fish and poultry constitute the major sources of EPA and DHA (Raper, Cronin, and Exler 1992).

LA is converted through a series of desaturation and elongation steps to arachidonic acid (AA) and LNA to EPA and DHA (Fig. 16-2). Elongation and desaturation of LNA to EPA and DHA occurs in human leukocytes and in the liver of both humans and rodents (e.g., rats) (deGomez Dumm and Brenner 1975; Emken et al. 1989). ω-3 and ω-6 fatty acids compete for the desaturation enzymes. But both δ-4 and δ-6 desaturases (the enzymes involved in desaturation) prefer the ω-3 to the ω-6 fatty acid (de Gomez Dumm and Brenner 1975; Hague and Christoffersen 1984, 1986). Retroconversion of DHA and of AA to shorter-chain fatty acids has been shown to occur by beta (β)-oxidation in humans (Fischer et al. 1987).

In rats the consumption of diets rich in cholesterol decreases both the activities of δ-5 desaturase and δ-6 desaturase, resulting in decreased formation of AA (20:4ω6) and in increased LA (18:2ω6). The amounts of the elongated polyunsaturated derivatives are dependent on the dietary source. Humans who consume large amounts of EPA and DHA, which are present in fatty fish and fish oils, have increased levels of these two fatty acids in their plasma and tissue lipids at the expense of LA and AA. Alternately, vegetarians, whose intake of LA is high, have more elevated levels of LA and AA and lower levels of EPA and DHA in plasma lipids and in cell membranes than omnivores. LA is the major PUFA in the Western diet.

Because most edible vegetable oils contain smaller amounts of LNA, their higher concentration of LA may depress synthesis of EPA and DHA from LNA, except selectively in tissues such as the retina (Anderson 1970), cerebral cortex (O'Brien and Sampson 1965), and testis (Poulos, Darin-Bennett, and White 1975), which are rich in DHA. DHA is one of the most abundant components of the brain's structural lipids, phosphatidylethanolamine (PE), phosphatidylcholine (PC), and phosphatidylserine (PS). As indicated earlier, DHA can be obtained directly from the diet by, for example, eating fish or it can be synthesized from dietary LNA. EPA and DHA are found in membrane phospholipids of practically all cells of individuals who consume ω-3 fatty acids. DHA is found mostly in phospholipids, whereas LNA is found mostly in triglycerides, cholesteryl esters, and in very small amounts in phospholipids. EPA is found in cholesteryl esters, phospholipids, and triglycerides.

Trans Fatty Acids

Trans fatty acids occur in nature in small amounts, but are formed in large quantities by the hydrogenation process which adds hydrogen atoms to double bonds. In the United States, the principal vegetable oil used for hydrogenation is LA (18:2ω6), and the main fatty acids produced from hydrogenation are oleic, elaidic, and stearic acids. Oleic acid (18:1) (Fig. 16-1) has one double bond of the cis configuration, in which the two hydrogen atoms attached to the double bond lie on the same side, making the molecule flexible and liquid. Elaidic acid is trans 18:1, its hydrogen atoms lie on opposite sides of the double bonds. It is a rigid molecule with a structure similar to that of saturated fats. Its presence solidifies fat. Other trans monounsaturated fatty acids are formed during the hydrogenation process (by moving the double bond up and down the LA molecule) that most likely behave biologically very much like elaidic acid. The third fatty acid resulting from the hydrogenation process is stearic acid (18:0), which is a saturated fatty acid and has no double bonds.

This change in structure of fatty acids that results from hydrogenation is depicted in Fig. 16-3 (Lees 1990). As early as 1969, Spritz and Mishkel suggested that many of the effects of unsaturated or polyunsaturated fatty acids were due to their cross-sectional area. The normal double bond bends the fatty acid at a 120° angle, and increases the area occupied by that chain. The presence of more double bonds increases the cross-sectional area even more. These authors further suggested that the cholesterol-lowering effects of PUFA were related to their "bulk" and that the ratio of the cross-sectional area of PUFA to that of the saturated dietary fatty acids controls should be directly related to the percent lowering of serum cholesterol. Spritz and Mishkel (1969) carried out feeding experiments and as expected, both LDL and HDL cholesterol were decreased following PUFA feedings. They further suggested and showed that the trans fatty acids should have the same effect as saturated fats in raising serum cholesterol concentrations. It was the first demonstration that changes in the structure of PUFA changed their biologic effects.

There are now many studies that have investigated the effects of trans fatty acids on serum lipid levels. In the aggregate, these studies show that trans fatty acids raise triglycerides in comparison to butterfat (Anderson, Grande, and Keys 1961) and raise LDL cholesterol in comparison to oleic acid, while they also lower HDL, whereas oleic acid does not lower HDL (Mensink and Katan 1990; Zock and Katan 1992). Nestel et al. (1992) showed also that a diet consisting of elaidic acid at 7% energy intake (about two times as high as the Australian diet, but within the range of the U.S. diet) increased LDL cholesterol, and significantly elevated lipoprotein (Lp) (a) compared to the other three diets: (1) enriched with butterfat (lauric, myristic, palmitic) (2) oleic acid rich, and (3) palmitic rich. The results of these studies suggest that margarines should not be promoted for the prevention of atherosclerosis and coronary heart disease, and trans fatty acids should not be recommended as substitutes for saturated fats (Table 16-1).

For the past 50 years there has been a great increase in the consumption of vegetable oils and particularly hydrogenated oils in the form of shortenings and margarines. The hydrogenation of vegetable oils came about as a result of technological advances. The change from tallow (which is 4–6% trans) to partially hydrogenated vegetable oils (which are 25–42% trans) for salads, cooking oils, and in frying foods has led to greater increases in trans fatty acids in the food supply (Enig et al. 1990). Foods containing trans fatty acids

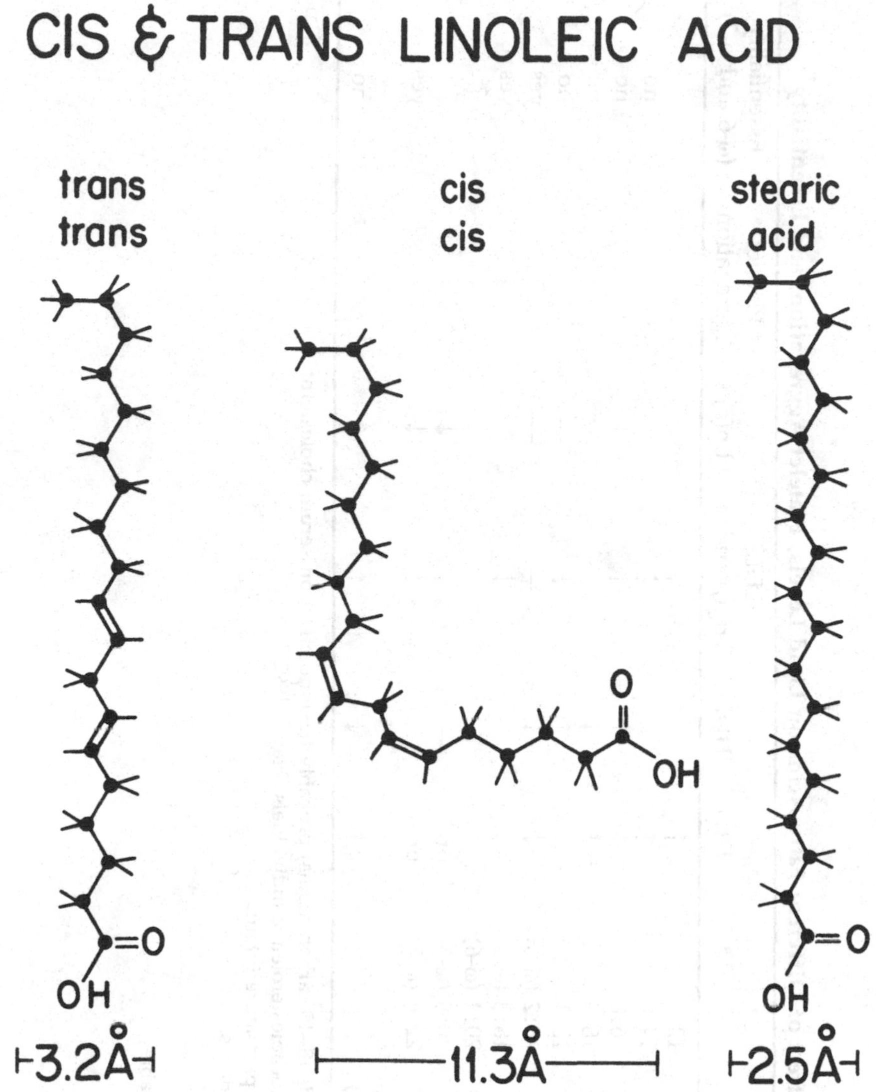

CIS & TRANS LINOLEIC ACID

trans trans | cis cis | stearic acid

O OH OH O OH

⊢3.2Å⊣ ⊢11.3Å⊣ ⊢2.5Å⊣

FIGURE 16-3. The cross-sectional area of two steroisomeric linoleic acids. The fully saturated analog stearic acid is included for comparison. The lateral cross-sectional area of the normally occurring cis-cis linoleic acid is much greater than that of the trans-trans linoleic acid produced by *in vitro* hydrogenation. The latter is not only structurally similar to stearic acid, but behaves metabolically (Reprinted from Lees 1990 pp. 1–38 by courtesy of Marcel Dekker Inc.)

Table 16-1. Effect of Different Fatty Acids on Lipid Levels, Platelet Aggregation, and Essentiality

Fatty Acid[a]		LDL	HDL	LDL Triglycerides	Lp(a)	Platelet Aggregation	Essentiality (ω-6 and ω-3)
Lauric acid	12:0	↑		↑			
Myristic acid	14:0	↑		↑			no
Palmitic acid[b]	16:0	↑	↑	↑	↓		no
Stearic acid	18:0	– ↑	↓			↑	no
Oleic acid (olive oil)	18:1	↓	↑	↓	—		no
Linoleic acid	18:2 (ω-6)	↓	↓	—	—	?	yes
Linolenic acid	18:3 (ω-3)	↓	—	↓		↓	yes
Arachidonic acid	20:4 (ω-6)					↑	?[c]
Eicosapentaenoic acid	20:5 (ω-3)	↓ or ↑[d]	—	↓	↓	↓	
Docosahexaenoic acid	22:6 (ω-3)	↓ or ↑[d]		↓	↓	↓	yes
Trans fatty acids (technologically developed)		↑	↓	↑	↑	↑	no

[a]Medium-chain fatty acids (8–10 carbon atoms) probably have no effect on serum cholesterol.
[b]May be neutral in normocholesterolemic individuals (Hayes 1992).
[c]May be essential in the premature infant.
[d]Decreases LDL at high doses.
↑ = increase
↓ = decrease
? = variable or questionable
– = no effect or neutral

include corn snacks, cheese/corn snacks, doughnuts, french fries, fried chicken, and fried fish. There is now convincing evidence that trans monounsaturated fatty acids raise LDL cholesterol levels as do saturated fats. In estimating the amount of saturated fats in the diet, we must include foods high in trans fatty acids. The 1987 reports from the edible oil industry indicate that salad oil target levels of trans fatty acids for soybean oil average 23.7% trans, and margarines base stock target levels average 51.7% trans (Mounts 1987). Obviously accurate knowledge of the composition of foods is needed. It is also essential that trans fatty acids and saturated fats that raise cholesterol or are thrombogenic be labeled as such. Evidence is accumulating that stearic acid does not raise cholesterol (Bonanome and Grundy 1988; Grundy 1990). However, stearic acid has thrombogenic properties and for that reason it should not be in the diet of persons with coronary heart disease or with a predisposition to coronary heart disease (Gautheron and Renaud 1972).

EVOLUTIONARY ASPECTS OF DIET

Studies of hunter-gatherer societies still in existence until very recently and estimates of foods available during man's evolution from paleolithic nutrition (Eaton and Konner 1985) indicate that man evolved on a diet that was low in saturated fat and the amounts of ω-3 and ω-6 fatty acids were about equal. Over the past 10,000 years with the development of agriculture, changes began to take place in the food supply, especially during the last 100–150 years, that led to increases in saturated fat from grain-fed cattle (Crawford, Gale, and Woodford 1969); increases in trans fatty acids from the hydrogenation of vegetable oils; and enormous increases in ω-6 fatty acids, about 30 g/day, due to the production of oils from vegetable seeds such as corn, safflower, and cotton (Fig. 16-4) (Leaf and Weber 1987). Increases in meat consumption have led to increased amounts of AA, about 200–1,000 mg/day (Phinney et al. 1990), whereas the amount of LNA is only 2.92 g/day (Raper, Cronin, and Exler 1992) and amounts of EPA and DHA are 48 and 72 mg/day, respectively. Thus a relative and absolute decrease in the amount of ω-3 fatty acids has led to an imbalance and increase in the ratio of ω-6:ω-3. It is now 20–30:1 in today's diet, whereas 1–4:1 was the ratio at the time when the human genetic code was established in response to diet.

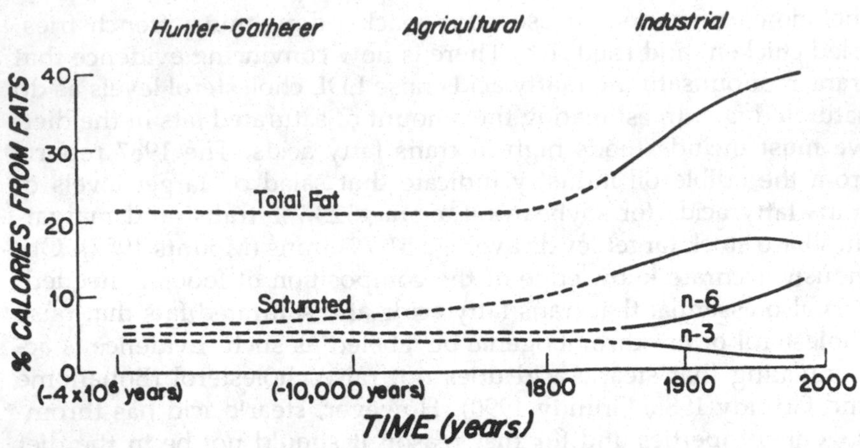

FIGURE 16-4. Scheme of the relative percentages of different dietary fatty acids (saturated fatty acids and ω-6 and ω-3 unsaturated fatty acids) and possible changes subsequent to industrial food processing, involving animal husbandry and hydrogenation of fatty acids (Reprinted from Leaf and Weber 1987, *Am. J. Clin. Nutr.* 45 (suppl):1048–53.)

Human milk is a good source of ω-3 fatty acids, but infants raised on infant formula do not receive any ω-3 fatty acids, as cow's milk does not contain any ω-3 (Carlson, Rhodes, and Ferguson 1986). The addition of LNA to formula may not adequately provide the DHA needs of the premature and possibly the full term infant (Uauy et al. 1990; Carlson and Salem 1991). Agribusiness and modern agricultural techniques with their emphasis on food production have further reduced the amount of ω-3 fatty acids in eggs and fish (Simopoulos and Salem 1989; van Vliet and Katan 1990).

BIOLOGICAL AND FUNCTIONAL EFFECTS OF ω-3 FATTY ACIDS

Eicosanoid Metabolism

EPA and AA metabolize to substances known as eicosanoids. These are prostanoids (prostaglandins and prostacylins) and leukotrienes (Fig. 16-5). Those derived from EPA are known as prostanoids of the 3-series and leukotrienes of the 5-series, whereas those derived from AA are known as prostanoids of the 2-series and leukotrienes

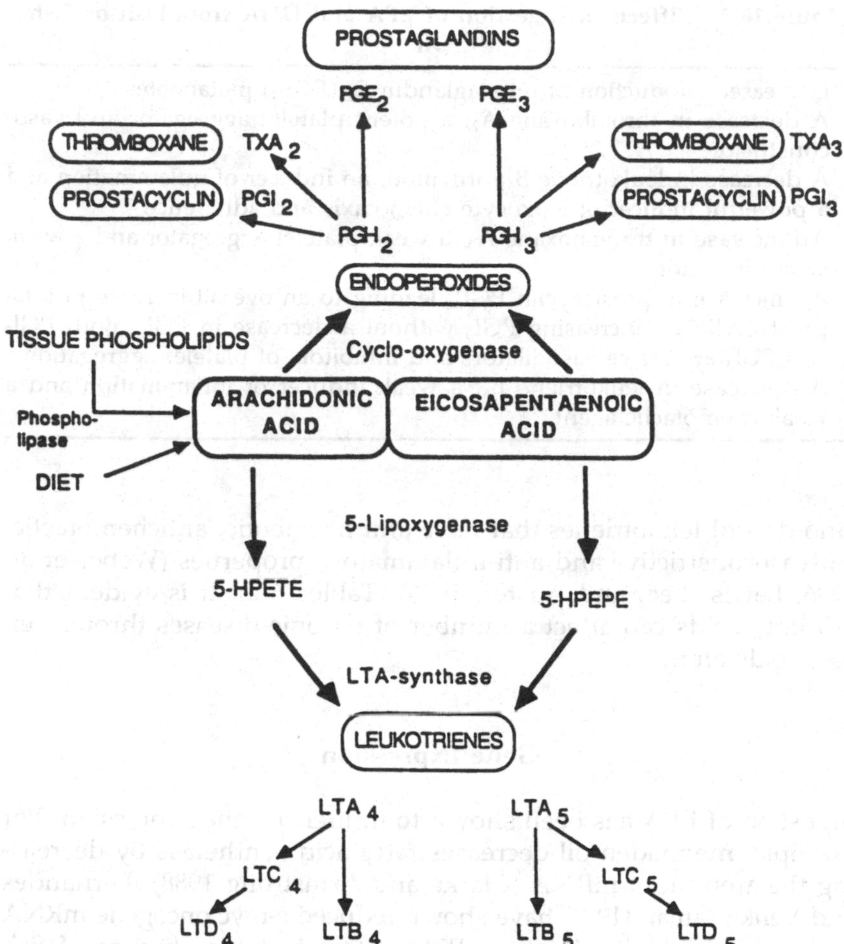

FIGURE 16-5. Oxidative metabolism of arachidonic acid and eicosapentaenoic acid by the cyclo-oxygenase and 5-lipoxygenase pathways. 5-HPETE denotes 5-hydroperoxyeicosatetraenoic acid; 5-HPEPE denotes 5-hydroxyeicosapentaenoic acid.

of the 4-series. Dihomo-gamma-linolenic acid (DGLA) is a precursor of the 1-series of prostaglandins. The metabolites of EPA and AA have competitive functions. Ingestion of EPA from fish or fish oil replaces AA from membrane phospholipids in practically all cells. Therefore, ingestion of EPA and DHA from fish or fish oil leads to a more physiologic state characterized by the production of pros-

Table 16-2. Effects of Ingestion of EPA and DHA from Fish or Fish Oil

- Decreased production of prostaglandin E_2 (PGE_2) metabolites
- A decrease in thromboxane A_2, a potent platelet aggregator and vasoconstrictor
- A decrease in leukotriene B_4 formation, an inducer of inflammation and a powerful inducer of leukocyte chemotaxis and adherence
- An increase in thromboxane A_3, a weak platelet aggregator and a weak vasoconstrictor
- An increase in prostacyclin PGI_3, leading to an overall increase in total prostacyclin by increasing PGI_3 without a decrease in PGI_2. Both PGI_2 and PGI_3 are active vasodilators and inhibitors of platelet aggregation
- An increase in leukotriene B_5, a weak inducer of inflammation and a weak chemotactic agent

tanoids and leukotrienes that have antithrombotic, antichemotactic, antivasoconstrictive and anti-inflammatory properties (Weber et al. 1986; Lewis, Lee, and Austen 1986) (Table 16-2). It is evident that ω-3 fatty acids can affect a number of chronic diseases through eicosanoids alone.

Gene Expression

Ingestion of EPA has been shown to influence gene expression. For example, menhaden oil decreases fatty acid synthetase by decreasing the amount of mRNA (Clarke and Armstrong 1988). Fernandes and Venkatraman (1991) have shown reduced c-myc oncogene mRNA levels and lower levels of ras-P21 protein in tumor tissues of fish oil-fed mice than in the tissues of either corn oil- or butterfat-fed mice.

Essentiality

The essentiality of LA was described by Burr and Burr (1929, 1930), and LA has been included in infant formula in amounts much larger than those found in human milk. The recognition of LNA as being essential for neural development of visual function and the cerebral cortex is very recent (Simopoulos 1990). Most of the research has taken place in the last eight years, although as early as 1973, Fiennes,

Table 16-3. The Differing Characteristics of ω-3 and ω-6 Essential Fatty Acid Deficiencies in the Infant Primate[a]

	ω-3	ω-6
Biochemical markers	Decreased 18:3ω3 and 22:6ω3 Increased 22:4ω6 and 22:5ω6 Increased 20:3ω9 (only if ω6 also low)	Decreased 18:2ω6 and 20:4ω6 Increased 20:3ω9 (only if ω3 also low)
Clinical features	Normal skin, growth and reproduction Reduced learning Abnormal electroretinogram Impaired vision Polydipsia	Growth retardation Skin lesions Reproductive failure Fatty liver Polydipsia

[a]Adapted from Connor, Neuringer, and Reisbick (1991), pp. 118–32, courtesy of Karger.

Sinclair, and Crawford (1973) described ω-3 fatty acid deficiency in the primate. Holman, Johnson, and Hatch (1982) described fatty acid deficiency in a five-year-old girl who was fed artificially for many months without the inclusion of ω-3 fatty acids in her tube feedings. She developed a neurologic condition that responded to the addition of LNA in her feedings.

Human milk contains both DHA and AA. Studies in animals and humans indicate that DHA is essential for normal visual and cerebral function in the premature infant and possibly in the full term infant (Neuringer and Connor 1989; Uauy et al. 1990; Connor, Neuringer, and Reisbick 1991; Carlson and Salem 1991; Salem and Carlson 1991; Farquharson et al. 1992). Table 16-3 shows the characteristics of ω-3 deficiency in the infant primate. Table 16-4 shows the characteristics of ω-3 fatty acid deficiency in adult humans. Table 16-5 lists the criteria for the diagnosis of ω-3 fatty acid deficiency in humans, and Table 16-6 shows the estimated ω-3 fatty acid requirement. It should be pointed out that the ESPGAN Committee on Nutrition (1991) has recommended the inclusion of DHA in infant formula of the premature infant. The Nordic countries have developed recommendations, and more recently the Canadian RDA (Scientific Review Committee 1990) include specific amounts of LNA and LA. In the United States the RDA, 10th edition, published in 1989, failed

Table 16-4. Symptoms Reported in ω-3 Fatty Acid Deficiency in Humans[a]

Biochemical	Clinical
Decreased 20:5ω3, 22:5ω3, and 22:6ω3	Imparied visual acuity
Increased 22:4ω6 and 22:5ω6	Abnormal retinograms
Hyperreactivity of lymphocytes	Learning deficiency
	Growth retardation
	Neurological dysfunction
	Hemorrhagic folliculitis
	Dermatitis

[a]Adapted from Bjerve (1991), pp. 66–77, courtesy of Karger.

to recognize the essentiality of LNA and particularly DHA for infants.

ω-3 Fatty Acids in Chronic Diseases

The important role of ω-3 fatty acids in disease states became evident as a result of epidemiologic studies and observations on Greenland Eskimos by Bang and Dyerberg (1972) and Bang, Dyerberg, and Hjorne (1976). Kromann and Green (1980) showed that cardiovascular disease, diabetes, and psoriasis in particular were rare in Eskimos despite the fact that the Eskimo diet was high in fat. How-

Table 16-5. Criteria for the Diagnosis of ω-3 Fatty Acid Deficiency in Humans[a]

(1) Long-term feeding on a completely defined diet supplying α-linolenic acid at less than approximately 50–100 mg/day, and long-chain ω-3 fatty acids less than 20–30 mg/day.

(2) Low intake of ω-3 fatty acids should be confirmed by low concentration of 20:5ω3 and 22:6ω3 in plasma and/or membrane phospholipids.

(3) Disappearance of clinical symptoms after supplementing with pure ω-3 fatty acids to an otherwise unchanged diet.

(4) Verify a concomitant increase in ω-3 fatty acid concentration in plasma and/or membrane phospholipids.

[a]Adapted from Bjerve (1991), pp. 66–77, courtesy of Karger.

Table 16-6. Estimates of Dietary Requirements of ω-3 Fatty Acids in Humans[a]

		Optimal Requirement	Minimal Requirement
Long-chain ω-3[b]	mg/day	350–400	100–200
	energy %	0.4	0.1–0.2
Linolenic acid[c]	mg/day	860–1,220	290–390
	energy %	0.54–1.2	0.2–0.3

The estimates are based on nutritional information and fatty acid changes observed in 10 patients with ω-3 fatty acid deficiency after supplementing with ω-3 fatty acids.
[a]Adapted from Bjerve (1991), pp. 66–77, courtesy of Karger.
[b]Dietary intake of linolenic acid below 100 mg/day.
[c]In the absence of dietary long-chain ω-3 fatty acids.

ever, the fatty acid composition of the Eskimos' diet varied enormously from that of the Danes. The Eskimos had high intakes of EPA, 6 g/day, and DHA, 9 g/day, and only 5 g/day of LA, whereas the Danes had 26 g/day of LA in their diet (Bang, Dyerberg, and Hjorne 1976).

The 1970s were auspicious times; the puzzle pieces seemed to come together. In 1978, two important concepts on eicosanoid formation were presented: (1) Moncada and Vane (1978) showed that the vascular wall could utilize endoperoxides released by adhering platelets for the formation of prostacyclin, and (2) Dyerberg, Bang, and Stoffersen (1978) proposed that EPA from oily fish or fish oil could, through the formation of the active analogue of prostacyclin, PGI_3, but inactive thromboxane, TXA_3, produce an antithrombotic state that protects against cardiovascular disease. In 1979, Needleman et al. showed that the eicosanoids from EPA differ biologically from those of AA. In the early 1980s a series of epidemiologic studies from Japan (Hirai et al. 1980; Hirai et al. 1984; Kagawa et al. 1982) indicated less cardiovascular disease and stroke, and a lower mortality from cardiovascular disease in fishing villagers versus non-fishing villagers. These were followed by a series of epidemiologic and intervention studies and studies on the mechanisms of function of ω-3 fatty acids and the development of standardized fish oil test material by the Department of Commerce for clinical and intervention studies and animal experiments (Simopoulos, Kifer, and Wykes 1991).

Hypolipidemic Effects

EPA and DHA in the form of fish or fish oil lower triglyceride levels consistently, by decreasing the synthesis of VLDL (Vascular LDL) (Tables 16-7 and 16-8) (Phillipson et al. 1985; Weber and Leaf 1991). Their effect on serum cholesterol has been inconsistent, from no effect, to increases in patients with types IIa and IV hyperlipidemia and in non-insulin-dependent diabetes mellitus (NIDDM) (Harris 1989). At high levels (20 g ω-3 fatty acids/day) ω-3 fatty acids lower serum cholesterol without lowering HDL, unlike ω-6 fatty acids, which consistently lower HDL cholesterol (Phillipson et al. 1985). In men given fish oil supplements, Abbey et al. (1990) noted significant changes in HDL characteristics. The proportion of HDL_2 rose, increasing the HDL_2/HDL_3 ratio, due partly to the decreased activity of lipid transfer protein, which normally transfers cholesteryl esters from HDL to other lipoproteins. The significance of this finding is a reduction in the redistribution of cholesterol from non-atherogenic to atherogenic proteins.

Antithrombotic Effects

The antithrombotic aspects of fish oil are due to decreases in platelet aggregation (Lorenz et al. 1983; Weber et al. 1986b; Weber 1989), a decrease in thromboxane A_2 (Weber et al. 1986b), increases in PGI_2 and PGI_3 production (Weber et al. 1986b), decreases in whole blood viscosity (Barcelli, Glas-Greenwalt, and Pollack 1985; Hostmark et al. 1988), and an increase in bleeding time (Lorenz et al. 1983; Weber 1989) (Table 16-8). Because of the increased amounts of ω-6 fatty acids in our diet, the eicosanoid metabolic products from AA, specifically prostaglandins, thromboxanes, leukotrienes, hydroxy fatty acids, and lipoxins, are formed in larger quantities than those formed from ω-3 fatty acids, particularly EPA. The eicosanoids from AA are biologically active in very small quantities and if they are formed in large amounts, they contribute to the formation of thrombus and atheromas; to allergic and inflammatory disorders, particularly in susceptible people; and to proliferation of cells (Weber and Leaf 1991). Thus a diet rich in ω-6 fatty acids shifts the physiological state to one that is prothrombotic and proaggregatory with increases in blood viscosity, vasospasm, and vasoconstriction and decreases in bleeding time. Bleeding time is decreased in groups of patients with hypercholesterolemia (Brox et al. 1983), hyperlipoproteinemia (Joist,

Table 16-7. Effects of ω-3 Fatty Acids on Factors Involved in the Pathophysiology of Atherosclerosis and Inflammation[a]

Factor	Function	Effect of ω-3 Fatty Acid
Arachidonic acid	Eicosanoid precursor; aggregates platelets; stimulates white blood cells	↓
Thromboxane A_2	Platelet aggregation; vasoconstriction; increase of intracellular Ca^{++}	↓
Prostacyclin ($PGI_{2/3}$)	Prevents platelet aggregation; vasodilation; increase cAMP	↑
Leukotriene (LTB_4)	Neutrophil chemoattractant; increase of intracellular Ca^{++}	↓
Tissue plasminogen activator	Increases endogenous fibrinolysis	↑
Fibrinogen	Blood clotting factor	↓
Red cell deformability	Decreases tendency to thrombosis and improves oxygen delivery to tissues	↑
Platelet activating factor (PAF)	Activates platelets and white blood cells	↓
Platelet-derived growth factor (PDGF)	Chemoattractant and mitogen for smooth muscles and macrophages	↓
Oxygen free radicals	Cellular damage; enhance LDL uptake via scavenger pathway; stimulate arachidonic acid metabolism	↓
Lipid hydroperoxides	Stimulate eicosanoid formation	↓
Interleukin 1 and tumor necrosis factor	Stimulate neutrophil O_2 free radical formation; stimulate lymphocyte proliferation; stimulate PAF; express intercellular adhesion molecule-1 on endothelial cells; inhibit plasminogen activator—thus procoagulants	↓

Table 16-7. Continued

Factor	Function	Effect of ω-3 Fatty Acid
Endothelial-derived relaxation factor (EDRF)	Reduces arterial vasoconstrictor response	↑
VLDL	Related to LDL and HDL level	↓
HDL	Decreases the risk for coronary heart disease	↑
Lp(a)	Lipoprotein(a) is a genetically determined protein that has atherogenic and thrombogenic properties	↓
Triglycerides and chylomicrons	Contribute to postprandial lipemia	↓

[a]Adapted from Weber and Leaf (1991), pp. 218–32, courtesy of Karger.
↑ = increase
↓ = decrease

Table 16-8. Functional Effects of ω-3 Fatty Acids in the Cardiovascular System[a]

- Decrease postprandial lipemia
- Reduce platelet aggregation
- Reduce blood pressure
- Decrease whole blood viscosity
- Reduce vascular intimal hyperplasia
- Reduce vasospastic response to vasoconstrictors
- Reduce cardiac arrhythmias
- Reduce albumin leakage in Type I diabetes mellitus

- Increase bleeding time
- Increase platelet survival
- Increase vascular (arterial) compliance
- Increase cardia beta-receptor function
- Increase postischemic coronary blood flow

[a]Adapted from Weber and Leaf (1991), pp. 218–32, courtesy of Karger.

Baker, and Schonfeld 1979), myocardial infarction, other forms of atherosclerotic disease, and diabetes (obesity and hypertriglyceridemia). Bleeding time is longer in women than in men and longer in young than in old people. There are ethnic differences in bleeding time that appear to be related to diet. In clinical trials (in both experimental and control groups) and in double-blind clinical trials (including restenosis trials and coronary artery graft surgery) there has never been increased evidence of clinical bleeding or increased perioperational blood loss despite the prolongation of bleeding time (DeCaterina et al. 1990).

Hypotensive Effects

EPA and DHA lower blood pressure in patients with mild to moderate hypertension, probably because of decreases in PGE_2 metabolites, the production of endothelial-derived relaxing factor (EDRF) (Vanhoutte, Shimokawa, and Boulanger 1991), and increased production of PGI_2 (Tables 16-7 and 16-8) (Knapp and FitzGerald 1989). It appears that there is a relationship between the amount of dietary fish intake and blood pressure. Bonaa et al. (1990), in a population-based intervention trial from Tromso, reported decreases of 6 mm Hg in systolic blood pressure and 3 mm Hg in diastolic blood pressure with fish-oil supplementation. These investigators monitored diet and assessed the concentration of plasma phospholipid fatty acids to determine the relation among diet, fatty acids, and blood pressure. Dietary supplementation with fish oil did not change mean blood pressure in the subjects who ate fish three or more times per week as part of their usual diet or in those who had a baseline concentration of plasma phospholipid ω-3 fatty acids >175.1 mg/L, suggesting that a relationship may exist between plasma phospholipid ω-3 fatty acid concentration and blood pressure. There was a lower blood pressure at baseline in subjects who habitually consumed larger quantities of fish, suggesting that supplementation with fish oils would be important from the primary prevention standpoint.

Antiatheromatous Effects

Evidence from a variety of studies suggests that a chronic intake of ω-3 fatty acids (Dyerberg, Bang, and Stofferson 1978; Hirai et al. 1980; Hirai et al. 1984; Kagawa et al. 1982) even in small amounts

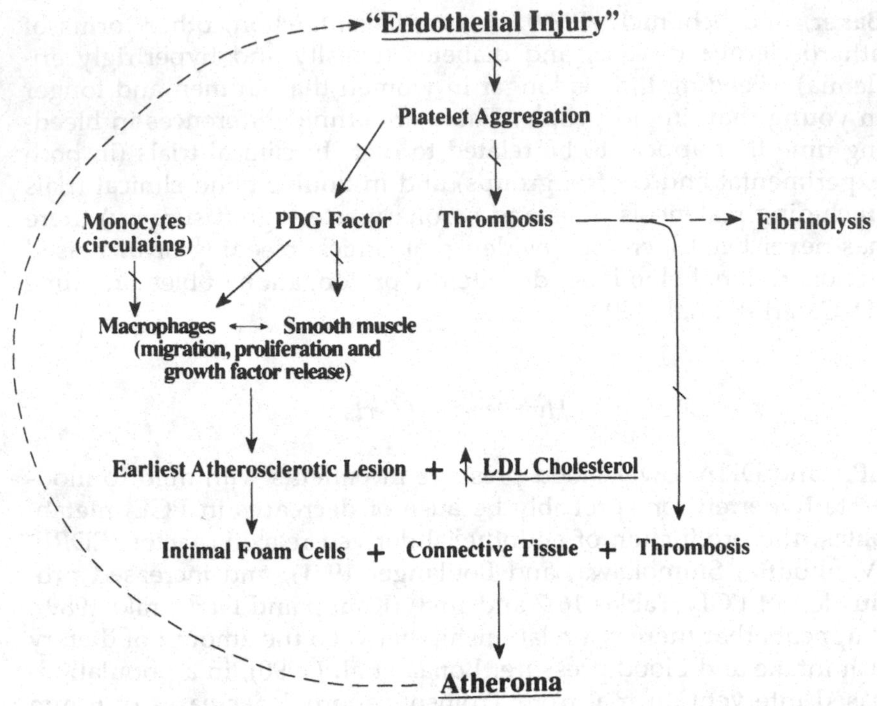

FIGURE 16-6. Sites of potential intervention for preventing development of atherosclerosis. Note that several of these possibilities would abort the disease before the concentration of plasma LDL cholesterol could contribute to the atherosclerotic process (Leaf and Weber 1987).

(Kromhout, Bosschieter, and Coulander 1985; Shekelle et al. 1985) are antiatherogenic. The proposed mechanism of this effect has focused on the role of EPA and DHA in eicosanoid metabolism (Table 16-2, Figs. 16-6 and 16-7). Recent data indicate that reductions in postprandial lipemia and increases in HDL_2 cholesterol levels may play a role in ω-3 fatty acids' ability to retard atherogenesis (Abbey et al. 1990). Harris and Windsor (1991) reported that 2.2 g/day of fish oil added to the regular American diet could lower postprandial lipid levels by lowering chylomicron triglycerides by 40%. The modern concepts on the development of coronary heart disease are based on the response to injury hypothesis developed by Ross (1986) and the theory of oxidation of LDL developed by Steinberg and his colleagues (1989). Thus the development of atheroma follows the process steps depicted in Fig. 16-6. The arrows indicate the locations

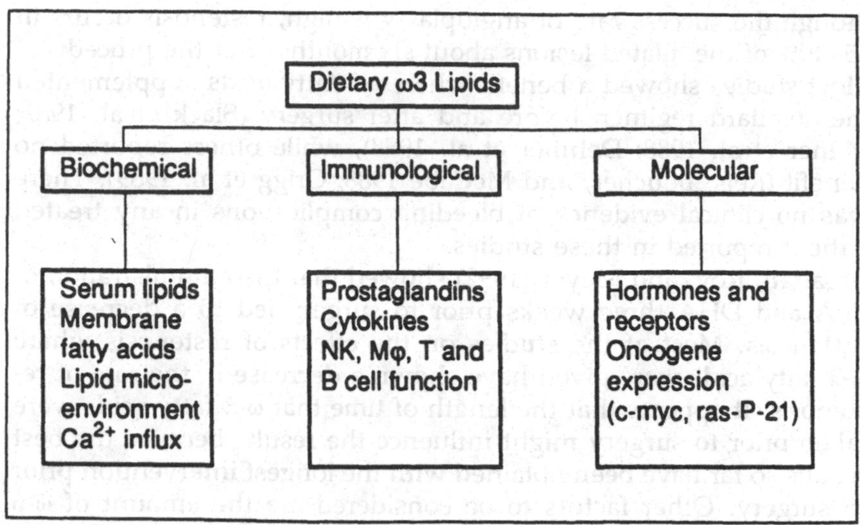

FIGURE 16-7. Possible mechanisms of ω-3 lipid action on modulating tumor growth (Fernandes and Venkatraman 1991) Courtesy of Karger.

or areas where ω-3 fatty acids supposedly could block the development of atherosclerosis. Recent work by Parthasarathy et al. (1990) shows that oleic acid prevents the oxidation of LDL in tissue culture, whereas LA promotes it.

Effects on Lipoprotein(a)

Lipoprotein(a) [Lp(a)] is a genetically determined lipoprotein whose structure is very similar to that of plasminogen. Lp(a) has both atherogenic and thrombogenic properties and is considered to be a risk factor for patients with familial hypercholesterolemia (Seed et al. 1990; Wiklund et al. 1990). A number of reports indicate that fish oils reduce Lp(a) and that dietary restriction and exercise potentiate the effects of EPA and DHA (Simopoulos 1989; Schmidt et al. 1991; Beil et al. 1991). As noted earlier, trans fatty acids increase Lp(a) (Nestel et al. 1992).

Prevention of Restenosis

Restenosis is a condition caused mainly by platelet aggregation, proliferation of smooth muscle cells, and coronary vasospasm. Although the success rate of angioplasty is high, restenosis occurs in 25–40% of the dilated lesions about six months after the procedure. Most studies showed a benefit when ω-3 fatty acids supplemented the standard regimen before and after surgery (Slack et al. 1987; Milner et al. 1988; Dehmer et al. 1988), while others reported no benefit (Reis, Boucher, and McCabe 1988; Grigg et al. 1989). There was no clinical evidence of bleeding complications in any treated patient reported in these studies.

Bairati, Roy, and Meyer (1992) showed that the administration of EPA and DHA three weeks prior to surgery led to a decrease of restenosis. Most of the studies on the effects of restenosis where ω-3 fatty acids were given have shown a decrease in the rate of restenosis. It appears that the length of time that ω-3 fatty acids were taken prior to surgery might influence the results because the best results so far have been obtained with the longest intervention prior to surgery. Other factors to be considered are the amount of ω-6 fatty acids and saturated fat in the diet. Saturated fat intake has been considered to prevent or interfere with the function of fish oil in decreasing the rate from coronary heart disease (Simonsen et al. 1987; Nordoy 1991).

Antiarrhythmic Effects

Animal studies have shown that ingestion of fish oils prevents or diminishes the development of arrhythmia in marmoset monkeys (Charnock 1991). Arrhythmia is associated with sudden death in humans. The findings by Burr et al. (1989) that fish or fish oil decreased the number of sudden deaths supports or is consistent with the animal studies (see also Burr in press).

Anti-inflammatory Effects

EPA and DHA compete with AA for the enzyme cyclo-oxygenase and thus decrease the production of the proinflammatory LTB_4 (Table 16-2, Fig. 16-5). LTB_4 is proinflammatory and chemotactic, and is increased in patients with arthritis (Kremer and Robinson 1991), psoriasis (Ziboh 1991), and ulcerative colitis (Stenson et al. 1992). The administration of fish oils in these conditions has diminished their severity in the majority of patients.

Arthritis

A number of studies (Sperling et al. 1987; Cleland et al. 1988; Kremer et al. 1990; Tempel et al. 1990) have shown that in rheumatoid arthritis patients, EPA and DHA decreased the production of LTB_4 and interleukin-1 (IL-1) while increasing LTB_5 and in some cases IL-2. Clinically these patients showed improvement; they had less pain and swelling in their joints and a decrease in the number of joints involved.

Psoriasis

In 1975 Hammarstrom et al. (1975) were the first to recognize that altered AA metabolism plays a role in the pathogenesis of cutaneous scaly disorders. They described abnormally high levels of AA and its lipoxygenase products LTB_4 and 12-HETE (Fig. 16-5) in the lesions (plaques) of patients with psoriasis. Neutrophils known to elaborate inflammatory mediators were also observed in these lesions. The findings suggested that the increased AA and its metabolites in the plaques of the psoriatic patients contribute at least in part to the pathogenesis of the disease. In patients with psoriasis, EPA and DHA decreased LTB_4 in the psoriatic lesion and an increase in the ratio of EPA and AA occurred in serum, granulocytes, and epidermis. There was a direct relationship between the level of EPA in the epidermis and the decrease in the erythema and itching of the psoriatic lesion. Only about 60% of patients with psoriasis appear to respond.

Ulcerative Colitis

Leukotriene B_4 and prostaglandin E_2, both products of AA metabolism, are increased in patients with ulcerative colitis. In ulcerative colitis LTB_4 is an important mediator of inflammation and has the ability to recruit additional neutrophils from the blood stream into the mucosa, exacerbating the disease process by the further increase of LTB_4. Because ω-3 fatty acids decrease the production of LTB_4, Stenson et al. (1992) investigated the effects of dietary supplementation with fish oil in patients with ulcerative colitis. Four months of diet supplementation with fish oils resulted in reductions in rectal dialysate LTB_4 levels, improvements in histologic findings, weight gain, and a reduction in the dose of prednisone. None of these changes occurred in the placebo group in which the dose of prednisone had to be raised during the period of the study. Encouraging results have been reported by others (McCall et al. 1989; Salomon, Kornbluth, and Janowitz 1990).

ω-3 Fatty Acids as an Adjuvant to Drug Therapy

In patients with arthritis (Kremer 1991), psoriasis (Allen 1991), ulcerative colitis (Stenson et al. 1992), and hypertension (Singer and Hueve 1991), EPA and DHA were added to the standard regimen and further improved the biochemical and chemical findings. In psoriasis, DHA prevented the hyperlipidemia that (usually) accompanies etretinate and prolonged the effects of phototherapy, indicating that ω-3 fatty acids have another function as adjuvants to drug therapy.

Cancer

Epidemiological studies have suggested an association between fat intake and incidence of breast cancer. However in many studies the effect of total energy intake was a confounding factor, as was body weight. Because Western diets are characterized by a high intake of fats and a low intake of fruits and vegetables (which are high in antioxidants), the question arises whether it was the absence of fruits and vegetables in the diet that was more important, as well as the type of fat (Pariza and Simopoulos 1987; Slater and Block 1991). Several animal studies have shown that high dietary fat intake, especially of ω-6 fatty acids, is related to the increased incidence of breast, prostate, and colon cancer (Carroll et al. 1986; Reddy 1986), whereas diets high in ω-3 fatty acids (EPA and DHA) have beneficial effects against several types of malignant tumors (Cave 1991). Because EPA and DHA inhibit both the cyclo-oxygenase and lipoxygenase pathways (Fig. 16-5), it is suggested that the eicosanoids produced by EPA and DHA modulate the induction and proliferation of tumor cells by altering immune function and oncogene expression, and possibly exerting a direct action on growth factors. Fig. 16-7 lists the various mechanisms of ω-3 fatty acid administration that may be involved in the modulation of tumor growth (Fernandes and Venkatraman 1991). In animal studies these authors showed that preliminary metastasis of MDA-MB231 cells was higher in corn oil-fed mice compared to a fish oil-fed group. Human trials are in progress to examine if EPA and DHA in fish oil can attenuate metastasis (Man-Fan Wan et al. 1991). Three epidemiological studies have shown that ω-3 fatty acid intake is inversely related to cancer prevalence (Kaizer et al. 1989; Willett et al. 1990; Dolecek and Grandits 1991).

CONCLUSIONS AND RECOMMENDATIONS

Epidemiologic studies, clinical investigations, animal experiments, tissue culture studies, biochemical advances in methods for identification of the various types of fatty acids and the development of purified forms of fatty acids for clinical trials and intervention studies have contributed to our knowledge and enhanced our understanding of the numerous biological and functional aspects of fatty acids.

At present we are standing at the threshold of the golden age of the nutritional aspects of fatty acids in terms of their role in growth and development and in health and disease. The chain length of the fatty acid molecule, degree of saturation (natural or as a result of hydrogenation that forms trans fatty acids), and the location of the first double bond appear to influence the biologic functions of fatty acids, in terms of their effects on serum lipids, platelet aggregation and thrombus formation, inflammation, cell-to-cell communication, gene expression, and essentiality.

Major changes have taken place in the U.S. food supply and in the Western diet in general since the turn of the century. These changes include increases in saturated fat intake from animal sources, and trans fatty acids from vegetable oils. But humans evolved on a diet that was low in saturated fat. Trans fatty acids occur in small amounts in nature, yet today contribute as much to the energy intake as saturated fats. Trans fatty acids and the hydrogenation process are new in terms of human evolutionary development because hydrogenation became popular only at the beginning of this century.

Saturated fats from animal sources raise cholesterol, and many dietary guidelines developed by governments around the world have recommended a decrease in saturated fat intake in order to lower serum cholesterol levels, and have recommended substituting polyunsaturated fatty acids for saturated ones. For this reason margarines and shortenings rich in trans fatty acids have increased rapidly in the U.S. food supply. Yet since 1969 it was pointed out by Spritz (see Spritz and Mishkel 1969) that the structure of trans fatty acids is similar to the structure of saturated fatty acids and the hydrogenation process changes polyunsaturates to saturates. Trans fatty acids, however, are worse than saturated fats because they not only raise LDL cholesterol, but also lower HDL cholesterol and raise Lp(a). We must learn our lesson from this. When technological modifications lead to structural changes in nutrients, their effects must be established prior to their introduction and wide dissemination in the food supply. The effects of fatty acids should not be limited to their effects on cholesterol, but also on platelet aggregation, prostaglandin and leukotriene metabolism, and other cellular functions, for fatty acids and lipids participate in many aspects of cellular metabolism.

Changes in the structure of fatty acids and in the ω-6/ω-3 balance of the food supply adversely affect human health. Again, humans evolved on a diet where they obtained about equal amounts of ω-6 and ω-3 fatty acids from their food. Both fatty acids are essential for growth and development, and DHA is particularly important for normal visual and cerebral function. Docosahexaenoic acid and arachidonic acid are found in human milk but not in cow's milk. Yet today we continue to produce infant formula devoid of both docosahexaenoic acid and arachidonic acid. The U.S. diet today is deficient in ω-3 fatty acids and has an enormous amount of ω-6 fatty acids, the result of changes in feeding practices of livestock, and the development of techniques that increase the production of vegetable oils rich in ω-6 fatty acids. Today the ratio of ω-6:ω-3 is about 20–30:1, instead of the 1:1 or 1–4:1 ratio estimated from studies of paleolithic nutrition and present day hunter-gatherer societies.

Coronary heart disease, hypertension, diabetes, arthritis, psoriasis, and ulcerative colitis are characterized by increased levels of prostaglandins and leukotrienes, products of AA metabolism. AA is present in such high amounts in our food supply that it has shifted human metabolism to a prothrombotic and proinflammatory state. Clinical investigations and animal experiments demonstrate that by increasing the dietary intake of EPA and DHA from eating fish or fish oil, we can alter plasma and cell membrane phospholipid fatty acid composition, the AA cascade, and the spectrum of biologically highly active eicosanoids. The alterations that occur from the increased ω-3 fatty acid intake induce an antithrombotic and anti-inflammatory state that is beneficial to health.

The changes recommended below are consistent with the diet during the time that human's genetic code was established in response to dietary changes over man's long evolutionary period. Today's diet was developed over the past 100–150 years and was not constructed on the basis of human physiological and metabolic needs. As a component of the human diet, trans fatty acids are the result of technological evolution and not the result of scientific evidence for their physiologic need. They should be kept as low as possible in the food supply.

- In calculating the total saturated fatty acid content of the diet, trans fatty acids should be added along with saturated fats from animal sources. Both saturated fats from animal sources and trans fatty acids should be reduced.
- Monounsaturated fatty acids do not have to be restricted and could be substituted for saturated fats and trans fatty acids.
- Polyunsaturated fatty acids should be considered separately as ω-6 and ω-3 fatty acids, as they compete in eicosanoid metabolism and for position in cell membranes.
- In order to return to a balanced ω-6/ω-3 ratio the ω-6 fatty acid content of the diet has to be reduced and the ω-3 fatty acid content will have to increase.
- ω-6 and ω-3 fatty acids should be provided in the diet in equal amounts. Substituting fish for meat 2–3 times per week and decreasing the intake of vegetable oils, shortening, and margarines is essential and should improve the ω-6/ω-3 balance.
- The proportion of ω-6 and ω-3 fatty acids should also be balanced in all solutions that are used intravenously and in all tube feedings.
- Infant formulas should contain DHA and AA in amounts equivalent to those found in human milk.

Advances in genetics and molecular biology indicate that genetic variation exists much more frequently than had been anticipated. Not everybody is susceptible to chronic diseases such as coronary heart disease, hypertension, diabetes, and cancer, or to the same degree. The identification of the individual at risk and the initiation of measures to prevent the disease or its expression will include dietary and pharmacologic management. We already know that ω-3 fatty acids influence gene expression and function as adjuvants to drug therapy. Therefore, we already have entered the period of Nutrients in the Control of Metabolic Disease(s) (Simopoulos and Childs 1990; Simopoulos, Herbert, and Jacobson 1993).

The importance of the essential role of fatty acids in health and disease and in growth and development has been recognized by the scientific community internationally. In 1991 the International Society for the Study of Fatty Acids and Lipids (ISSFAL) was established. The purpose of the society is to increase understanding, through research and education, of the role of dietary fatty acids and lipids in health and disease. Information about the society may be obtained from the author.

ISSFAL held its first international congress in Lugano, Switzerland in June 1993. The proceedings of the congress provide the most up-to-date information on fatty acids and lipids from cell biology to human disease (Galli, Simopoulos, and Tremoli 1994).

References

Abbey, M.; Clifton, P.; Kestin, M.; Belling, B.; and Nestel, P. 1990. Effect of fish oil on lipoproteins, lecithin: Cholesterol acyltransferase and lipid transfer protein activity in humans. *Arteriosclerosis* 10:85–94.

Adam, O. 1989. Linoleic and linolenic acids intake. In *Dietary ω3 and ω6 Fatty Acids: Biological Effects and Nutritional Essentiality*. Series A: Life Sciences Volume 171, C. Galli and A.P. Simopoulos, eds. New York: Plenum Press.

Allen, B.R. 1991. Fish oil in combination with other therapies in the treatment of psoriasis. In *Health Effects of ω3 Polyunsaturated Fatty Acids in Seafoods*. vol. 66, A.P. Simopoulos, R.R. Kifer, R.E. Martin, and S.M. Barlow, eds., pp. 436–45. Basel: Karger.

Anderson, R.E. 1970. Lipids of ocular tissues. IV. A comparison of the phospholipids from the retina of six mammalian species. *Exp. Eye Res.* 10:339–44.

Anderson, J.T.; Grande, R.; and Keys, A. 1961. Hydrogenated fats in the diet and lipids in the serum of man. *J. Nutr.* 75:388–94.

Bairati, I.; Roy, L.; and Meyer, F. 1992. Double-blind, randomized, controlled trial of fish oil supplements in prevention of recurrence of stenosis after coronary angioplasty. *Circulation* 85:950–56.

Bang, H.O., and Dyerberg, J. 1972. Plasma lipids and lipoproteins in Greenlandic West-coast Eskimos. *Acta Med. Scand.* 192:85–94.

Bang, H.O.; Dyerberg, J.; and Hjorne, N. 1976. The composition of food consumed by Greenland Eskimos. *Acta Med. Scand.* 200:69–73.

Barcelli, U.; Glas-Greenwalt, P.; and Pollack, V.E. 1985. Enhancing effect of dietary supplement with n-3 fatty acids on plasma fibrinolysis in normal subjects. *Thromb. Res.* 39:307–12.

Beil, F.U.; Terres, W.; Orgass, M.; and Greten, H. 1991. Dietary fish oil lowers lipoprotein(a) in primary hypertriglyceridemia. *Atherosclerosis* 90:95–97.

Bjerve, K.S. 1991. ω3 fatty acid deficiency in man: Implications for the requirement of alpha-linolenic acid and long-chain ω3 polyunsaturated fatty acids. In *Health Effects of ω3 Polyunsaturated Fatty Acids in Seafoods*. vol. 66, A.P. Simopoulos, R.R. Kifer, R.E. Martin, and S.M. Barlow, eds., pp. 66–77. Basel: Karger.

Bonaa, K.H.; Bjerve, K.S.; Straume, B.; Gram, I.T.; and Thelle, D. 1990. Effect of eicosapentaenoic and docosahexaenoic acids on blood pressure in hypertension. A population-based intervention trial from the Tromso Study. *N. Engl. J. Med.* 322:795–801.

Bonanome, A., and Grundy, S.M. 1988. Effect of dietary stearic acid on plasma cholesterol and lipoprotein levels. *N. Engl. J. Med.* 318:1244–48.

Brenner, R.R. 1989. Factors influencing fatty acid chain elongation and desaturation. In *The Role of Fats in Human Nutrition*. 2d. A.J. Vergroessen and M. Crawford. eds., pp. 45–79. London: Academic Press.

Brox, J.H.; Killie, J.E.; Osterud, B.; Holme, S.; and Nordoy, A. 1983. Effects of cod liver oil on platelets and coagulation in familial hypercholesterolemia (type IIa). *Acta Med. Scand.* 213:137–44.

Budiarso, I.T. 1990. Fish oil versus olive oil. *Lancet* 336:1313–1314.

Burr, G.O., and Burr, M.M. 1929. A new deficiency disease produced by the rigid exclusion of fat from the diet. *J. Biol. Chem.* 82:345–67.

———. 1930. On the nature and role of fatty acids essential in nutrition. *J. Biol. Chem.* 86:587–621.

Burr, M.L. (in press) Fish and ischaemic heart disease. In *Nutrition and Fitness in Health and Disease*. vol. 72, A.P. Simopoulos, ed. Basel: Karger.

Burr, M.L.; Fehily, A.M.; Gilbert, J.F.; Rogers S.; Holliday, R.M.; Sweetnam, P.M.; Elwood, P.C.; and Deadman, N.M. 1989. Effect of changes in fat, fish and fibre intakes on death and myocardial reinfarction: Diet and reinfarction trial (DART). *Lancet* ii:757–61.

Carlson, S.E.; Rhodes, P.G.; and Ferguson, M.G. 1986. Docosahexaenoic acid status of preterm infants at birth and following feeding with human milk or formula. *Am. J. Clin. Nutr.* 44:798–804.

Carlson, S.E.; and Salem, N., Jr. 1991. Essentiality of ω3 fatty acids in growth and development of infants. In *Health Effects of ω3 Polyunsaturated Fatty Acids in Seafoods*. vol. 66, A.P. Simopoulos, R.R. Kifer, R.E. Martin, and S.M. Barlow, eds., pp. 74–86. Basel: Karger.

Carroll, K.K.; Braden, L.M.; Bell, J.A.; and Kalamegham, R. 1986. Fat and cancer. *Cancer* 58(suppl):1818–25.

Cave, W.T. 1991. ω3 fatty acid diet effects on tumorigenesis in experimental animals. In *Health Effects of ω3 Polyunsaturated Fatty Acids in Seafoods*, vol. 66, A.P. Simopoulos, R.R. Kifer, R.E. Martin, and S.M. Barlow, eds., pp. 74–86. Basel: Karger.

Charnock, J.S. 1991. Antiarrhythmic effects of fish oils. In *Health Effects of ω3 Polyunsaturated Fatty Acids in Seafoods*. vol. 66, A.P. Simopoulos, R.R. Kifer, R.E. Martin, and S.M. Barlow, eds., pp. 278–291. Basel: Karger.

Clarke, S.D., and Armstrong, M.K. 1988. Suppression of rat liver fatty acid synthetase mRNA level by dietary fish oil. *FASEB J.* 2:A852.

Cleland, L.G.; French, J.K.; Betts, W.H.; Murphy, G.A.; and Elliot, M.J. 1988. Clinical and biochemical effects of dietary fish oil supplements in rheumatoid arthritis. *J. Rheumatol.* 15:1471–75.

Connor, W.E.; Neuringer, M.; and Reisbick, S. 1991. Essentiality of ω3 fatty acids: Evidence from the primate model and implications for human nutrition. In *Health*

Effects of ω3 Polyunsaturated Fatty Acids in Seafoods. vol. 66, A.P. Simopoulos, R.R. Kifer, R.E. Martin, and S.M. Barlow, eds., pp. 118–32. Basel: Karger.

Crawford, M.A.; Gale, M.M.; and Woodford, M.H. 1969. Linoleic acid and linolenic acid elongation production in muscle tissue of Syncerus caffer and other ruminant species. *Biochem. J.* 115:25–27.

DeCaterina, R.; Giannessi, D.; Mazzone, A.; Bernini, W.; Lazzerini, G.; Maffei, S.; Cerri, M.; Salvatore, L.; and Weksler, B. 1990. Vascular prostacyclin is increased in patients ingesting n-3 polyunsaturated fatty acids prior to coronary artery bypass surgery. *Circulation* 82:428–38.

de Gomez Dumm, I.N.T., and Brenner, R.R. 1975. Oxidative desaturation of alpha-linolenic, linoleic, and stearic acids by human liver microsomes. *Lipids* 10:315–17.

Dehmer, G.J.; Pompa, J.J.; van den Berg, E.K.; Eichhorn, E.J. 'ewitt, J.B.; Campbell, W.B.; Jennings, L.; Willerson, J.T.; and Schmitz, J.M. 1988. Reduction in the rate of early restenosis after coronary angioplasty by a diet supplemented with n-3 fatty acids. *N. Engl. J. Med.* 319:733–40.

Dolecek, T.A., and Grandits, G. 1991. Dietary polyunsaturated fatty acids and mortality in the Multiple Risk Factor Intervention Trial (MRFIT). In *Health Effects of ω3 Polyunsaturated Fatty Acids in Seafoods*. vol. 66, A.P. Simopoulos, R.R. Kifer, R.E. Martin, and S.M. Barlow, eds., pp. 205–16. Basel: Karger.

Dyerberg, J., and Bang, H.O. 1979. Haemostatic function and platelet polyunsaturated fatty acids in Eskimos. *Lancet* ii:433–35.

Dyerberg, J.; Bang, H.O.; and Hjorne, N. 1975. Fatty acid composition of the plasma lipids in Greenland Eskimos. *Am. J. Clin. Nutr.* 28:958–66.

Dyerberg, J.; Bang, H.O.; and Stoffersen, E. 1978. Eicosapentaenoic acid and prevention of thrombosis and atherosclerosis. *Lancet* ii:117–19.

Eaton, S.B., and Konner, M. 1985. Paleolithic nutrition. A consideration of its nature and current implications. *N. Engl. J. Med.* 312:283–89.

Emken, E.A.; Adolf, R.O.; Rakoff, H.; and Rohwedder, W.K. 1989. Metabolism of deuterium-labelled linolenic, linoleic, oleic, stearic and palmitic acid in human subjects. In *Synthesis and Applications of Isotopically Labelled Compounds 1988*, T.A. Baillie and J.R. Jones, eds., pp. 713–16. Amsterdam: Elsevier Science Publishers.

Enig, M.G.; Atal, S.; Keeney, M.; and Sampugna, J. 1990. Isomeric trans fatty acids in the U.S. diet. *J. Am. College Nutr.* 9:471–86.

ESPGAN Committee on Nutrition. 1991. Committee report. Comment on the content and composition of lipids in infant formulas. *Acta Pediatr. Scand.* 80:887–96.

Farquharson, J.; Cockburn, F.; Patrick, W.A.; Jamieson, E.C.; and Logan, R.W. 1992. Infant cerebral cortex phospholipid fatty-acid composition and diet. *Lancet* 340:810–813.

Fernandes, G., and Venkatraman, J.T. 1991. Modulation of breast cancer growth in nude mice by ω3 lipids. In *Health Effects of ω3 Polyunsaturated Fatty Acids in Seafoods*. vol. 66, A.P. Simopoulos, R.R. Kifer, R.E. Martin, and S.M. Barlow, eds., pp. 488–503. Basel: Karger.

Fiennes, R.N.T.W.; Sinclair, A.J.; and Crawford, M.A. 1973. Essential fatty acid studies in primates, linolenic acid requirements of capuchins. *J. Med. Prim.* 2:155–69.

Fischer, S.; Vischer, A.; Praec-Mursic, V.; and Weber, P.C. 1987. Dietary docosahexaenoic acid is retroconverted in man to eicosapentaenoic acid, which can be quickly transformed to prostaglandin I_3. *Prostaglandins* 34:367–73.

Galli, C., and Simopoulos, A.P., eds. 1989. *Dietary ω3 and ω6 Fatty Acids: Biological Effects and Nutritional Essentiality*. Series A: Life Sciences Volume 171. New York: Plenum Press.

Galli, C.; Simopoulos, A.P.; and Tremoli, E., eds. 1994. Fatty Acids and Lipids from Cell Biology to Human Disease. vol. 75. Basel: Karger.

Gautheron, P., and Renaud, S. 1972. Hyperlipidemia induced hypercoagulable state in rat. Role of an increased activity of platelet phosphatidylserine in response to certain dietary fatty acids. *Thromb. Res.* 1:353–70.

Grigg, L.E.; Kay, T.W.H.; Valetine, P.A.; Larkins, R.; Flower, D.J.; Malonal, E.G.; O'Dea, K.; Sinclair, A.J.; Hopper, J.L.; and Hunt, D. 1989. Determinants of restenosis and lack of effect of dietary supplementation with eicosapentaenoic acid on the incidence of coronary artery restenosis after angioplasty. *J. Am. Coll. Cardiol.* 13:665–72.

Grundy, S.M. 1990. Trans monounsaturated fatty acids and serum cholesterol levels. *N. Engl. J. Med.* 323:480–81.

Hague, T.A., and Christoffersen, B.O. 1984. Effect of dietary fats on arachidonic acid and eicosapentaenoic acid biosynthesis and conversion to C22 fatty acids in isolated liver cells. *Biochim. Biophys. Acta* 796:205–17.

———. 1986. Evidence for peroxisomal retroconversion of adrenic acid (22:4n6) and docosahexaenoic acid (22:6n3) in isolated liver cells. *Biochim. Biophys. Acta* 875:165–73.

Hammarstrom, A.; Hamberg, M.; Samuelsson, B.; Duell, E.A. §rawiski, M.; and Voorhees, J.J. 1975. Increased concentration of non-esterified arachidonic acid. 12L-hydroxy-5,8,10,14-eicosatetraenoic acid, prostaglandin E_2 and prostaglandin F_{2a} in epidermis of psoriasis. *Proc. Natl. Acad. Sci.* 72:5130–34.

Harris, W.S. 1989. Fish oils and plasma lipid and lipoprotein metabolism in humans: A critical review. *J. Lipid Res.* 30: 785–807.

Harris, W.S.; Connor, W.E.; Alam, N.; and Illingworth, D.R. 1988. Reduction of postprandial triglyceridemia in humans by dietary ω-3 fatty acids. *J. Lipid Res.* 29:1451–60.

Harris, W.S., and Windsor, S.L. 1991. N-3 fatty acid supplements reduce chylomicron levels in healthy volunteers. *J. Appl. Nutr.* 43:5–15.

Hayes, K.C. 1992. Chewing the fat. *Fat and Nutrition* 1:1–4.

Hirai, A.; Hamazaki, T.; Terano, T.; Nishikawa, T.; Tamura, Y.; and Kumagai, A. 1980. Eicosapentaenoic acid and platelet function in Japanese. *Lancet* ii:1132–33.

Hirai, A.; Terano, T.; Saito, H.; Tamura, Y.; Yoshida, S.; Sajiki, J.; and Kumagai, A. 1984. Eicosapentaenoic acid and platelet function in Japanese. In *Nutritional Prevention of Cardiovascular Disease*, Y. Goto and Y. Homma, eds., pp. 231–39. Orlando, FL: Academic Press.

Holman, R.T. 1960. The ratio of trienoic:tetraenoic acid in tissue lipids as a measure of essential fatty acid requirement. *J. Nutr.* 70:405–10.

Holman, R.T.; Johnson, S.B.; and Hatch, T.F. 1982. A case of human linolenic acid deficiency involving neurological abnormalities. *Am. J. Clin. Nutr.* 35:617–23.

Hostmark, A.T.; Bjerkedal, T.; Kierulf, P.; Flaten, H.; and Ulshagen, K. 1988. Fish oil and plasma fibrinogen. *Br. Med. J.* 297:180–81.

Joist, J.H.; Baker, R.K.; and Schonfeld, G. 1979. Increased in vivo and in vitro platelet function in type II- and type IV-hyperlipoproteinemia. *Thromb. Res.* 15:95–108.

Kagawa, Y.; Nishizawa, M.; Suzuki, M.; Miyatake, T.; Hamamoto, T.; Goto, K.; Motonaga, E.; Izumikawa, H.; Hirata, H.; and Ebihara, A. 1982. Eicosapolyenoic acids of serum lipids of Japanese Islanders with low incidence of cardiovascular disease. *J. Nutr. Sci. Vitaminol.* 28:441–53.

Kaizer, L.; Boyd, N.F.; Kriukov, V.; and Tritcher, D. 1989. Fish consumption and breast cancer risk: An ecological study. *Nutr. Cancer* 12:61–68.

Knapp, H.R., and FitzGerald, G.A. 1989. The antihypertensive effects of fish oil. A controlled study of polyunsaturated fatty acid supplements in essential hypertension. *N. Engl. J. Med.* 320:1037–43.

Kremer, J.M.; Lawrence, D.A.; Jubiz, W.; DiGiacomo, R.; Rynes, K.; Bartholomew, L.E.; and Sherman, M. 1990. Dietary fish oil and olive oil supplementation in patients with rheumatoid arthritis; clinical and immunological effects. *Arth. Rheumat.* 33:810–20.

Kremer, J.M., and Robinson, D.W. 1991. Studies of dietary supplementation with ω3 fatty acids in patients with rheumatoid arthritis. In *Health Effects of ω3 Polyunsaturated Fatty Acids in Seafoods*. vol. 66, A.P. Simopoulos, R.R. Kifer, R.E. Martin, and S.M. Barlow, eds., pp. 367–82. Basel: Karger.

Kromann, N., and Green, A. 1980. Epidemiological studies in the Uppernavik district, Greenland. *Acta Med. Scand.* 208:401–6.

Kromhout, D.; Bosschieter, E.B.; and Coulander, C. deL. 1985. The inverse relation between fish consumption and 20-year mortality from coronary heart disease. *N. Engl. J. Med.* 312:1205–9.

Leaf, A., and Weber, P.C. 1987. A new era for science in nutrition. *Am. J. Clin. Nutr.* 45(suppl):1048–53.

Lee, T.H.; Hoover, R.L.; Williams, J.D.; Sperling, R.I.; Ravalese III, J.; Spur, B.W.; Robinson, D.R.; Corey, E.J.; Lewis, R.A.; and Austen, K.F. 1985. Effect of dietary enrichment with eicosapentaenoic and docosahexaenoic acids on in vitro neutrophil and monocyte leukotriene generation and neutrophil function. *N. Engl. J. Med.* 312:1217–24.

Lees, R.S. 1990. Impact of dietary fat on human health. In *Omega-3 Fatty Acids in Health and Disease*, R.S. Lees and M. Karel, eds., pp. 1–38. New York: Marcel Dekker.

Lewis, R.A.; Lee, T.H.; and Austen, K.F. 1986. Effects of omega-3 fatty acids on the generation of products of the 5-lipoxygenase pathway. In *Health Effects of Polyunsaturated Fatty Acids in Seafoods*. A.P. Simopoulos, R.R. Kifer, and R.E. Martin, eds., pp. 227–38. Orlando, FL: Academic Press.

Lorenz, R.; Spengler, U.; Fischer, S.; Duhm, J.; and Weber, P.C. 1983. Platelet function, thromboxane formation and blood pressure control during supplementation of the western diet with cod liver oil. *Circulation* 67:504–11.

Man-Fan Wan, J.; Kanders, B.S.; Kowalchuk, M.; Knapp, H.; Szeluga, D.J.; Bagley, J.; and Blackburn, G.L. 1991. In *Health Effects of Polyunsaturated Fatty Acids in Seafoods*, A.P. Simopoulos, R.R. Kifer, and R.E. Martin, eds., pp. 477–87. Orlando, FL: Academic Press.

McCall, T.B.; O-Leary, D.; Bloomsield, J.; and O'Morain, C.A. 1989. Therapeutic potential of fish oil in the treatment of ulcerative colitis. *Aliment. Pharmacol. Ther.* 3:415–24.

Mensink, R.P., and Katan, M.B. 1990. Effect of dietary trans fatty acids on high-density and low-density lipoprotein cholesterol levels in healthy subjects. *N. Engl. J. Med.* 323:439–45.

Milner, M.R.; Gallino, R.A.; Leffingwell, A.; Pichard, A.D. ≤senberg, J.; and Lindsay, J. 1988. High dose omega-3 fatty acid supplementation reduces clinical restenosis after coronary angioplasty. *Circulation* 78(suppl 2):634.

Mishkel, M.A., and Spritz, N. 1969. The effects of trans isomerized trilinolein on plasma lipids of man. In *Advances in Experimental Medicine and Biology*, W.L. Holmes, L.A. Carlson, and R. Paoletti, eds., pp. 355–364. New York: Plenum Press.

Moncada, S., and Vane, J.R. 1978. Unstable metabolites of arachidonic acid and their role in haemostasis and thrombosis. *Br. Med. Bull.* 34:129–35.

Mounts, T.L. 1987. Alternative catalysts for hydrogenation of edible oils. In *Hydrogenation: Proceedings of an AOCS Colloquium*, R. Hastert, ed., pp. 47–64. Champaign, IL: Am. Oil. Chem. Soc.

National Research Council. 1988. *Designing Foods: Animal Product Options in the Marketplace*, pp. 367. Report of the Committee on Technological Options to Improve the Nutritional Attributes of Animal Products. Board on Agriculture. Washington, D.C.: National Academy Press.

Needleman, P.; Raz, A.; Minkes, M.S.; Ferrendeli, J.A.; and Sprecher, H. 1979. Triene prostaglandins: Prostacyclin and thromboxane biosynthesis and unique biological properties. *Proc. Natl. Acad. Sci. USA* 76:944–48.

Nestel, P.; Noakcs, M.; Belling, B.; McArther, R.; Clifton, P.; Janus, E.; and Abbey, M. 1992. Plasma lipoprotein lipid and Lp(a) changes with substitution of elaidic acid for oleic acid in the diet. *J. Lipid Res.* 33:1029–36.

Neuringer, M., and Connor, W.E. 1989. Omega-3 fatty acids in the retina. In *Dietary ω3 and ω6 Fatty Acids: Biological Effects and Nutritional Essentiality*. Series A: Life Sciences. vol. 171, C. Galli and A.P. Simopoulos, eds., pp. 177–90. New York: Plenum Press.

Nordoy, A. 1991. Cardiovascular II: Heart. In *Health Effects of ω3 Polyunsaturated Fatty Acids in Seafoods*. vol. 66, A.P. Simopoulos, R.R. Kifer, R.E. Martin, and S.M. Barlow, eds., pp. 34–38. Basel: Karger.

O'Brien, J.S., and Sampson, E.L. 1965. Fatty acid aldehyde composition of the major brain lipids in normal gray matter, white matter and myelin. *J. Lipid Res.* 6:545–51.

Pariza, M.W., and Simopoulos, A.P., eds. 1987. Calories and energy expenditure in carcinogenesis. *Am. J. Clin. Nutr.* 45(suppl):149–371.

Parthasarathy, S.; Khoo, J.C.; Miller, E.; Barnett, J.; Witztum, J.L.; and Steinberg, D. 1990. Low density lipoprotein rich in oleic acid is protected against oxidative modification: Implications for dietary prevention of atherosclerosis. *Proc. Natl. Acad. Sci.* 87:3894–3898.

Phillipson, B.E.; Rothrock, D.W.; Connor, W.E.; Harris, W.S.; and Illingworth, R. 1985. Reduction of plasma lipids, lipoproteins, and apoproteins by dietary fish oils in patients with hypertriglyceridemia. *N. Eng. J. Med.* 312:1210–16.

Phinney, S.D.; Odin, R.S.; Johnson, S.B.; and Holman, R.T. 1990. Reduced arachidonate in serum phospholipids and cholesteryl esters associated with vegetarian diets in humans. *Am. J. Clin. Nutr.* 51:385–92.

Poulos, A.; Darin-Bennett, A.; and White, I.G. 1975. The phospholipid bound fatty acids and aldehydes of mammalian spermatozoa. *Comp. Biochem. Physiol.* 46B:541–49.

Raper, N.R.; Cronin, F.J.; and Exler, J. 1992. Omega-3 fatty acid content of the US food supply. *J. Am. College Nutr.* 11:304–8.

Reddy, B.S. 1986. Diet and colon cancer: Evidence from human and animal models. In *Diet, Nutrition and Cancer; a Critical Evaluation.* vol. 1, B.S. Reddy and L.A. Cohens, eds., pp. 47–66. Boca Raton: CRC Press.

Reis, G.J.; Boucher, T.M.; and McCabe, C.H. 1988. Results of a randomized, double-blind placebo-controlled trial of fish oil for prevention of restenosis after PTCA. *Circulation* 78(supply 2):291.

Ross, R. 1986. The pathogenesis of atherosclerosis—an update. *N. Engl. J. Med.* 314:488–500.

Salem, N., and Carlson, S.E. 1991. Growth and development in infants. In *Health Effects of ω3 Polyunsaturated Fatty Acids in Seafoods.* vol. 66, A.P. Simopoulos, R.R. Kifer, R.E. Martin, and S.M. Barlow, eds., pp. 20–25. Basel: Karger.

Salomon, P.; Kornbluth, A.A.; and Janowitz, H.D. 1990. Treatment of ulcerative colitis with fish oil in 3 omega fatty acids: An open trial. *J. Clin. Gastroenterol.* 12:157–161.

Schmidt, E.B.; Klausen, I.C.; Kristensen, S.D.; Lervang, H-H.; and Foergeman, O. 1991. Effect of ω3 fatty acids on lipoprotein (a): In *Health Effects of ω3 Polyunsaturated Fatty Acids in Seafoods.* vol. 66, A.P. Simopoulos, R.R. Kifer, R.E. Martin, and S.M. Barlow, eds., pp. 529. Basel: Karger.

Scientific Review Committee. 1990. *Nutrition Recommendations.* (H49-42/1990E) Ottawa: Minister of National Health and Welfare, Canada.

Seed, M.; Hoppichler, F.; Reaveley, D.; McCarthy, S.; Thompson, G.R.; Boerwinkle, E.; and Utermann, G. 1990. Relation of serum lipoprotein (a) concentration and apolipoprotein (a) phenotype to coronary heart disease in patients with familial hypercholesterolemia. *N. Engl. J. Med.* 322:1494–99.

Shekelle, R.B.; Missell, L.V.; Oglesby, P.; Shryock, A.M.; and Stamler, J. 1985. Letter to the editor. *N. Engl. J. Med.* 313:820.

Simonsen, T.; Vartun, A.; Lyngmo, V.; and Nordoy, A. 1987. Coronary heart disease, serum lipids, platelets and dietary fish in two communities in northern Norway. *Acta Med. Scand.* 222:237–45.

Simopoulos, A.P. 1987. Terrestrial sources of omega-3 fatty acids: Purslane. In *Horticulture and Human Health: Contributions of Fruits and Vegetables,* B. Quebedeaux and F. Bliss, eds. Englewood Cliffs, N.J.: Prentice-Hall.

———. 1989. Executive summary. In *Dietary ω3 and ω6 Fatty Acids: Biological Effects and Nutritional Essentiality.* Series A: Life Sciences. vol. 171, C. Galli and A.P. Simopoulos, eds., pp. 391–402. New York: Plenum Press.

———. 1990. Omega-3 fatty acids in growth and development. In *Omega-3 Fatty Acids in Health and Disease,* R.S. Lees and M. Karel, eds., pp. 115–56. New York: Marcel Dekker.

———. 1991. Omega-3 fatty acids in health and disease and in growth and development. *Am. J. Clin. Nutr.* 54:438–63.

Simopoulos, A.P., and Childs, B., eds. 1990. *Genetic Variation and Nutrition.* vol. 63. Basel: Karger.

Simopoulos, A.P.; Herbert, V.; and Jacobson, B. 1993. *Genetic Nutrition. Designing a Diet Based on Your Family Medical History.* New York: MacMillan.

Simopoulos, A.P.; Kifer, R.R.; and Martin, R.E., eds. 1986. *Health Effects of Polyunsaturated Fatty Acids in Seafoods.* Orlando, FL: Academic Press.

Simopoulos, A.P.; Kifer, R.R.; Martin, R.E.; and Barlow, S.M., eds. 1991. *Health Effects of ω3 Polyunsaturated Fatty Acids in Seafoods.* vol. 66. Basel: Karger.

Simopoulos, A.P.; Kifer, R.R.; and Wykes, A.A. 1991. ω3 fatty acids: Research advances and support in the field since June 1985 (worldwide). In *Health Effects of ω3 Polyunsaturated Fatty Acids in Seafoods.* vol. 66, A.P. Simopoulos, R.R. Kifer, R.E. Martin, and S.M. Barlow, eds., pp. 51–71. Basel: Karger.

Simopoulos, A.P.; Norman, H.A.; Gillaspy, J.E.; and Duke, J.A. 1992. Common purslane: A source of omega-3 fatty acids and antioxidants. *J. Am. College Nutr.* 11:374–82.

Simopoulos, A.P., and Salem, N., Jr. 1989. n-3 fatty acids in eggs from range-fed Greek chickens. *N. Engl. J. Med.* 321:1412.

Singer, P., and Hueve, J. 1991. Blood pressure-lowering effect of fish oil, propranolol and the combination of both in mildly hypertensive patients. In *Health Effects of ω3 Polyunsaturated Fatty Acids in Seafoods.* vol. 66, A.P. Simopoulos, R.R. Kifer, R.E. Martin, and S.M. Barlow, eds., pp. 522. Basel: Karger.

Slack, J.D.; Pinkerton, C.A.; Van Tassel, J.; Orr, C.M.; and Nasser, W.K. 1987. Can oral fish oil supplement minimize restenosis after percutaneous transluminal coronary angioplasty? *J. Am. Coll. Cardiol.* 9(suppl):64a.

Slater, T.F., and Block, G., eds. 1991. Antioxidant vitamins and beta-carotene in disease prevention. *Am. J. Clin. Nutr.* 53(suppl):189S–396S.

Sperling, R.I.; Weinblatt, M.E.; Robin, J.L; Ravalese, J.; Hoover, R.L.; House, F.; Coblyn, J.S.; Fraser, P.A.; Spur, B.W.; Robinson, D.R.; Lewis, R.A.; and Austen, K.F. 1987. Effects of dietary supplementation with marine fish oil on leukocyte lipid mediator generation and function. *Arth. Rheumat.* 30:987–88.

Spritz, N., and Mishkel, M.A. 1969. Effects of dietary fats on plasma lipids and lipoproteins: An hypothesis for the lipid-lowering effect of unsaturated fatty acids. *J. Clin. Inves.* 48:78–86.

Steinberg, D.; Parthasarathy, S.; Carew, T.E.; Khoo, J.C.; and Witztum, J.L. 1989. Beyond cholesterol: Modifications of low-density lipoprotein that increase its atherogenicity. *N. Engl. J. Med.* 320:915–24.

Stenson, W.F.; Cort, D.; Rodgers, J.; Burakoff, R.; DeSchryver-Kecskemeti, K.; Gramlich, T.L.; and Beeken, W. 1992. Dietary supplementation with fish oil in ulcerative colitis. *Ann. Intern. Med.* 116:609–14.

Trevisan, M.; Krogh, V.; Freudenheim, J.; Blake, A.; Muti, P.; Panico, S.; Farinaro, E.; Mancini, M.; Menotti, A.; Ricci, G. and the Research Group ATS-RF2 of the Italian National Research Council. 1990. Consumption of olive oil, butter, and vegetable oils and coronary heart disease risk factors. *JAMA* 263:688–92.

Uauy, R.D.; Birch, D.G.; Birch, E.E.; Tyson, J.E.; and Hoffman, D.R. 1990. Effect of dietary omega-3 fatty acids on retinal function of very-low-birth-weight neonates. *Pediatr. Res.* 28:485–92.

van der Tempel, H.; Tulleken, J.E.; Limbuerg, P.C.; Muskiet, F.A.J.; and van Rijswijk, M.H. 1990. Effects of fish oil supplementation in rheumatoid arthritis. *Ann. Rheum. Dis.* 49:76–80.

van Vliet, T., and Katan, M.B. 1990. Lower ratio of n-3 to n-6 fatty acids in cultured than in wild fish. *Am. J. Clin. Nutr.* 51:1–2.

Vanhoutte, P.M.; Shimokawa, H.; and Boulanger, C. 1991. Fish oil and the platelet-blood vessel wall interaction. In *Health Effects of Polyunsaturated Fatty Acids in Seafoods*, A.P. Simopoulos, R.R. Kifer, and R.E. Martin, eds., pp. 233–44. Orlando, FL: Academic Press.

Weber, P.C. 1989. Clinical studies on the effects of n-3 fatty acids on cells and eicosanoids in the cardiovascular system. *J. Intern. Med.* 225(suppl):61–68.

Weber, P.C.; Fischer, S.; von Schacky, C.; Lorenz, R.; and Strasser, T. 1986. The conversion of dietary eicosapentaenoic acid to prostanoids and leukotrienes in man. *Prog. Lipid Res.* 25:273–76.

———. 1986a. Dietary omega-3 polyunsaturated fatty acids and eicosanoid formation in man. In *Health Effects of Polyunsaturated Fatty Acids in Seafoods*, A.P. Simopoulos, R.R. Kifer, and R.E. Martin, eds., pp. 49–60. Orlando, FL: Academic Press.

Weber, P.C., and Leaf, A. 1991. Cardiovascular effects of ω3 fatty acids. Atherosclerosis risk factor modification by ω3 fatty acids. In *Health Effects of ω3 Polyunsaturated Fatty Acids in Seafoods*. vol. 66, A.P. Simopoulos, R.R. Kifer, R.E. Martin, and S.M. Barlow, eds., pp. 218–32. Basel: Karger.

Wiklund, O.; Angelin B.; Olofsson, S-O.; Eriksson, M.; Fager, G.; Berglund, I.; and Bondjers, G. 1990. Apolipoprotein (a) and ischaemic heart disease in familial hypercholesterolaemia. *Lancet* 335:1360–63.

Willett, W.C.; Stampfer, M.J.; Colditz, G.A.; Rosner, B.A.; and Speizer, F.E. 1990. Relation of meat, fat, and fiber intake to the risk of colon cancer in a prospective study among women. *N. Engl. J. Med.* 323:1664–72.

Ziboh, V.A. 1991. ω3 polyunsaturated fatty acid constituents of fish oil and the management of skin inflammatory and scaly disorders. In *Health Effects of ω3 Polyunsaturated Fatty Acids in Seafoods*. vol. 66, A.P. Simopoulos, R.R. Kifer, R.E. Martin, and S.M. Barlow, eds., pp. 425–35. Basel: Karger.

Zock, P.L., and Katan, M.B. 1992. Hydrogenation alternatives: Effects of trans fatty acids and stearic acid versus linoleic acid on serum lipids and lipoproteins in humans. *J. Lipid Res.* 33:399–410.

Chapter 17

Phytochemicals and Antioxidants

Robert I-San Lin

INTRODUCTION: AN OVERVIEW

Plants synthesize a myriad of compounds, most of which are physiologically active when consumed. Many of them are essential to life and/or highly beneficial to health; many others are toxic. Plant foods provide an energy source that drives the metabolic process; the amino nitrogen that is the raw material for the synthesis of protein and nucleic acids; the essential fatty acids that form the cytomembranes and eicosanoids/docosanoids; the vitamins that function as coenzymes, metabolic regulators, and antioxidants; and minerals. All of them are well-known plant chemicals and their functions. Furthermore, there are additional arrays of compounds in plants, which in the 1950s and 1960s plant physiologists called secondary plant products or phytochemicals ("phyto" means plant). These include large families of phenolic compounds, alkaloids, derivatives of isoprene (e.g., terpenes, steroids, and carotenes), and numerous other types of chemicals. Their nutritional functions had been overlooked until recently, when their protective effects against some common degenerative diseases were investigated.

Through millions of years of evolution the human body has become well adapted to, or is quasi-optimized to the food sources that were prevalent before the establishment of agriculture about 20,000 years ago. The typical diets of our ancestors derived approximately three-quarters of its weight and two-thirds of its energy from plant sources, while the remaining portion was from animal sources. Fruits, leaves, seeds, tubers, and buds constituted the bulk of the diets that were consumed. These parts of plants are rich in various phytochemicals (Lin 1993a). Thus, it can be expected that the body has become responsive to many phytochemicals and uses them for regulatory and other functions. In contrast, modern diets are very rich in land animal products. From an evolutionary point of view, this imbalance is bound to cause certain undesirable impacts. The prevalence of cardiovascular diseases and cancer is at least partially attributable to the imbalance. For over three decades I have advocated reduced consumption of land animal-derived foods and increased consumption of plant-derived foods, particularly leafy vegetables, seeds, and nuts, in order to promote good health.

While Hippocrates said, "Let your food be your medicine; let your medicine be your food," the Chinese have been far more advanced in this approach. During the past 2,000 years, Chinese medical compendia have discussed the medicinal and nutritional properties of almost all foods. Foods are described with many pairs of opposite medicinal and/or nutritional attributes and with regard to the target organs to which a food contributes the most. The Chinese health-promoting philosophy emphasizes the balanced consumption of foods having opposite properties for maintaining health, and consumption of certain foods having properties that can counter the imbalances in one's body to prevent, treat, or cure diseases. Thus, the Chinese led the world in the concept of functional foods; we will draw some information from Chinese herbal medicine later on.

The term "phytochemical" originally had no specific definition other than that used by plant physiologists. However, in recent years many researchers have loosely referred to plant-derived substances that have anticancer and/or cardiovascular-protective properties as phytochemicals. In a position paper regarding functional foods, a U.S. Food and Drug Administration's expert has defined phytochemicals as "substances found in edible fruits and vegetables that may be ingested by humans daily in gram quantities, and that exhibit a potential for modulating human metabolism in a manner favorable for cancer prevention" (Jenkins 1993). For several reasons, I do not share this definition. Most phytochemicals that we will deal with exist in our diets at the milligram level. The term "phytochemicals" used in this chapter refers to *plant-derived substances that are nutritionally physiologically and/or medicinally highly active,* excluding starch, protein, common fatty acids, vitamins, and essential minerals.

Numerous plants and phytochemicals have been reported to be protective or potentially protective against various diseases. We will focus mainly on selected plants that have been used as foods and that appear to have some support for their efficacy in the prevention of some common degenerative diseases or in the promotion of general health. Plants and phytochemicals used for treating specific diseases of non-degenerative natures will be ignored here, as they may be considered as prescription drugs. For this reason, many very effective herbs and phytochemicals will not be reviewed here. The author writes this article not only from the viewpoint of a scientist, but also as a practitioner of Chinese herbal medicine for over 35 years.

CARDIOVASCULAR-PROTECTIVE PLANTS AND THEIR ACTIVE PRINCIPLES

Although bleeding from injuries might have been one of the major threats to life in the evolutionary past, occlusion of the blood circulation is the greatest threat to life among people living in modern affluent societies. *The following discussions will focus on the prevention of occlusive cardiovascular diseases only. Erroneously applying the information to other types of cardiovascular diseases may lead to dangerous situations.*

Several factors contribute to an occlusion of blood circulation in an otherwise healthy individual. These include atherosclerosis, thrombosis and/or embolism, and vasoconstriction/vasospasm. Phytochemicals that can mitigate these factors may potentially provide protection against occlusive cardiovascular diseases. When the safety and efficacy of a phytochemical or plant having these beneficial properties are supported by clinical studies, epidemiological studies, and/or a long history of use, then the evidence would be adequate for its use in food, provided that there is no stability problem. More than 400 foods and medicinal herbs have been known to have antiocclusive and antiatherosclerotic effects directly or indirectly. Four of them are discussed below.

Garlic and Thioallyl Compounds

Garlic (*Allium sativum L.*) is unique for containing a family of thioallyl compounds, $CH_2=CH\text{-}CH_2\text{-}X$, where X stands for a variety of organic structures (Lin 1989a, 1990; Lin, Abuirmeileh, and Qureshi 1990). In Chinese herbal medicine, concoctions of garlic and auxiliary herbs, garlic juice, garlic poultice, and garlic cloves that have been aged in vinegar or wine for two to three years are used to treat a broad spectrum of diseases that have a broad etiology, including occlusive circulatory disorders and neoplastic conditions. (Benign and malignant tumors were often not clearly distinguished.) Recent research on garlic focuses mainly on its cancer-protective and cardiovascular protective effects.

Hypolipidemic Effects

An aged garlic extract (ranging from 0.3–8.1% in a corn-based diet) and its active principle S-allyl cysteine (SAC; ranging from 15–405 ppm in the diet) were fed separately to normolipidemic and hyperlipidemic 2-week old female chickens for four weeks. In hyperlipidemic chickens, the aged garlic extract caused a dose-dependent reduction in serum total cholesterol (up to 30% reduction), low density lipoprotein (LDL) cholesterol (up to 50% reduction), apolipoprotein B (up to 19%), the activity of 3-hydroxy-3-methylglutaryl coenzyme A reductase (HMGR) (up to 26%), and the activity of 3-hydroxy-3-methylglutaryl coenzyme A synthase (up to 25%) (Qureshi et al. 1990). The latter two are the rate-limiting enzyme and the first key enzyme, respectively, in the biosynthesis of cholesterol. Similar but less drastic effects were observed in normolipidemic chickens. This supplementation of aged garlic extract also suppressed, in a dose-dependent manner, the activity of other enzymes of lipogenesis (acetyl CoA carboxylase and fatty acid synthetase), serum triglycerides level, and plasma thromboxane and platelet factor IV levels (Qureshi 1990). The latter two are important factors in platelet aggregation and subsequent thrombosis. Furthermore, platelet aggregation in platelet-rich plasma and in whole blood against collagen and against adenosine diphosphate was substantially inhibited. Garlic powder and garlic oil had similar effects.

Parallel to the study in chickens, feeding rats a diet containing 2% aged garlic extract led to a 30% reduction in serum triglycerides and a 15% reduction in total serum cholesterol. Radioactive ^{14}C-acetate and [2-^{3}H]-glycerol tracer studies with cultured rat hepatocytes demonstrated that the hypocholesterolemic and hypotriglyceridemic effects were due to suppression of cholesterol and fatty acid synthesis, but not due to suppression of esterification of fatty acids with glycerol.

These studies clearly demonstrate that, in laboratory animals, dietary garlic (aged garlic extract) supplementation can lower cholesterol and triglycerides levels in the blood, and reduce blood-clotting propensity. There are reports of similar effects of dietary garlic supplementation in humans. However, most of these reports are of very poor quality—in some cases amounting to advertisements for the sponsoring garlic supplement marketers. These reports cannot be used to support the efficacy and safety, in humans, of garlic supplementation on serum lipid levels. However, the results are not yet available. It is my opinion, based on valid data from biochemical, cell culture, and animal studies, as well as clinical experience, that garlic supplementation at about 10 to 15 g daily of cooked garlic, or the equivalent amounts of garlic oil or aged garlic extract, can lower serum cholesterol level by about 5–8% among the majority of hypercholesterolemic individuals. This range of reduction is significant but minor compared to the required degrees of reduction among many hypercholesterolemic individuals. The hypocholesterolemic function is only a minor part of the cardiovascular protective effects of garlic.

Antithrombosis, Antiadhesion, and Antiprolific Effects

Garlic has a substantial anticlotting effect. In addition to the abovementioned suppression of thromboxane B_2 formation, raw garlic and cooked garlic have been shown to enhance the blood's fibrinolytic activity, to suppress blood fibrinogen concentration, and to "thin" the blood through other mechanisms (Lin 1990; Makheja 1990). Sterile extracts of autoclaved garlic have a vasodilating effect in laboratory animals when given orally and intraperitoneally (Lin 1987, 1991), apparently, through its smooth muscle relaxing mechanism. Using a simulated arterial surface, aged garlic extract and its active compounds SAC and S-allyl mercaptocysteine (SAMC) have been demonstrated to have an antiplatelet adhesion effect (Steiner and Lin 1992). Platelet adhesion to the endothelial surface initiates the atherosclerotic process. SAMC has also been shown to inhibit the proliferation of rat aortal smooth muscle cells (Steiner and Lin 1993).

Although these were *in vitro* studies, they do strongly suggest that the aged garlic extract and its active principles may suppress platelet adhesion and smooth muscle cell proliferation *in vivo*, both important processes in the development of atherosclerosis. Smooth muscle cell proliferation is also the major factor in restenosis after angioplasty. It appears that the antiplatelet aggregation, the antiplatelet adhesion, and the antiproliferation properties of garlic (aged garlic extract) contribute more to the cardiovascular protection than the hypolipidemic properties of garlic.

In addition to the biochemical, cell culture, and animal studies that demonstrate the cardiovascular protective effects of garlic, there is epidemiological evidence of lowered cardiovascular incidence in the high garlic-consuming populations of the Mediterranean region and many Asian regions, as compared to populations who have similar life and dietary styles, but have low garlic consumption. Although such an epidemiological association in itself carries little weight in establishing a causal relationship, this does lend some support to the conclusion of laboratory studies that garlic consumption provides a certain degree of protection against occlusive cardiovascular disease.

There is only one clinical study on the effects of garlic supplementation on the end point cardiovascular event; i.e., a myocardial infarction or death (Bordia and Verma 1990). In the study, 432 cardiac patients were divided into a control group (210) and a garlic-supplemented group (222). The study states, "Garlic feeding reduced the mortality by 50% in the 2nd year and by about 66% in the 3rd year ($P < 0.01$). Similarly, the rate of re-infarction was reduced by 30% and 60% in the 2nd and 3rd year, respectively." Although the reported effects of garlic supplementation were phenomenal, only a small number of patients in both groups experienced the end event of death or re-infarction. A much larger scale study is needed to further substantiate and quantify the efficacy of garlic supplementation.

Garlic has been a popular subject of many writers. Unfortunately, most of popular books and articles on garlic have been written by unqualified writers and contain numerous mistakes, in addition to exaggerations. Some of the most common mistakes are the anti-infectious and bactericidal effects of garlic and garlic supplement pills. There is no evidence, both a long history of use and modern clinical studies, supporting garlic's efficacy in treating infectious diseases, when it is consumed at a level of ca. 20 g. (about six medium sized-cloves) or less daily, or equivalent amounts of garlic pills. This level is far above the levels commonly consumed by medium level-garlic consumers, which range from 5-10 g. daily. At levels substantially above this level, garlic consumption may provide some benefits against systemic fungal infections, but the safety and efficacy for these conditions have not been scientifically proven.

Active Principles of Garlic

Raw, intact garlic cloves contain about 1–2% thioallyl compounds, mainly in the form of alliin (S-allyl cysteine sulfoxide, $CH_2 = CH-CH_2-S(O)-CH_2-CH(NH_2)COOH$ and gamma glutamyl peptides of S-allyl cysteine. These compounds are odorless. When the cellular structures are ruptured, e.g., through cutting or crushing, alliin is brought into contact with an enzyme contained in the garlic cell, alliinase, which converts alliin into allicin, the odorous compound that is responsible for the odor of freshly crushed garlic. Allicin is unstable; it decomposes readily, following a quasi-first order kinetic. In a diluted solution in water, allicin is more stable. The decomposition fragments condense to form diallyl disulfide, diallyl trisulfide, diallyl sulfide, and dithiins, in addition to many minor thioallyls; these constitute the bulk of garlic essential oil. Diallyl sulfide and diallyl polysulfides are relatively more stable; they form a temporary pool of the transformation of garlic's thioallyl compounds. Thus, frying, excessive boiling, and steam distillation all lead to the transformation of garlic's thioallyls mainly into the diallyl sulfide/polysulfide forms and small quantities of dithiins and other sulfur compounds. Among the polysulfide homologues, there is a tendency of increasing bioactivity with an increasing number of sulfur atoms in the molecule. Due to instability of the thioallyl compounds, many garlic researchers have made erroneous conclusions; they often mistakenly assigned allicin to be the active principle of garlic or the garlic preparations that they studied. For more than a decade, allicin has been heavily promoted to be the active principle of garlic and garlic supplements by most garlic supplement vendors; each of them has been forceful in claiming that its garlic supplements contain the highest amount of allicin. The truth is that there is no truth in these allicin promotions and claims. Under normal physiological conditions, allicin, given as a part of freshly crushed garlic and given in pure form, does not pass through the digestive lining and enter the blood circulation. Injected allicin is rapidly destroyed in the blood. Similar situations occur with other oxygenated thioallyls, e.g., ajoene (see below). It appears that the body has low tolerance to oxygenated thioallyls (Lin 1993b). Garlic's thioallyls differ substantially in chemical structure; most of them share certain biological activities that are common among various garlic preparations. However, they differ in potency, the ability to generate garlic odor in the body, toxicity, stability, and bioavailability.

In the 1980s, an oxygenated thioallyl compound of garlic was identified as ajoene, which has a short term anti-coagulation effect in animals (Apitz-Castro and Jain 1990) and is capable of inhibiting human gastric lipase and may thus reduce fat absorption (Cudrey et al. 1990). Although ajoene can be produced during the aging of garlic macerates in oil suspension, it does not exist to a significant extent in freshly crushed garlic and garlic prepared under common cooking conditions. It is unstable and decomposes in the blood stream. These factors render ajoene unsuitable for food fortification.

As a result of the instability of the thioallyls, different conditions for preparing garlic lead to different families of thioallyls in different types of garlic products. Garlic powders made from raw garlic at temperatures below approximately 60° to 80° C contain alliin and gamma(γ)-glutamyl peptides of S-allyl cysteine as the main active principles. However, garlic powders may also contain trace amounts of diallyl sulfide/polysulfides due to the effects of processing. Garlic essential oils contain mainly diallyl sulfide/polysulfides and dithiins, in addition to small amounts of allyl and alkyl sulfides. However, commonly available garlic oils and garlic oil supplements contain highly diluted garlic essential oil, usually diluted 200- to 1000-fold with vegetable oils. A traditional way of preparing garlic is to age (soak) it in vinegar or wine for a long period of time, up to two to three years. The aging is usually accomplished in an enclosed jar or semi-enclosed system; thus, the end result is a reduction of alliin into SAC, with a small amount of SAMC also formed. Garlic oil macerates in vegetable oil contain mainly diallyl sulfides and smaller amounts of ajoene and dithiins. Cooked garlic contains thioallyls ranging from those similar to the ones in raw garlic to those similar to the ones in garlic essential oil, depending on the cooking conditions. In most cases, e.g., in garlic powders, garlic oil, and garlic oil macerates, the thioallyl composition changes constantly as the preparations are stored under ambient temperatures. Light can also cause changes in thioallyls, sometimes causing the ejection of sulfur atoms in the polysulfides, leading to the formation of sulfur powder.

The chemistry, pharmacology, efficacy, safety, and stability of garlic were discussed extensively during the First World Congress on the Health Significance of Garlic and Garlic Constituents, held in 1990 in Washington, D.C. Interested readers should read the Proceedings for further information (see Lin 1990).

Crataegus pinnatifida Bge. (Fruit)

The fruit of *Crataegus pinnatifida Bge.* has long been used to make candies in China, perhaps for its fruity flavor and tartness. This fruit has also long been used in Chinese herbal medicine to enhance circulation, treat swelling due to blood stasis, relieve pains associated with swelling, promote digestive function, and mitigate other conditions as well.

A study on 104 hypercholesterolemic patients demonstrated that a daily dose equivalent to 46 g of the fruit for 45 days caused normalization of cholesterol value in 75% of the patients, with an additional 15% of the patients experiencing a 20 milligrams/deciliter (mg/dl) reduction (Wu Han Health Department 1977). A similar treatment was reported to be effective in hypertriglyceridemia (Am Sun No. 1 Pharmaceutical Factory 1977). Daily supplementation of an extract of the fruit (equivalent to 15 g fresh fruit daily) for 12 weeks in 16 coronary artery patients with angina led to outstanding improvements in the conditions of most of the patients, including normalization of exertional electrocardiogram and resting electrocardiogram. There were also substantial reductions in serum triglycerides and cholesterol (Shen Yang Army Hospital 1975). Oral supplementation of extracts of the fruit has also been shown to be effective in lowering blood pressure in hypertensive patients (Liao Ling See Fung Disease Prevention Center 1974; Chau An County Health Department 1977), which was interpreted to be a result of peripheral vasodilation.

In addition to these human studies, many animal studies also demonstrated that the fruit and its extracts can reduce heart muscle fatigue, strengthen the heart muscles' contraction amplitude and pumping power, dilate the coronary artery, and enhance blood supply to the heart muscles. Tube feeding rabbits an extract of the fruit also lowered serum cholesterol and triglyceride levels, and reduced atherosclerotic lesions in the aorta, in comparison to animals fed a control diet. There was a significant reduction in the surface area and the thickness of the atherosclerotic mass, and in the number of lesions (Giang Su New Medical College 1975).

The Chinese have used mainly *C. pinnatifida* fruit as a food and medicine; however, several closely related species have also been used in China and Europe, e.g., *C. cuneata Sieb.*, *C. sanguinea Pall.*, *C. hupehensis Sarg.*, *C. monogyna*, *C. pentagyna*, and *C. oxyacantha*. The last one is common in Europe and is often called hawthorn berry. Extracts of hawthorn berry are sold in many European countries as a health food or cardiac tonic.

The fruit contains chlorogenic acid, caffeic acid, phlobaphene, L-epicatechol, choline, choline acetate, beta(β)-sitosterol, sorbitol, vitamin C, crategolic acid, hyperin, tartaric acid, citric acid and certain chromones. Some of the pharmacological activities, e.g., the hypotensive effects have been attributed to the chromones. Further research will undoubtedly uncover additional active principles, as neither these compounds nor combinations of them have the pharmacological properties that match adequately with the pharmacological properties of the whole fruit.

Ginkgo biloba L. (Seed and Leaf)

Ginkgo biloba seed is a food commonly consumed in China in small quantities. It contains certain toxic substances, vaguely known as ginkgotoxin, that deter excessive consumption. Ginkgo seeds are also used as a medicine for treating various conditions, including certain types of asthma, leukorrhea, and polyuria.

In the past three decades, ginkgo leaves have attracted much attention as an agent for improving circulation, particularly cerebral circulation, which may lead to improved mental function. In Europe and, to a lesser degree, in the United States, ginkgo leaf extracts are promoted heavily in the health food market and/or drug stores as popular dietary supplements. Sometimes the promotional literature purport or imply ginkgo leaf extracts as having been used for thousands of years in China. But before recent decades, traditional Chinese herbal medicine rarely used ginkgo leaves for treating occlusive circulatory disorders and other medical conditions. Many European studies reported that extracts of ginkgo leaves enhanced brain circulation (Larsen, Dupeyron, and Boulu 1978; Rapin and Le Poncin-Lafitte 1979; Le Poncin-Lafitte, Rapin, and Rapin 1980), had peripheral vasodilative and hypotensive effects (Auguet, De Feudis, and Clostre 1982), increased the brain's tolerance to hypoxia (Le

Poncin-Lafitte et al. 1982; Schaffler and Rech 1985), reduced venule thrombi formation (Borzeix, Labos, and Hartl 1980), increased venular and capillary blood flow and reactivated capillaries that were formerly shut off from circulation (Piovella 1973), protected retinal injury from free radicals (Clairambault et al. 1986), rendered the erythrocyte membrane more resistant to hemolysis (Etienne et al. 1982), enabled patients with peripheral arterial insufficiency to walk longer distances and improved their blood flow through the lower limbs (Bauer 1984), and improved cerebral hemodynamics and ameliorated mental deficit (Gessner, Voelp, and Klasser 1985; Haan et al. 1982; Eckmann and Schlag 1982). In some cases, the ginkgo leaf extracts out-performed modern drugs for mental insufficiency, e.g., ergot alkaloids and vincamine. Some researchers concluded that the effects were mediated by catecholamine release (Auguet, De Feudis, and Clostre 1982; Auguet et al. 1982; Brunello et al. 1985), and by calcium ion flux over smooth muscle cell membranes (Auguet and Clostre 1983). These effects share some similarities with those of theophylline and papaverine. Some of the studies and/or reports, however, were not quality works; therefore, better designed and controlled studies with more patients are needed to further support and quantify the efficacy.

The Chinese medical community followed the Europeans' lead and have engaged in research on ginkgo leaves in the past two decades. At a daily dose that is equivalent to 12 mg of total ginkgetin (ginkgetin + isoginkgetin), a three-month treatment caused significant improvements in coronary heart disease patients (Chez Giang Medical University 1972). In patients with mild cardiovascular disease, treatment with tablets made of a water extract of ginkgo leaves at a daily dose equivalent to 14 mg of ginkgetin for one to five months caused a reduction of serum cholesterol of 39 mg/dl, and improved the serum phospholipid to cholesterol ratio (Wu Han Army General Hospital 1973). However, these were not double-blind controlled studies. Serum cholesterol levels can be influenced by many factors, such as dietary intake of saturated fat, calories, and cholesterol, and the season of the year. Thus, the hypocholestolemic effect of ginkgo leaf extracts is yet to be further substantiated.

Active Principles of Ginkgo Leaves

The following ginkgo leaf compounds have been isolated and characterized: ginkgolic acid, hydroginkgolic acid, ginkgol, bilobol, gin-nol, ginkgotoxin, ginkgolides A, B, and C, and some flavonoids that are common to many other plants, including kaempferol guercetin, rutin, and isorhamnetin.

Compared to garlic and *Crataegus* berry, ginkgo leaves lack the support of a long history of effective and safe use.

Polygonatum Odoratum (Root)

Polygonatum odoratum (Mill.) Druce var. pluriforum (miq) Ohwi. root is commonly used in Chinese cooking and in Chinese herbal medicine. As a medicine, it has been used to treat "yin" deficiency, a syndrome well-defined in Chinese herbal medicine, but for which there is no counterpart in modern medicine. Certain types of hyperthyroidism and hypertension are results of a "yin" deficiency. Studies with animals demonstrated that extracts of the root lowered serum triglyceride and cholesterol levels, and slowed the progress of atherosclerosis. (Chin Chow Medical College 1977a, 1978). Extracts of the root have also been reported to be effective in treating hyperlipidemia and cardiac insufficiency due to coronary artery atherosclerosis (Chin Chow Medical College 1977b; Chauo Cho City Hospital 1972). In mice, injection of an extract of the herb substantially increased the animals' tolerance to hypo-oxygen conditions. Some constituents have been identified: convallamarin, convallarin, azetidine-2-carboxylic acid, kaempferol, and quercitol. However, few pharmacological studies have been done on them.

CANCER-PROTECTIVE PLANTS AND THEIR ACTIVE PRINCIPLES

Although cancers are a family of very complex diseases that have many causes, they share some common pathological characteristics. Most types of cancers are initiated by and go through the following steps or have the following properties: (a) activation of carcinogens, (b) initiation/transformation, (c) promotion, (d) progression, (e) capability of angiogenesis, (6)capability of immune evasion and resistance, and (7) capability of migration, adhesion, and invasion. Cancer cells can migrate through the blood vessels or the lymph ducts, adhere to the vessel surface, and then invade the underlying tissues, whereas many types of normal cells cannot. Plants and phytochemicals that can inhibit some of these steps or counteract these properties may have an anticancer effect.

Garlic and Thioallyl Compounds

Historically, herbal formulas containing garlic have been used in Chinese herbal medicine for treating neoplastic conditions. Recent work demonstrated that in dimethyl benz(a)anthracene (DMBA)-treated rats, an aged garlic extract inhibited DMBA-DNA adducts formation, reduced mammary tumor incidence and the number of tumors per animal, and delayed tumor appearance (Liu, Lin, and Milner 1992). DMBA is a potent carcinogen and is activated through oxidation by the cytochrome P_{450} enzymes to form DMBA diol epoxide. Covalent binding of the diol epoxide to DNA cause DNA lesions, initiates carcinogesis and transforms a normal cell into a cancerous one. The aged garlic extract also modulated the cell surface antigens and the growth of cultured human melanoma cells (Hoon et al. 1990). Cell surface antigens are associated with the invasiveness of cancer cells and their ability to resist detection/destruction by the immune cells; modification of cell surface antigens may have profound impacts on the course of the disease. A garlic extract also demonstrated a cancer-protective effect in DMBA-treated mice through suppression of initiation and promotion steps (Nishino 1990). A garlic-derived compound, diallyl sulfide (DAS) was reported to modulate carcinogen metabolism and detoxification (Wargovich, Sumiyoshi, and Baer 1990). In mice, injection of aged garlic extract suppressed cancer development of transplanted cancer cells (Lamm, Riggs, and DeHaven 1990). Various garlic preparations, including aged garlic extract, have been shown to suppress the formation of nitrosamines (carcinogens) in the stomach, inhibit the activation of polyarene carcinogens, enhance the excretion of metabolites of carcinogens, and reduce the rate of proliferation of cultured human breast cancer cells (Pinto, Tiwari, and Lin 1993). The antiprolific effects on arterial smooth muscle cells, as mentioned earlier in the section on cardiovascular-protective plants, also suggest that garlic may potentially inhibit angiogenesis in a cancer mass. Without blood vessels and blood perfusion, a cancer mass would not be able to grow large. These and other animal studies are consistent with epidemiological evidence that high garlic consumption is associated with low cancer incidence (Blot 1990). Thus, there is mounting evidence supporting the anticancer effect of appropriate consumption of garlic.

Soybean

Soybean (*Glycine max*) contains some potentially anticancerous compounds, including a protease inhibitor and phytic acid. Readers are urged to consult a recent review on this subject (Messina and Messina 1991).

Chlorophyllins

The evidence is overwhelming that high consumption of arachidonic acid-rich foods, and/or low consumption of leafy vegetables and whole grains is associated with high cancer incidence (Lin 1993c). All functional leaves contain large amounts of chlorophylls, which are essential for harvesting solar energy. Even leaves that are not green also often contain substantial amounts of chlorophylls. Chlorophyll molecules contain a heme chromophore to which a methyl and a long chain phytyl group are attached. These attached groups can be hydrolyzed and removed so that the resulting pigments have a higher water solubility. The modified chlorophylls are called chlorophyllins, and are used as food colors. Recently, chlorophyllins have been shown to reduce the bio-availability of chemical carcinogens (Bronzetti, Galli, and Croce 1990; Negishi et al. 1989), to inhibit aflatoxin B-DNA adducts formation, and to protect against mutagenicity of other carcinogens in a *Salmonella* assay (Dashwood, Breinholt, and Bailey 1991). These properties suggest that chlorophyllins and/or chlorophylls may provide anticancer benefits. However, it is my opinion that a reduction of a carcinogen's bioavailability alone may be insufficient in cancer prevention because cancer development is a multistep process. Much work needs to be done in this area to ascertain whether chlorophyllins or chlorophylls can indeed provide a significant level of protection against cancer.

Lignans

Lignans are diphenolic compounds derived from intestinal bacterial digestion of polymeric precursors in plants. These compounds have been reported to have estrogen properties and to exert some influence over estrogen regulation. For these reasons they are also called phytoestrogens. A high intake of lignans has been reported to reduce free estrogen levels (Adlercreutz et al. 1987). Thus, based on theory, increased consumption of lignan precursor-rich foods or lignan precursors may lower the risk of estrogen responsive cancers, such as certain types of breast cancer. Lignans have been shown to be inhibitory to the growth of cultured human breast cancer cells (Hirano et al. 1990). Many oil seeds and grains, such as flaxseed, rapeseed, soybean, triticale and wheat, are rich sources of lignan/lignan precursors. It is my opinion that lignans alone ineffective for cancer prevention, for several reasons. First, many types of cancer are not estrogen responsive. Second, the phytoestrogen theory is not well established. For example, if lignans are able to reduce the level of free circulating estrogen, then why do populations having a high intake of lignans/lignan precursors (such as certain Asian populations) have a lower osteoporosis incidence than populations having a low intake of lignan (such as the North American population)? If the phytoestrogen theory is correct, then a high intake of lignan would lower serum estrogen level and increase osteoporosis incidence. Although one could propose many hypotheses to interpret this discrepancy, there would be too many assumptions, weakening the validity of the entire theory. Third, epidemiological evidence supporting anti-cancer efficacy of lignans is deficient. Epidemiological evidence does support that high consumption of lignan precursor-rich foods is associated with low cancer risk. However, these foods are rich in other phytochemicals, such as flavonoids and other phenolic compounds, many of them also have anti-cancer properties. Fourth, growth suppression on cultured cancer cells is only a weak evidence for in vivo efficacy. Many compounds are cytostatic or cytocidal toward cultured cancer cells but are ineffective under in vivo conditions. Lignan's efficacy in vitro does not necessarily suggest its activity in vivo.

Cruciferous Vegetables

High intake of cruciferous vegetables is associated with low cancer risk (Kune et al. 1992). A study comparing the importance of organic content in drinking water and some dietary factors in the development of colon adenoma led to the conclusion that a low intake of cruciferous vegetables, among other factors, was far more important in the development of colorectal polyps (Hoff et al. 1992) than the water contaminants. Phenethyl isothiocyanate (PEITC) in these vegetables has been demonstrated to protect mice against nitrosamine-induced lung tumorigenesis (Chung, Morse, and Eklind 1992). In line with these and other observations, PEITC affects the activity of phase I and phase II xenobiotic-metabolizing enzymes, resulting in the inhibition of the oxidative activation of chemical carcinogens (Guo et al. 1992).

Tea

Tea (*Camellia sinensis O. Ktze.*; leaves and young branches) is one of the most commonly consumed beverages, particularly in China and Japan. An epidemiological study related a reduced gastric cancer risk to drinking 10 or more cups daily of green tea (Kono 1988). Many phenolic compounds of green tea, including catechins, have been shown to inhibit tumor formation in rats induced by N-methyl-N′-nitro-N-nitrosoguanidine (Jain et al. 1989) and mutation induced by benz(a)pyrene, aflatoxin, and alpha(α)-aminofluorene (Wang et al. 1989). The anticarcinogenic properties of these compounds have been attributed to their antioxidant capacity (Ruch, Cheng, and Klaunig 1989; Zhao et al. 1989). Many of these compounds are common to numerous other plants; thus, similar anticarcinogenic properties should be expected from these plants if these compounds are indeed effective. Tea merits further research, which may yield useful information that is applicable to other plant foods. Interestingly, while scientists in the West and promoters of green tea have focused on green tea, the Chinese believe that red tea is more beneficial, and they have demonstrated in rats some anticancer properties of an extract of the microorganisms that are used in making red tea through fermentation.

Turmeric

Turmeric is a spice isolated from the rhizome of the *Curcuma longa* plant. It enhanced the tumor suppressive effects of betel-leaf extract in hamsters treated with methyl(acetoxymethyl)nitrosamine (Azuine et al. 1992) and elevated the activity of carcinogen-detoxifying enzyme, glutathione-S-transferase in mice (Aruna and Sivaramakrishnan 1990). Some active principles of turmeric have been identified as yellow-colored phenolic compounds, curcumins I and III. These compounds inhibit carcinogenesis at the initiation, promotion, and progression stages (Nagabhusahan and Bhide 1992). The inhibitory effect of curcumin on a carcinogenic promotor 12-0-tetradecanoylphorbol-13-acetate (TPA)-induced tumor promotion in mouse epidermis has been suggested to be related to curcumin's anti-inflammatory property through its suppression of lipoxygenase and cyclooxygenase activities (Huang et al. 1991).

Interestingly, the rhizome is an important Chinese medicinal herb and is effective in treating certain circulatory disorders, such as angina pectoris, and blood clots from traumas. It is also used effectively to treat menstrual cramps and pains that are associated with insufficient flow, and certain neoplastic conditions. In these cases, it is often used with other herbs to enhance and/or modify its functions to fit individual needs. The rhizome contains many active principles, including turmerone, zingerene, phellandrene, 1,8-cineole, sabinene, borneol, and dehydroturmerone.

Licorice

Extracts of licorice (*Glycyrrhiza glabra L.*) root have been used both as a medicine and as a flavoring to sweeten foods, particularly candies. Licorice has many biological effects. In a recent study on rats, dietary supplementation with 3% licorice led to, among other effects, the appearance of 8 μg/ml of glycyrrhetic (glycyrrhetinic) acid (an active licorice compound) in the serum, suggesting the bioavailability of this active principle. Liver glutathione transferase, catalase, and protein kinase-C activity were elevated, suggesting a detoxification and potential anticancer effect of licorice, as glutathione transferase catalyzes the formation of glutathione conjugates of toxic substances for elimination. There were no untoward effects of the licorice supplementation (Webb et al. 1992). Parallel to these results, glycyrrhizin (glycyrrhizic acid, the sweet principle of licorice) is effective in the prevention of DMBA-DNA adducts formation and the subsequent tumorigenesis in mice treated with a carcinogen DMBA and a promoter 12-0-tetradecanoylphorbol-13-acetate (TPA) (Agarwal, Wang, and Mukhtar 1991).

Studies on licorice extracts from *G. glabra* varieties *glandulifera* and *violacea* have uncovered some active saponins, including glycyrrhizin (4–10%), which is a 3-0-diglucuronide of glycyrrhetic acid (Amagaya et al. 1985; Sticker and Soldati 1978). These compounds have been reported to have liver-protective effects through their anti-free radical properties (Kiso et al. 1984), a finding that is consistent with the proven traditional use of licorice root in China. These compounds have also been shown to have antibacterial and antiviral activities (Chandler 1985), as well as anti-inflammatory activity (Hikino 1985). Chromones (e.g., liquiritigenin, isoliquiritigenin), liquiritin, isoliquiritin, neoliquiritin, neoisoliquiritin, glycyrrhetol, glycyrrol (glycyrol), glycyrrhetol, licoricidin, liquoric acid, and licurazid have been found in various licorice extracts.

In contrast to the recent interest in licorice, the traditional Chinese herbal medicine uses the root of a related species, *G. uralensis fisch*, for detoxification, anti-inflammation, strengthening of digestive function, and numerous other purposes; it is almost always used together with some other herbs for these purposes. The Chinese have not used it for the prevention and/or treatment of neoplasia. Instead, they chose many other herbs for treating/preventing cancers; a few of them will be discussed below. Without the support of a long history of effective use, the efficacy demonstrated in laboratory animals and cultured cells and tissues has yet to be shown in human clinical trials.

Prunus Mume Sieb (Fruit)

This fruit is harvested just before ripening while it is still green, then baked dry at a low temperature. The finished product is black colored and extremely sour. The dried fruit is used in China to impart tartness and flavor in preparing beverage drinks. Concoctions of the fruit and some other herbs have been used to treat neoplasia-like conditions (Chen 1979). However, few well-controlled pharmacological studies have been conducted on this processed fruit. It is very rich in organic acids, flavonoids, other phenolic compounds, and phytosterols; the active compounds have not been identified.

Coix Lacryma-Jobi (Seed)

This spherical, white-colored seed is used as a food as well as a medicine in China. In combination with other herbs, the Coix seed has traditionally been used to treat lung and stomach cancers. A case report stated that a concoction of Coix seeds and certain herbs caused substantial improvement in digestive cancer patients (Yeh

1962). Some constituents of Coix seed have been identified, including coixenolide, coixol, and phytosterols. However, much work needs to be done on quantifying the efficacy of these compounds and the seed itself; none of these compounds comes close to matching the broad spectrum pharmacolological properties of the seed, and the protective mechanisms are mostly unknown.

Trichosanthes kirilowii Maximi and Related Species (Fruit and Root)

In Chinese herbal medicine, this plant's starchy root and squash-like fruit are the key herbs for treating neoplasia and many other conditions, particularly at early stages of cancers. Animal studies demonstrated a protein extract of the root to be effective against liver cancer (Kuo 1980). Extracts of the tuber have also been used intravenously for treating cancer with favorable results (Shanghai 2nd Medical College, Ree Zin Hospital 1972; Sunsee Zin Zung Region People's 2nd Hospital, Gynecology Department 1975; Wong Nan Medical College 1976). An extract of the fruit skin has been shown to be a potent cytocide against cultured cancer cells (Beijin Medical College Chinese Herbal Medicinc Research Group 1959). One of the active principles was identified as a small protein, trichosanthin. The fruit and root also contain many phenolic compounds and alkaloids. But little is known about their pharmacological properties and safety.

Taxus Species and Taxol

A diterpenoid derivative, taxol, isolated from the *Taxus* species, has demonstrated a potent anticancer property by stabilizing microtubules and inhibiting their depolymerization into tubulin (Vanhaelen 1992). A fungus growing in the bark has also been shown to synthesize a trace amount of taxol (Stierle, Strobel, and Stierle 1993). This compound is being actively studied worldwide as a drug for treating cancer. There are numerous publications on this subject. Interested readers should consult some of the recent ones (Chuang et al. 1993; Runowicz et al. 1993; Markman 1993; Lopes 1993). In China concoctions of the leaves of *Taxus cuspidata Sieb. et Zucc.* have been used to treat diabetes mellitus. Although the bark of the woody plant contains taxol, the leaf contains taxinines A, H, K, and L, which have an active structural element like that of the taxol molecule. There has been no report on its use in treating cancers.

IMMUNE-MODULATING PLANTS AND THEIR ACTIVE PRINCIPLES

Proper immune functions are important in maintaining health and preventing diseases, including cancer prevention. Theoretically, all the foods that we consume may have direct and/or indirect impacts of the immune functions. In recent years, numerous compounds have been found to stimulate immune cells in vitro. However, my experience shows that in vitro tests often lead to false hope, as in vivo tests often do not support the in vitro conclusions. Therefore, a long history of safe and effective use carries extra importance here. Five plants, which have a long history of effective use in herbal medicine and in tonic food preparations, will be discussed here.

Astragalus membranaceous Bge (Root)

The root of this plant is the most important tonic medicine used in Chinese herbal practice. It is used for enhancing the body's resistance to infections and for promoting wound healing with well-established efficacy. It is also used in preparing "tonic foods," where meat and *Astralagus* root, ginseng root, *Lycium* fruit (see the next section), and several other herbs are cooked together to make a soup. Both the meat and the soup are consumed as a tonic.

Animal studies demonstrated that extracts of the root have potent immune-enhancing properties, including improved phagocytotic activity in immune cells, strengthened resistance to tuberculosis and viral infections, enhanced interferon production, and increased total white cell count. Humans taking *Astragalus* extracts have high IgM and IgE counts. Certain *Astragalus*-containing herbal formulas are effective against various immune deficiency conditions, including chronic leukocytopenia, and other conditions such as chronic bronchitis, emphysema, and hepatitis. *Astragalus* is the key contributor to the efficacy of these formulas.

Astragalus contains many active compounds, including 2′,4′-dihydroxy-5,6-dimethoxy-isoflavone, kumatakenin, choline, and betaine. However, the chemical and pharmacological properties of many of its constituents are not known. In Chinese herbal medicine, this herb is routinely used in late stages of cancer, often with beneficial results. It appears to me that, in late stages of cancer, the anti-infection function of Astragalus is as important as its direct anticancer function. Both functions are related to its immune-enhancing properties. It is also used to treat uterine prolapse; it can enhance muscle tone of certain internal organs. It is contraindicated for asthma, except for very weak patients who suffers from concurrent infections, to whom the benefits of immune-enhancement may out-weight the risk of exacerbated asthmatic conditions. Even in this case, it is always used in small quantities and in conjunction with anti-asthmatic herbs. The long history of effective use of this herb for various conditions makes Astragalus a worthwhile subject for further studies.

Lycium Chinense Mill. (Fruit)

The bright red colored fruit of this plant has long been a key "cool-type" tonic herb in China. A "cool-type" herb is used to treat certain types of inflammatory conditions, hypertension, and hypermetabolic rate, among other conditions. It is also used in tonic soups and alcoholic drinks. Animal studies have demonstrated that extracts of *Lycium* fruits enhanced phagocytotic activity of immune cells and promoted white cell formation. The fruit contains 1% betaine and large quantities of carotenoids, folic acid, vitamin K, and other phenolic compounds. However, the immune-promoting compounds have not been identified.

Codonopsis pilosula (Franch.) Nannf. (Root)

The root of this plant has a long history as an effective tonic in China. It is often used in combination with other herbs, particularly with "cool-type" or "warm-type" tonic herbs for making concoctions, or for making tonic soups, in which the herbs are cooked with meat. A "warm-type" herb is used to treat certain types of hypotension, poor circulation, and hypometabolic rate, among other conditions. As a medicine, it is used for enhancing the immune function and the body's resistance to infections. In animal studies, extracts of Codonopsis root can enhance the phagocytic activity of immune cells. Some constituents of the root have been identified: scutellarein glucoside, codonopsine and codonopsinine; but the active principles are mostly unknown.

Ganoderma lucidum [Mushroom (Fruiting Body)]

Similar to the other immune-enhancing herbs we have mentioned, the *Ganoderma* mushroom is cooked with foods for preparing tonic soups, or is used with other herbs as a strengthening agent. The soup or extract is consumed as a tonic, but the mushroom is not, because of its woody texture. In animal studies, extracts of the mushroom enhanced the phagocytotic activity and improved the clearance of injected foreign protein. However, its immune-enhancing efficacy in humans is yet to be proven. The active principles are mostly unknown, although the phagocytotic-enhancing activity has been attributed to certain oligosaccharides.

Panax ginseng (Root)

This will be discussed in the next section on central nervous system-modulating plants.

In addition to immune-enhancing herbs, there are many immune-suppressing herbs, which may be effective in treating autoimmune diseases and suppressing rejection of transplanted organs. Some commonly used ones include *Sarcandra glabra (Thunb.) Nakai, Sophora subprostrata Chun et T. Chen, Gentiana manshurica Kitag.*, and *Andrographis paniculata (Burm f.) Nees.* Many immune-suppressing herbs are also anti-inflammative. However, due to space limitation we will not review them here.

CENTRAL NERVOUS SYSTEM-MODULATING PLANTS AND THEIR ACTIVE PRINCIPLES

Proper functioning of the central nervous system (CNS) is not only crucial to mental well-being but also important to other organ systems, because of neurohormonal regulations and other mechanisms. Therefore, some CNS-modulating plants and their constituents will be discussed here.

Panax ginseng C.A. Mey. (Root)

Ginseng has been a key Chinese medicinal herb. It is used mainly for the following purposes: (a) strengthening the body's vital force in a person who is weakened by chronic disease so that the body can cope with the disease, (b) improving the mind and treating various mental disorders including insomnia, neurasthenia, and mental deficiency, and (c) enhancing the chi (a function related to the circulation of body fluids and the ability of the muscles of the internal organs to contract and relax, for which there is no modern scientific term), such as in the case of cardiac insufficiency. In this case, ginseng can substantially enhance the cardiac output and the blood pressure. It is the main herb used to treat shock, often in combination with *Ophiopogon japonicus Ker-Gawl.* (a Chinese medicinal herb) and *Schisandra chinensis (Turcz.) Baill,* as in the "Pulse Generating Formula" (Lin 1989b).

Modern research supports the traditional use of ginseng (Sonnenborn and Proppert 1990; Brekhman and Dardymov 1969 a,b; Petkov and Mosharrof 1987; Liu and Xiao 1992; Singh, Agarwal, and Gupta 1984; Avakian et al. 1984). A ginseng-based herbal extract was used to treat limb paralysis from polio resulting in improved limb movement. White ginseng root is effective in certain insomnia that is refractory to barbiturates and other sleeping aids; however, chronic use diminishes its effectiveness. It has also been reported that ginseng enhanced learning ability in humans and animals (Wong 1977). In cultured chick embryonic ganglia, ginsenosides Rb_1 and Rd, and 20(S)-protopanaxydiol oligosides enhanced nerve fiber production induced by nerve growth factor (NGF) (Takemoto et al. 1984). Other studies demonstrated that ginseng or ginseng extracts protected animals from the toxicity of cocaine and strychnine; modulated brain metabolism of phenylalanine, cyclic AMP, lipid, protein, γ-aminobutyric acid (GABA), and dopamine; altered brain wave pattern; and reduced the effects of central nervous system-suppressing drugs. These are only some of many reported effects of ginseng on the central nervous system; some of these effects are consistent with the traditional use of ginseng as a brain tonic.

In several species of animals, ginseng and ginseng extracts protected liver against xenobiotic-induced damage, stimulated adrenal synthesis of corticosteroids, and increased white cell production; the latter effect suggests an immune-enhancing function of ginseng. Interestingly, a recent study showed that ginsenosides Rb_1, Rb_2, and Rb_3 protected cultured rat myocardiocytes against superoxide radical (Jiang et al. 1993). It suggests that the steroid ginsenosides may turn on the genes that are responsible for certain cellular antioxidative defense mechanisms, because ginsenosides themselves are not effective antioxidants. A polyacetylenic alcohol of ginseng, panaxynol, has been shown to inhibit the growth of various kinds of cultured tumor cells in a dose-dependent manner. The tumor cell rapidly consumed the compound (Saita et al. 1993).

Ginseng contains about 30 ginsenosides, which are hormone-like steroid glycosides (Shibata et al. 1985), and acetylenic compounds: panaxydiol, panaxynol, panaxytriol (Hansen and Boll 1986; Hirakura et al. 1991, 1992; Shim, Koh and Han 1983), and many other active compounds, including panacene, panacon, panaquilon, and panax acid. Some peptide glycans have also been identified (Tomoda et al. 1984, 1985; Konno et al. 1984). Although these glycans have been reported to have hypoglycemic effects, ginseng is seldom used for treating diabetics by the Chinese herbal practitioners.

Schisandra chinensis (Fruit)

In China, the dried, purplish-black berry of *Schisandra chinensis*, about a few millimeters in diameter, is used both as a flavoring agent in foods and beverages, and as a medicine. In the latter case, *Schisandra* berries are used, almost always in combination with other herbs, to treat excessive sweating, certain types of bronchitis and asthma, and other conditions. Modern studies suggest that it has CNS-stimulating effects in animals. The berry contains many active compounds: schizandrin, deoxyschizandrin, gamma(γ)-schizandrin, pseudo-γ-schizandrin, schizandrol, alpha-chamigrene, chamigrenal, phytosterols, citral, and vitamins C and E. Little is known about the mechanisms of the active principles.

In last decade, tablets and powders of *Schisandra* berry have been promoted widely in the United States and some other countries as nutritional supplements or medicines for promoting longevity, with annual sales over ten million U. S. dollars. Promoters often purport that the berry has been used in China for promoting longevity. However, the Chinese have never used it for this purpose; neither is there valid scientific evidence supporting the claim.

Ginkgo biloba

This plant was discussed in the section on cardiovascular protective plants.

Passion Flower

The *Passiflora incarnata L.* plant (excluding the root) has long been used as a sedative. Maltol and many flavonoids (vitexin, isovitexin, iso-orientin, and others) have been isolated from the plant, but none of these compounds has sedative properties (Qimin, van den Heuvel, and Delorenzo 1991; Congora, Proliac, and Raynaud 1986; Glotzbach and Rimpler 1968; Aoyagi, Kimura, and Murata 1974; Jaspersen-Schib 1990; Brasseur and Angenot 1984; Speroni and Minghetti 1988). Further research is needed to ascertain the active principles and to quantify the efficacy and safety of this plant.

LIVER-PROTECTIVE PLANTS AND THEIR ACTIVE PRINCIPLES

The liver is an important organ where serum proteins and lipids are produced, and xenobiotics are metabolized, mainly through cytochrome P_{450} enzymes. Some cytochrome P_{450} enzymes oxidize xenobiotics, which are then detoxified and eliminated through conjugate formation, while some other cytochrome P_{450} enzymes may oxidize and activate carcinogens. Thus, protecting the liver's functions, promoting its detoxifying ability, and suppressing its carcinogen-activating activity are all important. Free radicals are generated in the cytochrome P_{450} oxidation of xenobiotics, which may cause damage in liver cells and may lead to fatty liver and cirrhosis. Numerous plants are rich in free radical quenchers that have liver-protective properties.

Schisandra chinensis. Fruit

This fruit was discussed in the section on central nervous system-modulating plants. One of the most important modern uses of this berry is in the treatment of bacterial and viral chronic hepatitis and xenobiotic-induced hepatitis. Some of the constituents have been mentioned above.

Artemisia capillaris Thunb (Young Plant)

In China, the young plant is mainly used for treating various forms of hepatitis, although small quantities of it have also been used as a bitter flavoring. It is one of the most effective herbs that I routinely use to treat hepatitis, always in combination with other herbs to balance their cool/warm and other properties. Some of its constituents have been identified: capillin, capillone, capillene, capillarin, capillarisine, β-pinene, chlorogenic acid, caffeic acid, and several chromones. Many of these compounds demonstrate hepato-protective properties in cultured hepatocytes and animals.

In addition to the herbs I have discussed, aged garlic extract, astragalus, ginseng, and licorice also have substantial hepato-protective functions.

HEMOPOIESIS-PROMOTING PLANTS AND THEIR ACTIVE PRINCIPLES

Proper hemopoietic functions are important to the maintenance of general health. There are many herbs that enhance the hemopoietic functions, including ginseng and codonopsis; these have been discussed before. Two additional herbs will be discussed here.

Angelica sinensis (Root)

Angelica root is the most commonly used hematinic tonic throughout Chinese history. It is also used as a flavoring in China. For hematinic purposes, it is almost always used in combination with *Rehmannia glutinosa* (see below). Angelica has a broad spectrum of effects on the circulatory system, including antithrombotic effects, modulation of myocardial muscular contraction and relaxation, dilation of coronary arteries and increasing of coronary flow, and modulation of uterine muscular activity. Many compounds have been isolated from *Angelica* root, including: ligustilide, butylidene phthalide, n-valerophenone-O-carboxylic acid, and dihydrophthalic acid. Little is known about their pharmacological properties.

Rehmannia glutinosa (Gaertn.) Libosch. (Root)

This herb is used in conjunction with *Angelica* as a hematinic; frequently both are cooked together with a few additional herbs and meat to make a tonic soup. The soup with the meat is consumed as a hematinic tonic. Some constituents of *Rehmannia* have been identified: rehmannin, catalpol, and campesterol. Little is known about the nutritional and pharmacological properties of this herb and its constituents other than the traditional use of it.

ANTIOXIDANT PHYTOCHEMICALS AND THEIR HEALTH SIGNIFICANCE

Oxidation is a normal bodily process in which heat and free energy are released for maintaining body temperature, constructing and repairing cellular structures, degrading and eliminating unwanted ones, and other metabolic processes. However, undesirable oxidation often occurs, causing damage to cells and tissues. Many conditions can exacerbate oxidative damage, including:

(a) Excessive exposure to xenobiotics,

(b) Presence of improperly sequestered and/or excessive amounts of semiquinones and transitional metals, which can cause one-electron reduction of molecular oxygen to form superoxide radicals,

(c) Photo-induced lysis of chemical bonds (or electron redistribution) to form various free radicals, which eventually lead to, in the presence of molecular oxygen, the formation of oxygen centered radicals and peroxidation,

(d) Photosensitized formation, in the presence of chromophores, of singlet oxygen, which attacks unsaturated centers of an organic molecule and abstracts a hydrogen atom, leading to the formation of hydroperoxides and peroxidation,

(e) Infections.

Even in the absence of these abnormal conditions, erroneous oxidation that damages cellular constituents may occur in the normal oxidation pathways, as dictated by the uncertainty principle of quantum mechanics. Thus the body has evolved to have arrays of anti-oxidative protective and repair mechanisms. The body's arrays of antioxidative devices include substances that can quench singlet oxygen and free radicals (superoxide ion, hydroxy radical, and other radicals), inactivate hydrogen peroxide and hydroperoxides, and sequester transitional metal ions. They range from small antioxidant molecules, such as ascorbate, tocophorols, and glutathione, to complex antioxidant enzymes, such as superoxide dismutase, glutathione peroxidase, peroxidases, and catalase.

In addition to these arrays of protective molecules, physical segregation of the lipid component (which are most liable to oxidation) of the cell into small structures, such as membranes and small lipid globules, also slows the propagation of lipid auto-oxidation. This is because when a large continuous lipid phase is interrupted by more stable molecules the auto-oxidation chains are interrupted.

There are numerous types of antioxidants in plants; the most important ones are tocopherols, ascorbate, phenolic compounds, and thiols (and their precursors). Tocopherols and ascorbate are outside the focus of this chapter and will not be discussed here. Phenolic compounds, such as flavonoids, chromones, and lignans, are abundant in plants. Some of them have been discussed in previous sections. Garlic, which contains many precursors of thiol-compounds, has also been previously discussed. β-carotene and ubiquinones/ubiquinols, both believed to play important antioxidant roles in the body, will be discussed here.

Beta(β)-carotene

β-carotene and other carotenoids exist widely among plants, where they function partly as antioxidants by quenching singlet oxygen that may accidentally be generated from transfer of the triplet energy of photo-excited chlorophyll molecules to ground state triplet oxygen molecules. Usually, the richer in green the color, the more chlorophylls and carotenoids the plant contains. The love affair with β-carotene started in the late 1960s when it was discovered that β-carotene is a very potent singlet oxygen quencher (Krinsky 1968). In 1975, β-carotene was shown to inhibit the *initiation* but not the *propagation* phase of auto-oxidation of cooking oils and fried foods. In these products, β-carotene served as an antioxidant through its singlet oxygen quenching ability. In animal studies, β-carotene was found to bind to various cellular compounds and nuclear proteins, suggesting that it performs regulatory functions in the body, in addition to serving as a precursor of vitamin A (Lin 1975).

Since 1982 the United States National Academy of Sciences and some governmental agencies have repeatedly emphasized the importance of consuming carotene-rich foods in the prevention of cancer. In recent years, β-carotene has come into the limelight, because of new discoveries that imply its health benefits and heavy commercial promotion. At every medical school where I have lectured in recent years, professors and students asked about the functions and benefits of β-carotene; many of them also take β-carotene supplements. Statements such as "There is justified hope that beta-carotene will be proven efficacious in the prevention and/or delaying of the onset of these chronic diseases" (referring to cancer and cardiovascular disease) (Muggli 1992) are commonly made. During the last decade, numerous articles on β-carotene have appeared in scientific and popular publications; it appears to have become the molecule of the decade.

Evidence supporting β-carotene's anticancer properties comes from epidemiological studies, animal studies, and studies on cultured cells, tissues, and organs.

β-carotene versus Cancer: Epidemiological Studies

Various degrees of association of low intake of vegetables, fruits, and/or carotenoids with increased cancers of various sites have been demonstrated in several prospective epidemiological studies (Hirayama 1979, 1985; Shekelle et al. 1981; Kvale, Bjelke, and Gart 1983; Colditz et al. 1985; Paganini-Hill et al. 1987). In a prospective study, information about the dietary intake and body fluid analysis of a seemingly healthy population sample is obtained, and this information is used at a future date to compare those who develop a disease with those who do not develop the disease, to ascertain whether there is an association between dietary intake and the disease.

Similar correlations have also been obtained in many retrospective population studies (MacLennan et al. 1977; Hinds et al. 1984; Ziegler et al. 1984, 1986; Wu et al. 1985; Samet et al. 1985; Pisani et al. 1986; Pastorino et al. 1987; Bond, Thompson, and Cook 1987; Byers et al. 1987). A retrospective study refers to a study in which the information of dietary intake (and body fluids) of a diseased population sample taken around the time of diagnosis is compared to a comparable control sample of un-diseased individuals, so that correlations between dietary intake (and blood indices) and the disease can be ascertained.

A more recent study found that cancer patients and their immediate families had substantially lower serum β-carotene levels as compared to the levels of non-cancer individuals and their families (Smith and Waller 1991). The association of low serum β-carotene with increased cancer incidence can be found in populations worldwide, in Switzerland (Stahelin 1984), England, the United States (Comstock, Helzlsouer, and Bush 1991; Le Gardeur, Lopez, and Johnson 1990; Palan et al. 1991), Finland (Knekt et al. 1991), and Italy (Buiatti et al. 1990), and across racial boundaries among American blacks (Gridley et al. 1990), and Japanese Hawaiians (Nomura et al. 1985).

β-carotene Versus Cancer: Animal and in vitro Studies

In animal studies, β-carotene (or carrot, which is a rich source of β-carotene) given through various routes, including oral and intraperitoneal routes, has demonstrated antitumor activities against chemically induced and ultraviolet light-induced tumors (Matthews-Roth et al. 1977; Matthews-Roth 1982; Dorogokupla et al. 1973; Santamaria et al. 1981; Epstein 1977; Rieder, Adamek, and Wrba 1983; Rettura, Levenson, and Seifter 1984; Matthews-Roth and Krinsky 1985, 1987; Temple and Basu 1987; Alam and Alam 1987; Suda, Schwartz, and Shklar 1986; Rettura et al. 1982). Feeding rats with 30 mg/kg β-carotene was reported to reduce oxidative liver damage and red blood cell membrane peroxidation (Zamora, Hidalgo, and Tappel 1990). In cultured cells and tissues derived from various animals, β-carotene protected the cells against damage to the cell nuclei induced by potent carcinogens, including N-nitrosodiethylamine and N-methylnitrosurea (Manoharan and Bennerjee 1985), superoxide ion (Weitberg et al. 1985), and genotoxic chemicals (Stich and Dunn 1986; Peng et al. 1988). These findings are consistent with the purported anticancer effect of β-carotene, as nuclear damage may lead to the transformation of normal cells into cancerous cells. Palm oil is rich in carotenoids, including β-carotene. Dietary supplementation of palm oil showed a dose-dependent antitumor activity in mice (Azuine et al. 1992).

Most of the authors attributed the anticancer effect to antioxidant properties of carotenoids, although a recent report indicates that exposure to a chemical carcinogen (benzopyrene) induced a tissue depletion of vitamin A, and that high intake of β-carotene prevented the depleting effect of the carcinogen (Edes et al. 1991). This observation suggests that β-carotene may exert an anticancer effect through its conversion to vitamin A, which is a classical function of the carotene. Many other studies demonstrate that β-carotene's anticancer activity is independent of its function as a precursor of vitamin A.

Recently, interest in β-carotene as a protective nutrient has extended into protection against cardiovascular disease. It was found that low plasma levels of vitamins A, C, and E and β-carotene, individually and as a group, were associated with an increased risk of angina (Riemersma et al. 1991). The Basel Study, which involved a 12-year follow-up of 2,974 participants with 132 ischemic heart disease deaths, 31 stroke deaths, and 204 cancer cases, showed that low β-carotene status increased mortality of ischemic heart disease, cerebral vascular diseases, and cancer (Eichholzer, Stahelin, and Gey 1992). These authors attributed the protective effects to the antioxidant and free radical quenching ability of β-carotene.

β-carotene does not appear to act as an antioxidant through singlet oxygen quenching because singlet oxygen does not occur commonly in the body. A new theory has been advanced to explain the molecular mechanisms of β-carotene's bioactivity. The theory states that β-carotene acts as an antioxidant ". . . especially at the lowest partial pressure of O_2(15 torr)." (Burton 1988; Burton and Ingold 1984). Statements such as "many carotenes . . . are excellent antioxidants and radical trapping agents, especially for peroxyl and hydroxy radicals." (Blumberg 1992) are commonly made by many leading scientists. The array of publications, the reputation of the authors, and the arguments are indeed impressive. No wonder many nutritionists, physicians, and the informed public have stampeded to the rescue of β-carotene by taking β-carotene supplements.

However, I found several mistakes and/or misleading arguments in many of these publications. First, under physiological conditions, where partial pressure of oxygen is between 110 torrs in the arterial blood and 40 torrs in the venous blood, β-carotene cannot be an effective antioxidant. Second, the typical molar ratio of β-carotene to tocopherols in tissues is approximately 1 to 50; so approximately 2% of the antioxidant task would be performed by β-carotene, assuming that β-carotene and tocopherols are equally effective free radical quenchers on a per-molecule basis. The molar concentration of β-carotene in most human tissues ranges approximately from 6 to 60 micromolar; this is too low a range for it to be effective as an antioxidant. It is clear that tocopherols and other antioxidants contribute the bulk of antioxidative protection. Thus, the antioxidant theory is faulty. (However it cannot be ruled out that β-carotene may play an unlikely, unconventional role in the cellular antioxidant defense through cross-membrane charge transfer; i.e., the long conjugated β-carotene molecule forms a bridge across a cytomembrane and gives up a hydrogen atom or an electron on one side of the membrane to quench a radical; meanwhile it picks up a proton or an electron from the other side of the membrane, with subsequent rearrangement of the electrons and protons.) Third, although at extremely high intake levels, i.e., about several hundred to a thousand times human intake level, β-carotene has been reported to reduce lipid peroxidation and to protect liver damage in animals, these may be secondary effects, e.g., through induction of protective enzymes. Fourth, most researchers equate β-carotene-rich foods to β-carotene. β-carotene rich foods usually contain numerous active plant compounds, e.g., phenolic compounds and chlorophylls. Thus, the observed beneficial properties of these foods may be at least partially attributed to these compounds. Fifth, the correlation of high plasma β-carotene level with the reduced risks does not necessarily prove that β-carotene does indeed provide the benefits; the high plasma β-carotene level also indicates that the individuals consume more β-carotene rich foods, which provide many other active compounds. Finally, the efficacy of β-carotene observed in laboratory animals may not be obtainable in humans. This is because the laboratory animals are raised on diets specified by the U.S. National Academy of Sciences/National Research Council, or similar diets. While these diets are excellent for maintaining the health of animals raised under simple and protected laboratory environments, they are far simpler than the diets that the animals would have consumed under natural conditions. These simple laboratory diets are likely to lack phytochemicals that may protect the animals against exposure to certain adverse environments, such as a carcinogen challenge. Under these conditions, β-carotene is more likely to have prominent effects. The effective doses in animals, usually 1–2% in the diets, is about 1,000 times higher than the β-carotene content in human diets. It is simply impossible for humans to consume daily 50 to 100 kilograms of β-carotene-rich leaves and fruits in order to reach the intake levels that have been shown to have some effect in animals. The benefits seen in the animal studies should be considered nontypical pharmacological effects rather than nutritional effects. Similar arguments apply to the cell and tissue culture studies.

Therefore, in the height of β-carotene worship, I must voice a lone opinion that the pursuit of cancer prevention and cardiovascular protection through β-carotene supplementation alone will result in something that is only slightly better than the pursuit of a mirage. Nevertheless, I personally do take a nutritional supplement that contains β-carotene, and recommend the incorporation of β-carotene in nutritional supplements simply because there is evolutionary and laboratory evidence suggesting that β-carotene supplementation provides other benefits. For example, the eye is susceptible to photo-sensitized damage and adequate intake of carotenoids can provide some protection against photo-injury. In this connection, it is interesting to note that Chinese herbal medicine uses the carotene-rich *Lycium* fruit (see the section on immune-modulating plants) as an eye tonic.

Ubiquinones (Coenzyme Q's) and Ubiquinols

Ubiquinones are a family of compounds having a repeated side chain (polyprenyls) ranging from 1 to 12 units. A common one that exists among higher plants and animals has 10 repeating units and is called ubiquinone-10, coenzyme Q-10, or CoO_{10}. Corn oil and many other vegetable oils contain significant amounts of ubiquinones. Ubiquinones perform redox function and electron transport in the mitochondrial membrane, where electrons are stripped from respiratory substrates and transported to oxygen molecules during cellular respiration. Ubiquinols, the reduced forms of ubiquinones, may function as antioxidants. The antioxidant efficacy of ubiquinol-10 at physiological concentrations in the lipid components of cells has been discussed (Frei, Kim, and Ames 1990).

CoQ_{10}'s antioxidative, electron transport, and membrane-stabilizing properties have recently been investigated aiming at prevention of and/or treatment of various cardiovascular diseases, including prevention of cellular damage during reperfusion, angina pectoris, hypertension, myocardial ischemia, and congestive heart failure (Greenberg and Frishman 1990).

In patients with mitochondrial encephalomyopathy, CoQ_{10} treatment, 150 mg/day, increased mitochondrial functions and exercise performance, and reduced the acidosis associated with exercise (Bendahan 1992). This finding is consistent with a model that CoQ_{10} participates in electron transport in the mitochondrial membranes, and in the biological oxidation of cellular fuels for energy generation.

CoQ_9 deficiency was reported to be associated with a viral infection, and supplementation of CoQ_{10} in acquired immune deficiency syndrome (AIDS) patients resulted in enhanced macrophage activity and increased serum level of IgG. CoQ_{10} treatment has been reported to provide some benefits in cancer patients, and enhanced hematopoietic activity in monkeys, rabbits, poultry, and malnourished (Kwashiorkor) children (Folkers et al. 1993). All the evidence suggests that ubiquinones may play certain roles in maintaining and promoting health, under normal and abnormal conditions.

PHARMACOLOGICALLY ACTIVE PRINCIPLES OF SOME FLAVORING EXTRACTS OF PLANTS

Many flavoring extracts permitted for food use contain active compounds that merit further investigation to ascertain whether they can play a role in the prevention of degenerative diseases. Some of them will be reviewed here.

Ginger

Ginger (*Zingiber officinale Rosc.*) contains many active compounds. A series of gingerols and shogaols, differing in side chain length and unsaturation (4, 6, 8, methylene), have been found to have profound effects on the digestive system (Suekawa et al. 1984). Among the effects are the enhanced peristaltic movement and muscle tone of intestinal muscles. These properties parallel the traditional use, both in the West and the East, of ginger for treating poor appetite and certain types of indigestion. A recent study demonstrated that ginger reduces the incidence of nausea and vomiting, which often occur subsequent to major surgeries (Bone et al. 1990). Many other compounds are known to be present in ginger and ginger extracts, including zingiberol, zingiberene, phellandrene, camphene, citral, linalool, methylheptenone, nonyl aldehyde, d-borneol, zingerone zingiberone, and 2-pipecolic acid.

Gentian

Extracts of gentian (*Gentiana lutea*) are commonly used as a bitter flavoring. Several phenolic compounds of gentian, e.g., gentisin and amarogentin, are intensively bitter and can induce saliva and gastric secretion. Gentian extract is included in the group "natural flavoring substances and natural substances used in conjunction with flavors" and is permitted for use in alcoholic beverages under U. S. Code of Federal Regulations (21 CFR 172.510).

Golden Seal

Golden seal (*Hydrastis canadensis*) root is a key medicinal herb traditionally used by American Indians for various conditions, including hemorrhages and bacterial infections. This root produces an active alkaloid, hydrastine, which can cause vasoconstriction in widespread organ systems, including the reproductive, respiratory, and nervous systems (Genest and Hughes 1969). Other alkaloids have also been isolated from the root of this plant, including berberine and canadine, which stimulate uterine muscles. (Shideman 1950). The antihemorrhagic effect and the uterine-stimulating effect underscore its traditional use by the Indians for treating menstrual problems.

Guaiacum Extract

Guaiacum (*Guaiacum officinale L.*) extracts are used as a flavoring agent permitted under the U.S. Federal Regulations (21 CFR 172.510). They contain many phenolic compounds, including guaiacic acid, guaiaretic acid and α- and β-guaiaconic acids. These compounds have anti-inflammatory properties. Guaiacum extracts have been used for treating rheumatic conditions (Schrecker 1957; King and Wilson 1964; Kratochvil et al. 1971).

Marshmallow Extract

Extracts of marshmallow root (*Althaea officinalis L.*) are used as flavorings in foods under U.S. Federal Regulation, 21 CFR 172.510. They contain many phenolic compounds and flavonoids, such as hypolaetin-8-glucoside, isoscutellarein 4′-methyl ether 8-glucoside-2′-sulfonate, caffeic acid, p-coumaric acid, ferulic acid, p-hydroxybenzoic acid, vanillic acid, syringic acid, and p-hydroxyphenylacetic acids (Gudej 1991). These compounds have some anti-inflammatory properties.

Chamomile Extracts

Extracts of German or Hungarian chamomile (*Matricaria chamomilla L.*) flower are used as flavoring under 21 CFR 182.10 and 182.20. They contain anti-inflammatory compounds, including α-bisabolol, chamazulene, matricin, herniarin, umbelliferone, and apigenin (Issac 1979; Glowania, Raulin, and Swoboda 1987; Jakovlev et al. 1979). Some polysaccharides of the flower have immune-stimulating effects (Wagner et al. 1985).

Roman or English chamomile (*Anthemis nobilis L.*) flower is a common medicinal herb used in Europe for treating a wide variety of disorders, including dyspepsia and nausea. Extracts of the herb are permitted for use as a flavoring (21 CFR 182.10 and 182.20) and are known to contain flavonoids (apigenin and luteolin), nobilin, anthemene, anthemic acid, anthesterol, and n-butyl angelate (Rüker, Mayer, and Lee 1989; Melegari et al. 1988; Tosi et al. 1991; Hall et al. 1980).

Both plants belong to the *Asteraceae* family; members of the family contain a poison helenalin, an α-methylene-γ-lactone. This type of lactone exists in many other plants and can readily add to thiols in the cellular environments to cause widespread alterations in cellular physiology (Hladon and Twardowski 1979). It inhibits prostaglandin synthesis, modulates cyclic AMP level, prevents lysosomal rupture, suppresses RNA and protein synthesis, disrupts the genetic apparatus, and arrests cell division. The last two properties have led to some interests in studying it for cancer treatment. In animals it can cause paralysis of voluntary and cardiac muscles, leading to cardiac arrest and death.

Parsley Plant

Parsley plant (*Petroselinum crispum (Mill.) Manst.*) is consumed as a food or used as a flavoring agent (21 CFR 182.10 and 182.20). Some compounds isolated from the plant have been reported to have diuretic, vasodilating, and spasmolytic functions: myristicin (which is recognized as a naturally occurring toxicant in food), apiole, apiin, bergapten, 8-methoxypsoralen, isopimpinellin, and psoralen (Bézanger-Beauquesne et al. 1980; Chaudhary et al. 1986; Innocenti, Dall'Acqua, and Carporale 1976).

Peppermint

Extracts of peppermint (*Mentha piperita L.*) are used as flavoring agents (21 CFR 182.10 and 182.20.) The essential oil of peppermint contains menthol, neomenthol, isomenthol, menthone, menthofuran, cineole, limonene, and viridoflorol (Clark and Menary 1981). Many of these compounds and peppermint's flavonoids have been reported to have muscle-relaxing effects (Steinegger and Hänsel 1988; Taddei et al. 1988; Leicester and Hunt 1982; Rees, Evans, and Rhodes 1979).

Prickly Ash Bark Extract

Extracts of prickly ash bark (*Zanthoxylum clava-herculis L.* or *Xanthoxylum americanum Mill.*) are used as flavoring agents (21 CFR 182.20). They contain alkaloids chelerythrine, herclavin, neoherculin, nitidine, and temberatine (Fish et al. 1975; Crombie 1952, 1955; Fish and Waterman 1973; Crombie and Taylor 1957). The extracts also contain many phenolic compounds. These compounds have been reported to stimulate blood circulation.

Rhubarb Extract

Extracts of rhubarb (*Rheum palmatum L.* or *R. officinale Baill.* and related species) are used as a bitter flavoring (21 CFR 172.510). They contain extremely bitter glycosides of anthraquinones (rhein, aloe-emodin, chrysophanol, emodin, physcion), heterodianthrones (palmidins A, B, C, and D), rheidins, A, B, and C; sennidins A, B, C, and D; and dianthrones of aloe-emodin, chrysophanol, emodin, and physcion (Verhaeren, Dreessen, and Lemli 1982; Oshio et al. 1974; Yamagishi et al. 1987). They also contain many flavonoids, catechin, and numerous other phenolic compounds. Although rhubarb extracts are well known for their laxative properties, traditional Chinese herbal medicine also uses rhubarb in conjunction with other herbs for treating inflammatory conditions. In my practice I have consistently found rhubarb-containing herbal formulas to be efficacious for certain inflammatory conditions.

Sarsaparilla Extract

Extracts of sarsaparilla (*Smilax aristolochiaefolia Miller*) are used as flavoring agents in the United States (21 CFR 172.510). Sometimes closely related species are also used, e.g., *S. ornata and S. febrifuga Kunth.* The steroidal saponins in the extracts have anti-inflammatory and hepato-protective effects. Some of these compounds have been characterized: sarsaparilloside, sarsasapogenin, smilagenin, and parillin (Ageel et al. 1989; Rafatullah et al. 1991; Devys et al. 1969; Thurmon 1942).

Senna Leaf Extract

Extracts of senna leaves (Khauton senna or Alexandrian senna, *Cassia senna L. or C. acutifolia Dellile*) are used as flavoring agents (21 CFR 172.510). They contain potent laxatives, which are hydroxyanthracene glycosides (sennosides A, B, C, and D) and glycosides of other phenolic compounds, including tinnevellin glucoside and 6-hydroxymusizin-8-glucoside (Lemli 1986; Nakajima, Yamauchi, and Kuwano 1985; Leng-Peschlow 1986).

Valerian Extract

Extracts of valerian (*Valeriana officinalis L.*) root are used as a flavoring agent in the United States (21 CFR 172.510). These extracts have mild sedative effects and contain valerianol, valeranone, valerenone, valerenal, valerenic acid, bornyl isovalerate, borneol, camphene, α-pinene, d-terpineol, l-limonene, pyrryl-α-methyl ketone, α-fenchene, myrcene, phellandrene, γ-selinene, alloaromadendrene, γ-terpinene, terpinolene, candinene, eremophilene, l-caryophyllene, hesperitinic acid, behenic acid, isovaleric acid, valerenolic acid, maali alcohol, l-bornyl acetate, l-myrtenol, (S)-(-)-actinidine, curcumenes, β-bisabolene, ledol, valerianine, valerosidatum, isovaleroxy-hydroxy dihydrovaltrate, valerine, valerianine, chatinine, tannin, and valepotriates (Thies 1968; Bos et al. 1986; Titz et al. 1982, 1983), among others. The root is used in China as a sedative and an antispasmodic drug (especially for intestinal spasms). It is also used to enhance menstrual flow and to treat arthritis.

Yarrow Extract

Extracts of yarrow (which is the aerial part of *Achillea millefolium L.*) are used for flavoring purposes (21 CFR 172.510). They contain volatile compounds (linalool, camphor, chamazulene), sesquiterpenes (achillin, achillicin, achilleine, leucodin), and flavonoids (apigenin, luteolin, isorhamnetin, rutin, artemetin, casticin), and many other phenolic compounds (Hofmann et al. 1992; Ulubelen, Öksüz, and Schuster 1990). This herb has also been used for curing colds through its sweating effect.

Hops Extract

Extracts of dried flowers of hops (*Humulus lupulus L.*) are used as flavoring agents under the U.S. regulation 21 CFR 182.20. These extracts contain many phenolic compounds, including xanthohumol, lupulone, and humulone (Hänsel and Schulz 1986a,b; McMurrough 1981; Song-San, Watanabe, and Saito 1989), humuladienone, humulenone-II, α-corocalene, γ-calacorene, myrcene, humulene, astragalin, isoquercitrin, quercetin, leucocyanidin, leucodelphinidin, and kaempferol. Extracts of hops have been reported to have a subtle sedative effect (Bravo et al. 1974). The fresh flower contains estrogen-like substances and antispasmodic substances. The herb has been used to treat inflammation of bladder and to induce diuresis.

SCIENTIFIC, TECHNOLOGICAL, AND LEGAL ASPECTS OF FOOD FORTIFICATION WITH PHYTOCHEMICALS

Scientific and Technological Aspects

From a scientific and technological point of view, safety, stability, and efficacy are the main concerns of food scientists, food companies, and governments. A pure phytochemical and a plant containing the same amount of the phytochemical may differ substantially in safety, stability, and efficacy. Consumption of a certain quantity of pure phytochemical may not be safe, even if consumption of the whole plant that contains the same quantity of the phytochemical is safe (and vice versa). This is so because the whole plant contains many compounds, which may counteract each other and reduce the toxicity. The United States National Academy of Sciences, governmental agencies, and many leading scientists have repeatedly extrapolated certain toxicity data obtained from pure phytochemicals and concluded or implied that it is unsafe to consume certain plant foods that contain the phytochemicals, and/or have claimed that the hazard of naturally occurring toxicants in plants is as great as that of pure synthetic toxicants (Ames and Gold 1990, 1992; Ames 1991; Gold et al. 1991; Gold, Manley, and Ames 1992; Ames, Profet, and Gold 1990a,b). The mistake here is a failure to consider the interactions of plant constituents when they are consumed together. For example, a whole plant contains many antioxidants. When these antioxidants are consumed together with a compound that is contained in the plant and that can cause free radical/oxidative damage in the liver, the end result is a reduced toxicity of the plant in comparison to the pure compound. Many plant constituents can also modulate certain liver cytochrome P_{450} enzymes, so that when these constituents are consumed together with a plant toxicant, the toxicant may be detoxified faster because of the elevated enzymatic activity.

A phytochemical must be tested for safety more rigorously than a drug would be, before the pure compound can be used for food fortification. A reason for a more stringent test is that the consumers' exposure to phytochemicals is expected to be greater than their exposure to drugs. Another reason is that the benefit of a phytochemical is expected to be lower than that of a drug, at least in the short term. Therefore, the use of a phytochemical must have a very low safety risk in order to achieve an acceptable benefit/risk ratio.

A more difficult and serious problem related to functional foods is that a plant/phytochemical may reduce the risk of certain diseases in some individuals, but may increase the risk of other diseases in some other individuals, or even in the same individuals. As discussed in the Introduction, many foods have an effect of counterbalancing certain imbalanced physiological states, so that the risk of certain diseases associated with the imbalance may be mitigated. A phytochemical may enhance the balance in some, but may cause excess in others. Typical examples: antiocclusive phytochemicals can reduce the risk of occlusive cardiovascular diseases in the majority of people, but they may also increase the risk of hemorrhages in a certain segment of the population; an immune-promoting phytochemical may enhance resistance to infection and cancer, but may also increase the risk of autoimmune diseases and/or allergic conditions in the same individuals, as in the case of *Astragalus*, which increases both IgG and IgE. All the Chinese herbs have contraindications; i.e., none of the herbs, no matter how beneficial it is, is suitable for all people. Looking at this issue from another angle: How far can a functional food go without entering the prescription arena? Furthermore, no nutrient should be taken in substantial excess over a long period of time; phytochemicals are no exception.

Stability can be another problem. I found that some Chinese herbs retain their potency and safety for years, even for decades, when they are stored under ambient conditions. But, in most cases, the isolated active principles themselves are chemically unstable; they often change within months, even within days. The most commonly occurring reaction is oxidative polymerization of phenolic compounds.

Proving efficacy is perhaps the most difficult aspect of the phytochemical matter. In many cases, the required degree of proof for drugs is not obtainable with phytochemicals, simply because the risk/reward ratio is too great to justify subjecting seemingly healthy humans to the testing conditions of a drug test. In drug clinical trials, the expected relief from suffering and/or the expected reduction of risk of death from a disease may justify the risk of trying an unproven drug; but in the case of phytochemicals and/or functional foods there is a lack of strong justifications for risk taking. My opinion is that the purported benefits of a plant (or a phytochemical) must be supportable by a long history of use and/or by *sound extrapolation* of established science, which may include the principles and theories of physiology, nutrition, and medicine. The soundness of extrapolation is emphasized. The mistakes of overestimating the hazard of many naturally occurring toxicants in foods by the scientific community and governments, and of the carotene matter, as discussed above, are examples of unsound extrapolation of established science. Results of modern scientific studies should be subject to extensive and careful scrutiny because the simple, controlled laboratory and/or clinical conditions are far removed from the complex real world situations. In the field of herbs and health, often modern studies should be considered as footnotes that support the traditional use. It is not uncommon that laboratory tests of an active principle demonstrated or suggested certain medicinal use of it that are radically different from the traditional use of the herb from which the active principle is derived.

In vitro studies, such as effects on cultured cells and tissues, should not be given much importance because a substance that is active and nontoxic *in vitro* may become ineffective and/or toxic when taken orally because of digestion and alteration by the liver. Furthermore, it is not likely that food fortification with a single phytochemical can have a major impact on reducing risks of cancer and/or occlusive cardiovascular diseases.

Legal Aspects

In most countries, there are two sets of laws and regulations that govern foods and drugs. In the United States, there are four statutory definitions of a drug. The one that is most closely related to phytochemicals is: "The term 'drug' means . . . articles intended for use in the diagnosis, cure, mitigation, treatment, or prevention of disease in man or other animals." [21. U.S.C. 321(g)(1)(B).] Phytochemicals that are intended for disease prevention are drugs under the statutory definition, and would require drug approval before marketing, unless another law/regulation specifically exempts them from drug status. Governments have taken actions against many food items that make drug claims (i.e., claiming or implying to be effective for prevention or treatment of certain diseases), and the health food industry has also taken counteractions against food and drug authorities in various countries. The Nutrition Labeling and Education Act of 1990 of the U.S. appears to take a piece-meal approach in dealing with food and disease prevention. A better law dealing with the entire scope of functional (medicinal) food is needed. To resolve the problem of a food that has disease preventing and/or mitigating properties being classified as a drug, I have long proposed a logical, scientific, and simple way of adding the phrase "other than food" in the drug definition cited above, and converting the definition to "The term 'drug' means . . . articles, other than food, intended for use in the diagnosis, cure, mitigation, treatment, or prevention of disease in man or other animals."

The U.S. Food and Drug Administration has recently proposed banning 415 ingredients from seven categories of nonprescription drugs, because they have not been shown to be safe and effective for their stated claims (U.S. Department of Health and Human Services 1992). The majority of them are plants, plant extracts, and phytochemicals, including chamomile flower, asafetida, ginger, peppermint, rhubarb fluid extract, senna, allyl isothiocyanate, camphor, gentian, licorice, golden seal, and parsley. Many of them have been discussed in this chapter.

In the United States, the Federal Trade Commission (FTC) requires the substantiation of an advertisement. An advertisement of a phytochemically fortified food will not be allowed to make a benefit claim without acceptable proof of efficacy. Unless the government and the society accept "long history of use and/or sound extrapolation of established science" as supporting evidence, it might be difficult for some phytochemically fortified foods to advertise their benefits. In my opinion, the substantiation of the safety of phytochemicals must be more stringent than that of drugs, and the substantiation of the efficacy of phytochemicals must be more relaxed than that of drugs.

Another important aspect of functional foods is the prevention of false and/or misleading claims. In the health food industry, numerous unscrupulous operators market over 10,000 items with myriads of claims from cancer drugs and aphrodisiacs to magic pills which reduce body weight while sleeping. Most of the pills are ineffective for the claimed conditions. A consumer may suffer direct and/or indirect injures, in addition to economical losses. A direct injury refers to the toxicity caused by the consumption of a product; an indirect injury refers to the situation where a consumer is mislead to depend on the ineffective health food product for the treatment of his conditions, leading to the worsening of the conditions. Currently, governmental authorities in most countries have not taken adequate actions against these unscrupulous operators. The functional food movement is bound to further stimulate some unscrupulous operators. Thus, governments, and consumer and industrial organizations must strive to formulate new regulations and allocate additional resources to prevent unscrupulous exploitation of functional foods.

CONCLUSION

Plants contain many physiologically and/or pharmacologically active compounds; their consumption may reduce the risk of some common degenerative diseases, including cardiovascular disease and cancer, in the majority of people. Food fortification with these plants or their active principles is a fertile field for research and development.

References

Adlercreutz, H.A.; Hockerstad, K.; Bannwart, C.; Bloigu, S.; Hamalainen, E.; Fotsis, T.; and Ollus, A. 1987. Effect of dietary components, including lignans and phytoestrogens, on enterohepatic circulation and live metabolism of estrogens and on sex hormone binding globulin (SHBG). *J. Steroid Biochem.* 27:1135–1144.

Agarwal, R.; Wang, Z.Y.; and Mukhtar, H. 1991. Inhibition of mouse skin tumor-initiating activity of DMBA by chronic oral feeding of glycyrrhizin in drinking water. *Nutr. Cancer* 15:187–193.

Ageel, A.M.; Mossa, J.S.; Al-Yahya, M.A.; Al-Said, M.S.; and Tariq, M. 1989. Experimental studies on anti-rheumatic crude drugs used in Saudi traditional medicine. *Drugs Exptl. Clin. Res.* 15:369–372.

Alam, B.S., and Alam S.Q. 1987. The effect of different levels of dietary beta-carotene on DMBA-induced salivary gland tumors. *Nutr. Cancer* 9:93–101.

Am Sun No. 1 Pharmaceutical Factory. 1977. *Journal of Traditional Chinese Medicine.* 57 (11):677. (in Chinese.)

Amagaya, S.; Sugishita, E.; Ogihara, Y.; Ogawa, S.; Okada, K.; and Aizawa, T. 1985. Separation and quantitative analysis of 18, α-glycyrrhizae radix by gas-liquid chromatography. *J. Chromatogr.* 320:430–434.

Ames, B.N. 1991. Natural carcinogens and dioxin. *Sci. Total Environ.* 104:159–166.

Ames, B.N., and L.S. Gold. 1990. Carcinogenesis debate. *Science* 250:1498–1499.

———. 1992. Animal cancer tests and cancer prevention. *Monographs/Natl. Cancer Inst.* pp. 125–132.

Ames, B.N.; Profet, M.; and Gold, L.S. 1990a. Dietary pesticides (99.99% all natural). *Proc. Natl. Acad. Sci. USA* 87:7777–7781.

———. 1990b. Nature's chemicals and synthetic chemicals: Comparative toxicology. *Proc. Natl. Acad. Sci. USA* 87:7782–7786.

Aoyagi, N.; Kimura, R.; and Murata, T. 1974. Studies on *Passiflora incarnata* dry extract. I. Isolation of maltol and pharmacological action of maltol and ethyl maltol. *Chem. Pharma. Bull.* 22:1008–1013.

Apitz-Castro, R., and Jain, M.K. 1990. Anti-thrombotic action of ajoene, a well characterized compound derived from garlic. In *Garlic in Biology and Medicine: Proceedings of the First World Congress on the Health Significance of Garlic and Garlic Constituents.* Nutrition International Co., P.O. Box 50632, Irvine, CA 92619-0632.

Aruna, K., and Sivaramakrishnan, V.M. 1990. Plant products as protective agents against cancer. *Indian J. Exptl. Biol.* 28:1008–1011.

Auguet, M., and Clostre, F. 1983. Effects of an extract of *Ginkgo biloba* and diverse substances on the phasic and tonic components of the contraction of an isolated rabbit aorta. *Gen. Pharmacol.* 14:277.

Auget, M.; De Feudis, V.; and Clostre, F. 1982. Effects of *Ginkgo biloba* on arterial smooth muscle responses to vaso-active stimuli. *Gen. Pharmacol.* 13, 169–172.

Auguet, M.; De Feudis, V.; Clostre, F.; and Deghenghi, R. 1982. Effects of an extract of *Ginkgo biloba* on rabbit isolated aorta. *Gen. Pharmacol.* 13:225.

Avakian, E.V.; Sugimoto, R.B.; Taguchi, S.; and Horvath, S.M. 1984. Effects of *panax ginseng extract* on energy metabolism during exercise in rats. *Planta Medica* 50:151–154.

Azuine, M.A., and Bhide, S.V. 1992. Protective single/combined treatments with betel leaf and turmeric against methyl (acetoxymethyl) nitrosamine-induced hamster oral carcinogenesis. *Intl. J. Cancer* 51:412–415.

Azuine, M.A.; Goswami, U.C.; Kayal, J.J.; and Bhide, S.V. 1992. Anti-mutagenic and anticarcinogenic effects of carotenoids and dietary palm oil. *Nutr. Cancer* 17:287–295.

Bauer, U. 1984. 6-month double-blind randomised clinical trial of *Ginkgo biloba* extract versus placebo in two parallel groups in patients suffering from peripheral arterial insufficiency. *Arzneim. Forsch.* 34, 716.

Beijin Medical College Chinese Herbal Medicine Research Group. 1959. J. Beijin Medical College 1:104 (in Chinese).

Bendahan, D., Desnuelle, C., Vanuxem D., Confort-Gouny, S., Figarella-Branger, D., Pellisier, J.F., Kozak-Ribbens, G., Pouget, J., Serratrice, G., and Cozzone, P.J. 1992. ^{31}P NMR spectroscopy and ergometer exercise test as evidence for muscle oxidative performance improvement with coenzyme Q in mitochondrial myopathies. *Neurology* 42:1203–1208

Bézanger-Beauquesne, L.; Pinkas, M.; Torck, M.; and Trotin, F. 1990. Plantes Médicinales des Régions Tempérées. Paris, Maloine.

Blot, W.J. 1990. Garlic in relation to cancer in human populations. In *Garlic in Biology and Medicine: Proceedings of the First World Congress on the Health Significance of Garlic and Garlic Constituents*. Nutrition International Co., P.O. Box 50632, Irvine, CA 92619-0632.

Blumberg, J.B. 1992. The role of antioxidants in the prevention of cancer. In Proceedings of *Nutrition in Disease Prevention*. Sydney. The Australian Council For Responsible Nutrition, Inc.

Bond, G.G., Thompson, F.E. and Cook, R.R. 1987. Dietary vitamin A and lung cancer: Results of a case-control study among chemical workers. *Nutr. Cancer* 9:109–121.

Bone, M.E.; Wilkinson, D.J.; Young, J.K.; McNeil, J.; and Charlton, S. 1990. Ginger root: A new antiemetic. The effect of ginger root on postoperative nausea and vomiting after major gynecological surgery. *Anaesthesia* 45:669–671.

Bordia, A., and Verma, S.K. 1990. Effect of three years treatment with garlic on the rate of re-infarction and mortality in patients with coronary artery disease. In *Garlic in Biology and Medicine: Proceedings of the First World Congress on the Health Significance of Garlic and Garlic Constituents*. Nutrition International Co., P.O. Box 50632, Irvine, CA 92619-0632

Borzeix, M.G.; Labos, M.; and Hartl, C. 1980. Recherches sur l'action antiagrégant de l'extrait de *Ginkgo biloba*. Activité au niveau des artères et des veines de la piemère chez le lapin. *Sem. Hôp. (Paris) 56:393.*

Bos, R.; Hendricks, H.; Bruins, A.P.; Kloosterman, J.; and Sipma, G. 1986. Isolation and identification of valerenane sesquiterpenoids from valeriana officinalis. Phytochem. 25:133–135.

Brasseur, T., and Angenot, L. 1984. Contribution to the pharmacognostical study of Passion flower. *J. Pharm. Belg.* 39:15–22.

Bravo, L.; Cobo, J.; Fraile, A.; Jimenez, J.; and Villar, A. 1974. Estudio farmacodinamico del lupulo (Humulus lupulus L.). Action transquilizante. *Boll. Chim. Farm* 113:310–315.

Brekhman, I.I., and Dardymov, I.V. 1969a. New substances of plant origin which increase non-specific resistance. *Ann. Rev. Pharmacacol.* 9:419–430.

———. 1969b. Pharmacological investigation of glycosides from ginseng and *Eleutherococcus*. *Lloydia* 32:46–51.

Bronzetti, G.; Galli, A.; and Croce, C.D. 1990. Antimutagenic effects of chlorophyllin. In: *Antimutagenesis and Anticarcinogenesis Mechanisms II*. Y. Kuroda, D.M. Shankel, and M.D. Waters. eds., pp. 463–468. New York: Plenum Press.

Brunello, N., Racagni, G., Clostre, F., Drieu, K., Braquet, P. 1985. Effects of an extract of *Ginkgo biloba* on noradrenergic systems of rat cerebral cortex. *Pharm. Res. Commun.* 17, 1063.

Buiatti, E.; Palli, D.; Decarli, A.; Amadori, D.; Avellini, C.; Bianchi, S.; Bonaguri, C.; Cipriani, F. %cco, P.; Giacosa, A.; Marubini, E.; Miracci, C.; Puntoni, R.; Russo, A.; Vindigni, C.; Fraumeni, J.F.; and Blot, W.J. 1990. A case-control study of gastric cancer and diet in Italy, II. Association with nutrients. *Int. J. Cancer* 45:896–901.

Burton, G.W. 1988. Antioxidant action of carotenoids. *J. Nutr.* 118:109–111.

Burton, G.W., and Ingold, K.U. 1984. Beta-carotene: An unusual type of lipid antioxidant. *Science*. 224:569.

Byers, T.E., Graham, S., Haughey, B.P., Marshall, J.R. and Swanson, M.K. 1987. Diet and lung cancer risk: Findings from the Western New York Diet Study. *Am. J. Epidemiol.* 125:351–363.

Chandler, R.F. 1985. Licorice, more than just a flavor. *Can. Pharm. J.* 118:420–424.

Chau An County Health Department. 1977. *Data Base on Canton Medicines*. 11:44.

Chaudhary, S.K.; Ceska, O.; Tetu, C.; Warrington, P.J.; Ashwood-Smith, M.J.; and Poulton, G.A. 1986. Oxypeucedanin, a major furocoumarin in parsley, Petroselinum Crispum. *Planta Medica* 52:462–464.

Chauo Cho City Hospital. 1972. *Scientific Communications*. 1:26. (in Chinese.)

Chen, Y.S. 1979. *Chunking Senior Herbal Practitioners Data Base*. 4:8. (in Chinese).

Chez Giang Medical University. 1972. Medical Science Report, Pharmaceutical Section 6:13.

Chin Chow Medical College. 1977a. *Chin Chow Medical Science*. 6:19. (in Chinese.)

———. 1977b. *Chin Chow Medical Science*. 1977. 6:35 (in Chinese.)

———. 1978. *Chin Chow Medical Science*. 9:21 (in Chinese.)

Chuang, L.T.; Lotzova, E.; Cook, K.R.; Cristoferoni, P.; Morris, M.; and Wharton, J.T. 1993. Effect of new investigational drug Taxol on oncolytic activity and stimulation of human lymphocytes. *Gyne. Oncol.* 49:291–8.

Chung, F.L.; Morse, M.A.; and Eklind, K.I. 1992. New potential chemopreventive agents for lung carcinogenesis of tobacco-specific nitrosamine. *Cancer Res.* 52:2719S–2722S.

Clairambault, P.; Magnier, B.; Droy-Lefaix, M.T.; Magnier, M.; and Pairault, C. 1986. Effet de l'extrait de *Ginkgo biloba* sur les lesions induites par une photocoagulation au laser à l'argon sur la rétine de lapin. *Sem. Hôp. Paris* 62, 57.

Clark, R.J., and Menary, R.C. 1981. Variations in composition of peppermint oil in relation to production areas. *Econ. Botany* 35:59–69.

Colditz, G.A.; Branch, L.G.; Lipnick, R.J.; Willett, W.C.; Rosner, B.; Posner, B.M.; and Hennekens, C.H. 1985. Increased green and yellow vegetable intake and lowered cancer deaths in an elderly population. *Am. J. Clin. Nutr.* 41:32–36.

Comstock, G.W.; Helzlsouer, K.J.; and Bush, T.L. 1991. Prediagnostic serum levels of carotenoids and vitamin E as related to subsequent cancer in Washington County, Maryland. *Am. J. Clin. Nutr.* 53:260S–264S.

Congora, C.; Proliac, A.; and Raynaud, J. 1986. Isolation and identification of two mono-C-glucosyl-luteolins and of the di-C substituted 6,8-diglucosyl-luteolin from the leafy stalks of *Passiflora incarnata L. Helv. Chim. Acta* 69:251–253.

Crombie, L. 1952. Amides of vegetable origin. Part I. Stereoisomeric N-isobutyl-undeca-1:7-diene-1-carboxyamides and the structure of Herculin. *J. Chem. Soc.* 2997–3008.

———. 1955. Amides of vegetable origin. Part II. Structure and stereochemistry of neoherculin. *J. Chem. Soc.* 995–998.

Crombie, L., and Tayler, J.L. 1957. Amides of vegetable origin. Part III. The constitution and configuration of the Sanshoöls. *J. Chem. Soc.* 2760–2766.

Cudrey, C.; Gargouri, Y.; Moreau, H.; Carriere, F.; Lauwereys, M.; Matthyssens, G.; and Verger, R. 1990. Ajoene prevents fat digestion by human gastric lipase in vitro (abstract). In *Garlic in Biology and Medicine: Proceedings of the First World Congress on the Health Significance of Garlic and Garlic Constituents*. Nutrition International Co., P.O. Box 50632, Irvine, CA 92619-0632.

Dashwood, R.H.; Breinholt, V.; and Bailey, G.S. 1991. Chemopreventive properties of chlorophyllin: Inhibition of aflatoxin B1 (AFB_1)-DNA binding in-vivo and anti-mutagenic activity against AFB_1 and two-heterocyclic amines in the Salmonella assay. *Carcinogenesis* 12:939–942.

Devys, M.; Alcaide, A.; Pinte, F.; and Barbier, M. 1969. Pollinastanol dans la fougére polypodium vulgare L. et la salsapareille smilax medica schlecht et Cham. *C.R. Acad. Sc. Paris Série D.* 269:2033–2034.

Dorogokupla, A.G.; Postolnikov, S.F.; Troitskaya, E.G.; and Adilgireeva, L.K. 1973. Role of carotene-containing foods in the occurrence of experimental tumors. In *Inst. Klin. Eksp. Khir*, S.R. Karynbaev, ed., pp. 153–155. Moscow: Alma-Ata.

Eckmann, F., and Schlag, H. 1982. Kontrollierte Doppelblind—Studie zum Wirksamkeitsnachweis von Tebonin forte bei Patienten mit zerebrovaskularer Insuffizienz. *Fortschr. Med.* 100, 474.

Edes, T.E.; Gysbers, D.G.; Buckley, C.S.; and Thornton, W.H. 1991. Exposure to the carcinogen benzopyrene depletes tissue vitamins A: Beta-carotene prevents depletion. *Nutr. Cancer* 15:159–166.

Eichholzer, M.; Stahelin, H.B.; and Gey, K.F. 1992. Inverse correlation between essential antioxidants in plasma and subsequent risk to develop cancer, *ischemic heart disease and stroke*, respectively: 12 year follow-up of the prospective Basel study. *Exs.* 62:398–410.

Epstein, J.H. 1977. Effect of beta-carotene on ultraviolet-induced cancer formation in the hairless mouse skin. *Photochem. Photobiol.* 25:211–213.

Etienne, A.; Hecquet, F.; Clostre, F.; and De Feudis, F.V. 1982. Comparison des effets d'un extrait de *Ginkgo biloba* et de la chloropromazine sur la fragilité osmotique, in vitro, d'erythrocytes de rat. *J. Pharmacol. (Paris)* 13:291.

Fish, F.; Gray, A.I.; Waterman, P.G.; and Donachie, F. 1975. Alkaloids and coumarins from North American *Zanthoxylum* species. *Lloydia* 38:268–270.

Fish, F., and Waterman, P.G. 1973. Alkaloids in the bark of *Zanthoxylum* clava-herculis. *J. Pharm. Pharmac.* 25:(Suppl.) 115–116.

Folkers, K.; Brown, R.; Judy, W.V.; and Morita, M. 1993. Survival of cancer patients on therapy with Coenzyme Q_{10}. *Biochem. Biophys. Res. Comm.* 192:241–245.

Frei, B.; Kim, M.C.; and Ames, B.N. 1990. Ubiquinol-10 is an effective lipid-soluble antioxidant at physiological concentrations. *Proc. Natl. Acad. Sci. USA* 87:4879–4883.

Genest, K., and Hughes, D.W. 1969. Natural products in Canadian Pharmaceuticals IV. Hydrastis canadensis. *Can. J. Pharm. Sci.* 4:41–45.

Gessner, B.; Voelp, P.; and Klasser, M. 1985. Study of the long-term action of a *Ginkgo biloba* extract on vigilance and mental performance as determined by means of quantitative pharmaco-EEG and psychometric measurements. *Arzneim. Forsch.* 35:1459.

Giang Su New Medical College. 1975. *Pathological Data Base*. Vol. 2, p. 76. (in Chinese.)

Glotzbach, B., and Rimpler, H. 1968. Die Flavonoide von *Passiflora incarnata L.*, *Passiflora quadrangularis L.*, und *Passiflora pulchella H.B.K.* Eine chromatographische Untersuchung. *Planta Medica* 16:1–7.

Glowania, H.J.; Raulin, C.; and Swoboda, M. 1987. The effect of chamomile on wound healing—A controlled clinical experimental double-blind trial. *Z. Hautkr.* 62:1262–1271.

Gold, L.S.; Manley, N.B.; and Ames, B.N. 1992. Extrapolation of carcinogenicity between species: Qualitative and quantitative factors. *Risk Anal.* 12:579–588.

Gold, L.S.; Slone, T.H.; Manley, N.B.; Garfinkel, G.B.; Hudes, E.S.; Rohrbach, L.; and Ames, B.N. 1991. The carcinogenic potency database: Analysis of 4,000 chronic animal cancer experiments published in the general literature and by the U.S. National Cancer Institute/National Toxicology Program. *Environ. Health Persp.* 96:11–15.

Greenberg, S., and Frishman, W.H. 1990. Co-enzymes Q_{10}: A new drug for cardiovascular disease. *J. Clin. Pharmacol.* 30:596–608.

Gridley, G.; McLaughlin, J.K.; Block, G.; Blot, W.J.; Winn, D.M.; Greenberg, R.S. *hoenburg, J.B.; Preston-Martin, S.; Austin, D.F.; and Fraumeni, J.F. 1990. Diet and oral and pharyngeal cancer among blacks. *Nutr. Cancer* 14:219–225.

Gudej, J. 1991. Flavonoids, phenolic acids and Coumarins, from roots of *Althaea officinalis*. *Planta Medica* 57:284–285.

Guo, Z.; Smith, T.J.; Wang, E.; Sadrieh, N.; Ma, Q.; Thomas, P.E.; and Yang, C.S. 1992. Effects of phenethyl isothiocyanate, a carcinogenesis inhibitor, on xenobiotic-metabolizing enzymes and nitrosamine metabolism in rats. *Carcinogenesis* 13:2205–2210.

Haan, J.; Reekermann, V.; Welter, F.L.; Sabin, G.; and Müller, E. 1982. *Ginkgo biloba* Flavonglykoside. Therapiemoglichkeit der zerebralen Insuffizienz. *Med. Welt* 33, 1001.

Hall, I.H.; Starnes, C.O.; Lee, K.H.; and Waddell, T.G. 1980. Mode of action of sesquiterpene lactones as anti-inflammatory agents. *J. Pharm. Sci.* 68:537–543.

Hänsel, R., and Schulz, J. 1986a. Hopfen und Hopfenpräparate. Fragen zur Pharmazeutischen Qualität. *Deutsch Apoth. Ztg.* 126:2033–2037.

———. 1986b. Hopfenzapfen (Lupuli strobulus). Dunnschicht-chromatographische Prüfung auf Identität. *Deutsch Apoth. Ztg.* 126:2347–2348.

Hansen, L., and Boll, P.M. 1986. Polyacetylenes in Araliaceae: Their chemistry, biosynthesis and biological significance. *Phytochem.* 25:285–293.

Hikino, H. 1985. Recent research on Oriental medicinal plants. In *Economic and Medicinal Plant Research* 1, pp. 53–85. New York: Academic Press.

Hinds, M.W.; Kolonel, L.N.; Hankin, J.H.; and Lee, J. 1984. Dietary vitamin A, carotene, vitamin C and risk of lung cancer in Hawaii. *Am. J. Epidemiol.* 119:227–237.

Hirakura, K.; Morita, M.; Nakajima, K.; Ikeya, Y.; and Mitsuhashi, H. 1991. Polyacetylenes from the roots of *Panax ginseng*. *Phytochem.* 30:3327–3333.

———. 1992. Three acetylenic compounds from roots of *Panax ginseng*. *Phytochem.* 31:899–903.

Hirano, T.; Fukuoka, K.; Oka, K.; Naito, T.; Hosaka, K.; Mitsuhashi, H.; and Matsumoto, Y. 1990. Antiproliferative activity of mammalian lignan derivative against human breast carcinoma cell line, ZR-75-1. *Cancer Invest.* 8:595–601.

Hirayama, T. 1979. Diet and Cancer. *Nutr. Cancer* 1:68–81.

———. 1985. A large scale cohort study on cancer risks with special reference to the risk reducing effects of green-yellow vegetable consumption. *Int. Symp. Princess Takamatsu Cancer Res. Fund* 16:41–53.

Hladon, B. and Twardowski, T. 1979. Sesquiterpene lactones, XXV. Studies on the mode of action. The cellular and molecular basis of cytostatic action. *Pol. J. Pharmecol.* 31:35–43.

Hoff, G.; Moen, I.E.; Mowinckel, P.; Rosef, D.; Nordbo, E.; Sauar, J.; Vatn, M.H.; and Torgrimsen, T. 1992. Drinking water and the prevalence of colorectal adenomas: An epidemiologic study in Telemark, Norway. *Europ. J. Cancer Prevent.* 1:423–428.

Hofmann, L.; Fritz, D.; Nitz, S.; Kollmannsberger, H.; and Drawert, F. 1992. Essential oil composition of three polyploids in the Achillea Millefolium "Complex." *Phytochem.* 31:537–542.

Hoon, D.S.B.; Sze, L.; Lin, R.I.S.; and Irie, R.F. 1990. Modulation of cancer antigens and growth of human melanoma by aged garlic extract. In *Garlic in Biology and Medicine: Proceedings of the First World Congress on the Health Significance of Garlic and Garlic Constituents.* Nutrition International Co., P.O. Box 50632, Irvine, CA 92619-0632.

Huang, M.T.; Lysz, T.; Ferraro, T.; Abidi, T.F.; Laskin, J.D.; and Conney, A.H. 1991. Inhibitory effects of curcumin on in vitro lipoxygenase and cyclooxygenase activities in mouse epidermis. *Cancer Res.* 51:813–819.

Innocenti, G.; Dall' Acqua, F.; and Caporale, G. 1976. Investigations of the content of furocoumarins in *Apium graveolens and Petroselinum sativum. Planta Medica* 29:165–170.

Issac, O. 1979. Pharmacological investigation with compounds of chamomile. I. On the pharmacology of (-)-α-bisabolol and bisabolol oxides. *Planta Medica* 35:118–124.

Jain, A.K.; Shimoi, K.; Nakamura, Y.; Kada, T.; Hara, Y.; and Tomita, I. 1989. Crude tea extract decreases the mutagenic activity of N-methyl-N'-nitro-N-nitrosoguanidine in vitro and in intragastic tract of rats. *Mutation Res.* 210:1–8.

Jakovlev, V.; Isaac, O.; Thiemer, K.; and Kunde, R. 1979. Pharmacological investigations with compounds of chamomile. II. New investigations on the antiphlogistic effects of (-)-α-bisabolol and bisabolol oxides. *Planta Medica* 35:125–140.

Jaspersen-Schib, R. 1990. Sédatifs à base de plantes. *Schweiz. Apoth. Ztg.* 128:248–251.

Jenkins, M.L.Y. 1993. Research issues in evaluating "Functional Foods." *Food Technology* 47(5):76–79.

Jiang, Y., Zhong, G.G., Chen, L., and Ma, X.Y. 1993. Influence of ginsenosides Rb1, Rb2, and Rb3 on electric and contractile activities of normal and damaged cultured myocardiocytes. *Chung-kou Yao Li Hsueh Pao (Acta Pharmacologica Sinica)* 13:403–6.

King, F.E., and Wilson, J.G. 1964. The chemistry of extractives from hardwoods. Part XXXVI. The lignans of *Guaiacum officiale L. J. Chem. Soc.* 4011–4024.

Kiso, Y.; Tohkin, M.; Hikino, H.; Hattori, M.; Sakamoto, T.; and Namba, T. 1984. Mechanism of antihepatotoxic activity of glycyrrhizin. I. Effect on free radical generation and lipid peroxidation. *Planta Medica* 50:298–302.

Knekt, P.; Jarvinen, R.; Seppanen, R.; Rissanen, A.; Aromaa, A.; Heinonen, O.P.; Albanes, D.; Heinonen, M.; Pukkala, E.; and Teppo, L. 1991. Dietary antioxidants and the risk of lung cancer. *Am. J. Epidemiol.* 134:471–479.

Konno, C.; Sugiyama, K.; Kano, M.; Takahashi, M.; and Hikino, H. 1984. Isolation and hypoglycemic activity of Panaxans A, B, C, D, and E, glycans of *Panax ginseng* roots. *Planta Medica* 50:434–436.

Kono, S.; Ikeda, M.; Tokudome, S.; and Kuratsune, M. 1988. A case-control study of gastric cancer and diet in northern Kyushu, Japan. *Japan J. Cancer Res.* 79:1067–1074.

Kratochvil, J.F.; Burris, R.H.; Seikel, M.K.; and Harkin, J.M. 1971. Isolation and characterization of α-guaiaconic acid and the nature of Guaiacum Blue. *Phytochem.* 10:2529–2531.

Krinsky, N.I. 1968. The protective function of carotenoid pigments. In *Photophysiology*. vol. 3, A.C. Giese, ed., pp. 123–196. London: Academy Press.

Kune, G.A.; Bannerman, S.; Field, B.; Watson, L.F.; Cleland, H.; Merenstein, D.; and Vitetta, L. 1992. Diet, alcohol, smoking, serum beta-carotene, and vitamin A in male nonmelanocytic skin cancer patients and controls. *Nutr. Cancer* 18:237–244.

Kuo, F. 1980. *Clinical and Applied Immunology* 1:15 (in Chinese).

Kvale, G.; Bjelke, E.; and Gart, J.J. 1983. Dietary habits and lung cancer risk. *Int. J. Cancer* 31:397–405.

Lamm, D.L.; Riggs, D.R.; and DeHaven, J.I. 1990. Intralesional immunotherapy of murine transitional cell carcinoma using garlic extract. In *Garlic in Biology and Medicine: Proceedings of the First World Congress on the Health Significance of Garlic and Garlic Constituents*. Nutrition International Co., P.O. Box 50632, Irvine, CA 92619-0632.

Larsen, R.G.; Dupeyron, J.P.; and Boulu, R.G. 1978. Modéle d'ischémie cérébrale expérimentale par microsphères chez le rat. Etude de l'effet de deux extraits de *Ginkgo biloba* et du naftidrofuril. *Thérapie* 33:651.

Le Gardeur, B.Y.; Lopez, A.; and Johnson, W.D. 1990. A case-control study of serum vitamins A, E, and C lung cancer patients. *Nutr. Cancer.* 14:133–140.

Le Poncin-Lafitte, M.; Rapin, J.; and Rapin, J.R. 1980. Effects of *Ginkgo biloba* on changes induced by quantitative cerebral microembolization in rats. *Arch. Int. Pharmacodyn.* 243:236.

Le Poncin-Lafitte, M.; Martin, P.; Lespinasse, P.; and Rapin, J.R. 1982. Ischèmie cérébrale après ligature non simultanée des artéres carotides chez le rat: Effet de l'extrait de *Ginkgo biloba*. *Sem. Hôp. Paris* 58:403.

Leicester, R.J., and Hunt, R.H. 1982. Peppermint oil to reduce colonic spasm during endoscopy. *Lancet* II 989.

Lemli, J. 1986. The chemistry of senna. *Fitoterapia* 57:33–40.

Leng-Peschlow, E. 1986. Dual effect of oral administered sennosides on large intestines transit and fluid absorption in the rat. *J. Pharm. Pharmacol.* 38:606–610.

Liao Ling See Fung Disease Prevention Center. 1974. *Tieh Ling Medicine.* 2:33.

Lin, R.I.S. 1990. Introduction: An overview of the nutritional and pharmacological properties of garlic. In *Garlic in Biology and Medicine: Proceedings of the First World Congress on the Health Significance of Garlic and Garlic Constituents.* Nutrition International Co., P.O. Box 50632, Irvine, CA 92619-0632.

Lin, R.I.S. 1975. *Beta-carotene as a singlet oxygen quencher in foods and its physiological significance.* Frito-Lay Company Research and Development Report. Dallas.

———. 1987. Lipid Nutrition and Cardiovascular Diseases. In: Trends in Nutrition and Food Policy (*Proceedings of the 7th World Congress of Food Science and Technology.*) A.H. Gee ed. Singapore. Singapore Institute of Food Science and Technol ogy.

———. 1989a. *Garlic in Nutrition and Medicine.* Nutrition International Co., P.O. Box 50632, Irvine, CA 92619-0632.

———. 1989b. *Ginseng and Herbs.* Nutrition International Co., P.O. Box 50632, Irvine, CA 92619-0632.

———. 1991. Cardiovascular Protective Effects of Garlic. Paper read at National Conference on Cholesterol and High Blood Pressure Control, May 8–10, 1991, Washington, D.C.

———. 1993a. *Evolutional Aspects of Diet, Nutrition, and Disease.* Nutrition International Co., P.O. Box 50632, Irvine, CA 92619-0632.

———. 1993b. *Bioavailability and Transformation of Garlic's Thioallys in Animals.* Nutrition International Co., P.O. Box 50632, Irvine, CA 92619-0632.

———. 1993c. *Garlic and Health: Recent Advances in Research.* Nutrition International Co., P.O. Box 50632, Irvine, CA 92619-0632.

Lin, R.I.S.; Abuirmeileh, N.; and Qureshi, N. 1990. Inhibition of cholesterol synthesis by kyolic (aged garlic extract) and S-allyl cysteine in a hypercholesterolemic model. In *Garlic in Biology and Medicine: Proceedings of the First World Congress on the Health Significance of Garlic and Garlic Constituents.* Nutrition International Co., P.O. Box 50632, Irvine, CA 92619-0632.

Liu, C.X., and Xiao, P.G. 1992. Recent Advances on Ginseng Research in China. *J. Ethnopharmacol.* 36:27–38.

Liu, J.; Lin, R.I.S.; and Milner, J.A. 1992. Inhibition of 7,12-dimethylbenz(a)anthracene-induced mammary tumors and DNA adducts by garlic powder. *Carcinogenesis* 13:1847–1851.

Lopes, N.M.; Adams, E.G.; Pitts, T.W.; and Bhuyan, B.K. 1993. Cell kill kemetics and cell cycle effects of taxol on human and hamster ovarian cell lines. *Cancer Chemother. Pharm.* 32:235–42.

MacLennan, R.; DaCosta, J.; Day, N.E.; Law, C.H.; Ng, Y.K.; and Shanmugaratnam, K. 1977. Risk factors for lung cancer in Singapore Chinese, a population with high female incidence rates. *Int. J. Cancer* 20:854–860.

Makheja, A.N. 1990. Anti-platelet constituents of garlic and their effects on platelet aggregation and cyclo- and lipoxygenase enzymes. In *Garlic in Biology and Medicine: Proceedings of the First World Congress on the Health Significance of Garlic and Garlic Constituents.* Nutrition International Co., P.O. Box 50632, Irvine, CA 92619-0632.

Markman, M. 1993. Report from the Second National Cancer Institute Workshop on Taxol and Taxus. *J. Cancer Res. Clin. Oncol.* 119:439–40.

Matthews-Roth, M.M. 1982. Antitumor activity of beta-carotene, canthaxanthin and phytoene. *Oncology* 39:33–37.

Matthews-Roth, M.M., and Krinsky, N.I. 1985. Carotenoid dose level and protection against UV-B induced skin tumors. *Photochem. Photobiol.* 42:35–38.

———. 1987. Carotenoids affect development of UV-B induced skin cancer. *Photochem. Photobiol.* 46:507–509.

Matthews-Roth, M.M.; Pathak, M.A.; Fitzpatrick, T.B.; Harber, L.H.; and Kass, E.H. 1977. Beta-carotene therapy for erythropoietic protoporphyria and other photosensitivity diseases. *Arch. Dermatol.* 113:1229–1232.

McMurrough, I. 1981. High-performance liquid chromatography of flavonoids in barley and hops. *J. Chromatogr.* 28:683–693.

Melegari, M.; Albasini, A.; Pecorari, P.; Vampa, G.; and Rinaldi, M. 1988. Chemical characteristics and pharmacological properties of the essential oils of *Anthemis nobilis*. *Fitoterapia* 59:449–455.

Messina, J., and Messina, V. 1991. Increasing use of soyfoods and their potential role in cancer prevention. *J. Am. Diet. Assoc.* 91:836–840.

Manoharan, K., and Bannerjee, M.R. 1985. Beta-carotene reduces sister chromatid exchange induced by chemical carcinogens in mouse mammary cells in organ culture. *Cell Biol. Int. Rep.* 9:783–789.

Muggli, R. 1992. Beta-carotene and disease prevention. *J. Nutr. Sci. Vitaminol.* Spc. No. 560–563.

Nagabhushan, M., and Bhide, S.V. 1992. Curcumin as an inhibitor of cancer. *J. Am. College Nutr.* 11:192–198.

Nakajima, K.; Yamauchi, K.; and Kuwano, S. 1985. Isolation of a new aloe-emodin dianthrone diglucoside from senna and its polentiating effect on the purgative activity of Sennoside A in mice. *J. Pharm. Pharmacol.* 37:703–706.

Negishi, T.; Arimoto, S.; Nishizaki, C.; and Hayatsu, H. 1989. Inhibitory effect of chlorophyll on the genotoxicity of 3-amino-1-methyl-5H-pyrido (4,3-b) indole (Trp-P-2). *Carcinogenesis* 10:145–149.

Nishino, H. 1990. Inhibition of tumorigenesis by garlic constituents. In *Garlic in Biology and Medicine: Proceedings of the First World Congress on the Health Significance of Garlic and Garlic Constituents*. Nutrition International Co., P.O. Box 50632, Irvine, CA 92619-0632.

Nomura, A.M.Y.; Stemmerman, G.N.; Heilbrun, L.K.; Salkeld, R.M.; and Vuilleumier, J.P. 1985. Serum vitamin levels and the risk of cancer of specific sites in men of Japanese ancestry in Hawaii. *Cancer Res.* 45:2369–2372.

Oshio, H.; Imai, S.; Fujioka, S.; Sugawara, T.; Miyamoto, M.; and Tsukui, M. 1974. Investigation of rhubarbs. III. New purgative constituents, Sennosides E and F. *Chem. Pharma. Bull.* 22:823–831.

Paganini-Hill, A.; Chao, A.; Ross, R.K.; and Henderson, B.E. 1987. Vitamin A, beta-carotene, and the risk of cancer: A prospective study. *J. Natl. Cancer Inst.* 79:443–448.

Palan, P.R.; Mikhail, M.S.; Basu, J.; and Romney, S.L. 1991. Plasma levels of antioxidant beta-carotene and alpha-tocopherol in uterine cervix dysplasia and cancer. *Nutr. Cancer* 15:13–20.

Pastorino, U.; Pisani, P.; Berrino, F.; Andrecoli, C.; Barbieri, A.; Costa, A.; Mazzoleni, C.; Gramegna, G.; and Marubini, E. 1987. Vitamin A and female lung cancer: A case-control study on plasma and diet. *Nutr. Cancer* 10:171–179.

Peng, A.; Rundhaug, J.E.; Yoshizawa, C.N.; and Bertram, J.S. 1988. β-carotene and canthaxanthin inhibit chemically-and physically-induced neoplastic transformation in 10T1/2 cells. *Carcinogenesis* 9:1533–1539.

Petkov, V.D., and Mosharrof, A.H. 1987. Effects of standardized ginseng extract on learning, memory and physical capabilities. *Am. J. Clin. Med.* 15:19–29.

Pinto J.; Tiwari, R.; and Lin, R.I.S. 1993. Effects of aged garlic extract on cultured human breast cancer cells. (Manuscript in preparation.)

Piovella, C. 1973. Effetti della *Ginkgo biloba* sui microvasi della congiuntiva bulbare. *Min. Med.* 64, 4179.

Pisani, P.F.; Berrino, F.; Macaluso, M.; Pastrino, U.; Crosignani, P.; and Baldasseroni, A. 1986. Carrots, green vegetables, and lung cancer: a case control study. *Int. J. Epidemiol.* 15:463–468.

Qimin L.; van den Heuvel, H.; and Delorenzo, O. 1991. Mass spectral characterizations of c-glycosidic flavonoids isolated from a medicinal plant (*Passiflora incarnata*). *J. Chromatogr.* 562:415–446.

Qureshi, N.; Lin, R.I.S.; Abuirmeileh, N.; and Qureshi, A.A. 1990. Dietary kyolic (aged garlic extract) and S-allyl cysteine reduces the levels of plasma triglycerides, thromboxane B_2 and platelet aggregation in hypercholesterolemic model. In *Garlic in Biology and Medicine: Proceedings of the First World Congress on the Health Significance of Garlic and Garlic Constituents*. Nutrition International Co., P.O. Box 50632, Irvine, CA 92619-0632.

Rafatullah, S.; Mossa, J.S.; Ageel, A.M.; Al-Yahya, M.A.; and Tariq, M. 1991. Hepatoprotective and safety evaluation studies on sarsaparilla. *Int. J. Pharmacognosy* 29:296–301.

Rapin, J.R.; and Le Poncin-Lafitte, M. 1979. Modèle expérimental d'ischémie cérébrale. Action preventive de l'extrait de Ginkgo. *Sem. Hôp. Paris* 55:2047.

Rees, W.D.W.; Evans, B.K.; and Rhodes, J. 1979. Treating irritable bowel syndrome with peppermint oil. *Br. Med. J.* 2:835.

Rettura, G.; Levenson, S.M.; and Seifter, E. 1984. Beta-carotene prevents dimethylbenz(a)anthracene-induced tumors. In: *Vitamins, Nutrition, and Cancer*, K.N. Presad, ed. pp. 296–297. New York: Karger, S.

Rettura, G.; Stralford, F.; Levenson, S.M.; and Seifter, E. 1982. Prophylactic and therapeutic actions of supplemental beta-carotene in mice inoculated with C3HBA adenoma carcinoma cells: Lack of therapeutic actions of supplemental ascorbic acid. *J. Natl. Cancer Inst.* 69:73–77.

Rieder, A.; Adamek, M.; and Wrba, H. 1983. Delay of diethylnitrosamine-induced hepatoma in rats by carrot feeding. *Oncology* 40:120–123.

Riemersma, R.A.; Wood, D.A.; MacIntyre, C.C.A.; Elton, R.A.; Gey, K.F.; and Oliver, M.F. 1991. Risk of angina pectoris and plasma concentrations of vitamins A, C, and E and carotene. *Lancet* 337:1–5.

Ruch, R.J.; Cheng, S.J.; and Klaunig, J.E. 1989. Prevention of cytotoxicity and inhibition of intercellular communication by antioxidant catechins isolated from Chinese green tea. *Carcinogenesis* 10:1003–1006.

Rüker, G.; Mayer, R.; and Lee, K.R. 1989. Peroxides as plant constituents. VI. Hydroperoxides from blossoms of Roman chamomile, *Anthemis nobilis L. Arch. Pharm.* 322:821–826.

Runowicz, C.D.; Wiernick, P.H.; Einzig, A.I.; Goldberg, G.L.; and Horwitz, S.D. 1993. Taxol in ovarian cancer. *Cancer* 71:1591–6.

Saita, T.; Katano, M.; Matsunaga, H.; Yamamoto, H.; Fujito, H.; and Mori, M. 1993. The first specific antibody against cytotoxic polyacetylenic alcohol, panaxynol. *Chem. Pharma. Bull.* 41:549–552.

Samet, J.M.; Skipper, B.J.; Humble, C.G.; and Pathak, D.R. 1985. Lung cancer risk and vitamin A consumption in New Mexico. *Ann. Rev. Respir. Dis.* 131:198–202.

Santamaria, L.; Bianchi, A.; Arnaboldi, A.; and Andreoni, L. 1981. Prevention of the benz(α)pyrene photocarcinogenic effect by beta-carotene and canthoxanthin. *Med. Biol. Environ.* 9:113–120.

Schaffler, K., Rech, P.W. 1985. Doppellblind studie zur hypoxieprotektiven Wirkung eines standardieserten *Ginkgo-biloba* Praeparates nach Mehrfachverabreichung au gesunden Probanden. *Arzneim. Forsch.* 35, 1283.

Schrecker, A.W. 1957. Meso-dihydrogaiaretic acid and its derivatives. *J. Am. Chem. Soc.* 79:3823–3827.

Shanghai 2nd Medical College, Ree Zin Hospital. 1972. *Medical Science Data Base* 16:30 (in Chinese).

Shekelle, R.B.; Lepper, M.; Liu, S.; Maliza, C.; Raynor, W.J.; Rossof, A.H.; Paul, O.; Shyrock, A.M.; and Stamler, J. 1981. Dietary vitamin A and risk of cancer in the Western Electric study. *Lancet* 2:1185–1190.

Shen Yang Army Hospital. 1975. *Medical Data Base*. vols. 1 and 2, p. 20. (in Chinese.)

Shibata, S.; Tanaka, O.; Shoji, J.; and Saito, H. 1985. Chemistry and pharmacology of panax. In *Economic and Medicinal Plant Research*, 1:217–284, H. Wagner, H. Hikino, and N.R. Farnsworth, eds. New York: Academic Press.

Shideman, F.E. 1950. A review of the pharmacology and therapeutics of hydrastis and its alkaloids hydrastine, berberine, and canadine. *Bull. National Formulary Comm.* 18:3–19.

Shim, S.C.; Koh, H.Y.; and Han, B.H. 1983. Polyacetylenes from *Panax ginseng* roots. *Phytochem.* 22:1817–1818.

Singh, V.K.; Agarwal, S.S.; and Gupta, B.M. 1984. Immuno-modulatory activity of *Panax ginseng* extract. *Planta Medica* 50:462–465.

Smith, A., and Waller, K. 1991. Serum beta-carotene in persons with cancer and their immediate families. *Am. J. Epidemiol.* 133:661–671.

Song-San; Watanabe, S.; and Saito, T. 1989. Chalcones from *Humulus lupulus*. *Phytochem.* 28:1776–1777.

Sonnenborn, U., and Proppert, Y. 1990. Ginseng (Panax Ginseng C.A. Meyer). *Z. Phytotherapie* 11:35–49.

Speroni, E., and Minghetti, A. 1988. Neuro-pharmacological activity of extracts from *Passiflora incarnata*. *Planta Medica* 54:488–491.

Stahelin, H.B.; Rosel, F.; Buess, E.; and Brubacher, G. 1984. Cancer, vitamins, and plasma lipids: Prospective Basel study. *J. Natl. Cancer Inst.* 73:1463–1468.

Steiner, M., and Lin, R.I.S. 1992. Anti-platelet adhesion effects of thioallyl compounds. *J. Am. College Nutr.* 11:56.

———. 1993. Inhibition of Smooth Muscle Cell Proliferation by Thioallyl Compounds Derived from Garlic. (manuscript in preparation.)

Stich, H.F., and Dunn, B.P. 1986. Relationship between cellular levels of beta-carotene and sensitivity to genotoxic agents. *Int. J. Cancer* 38:713–717.

Sticker, O., and Soldati, F. 1978. Glycrrhizin-säure-bestimmung in Radix liquiritae mit Hochleistungs-Flüssigkeitschromatographie (HPLC). *Pharm. Acta Helv.* 53:46–52.

Stierle, A.; Strobel, G.; and Stierle, D. 1993. Taxol and Taxane production by *Taxomyus andreanae*, an endophytic fungus of Pacific yew. *Science* 260:214–215.

Suda, D.; Schwartz, J.; and Shklar, G. 1986. Inhibition of experimental oral carcinogenesis by topical beta-carotene. *Carcinogenesis* 7:711–715.

Suekawa M.; Ishige, A.; Yuasa, K.; Sudo, K.; Aburoda, M.; and Hosoya, E. 1984. Pharmacological studies on ginger. 1. Pharmacological actions of pungent constituents [6]- gingerol and [6]-shogaol. *J. Pharm. Dyn.* 7:836–48.

Sunsee Zin Zung Region People's 2nd Hospital, Gynecology Department 1975. *Neoplasia Treatment and Prevention Communications* 1:31. (in Chinese.)

Taddei, L.; Giachetti, D.; Taddei, E.; Mantovani, P.; and Bianchi, E. 1988. Spasmolytic activity of peppermint, sage, and rosemary essences and their major constituents. *Fitoterapia* 59:463– 468.

Takemoto, Y., Ueyama, T., Saito, H., Horio, S., Sanada, S., Shoji, J., Yahara, S., Tanaka, O., and Shibata, S. 1984. Potentiation of nerve growth factor mediated nerve fiber production in organ cultures of chicken embryonic ganglia by ginseng saponins: structure-activity relationship. *Chem. Pharm. Bull.* 32:3128–3131.

Temple, N.J., and Basu, T.K. 1987. Protective effect of β-carotene against colon tumors in mice. *J. Natl. Cancer Inst.* 78:1211–1214.

Thies, P.W. 1968. Die Konstitution der Valepotriate. Mitteilung über die Wirkstoffe des Baldrians. *Tetrahedron* 24:313–347.

Thurmon, F.M. 1942. The treatment of psoriasis with a sarsaparilla compound. *N. Engl. J. Med.* 227:128–133.

Titz, W.; Jurenitsch, J.; Fitzbauer-Busch, E.; Wicho, E.; and Kubelka, W. 1982. Valepotriates and essential oil of morphologically and karyologically defined types of *Valeriana officinalis* s.1. I. Comparison of valepotriate content and composition. *Sci. Pharm.* 50:309–324.

Titz, W.; Jurenitsch, J.; Gruber, J.; Schabus, I.; Titz, E.; and Kubelka, W. 1983. Valepotriates and essential oil of morphologically and karyologically defined types of *Valeriana officinalis* s.1. II. Variation of some characteristic compounds of the essential oil. *Sci. Pharm.* 51:63– 86.

Tiwari, R.K.; Pinto, J.; Qiao, C.H.; Lin, R.I.S.; and Osborne, M.P. 1992. Chemopreventive effects of garlic compounds in human breast carcinoma cells. *Breast Cancer Res. and Treat.* 23–171.

Tomoda, M.; Shimada, K.; Konno, C.; and Hikino, H. 1985. Structure of Panaxan B, a hypoglycemic glycan of *Panax ginseng* roots. *Phytochem.* 24:2431–2433.

Tomoda, M.; Shimada, K.; Konno, C.; Sugiyama, K.; and Hikino, H. 1984. Partial structure of Panaxan A, a hypoglycemic glycan of *Panax ginseng* roots. *Planta Medica* 50:436–438.

Tosi, B.; Bonora, A.; Dall' Olio, G.; and Bruni, A. 1991. Screening for toxic thiophene compositae used in northern Italy. *Phytother. Res.* 5:59–62.

U.S. Department of Health and Human Services. 1992. *HHS News*. pp. 92–97.

Ulubelen, A.; Öksüz, S.; and Schuster, A. 1990. A sesquiterpene lactone from *Achillea millefolium* subsp. *millefolium*. *Phytochem.* 29:3948–3949.

Vanhaelen, M.H. 1992. Taxol and related diterpenoids of *Taxus* sp: Important acquisitions in the treatment of cancer. (translated) *J. de Pharm. de Belgique* 47:417–424.

Verhaeren, E.H.C., Dreessen, M., Lemli, J. 1982. Studies in the field of drug containing anthracene derivatives. XXXIII. Variation of the ratio of reduced and oxidized forms of anthranoids in the root of *Rheum officinale* during a year cycle. *Planta Medica* 45:15–19.

Wagner, H., Proksch, A., Reiss-Maurer, I., Vollmar, A., Odenthall, S., Stuppner, H., Jurcic, K., le Turdu, M., and Fang, J.N. 1985. Immunostimulating polysaccharides (heteroglycans) of higher plants. *Arzneim. Forsch.* 35:1069–1075.

Wang, Z.T.; Cheng, S.J.; Zhou, Z.C.; Athar, M.; Khan, W.A.; Bickers, D.R.; and Mukhtar, H. 1989. Antimutagenic activity of green tea polyphenols. *Mutation Res.* 223:273–285.

Wargovich, M.J.; Sumiyoshi, H.; and Baer, A. 1990. Chemoprevention studies in animal models for colon and esophageal cancer using diallyl sulfide. In *Garlic in Biology and Medicine: Proceedings of the First World Congress on the Health Significance of Garlic and Garlic Constituents*. (in press.) Nutrition International Co., P.O. Box 50632, Irvine, CA 92619–0632.

Webb, T.E.; Stromberg, P.C.; Abou-Isaac, H.; Curley, R.W.; and Moeschberger, M. 1992. Effect of dietary soybean and licorice on the male F344 rat: An integrated study of some parameters relevant to cancer chemoprevention. *Nutr. Cancer* 18:215:230.

Weitberg, A.B.; Weitzman, S.A.; Clark, E.P.; and Stossel, T.P. 1985. Effects of antioxidants on oxidant-induced sister chromatid exchange formation. *J. Clin. Invest.* 75:1835–1841.

Wong Nan Medical College 1976. *Wong Nan J. Medicine* 1:14. (in Chinese.)

Wong, M. 1977. *Studies on Ginseng,* Shanghai, China: Shanghai Chinese Medicine College.

Wu Han Army General Hospital. 1973. *New Medical Science* 1:13. (in Chinese.)

Wu Han Health Department 1977. Clinical studies on *Crataegus pinnatifida.* In *Health and Medicine Data Base.* vol. 40. (in Chinese.)

Wu, A.H.; Henderson, B.E.; Pike, M.C.; and Yu, M.C. 1985. Smoking and other risk factors for women. *J. Natl. Cancer Inst.* 74:747–751.

Yamagishi, T.; Nishizawa, M.; Ikura, M.; Hikiehi, K.; Nonako, G.; and Nishioka, I. 1987. New laxative constituents of rhubarb. Isolation and characterization of Rheinosides A, B, C. and D. *Chem. Pharm. Bull.* 35:3132–3138.

Yeh, G.C. 1962. *Jiang Su Chinese Medicine* 1:29. (in Chinese.)

Zamora, R.; Hidalgo, F.J.; and Tappel, A.L. 1990. Comparative antioxidant effectiveness in dietary beta-carotene, vitamin E, selenium, and coenzyme Q_{10} in rat erythocytes and plasma. *J. Nutr.* 121:50–56.

Zhao, B.L.; Li, X.J.; He, R.G.; Cheng, S.J.; and Xin, W.J. 1989. Scavenging effect of extracts of green tea and natural antioxidants on active oxygen radicals. *Cell Biophysics* 14:175–185.

Ziegler, R.G.; Mason, T.J.; Stemhagen, A.; Hoover, R.; Schoenberg, J.B.; Gridley, G.; Virgo, P.W.¸tman, R.; and Fraumeni, J.F. 1984. Dietary carotene and vitamin A and risk of lung cancer among whites in New Jersey. *J. Natl. Cancer Inst.* 73:1429–1435.

———. 1986. Carotenoid intake, vegetables, and the risk of lung cancer among white men in New Jersey. *Am. J. Epidemiol.* 123:1080–1093.

Part IV

Market and Competition

Chapter 18

Functional Foods in Japan

Tomio Ichikawa

INTRODUCTION

Japan is viewed as the world pioneer leader in the development of functional foods, in part perhaps for reasons grounded in the Japanese culture and also because the government has been proactive toward this project. In Japan functional foods are being very seriously considered as a major new product opportunity for the food industry.

The functional foods project started in 1984 when the Ministry of Education, Science, and Culture sponsored a research project on "A statistical analysis and an outlook on food nutrition." The Ministry completed an initial report by 1986, in which it was concluded that foods, according to the functional foods concept, have three functions in the body. The primary function is the food's nutritive value. The secondary function is the food's sensory appeal, or organoleptic properties. The tertiary function refers to the food's physiological aspects, such as neutralizing harmful substrates, regulation of body functions and physical conditions, preventing diseases, and promoting recovery and general good health.

Functional foods, which offer additional nutritional or health benefits to the daily diet, became popular in Japan. There are many reasons for this phenomenon—cultural, historical, and philosophical. However, most important is the concern of the Japanese government about the aging of the country's population (the Japanese people have the longest life span in the world) and the resultant cost of healthcare. Of the three Ministries involved in functional foods (Ministry of Health and Welfare (MHW), Ministry of Agriculture, Forestry, and Fisheries (MAFF), and Ministry of Education, Science, and Culture (MESC)), the most interested and influential is the Ministry of Health and Welfare. The MHW recognized the beneficial connection between diet and health to an aging population and was the main catalyst in launching a program recently to promote the development of functional foods in Japan. In August 1986, the Ministry established the Functional Food Forum of six experts from food and nutrition departments of several universities. Their role was to propose ways of improving the population's health (PA Consulting Group 1990). This attitude of the government was favorably accepted by the Trade Association of Interested Food and Drink Manufacturers, which was formed in 1987, and by the health food industry. The resulting flurry of new product activity (over 100 products) has given rise to a whole new classification of foods called functional foods.

In 1988 the Japanese Health Food and Nutritional Food Association (JHFNFA) and the Research Board of Technology for Future Foods were established, and the JHFNFA formed its Functional Foods Committee. While the MHW was promoting the development of functional foods, and the food manufacturers were expecting to market functional foods with health claims, there was a disagreement within the MHW regarding this issue. While the General Environmental Health Bureau apparently was pushing for legal recognition and designation of certain products as so-called functional foods, the general Pharmaceutical Affairs Bureau reportedly did not want to allow indications of effectiveness to be displayed on the labels of these products (Anon. 1990). According to article 2, section 1(3) of the Pharmaceutical Affairs Law, drugs (or medicines) are defined as materials for the diagnosis, cure, and prevention of diseases, which affect the structure or function of the human body. On the other hand, the Food Sanitary Law defines food as material that penetrates the human body through the mouth and is not classified as drugs (or medicine). In addition, the situation was further complicated by the involvement of the two other ministries, MAFF and

MESC. However, in April 1989 it was agreed that the manufacture of functional foods should be embodied in the Nutrition Improvement Law and that functional foods should be interpreted as foods with health claims, according to article 12 of this law. In 1989 the government established the Functional Foods Liaison Board to serve as the sole intermediary with industry. Individual subcommittees were formed to deal with each potential food ingredient (Kamewada et al. 1988). Until September 1, 1991 the term "functional foods" was widely used in Japan. Since then the term "foods for specified health use" has been used and has become the official term. Both terms will be used in this chapter for reasons of convenience.

THE JAPANESE APPROVAL PROCESS FOR FOODS FOR SPECIFIED HEALTH USE

As a result of the above-mentioned organizational effort, a complex network of committees and councils was set up to provide a legal framework for monitoring and controlling health claims to be made by functional foods and to enable the manufacture and promotion of foods with specific health claims (Fig. 18-1) (Potter 1990). The Ministry of Health and Welfare receives applications for approval from manufactures and processes these applications rapidly. We will elaborate later in this chapter on the various components comprising the network of committees.

The Definition of Foods for Specified Health Use (Functional Foods)

Foods for specified health use, or functional foods, have been defined by the MHW as "processed foods containing ingredients that aid specific bodily functions in addition to being nutritious." These include foods that aid in enhancement of the body's defense mechanisms, prevention of diseases like hypertension, diabetes, and congenital metabolic disorders, recovery from disease, control of physical conditions, and slowing down of aging (Cole 1991). The Ministry has highlighted three conditions that define food for specified health use: (a) it is a food, not in the form of a capsule or a powder, and is derived from naturally occurring ingredients, (b) it can and should be consumed as part of the daily diet, and (c) it has a particular

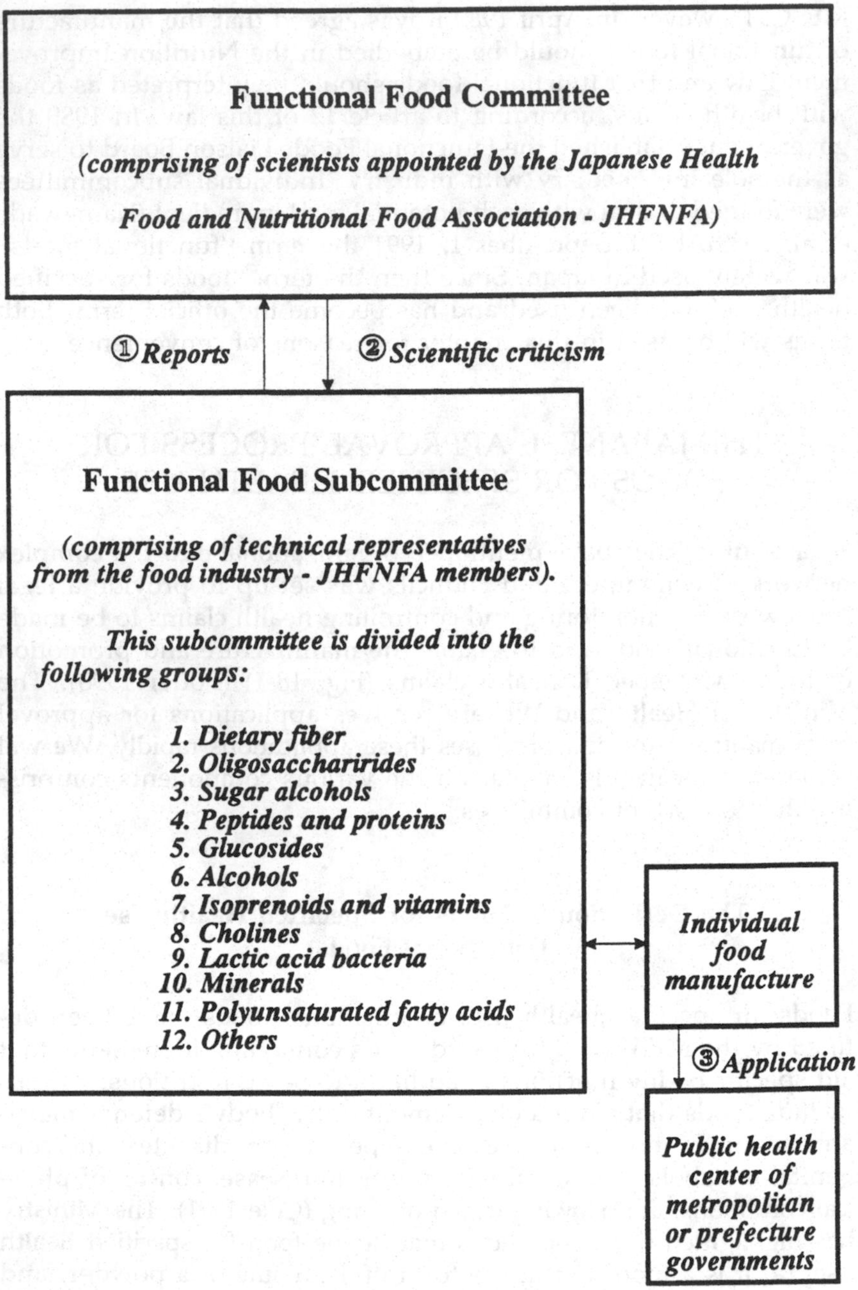

FIGURE 18-1. The Japanese Legislation Process for Foods for Specified Health

health function, when ingested, as specified above. This definition grants such products individual health claims for specific health needs based on approved and sufficient research data.

Because the health benefits that can be offered by the food products are limited only by the ingredients used, the MHW is currently considering 12 different classes of health-enhancing ingredients for use in foods for specified health use (Fig. 18-1). These include dietary fiber, oligosaccharides, sugar alcohols, peptides and proteins, glycosides, alcohols, vitamins, cholines, lactic acid bacteria, minerals, fatty acids, and a miscellaneous collection of unclassified ingredients for which health benefits are claimed (e.g., phytochemicals and antioxidants) Several of these classes include ingredients and products that are already being manufactured or researched within the food industry. The listed ingredients appear in a full menu of products—foods ranging from breakfast cereals to salad dressing.

Legislation Guidelines for Foods for Specified Health Use

The following guidelines were determined by the MHW for legislation of foods for specified health use.

1. The food should improve the diet and health.
2. The health and nutritional benefits of the food or the specific ingredient must have a solid scientific basis.
3. The appropriate daily intake quantity of the food or the ingredient must be established by medical and nutrition experts for approval.
4. The food or the ingredient should be safe with regard to a balanced diet.
5. The ingredient should be characterized by: (a) its physical and chemical properties; detailed analytical methods for its characterization should be given, and (b) its quantitative and qualitative determination in the food.
6. The ingredient should not reduce the nutritive value of the food.
7. The food must be consumed in the normal way.
8. The food should not be in the form of tablets, capsules, or powder.
9. The ingredient should be a natural compound.

Documents Required for the Approval Process

The following documents must be submitted in order to obtain the necessary approval for labeling foods for specified health use:

1. A document containing extensive information and clarifying the enhancing effects of the food or the specific food ingredient on health on the basis of medical and nutritional scientific data.
2. A document concerning the daily intake quantity required to produce such health-enhancing effects, based on medical and nutritional scientific data.
3. Disclosure of reports of investigations concerning the safety for use of the food or ingredient.
4. Data concerning the stability of the food or the ingredient.
5. Documents regarding the physical and chemical identity and composition of the ingredient, together with the analytical methods used.
6. Reports on the qualitative and quantitative analyses of the ingredient in the food, together with detailed protocols for analyses.

The scientific data may include those articles that were published in the scientific literature. Analytical data should be obtained by national, regional, or private institutes of food sanitation that are authorized by the MHW. The documents describing these data should be signed by the person authorized by the institute. In the case where this information is common knowledge to the medical and nutrition communities, and when these documents have been authorized by medical and nutrition scientists, these documents do not have to be submitted for approval.

The Process for the Approval of Foods for Specified Health Use

The process established for the approval of foods for specified health use is described in Fig. 18-1 and 18-2. Before a food is presented to MHW to get the approval of foods for specified health use, it and its special component are usually evaluated by the Functional Food Committee appointed by the Japanese Health Nutrition Food Association (JHFNFA, see Fig. 18-1). When the food is recognized to be able to apply as foods for specified health use by The Functional

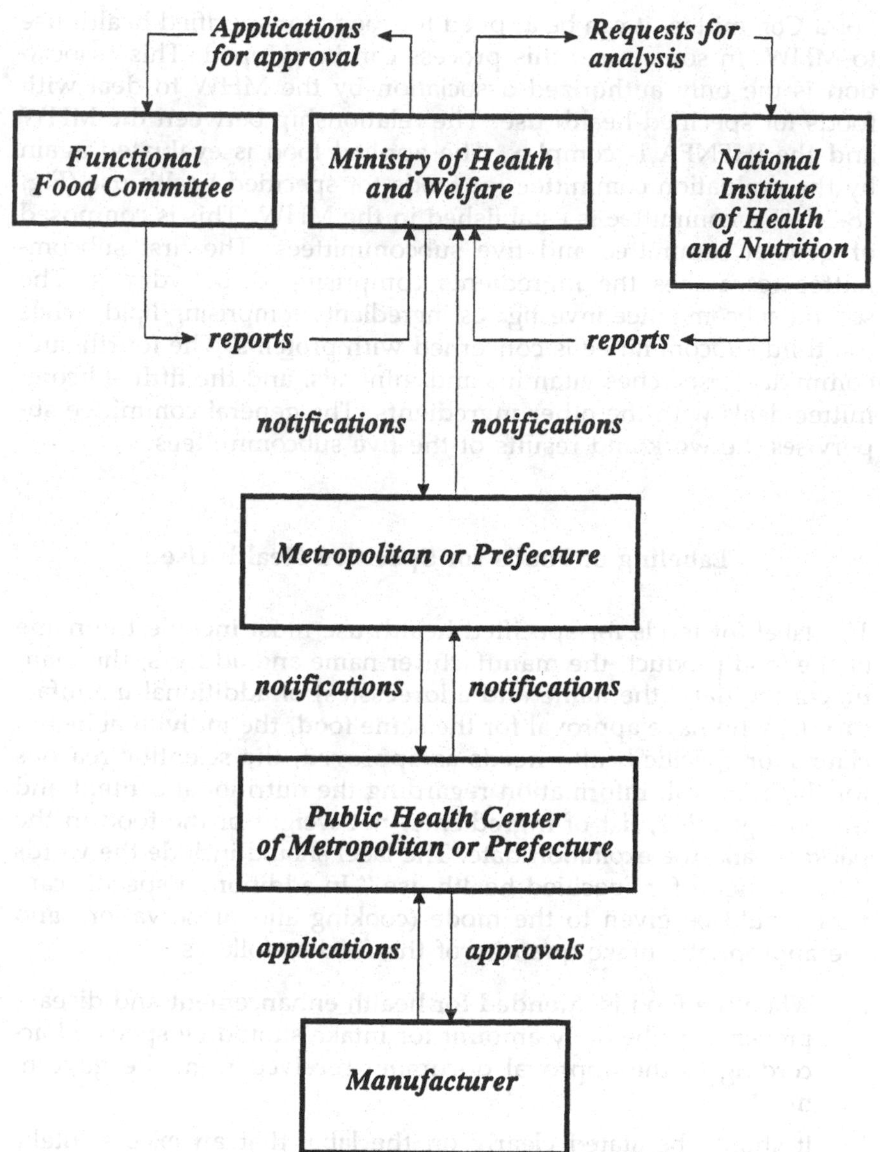

FIGURE 18-2. A System for Approving Foods for Specified Health

Food Committee, it can be applied for foods for specified health use to MHW. In some case, this process can be skipped. This association is the only authorized association by the MHW to deal with foods for specified health use. The relationship between the MHW and the JHFNFA is complex. The applied food is evaluated again by the evaluation committee for foods for specified health use (Fig. 18-2). This committee is established in the MHW. This is composed of general committee and five subcommittees. The first subcommittee researches the ingredients comprising carbohydrates. The second subcommittee investigates ingredients comprising lipid, while the third subcommittee is concerned with proteins. The fourth subcommittee researches vitamins and minerals, and the fifth subcommittee deals with the other ingredients. The general committee supervises the work and results of the five subcommittees.

Labeling of Foods for Specified Health Use

The label for foods for specified health use must include the name of the food product, the manufacturer name and address, the manufacturing date, the name and addreess(es) of additional manufacturer(s) who have approval for the same food, the individual health claims for specific health needs as approved, the scientific reasons for the approval, information regarding the nutritional content and the energy value, list of ingredients, net weight of the food in the package, and the expiration date. The label should include the words "this is a food for specified health use." In addition, a specific caution should be given to the mode (cooking and preservation) and the appropriate intake quantity of the food as follows.

a. When the food is intended for health enhancement and disease prevention, the daily amount for intake should be specified according to the approval document received from the government.
b. It should be stated clearly on the label that an excess intake quantity of the food would not result in a further improvement of health.
c. When health damages caused by consuming an excess of the specific food are known or are possible according to the documents submitted for approval, they should be stated on the label.

d. Other statements should be in accordance with the Nutrition Improvement Law.

MARKETING OF FOODS FOR SPECIFIED HEALTH USE

The Market and Products

So far more than 100 Japanese companies have invested in the development and marketing of functional foods. The scope of food fortification is enormous, and "functional food" products have been launched in the following areas: soft drinks, ready meals, breakfast cereals, biscuits, confectionery, milk products, ice cream, and salad dressing. The current Japanese market for functional foods, even without manufacturers being able to make explicit health claims, is estimated at $3 billion and is expected to reach a level of $7 billion (PA Consulting Group 1990).

We can mention here only a few companies who recently have expanded their interest in developing foods for specified health use.

Nissin Food Products Co., Ltd. and the food wholesaler Toshoku, Ltd. have jointly acquired the milk manufacturing firm Yoke Honsha K.K. in a bid to expand their sales of functional foods and beverages. Yoke Honsha, established in 1969, was the first company in Japan to introduce a yogurt drink (Anon. 1990a).

Takeda Chemical Industries, Ltd. is planning to expand its range of health foods and beverages. The company is now planning to develop more specialized functional foods and beverages aimed at health-conscious consumers via the Takeda Health Care Division, which has been marketing vitamin-based health drinks targeted as soft drinks for the general public (Anon. 1990b).

Hayashibara Biochemical Laboratories Inc. have licensed the manufacturing technology of water soluble rutin to Toyo Sugar Refining Co., Ltd., which will target this product mainly for the food industry as an ingredient for functional foods. Rutin (quercetin-3-rutinoside), a flavonol glycoside and a natural yellow pigment found in the bean family and buckwheat, is used as vitamin P to strengthen damaged capillaries, especially those damaged by hypertension or radiation. It also can protect against damage to tissues from oxidation and ultraviolet radiation. It is used in preventing and treating cerebral thrombosis and lung hemorrhage, but because of its low solubility only tablets are available. Hayashibara Biochemical Laboratories, Inc. have developed a modified form of rutin that readily dissolves in water, and thus has given it a wide variety of uses, including the production of functional foods (Anon. 1990c,d,e).

Nisshin Oil Mills, Ltd., has set up an in-house Functional Food Development Promotion Team in order to follow up on administrative and legislative developments in the functional foods field. The team will choose from among the company's products those types that can be classified as functional foods and will then decide which type of products to develop in the future (Anon. 1989).

Yamanouch Pharmaceutical Co., Ltd. has established a Health Science Research Laboratory at its Tokyo head office, which will conduct research on health foods, functional foods, and health care-related products. A Yamanouchi subsidiary, Yamanouchi Shoji Co., Ltd., markets health food products, including dietary fiber and calcium supplements (Anon. 1989a).

Hitachi Zosen Corp. has decided to strengthen its business activities in the functional foods area. The company currently markets health beverages, including the tea product Toshu Cha containing calcium and other "healthy" ingredients. The company plans to expand its range of functional foods products to include foodstuffs and confections (Anon. 1989b).

These are only a few randomly chosen examples to demonstrate the broad interest of companies in this fast-developing segment of the food industry.

Many functional food products were developed and are marketed (for a list see PA Consulting Group 1990). Here I will discuss only a few random examples to show the diversity of potential functional food products and active ingredients. These products are not foods for specified health use; they are only intended to have health-enhancing effects (see below).

Asahi Chemical Industry Co., Ltd. and its fully owned subsidiary Asahi Food Corp. has developed Fibermix 391, which is a wafer cookie with a high (39%) fiber content, and is being marketed as a constipation preventative. The wafers contain four types of natural fiber, including soybean and corn fiber, processed with wheat flour. This company developed another functional food called Hemace, which is an iron supplement candy containing heme purified from animal blood. The heme reportedly is absorbed by the intestine 30 times more efficiently than iron components obtained in other types of iron supplemental agents (Anon. 1988).

Shikibo Ltd. has entered the market for functional foods with Mermaid Fiber, a foodstuff high in dietary fiber obtained from the seed husks of isabgol. This is a plant product in bulk form used by food manufacturers for incorporation into various foodstuffs, including beverages and ice cream (Anon. 1989c).

Over 70% of the functional foods developed in Japan are in liquid form. Soft drink production in Japan for 1989 was 2.4 billion gallons, up 9.4% from the previous year (Anon. 1991a). Health drinks are becoming more popular as consumers become increasingly health conscious. Originally health drinks were developed as sport drinks. The idea behind sport drinks in Japan is that they replace nutrients lost during sweating. This idea started the functional food craze in Japan, and was promoted by a company selling isotonic drinks for athletes ("anti fatigue tonic"—Oronamin C produced by Otsuka). These beverages comprise a 9.1% market share, and this category of functional drinks is currently the most active category in the Japanese beverage market. In 1988 Otsuka Pharmaceuticals, one of the biggest producers of soft drinks, launched FibeMini as a dietary fiber supplement. The company claims that this, Japan's best selling soft drink, contains all the minerals, vitamins, and fiber the body needs in a day. Another drink Otsuka introduced in 1990 is Pocari Sweat Stevia, which, as its middle name implies, is an isotonic sport drink containing the low calorie sweetener stevia, as well as fructose and sugar, acids, sodium citrate, sodium chloride, vitamin C, potassium chloride, calcium lactate, amino acids, magnesium chloride, and flavors. In some places Sweat machines outsell Coke machines. Coca-Cola Japan Co. Ltd. produces Fibi, a fiber-rich soft drink aimed at general digestive health. It contains high fructose corn syrup, fiber from guar gum, caramel color, flavors, citric acid, and vitamin C.

Other examples of functional drinks are Yakult, a fermented milk and Bifiel, a fermented yogurt produced by Yakult. Yakult (containing lactic acid bacteria as its key active ingredient; its health benefits may result from modifying the microflora in the human gut) is sold in 50-ml bottles at the rate of 1 million a day by "Yakult Aunties" off trolleys wheeled through office blocks. It is targeted at office employees (Anon. 1991). Bifiel brand was launched in 1989 by Yakult and is targeted at young women, for whom the manufacturer claims it prevents aging and maintains skin elasticity. Ingredients include calcium, vitamins, and oligosaccharides that are claimed to aid digestion. The low calorie content of this drink is favored by these consumers.

Vege Take emphasizes vegetable nutrients, as the name implies. It contains fructose/glucose liquid sugar, dextrin, vegetable extract, flavor, acid, calcium lactate, soy peptide, calcium carbonate, gardenia color, potassium chloride, beta(β)-carotene, and vitamins. This product is manufactured by Kirin Brewery Co.

Meiji Seika Co manufactures a cereal with added fructo-oligosaccharide and dietary fiber, targeted at young adults. The cereal comes in three varieties: milk corn flakes, which is slightly salty; multi-balanced flakes 30, which is made from 30 ingredients including fruits, vegetables, grains, honey, and small dried sardines; and nuts flakes and fruit corn, a bittersweet flake, with a base of peanuts, coconut, and fruit chips (Cole 1991).

Based on the various functional foods identified, the segmentation of products by candidate ingredients is estimated as dietary fiber: 40%, calcium: 20%, oligosaccharides: 20%, lactic acid bacteria: 10%, and other: 10% (PA Consulting Group 1990).

Problems in Marketing Food for Specified Health Use

It should be pointed out that until legislation is passed, no foods for specified health use (functional foods) developed and marketed so far can be officially labeled with health claims. The Japanese manufacturers still remember the recent imprisonment of Kisaburo Shiko, President of V-Call, who attributed specific medical benefits to his health drinks. The product contained vitamins B, C, and E plus other nutrients which were said to be effective against high blood pressure and diabetes (Lake 1990).

As was mentioned earlier, in Japan the Pharmaceutical Affairs Law prohibited health claims on foods, and when a material is labeled with health claims it is regarded as a drug or a medicine. However according to the Nutrition Improvement Law, a certain health claim can be labeled under limited conditions; i.e., the food for specified health use should also be a food for special nutritive use as is listed in Fig. 18-3.

The procedure required for approval of health- and nutrition-enhancing foods, according to the Nutrition Improvement Law, is similar to that required for the approval of any food additive or food used for a specific nutritional purpose. However, the scientific evaluation required for the approval of functional foods will be carried out by the Functional Food Committee, nominated by the JHFNFA (see Fig. 18-2). At present, too many documents are required, which in some cases are difficult to obtain. This is in contrast to food additives or foods used for a specific nutritional purpose; for these the documents related to their health effects are not needed. According

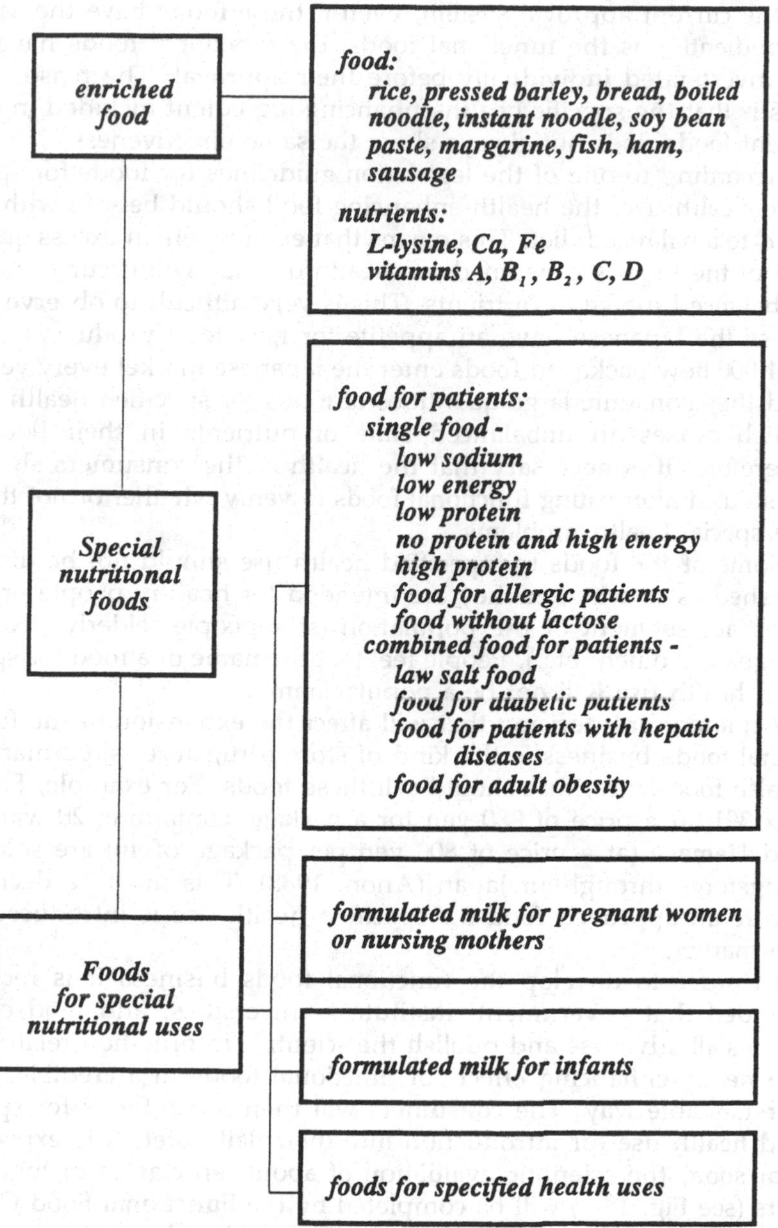

FIGURE 18-3. Special Nutritional Foods

to the current approval system, even if these foods have the same ingredient(s) as the functional foods, the functional foods have to be investigated individually before their approval. The reason for this is that the specific health-enhancing ingredient included in different foods does not always show the same effectiveness.

According to one of the legislation guidelines for foods for specified health use, the health-enhancing food should be safe with regard to a balanced diet. This means that even when an excess quantity of the food is consumed, no health damage will occur from an unbalanced intake of nutrients. This is very difficult to observe because the Japanese have an appetite for new food products (3,500 to 4,000 new packaged foods enter the Japanese market every year), and they consume large quantities of foods for specified health use which causes an unbalanced state of nutrients in their bodies. Therefore, it is necessary that the health of the consumers should be studied after eating functional foods to verify whether or not there are special health problems.

Some of the foods for specified health use should not be distinguished as to whether they are intended for healthy people or for a certain segment of the population (sick people, elderly people, pregnant women, etc.). People feel that the name of a food for specified health use will not be a popular name.

An important decision that will affect the expansion of the functional foods business is the kind of store (drugstore, supermarket, health food store) that should sell these foods. For example, Fibermix 391 (at a price of 880 yen for a package containing 20 wafers) and Hemace (at a price of 800 yen per package of 40) are sold in drugstores throughout Japan (Anon. 1988). This must be decided before an approved food for specified health use is introduced to the market.

In order to develop the functional foods business it is recommended that government, institutes, universities, and food companies all advertise and publish the scientific information related to the health-enhancing effects of functional foods in a credible, understandable way. The consumers will then select foods for specified health use for introduction into their daily diet. It is expected that soon, the scientific evaluation of about ten classes of ingredients (see Fig. 18-1) will be completed by the Functional Food Committee, especially for dietary fiber, oligosaccharides, calcium, and heme-iron. This will pave the way for the introductions of the first foods for specified health use with health claims in the market by the beginning of 1993. Certainly the food industry expects that the demand and the profits for foods for specified health use will be high enough to justify the enormous effort required for their approval.

cf. how people can see about 15 foods for specific health use including the oligosaccharite containing food and so on, at the market in Japan.

References

Anon 1988. Comline Biotechnology and Medical. December 2, p. 4.

Anon 1989. Comline Biotechnology and Medical. September 6, p. 4.

Anon 1989a. Comline Biotechnology and Medical. July 10, p. 6.

Anon 1989b. Comline Biotechnology and Medical. June 27, p. 6.

Anon 1989c. Comline Biotechnology and Medical. July 6, p. 6.

Anon 1990. Comline Biotechnology and Medical. January 24, p. 5.

Anon 1990a. Comline Biotechnology and Medical. July 9, p. 4.

Anon 1990b. Comline Biotechnology and Medical. May 9, p. 3.

Anon 1990c. Comline Biotechnology and Medical. February 27, p. 1.

Anon 1990d. Japan Report Medical Technology. February.

Anon 1990e. New Technology Japan. February, p. 34.

Anon 1991. Food Europe, Spring pp. 16–19.

Anon 1991a. Prepared Foods New Products Annual. pp. 30–31.

Cole, M. 1991. When food meets medicine. *Food Rev.* June/July, pp. 15–17.

Kamewada, M.; Oyama, C.; Inaba, H.; Iwamoto, M.; Ota, M.; Kan, T.; Kobayashi, Y.; Shibata, T.; and Kanno, T. 1988. *Developmental Studies of Functional Foods*, p. 309. Tokyo, Japan: CMC Co., Ltd.

Lake, C. 1990. Up in the air over functional foods. *Food FIPP*. June.

PA Consulting Group. 1990. Functional foods: a new global added value market? London, England.

Potter, D. 1990. Functional foods—a major opportunity for the dairy industry? *Dairy Ind. Int.* 55:32–33.

Chapter 19

The Development of the Functional Food Business in the United States and Europe

Jeffrey C. Gardner

INTRODUCTION

Functional foods are loosely defined as products that offer a physiological benefit beyond the fulfillment of daily nutritional requirements. Because of this loose definition, it is difficult to chart the evolution of these products. Further compounding the problem, the definition of what constitutes a functional food product varies greatly by geographic region and consumer demand. For example, the Japanese, widely credited with developing the functional food industry, often define functional food products based on their utilization or inclusion of natural ingredients (see Chapter 18). These natural ingredients define the functional characteristics of the product as well as, at times, the marketing strategy for the finished product. A similar viewpoint is shared by many European functional food producers. In Europe, functional foods tend to focus on physiological benefits offered by the natural ingredient. Greater emphasis is placed on the utilization of natural ingredients that offer the physiological benefit. On the other hand, in the United States, functional food products tend to be classified by their design to enhance the desired functional characteristics. This design may be based on the functional characteristic of the ingredients, but does not necessarily rely on these ingredients being of natural origin. In the United States, functional food products are often based on designed (or "designer") ingredients. Recent examples of specific designed ingredients include Monsanto's Simplesse and Procter & Gamble's Olestra (Mentzer-Morrison 1990; Jandacek 1991) Although these ingredients may not impart a functional characteristic to the finished product, they do serve an important function in the product's organoleptic qualities, and therefore, are important advances in functional ingredients.

In recent years, there has been a significant increase in consumer demand and acceptance of functional food products worldwide. The nature of this demand varies by geographic region. In Europe, the demand for functional food products has closely tracked increasing consumer interest in homeopathic and herbal/botanical medicines. In the United States, consumer interest in functional foods has followed the upswing in interest in physical fitness and overall physical well-being. There has also been an increasing demand for products that are claimed to be of natural origin, closely following existing trends in Europe and elsewhere.

The functional food product industry has developed differently in Europe and the United States because of differences in:

- The therapeutic or functional property desired for the specific food material
- Regulatory status of the ingredient and/or final product
- Perceived or actual consumer acceptance of that ingredient/product.

In all cases, the most important deciding factor influencing the development of the functional food product is actual consumer perception and acceptance. Regardless of the country of origin, the therapeutic claim, or the acceptance of an ingredient by the governing regulatory body, if the consumer is not interested or does not believe in the product's ability to provide the stated benefit the product will not succeed.

EUROPE

In Europe, the average daily diet has been traditionally low in saturated fat, low in sodium, and high in dietary fiber. Europeans place great emphasis on the consumption of vegetables, grains, and other products that provide a relatively healthy dietary intake. Consumer preferences for products of natural origin and, at times, a suspicion of processed food products resulted in the development of functional food products to meet this demand. In recent decades, processed and/or shelf-stable foods have become an increasing part of European diets. Within the last few years, this interest has been offset by increased pressures from consumers for processed foods that offer both convenience and improved nutrition. Concern has been expressed over the "Americanization" of the traditional European diet through the increased popularity of "fast foods" and convenience products. Many nutritional experts are calling for a return to the more traditional diet (more vegetables and grains and less processed food products). This may provide an impetus to the further development of the functional food industry in Europe.

The high level of consumer acceptance and resultant demand for homeopathic and/or herbal/botanical medicines is closely related. In European countries, these are mainstay consumer self-medication products. The level of acceptance of homeopathic or herbal medicines is enough to influence other areas of consumer products, including health and beauty products and foodstuffs. The long history of availability of these medicinal products has resulted in enormous individual awareness of the therapeutic and physiological benefits of herbs, plants, and their respective derivatives and extracts. Thus, when a traditional medicinal ingredient, such as garlic, is used in a functional food product, the average consumer is much more likely to understand and accept any therapeutic claim made, explicitly or implicitly, for the product. The homeopathic and herbal medicine industries have helped to shape the functional food industry by fostering consumer knowledge and acceptance of natural products, and they have influenced the form many functional food products take.

Recently there has been increased interest in orthomolecular medicine in Europe. This utilization of plant and plant products that have been specifically bred to offer increased levels of vitamins, amino acids, and other nutritional materials has direct relevance to the functional food industry. The plants are being produced through specialized breeding programs and advanced cultivation methods and procedures. Orthomolecular products offer a means to increase the therapeutic benefits of many traditional herbal and botanical extracts and derivatives while maintaining their high level of acceptance as natural products.

The main focus of functional food products in Europe at this time is the use of garlic, omega (ω)-fatty acids, oligosaccharides, and beta (β)-carotene. Garlic is often viewed as a mainstay homeopathic medicinal ingredient, but in recent times its importance as a food ingredient or dietary supplement has greatly increased. In the case of the ω-fatty acids and β-carotene, their importance in European diets has been long established, but increased attention within the scientific community on the therapeutic benefits of these compounds has spurred greater consumer interest and demand. Oligosaccharides, on the other hand, are not as widely known as functional food ingredients, but represent an area of strong growth potential within the industry. The interest in other types of functional and designed ingredients, such as those seen in the United States, has not developed in Europe. For example, fat replacements or substitutes, quite popular in the United States as functional food ingredients, are far less important in Europe. This is directly related to consumer preferences and acceptance of such products.

For the most part, the European community favors the development of food products that are based on natural ingredients proven to be safe. While the EC (European Community) and independent nations have not overtly reacted to the establishment and development of the functional food business, limitations are placed on the therapeutic claims that can be made. Functional food products and ingredients cannot make explicit therapeutic claims, and thus must be proven safe for the average consumer. Despite widespread consumer distrust of biotechnologically derived products in Europe, the regulatory bodies have not moved to prohibit the use of such ingredients within functional food products. In some cases, however, the regulatory agencies have acted in such a way as to discourage the development of biotechnological products. This, in turn, fuels consumer skepticism and concern over the apparent safety of such products.

Most of the regulations governing food ingredients have been in place for decades and have received only minor revisions in subsequent years. There is no clear indication that the EC will move to follow the United States in the development of new legislation focused on increasing consumer awareness and industry standardization. Rather, the EC is likely to continue to act to preserve consumer safety while not limiting pan-European trade.

UNITED STATES

The functional food product industry in the United States evolved differently than in Europe. Major differences relate to:

- Diets and lifestyles
- Consumer marketing forces
- The lack of an established and mainstream homeopathic medicine market.

Much attention has been paid to the degeneration of the American diet. The widespread availability of and demand for fast foods, convenience products, and refined or processed foods has had a deleterious effect on the health of the general public. These products are generally relatively high in saturated fats, high in sodium, and low in dietary fiber. In addition, these foods often displace vegetables, grains, and other healthier foodstuffs in an average American diet.

In the late 1970s, the trend toward improved physical fitness and overall well-being began. At that time scientific evidence, as reported in the popular media, stressing the importance of a balanced diet low in saturated fat, sodium, and cholesterol began to be accepted. By the mid-1980s, the food industry began to respond to these trends by introducing products that were (claimed to be) lower in fat, sodium, and cholesterol (see Chapter 1). While many of these products could not be classified as functional in nature, they did represent a milestone in the introduction of functional food products to the general public.

As mentioned earlier, the loose definition used to identify functional food products makes it difficult to follow product evolution, particularly in the United States. One of the first generations of products designed specifically to offer consumers a therapeutic effect was the class of products called isotonic sports beverages, such as Gatorade. These products were marketed as offering a therapeutic and physiological benefit to the physically active consumer, and were initially introduced to the professional athletic market. With the physical fitness trend, these products achieved widespread consumer acceptance. Of course, the therapeutic/physiological claims of these products are very limited, but they did represent the first functional food products that made claims and achieved a very high level of consumer acceptance.

An additional market factor that has affected the development of functional foods in the United States is the health food market. In the United States, the botanical and herbal products common in many European countries reached a limited, but educated, market via the health food industry. The health food market reached its peak in the late 1960s and early 1970s and served as a "healthier" alternative to the developing processed and packaged food industries. In the 1970s, the health food industry lost a significant portion of its market as consumer acceptance and desire for the convenience and abundance of processed and packaged foods grew.

This trend began to reverse in the early 1980s, as many Americans became interested in improving their diet. The early 1990s have seen an even greater increase in consumer interest in products offering real or perceived health benefits. The health food industry moved to meet this increased demand for healthier products and to provide information on the benefits of traditional botanical and herbal ingredients and products. In many cases, the products being offered through the health food stores were imported from Europe or produced as copies of existing European brands. This, in turn, led to an increased consumer awareness and interest in homeopathic and traditional medicines; a market which until recently has had extremely limited appeal to the general public.

These factors have combined to lead to a high level of consumer acceptance of products that offer improved nutrition. Unlike its European counterpart, the mainstream U.S. food industry is strictly limited in its abilities to make therapeutic claims for its products. Many advocates of functional foods had hoped that the Nutritional Labeling and Education Act (NLEA) of 1990 would provide a means to gain acceptance of these products and their claims in the United States. In fact, the Food and Drug Administration (FDA) has moved to examine the possible connections between nutritional products and a diseased state, including:

- Calcium and osteoporosis
- Sodium and hypertension
- Lipids and cardiovascular disease
- Lipids and cancer
- Dietary fiber and cardiovascular disease
- Zinc and immune function in the elderly
- Antioxidant vitamins and cancer.

Provisions have already been made to allow food processors to make limited claims regarding the role of dietary fiber and reduced sodium in overall dietary intake, but these guidelines still do not allow individual product claims to be made.

With the publication of the final NLEA guidelines in the January 6, 1993 edition of the *Federal Register* (Volume 58, Number 3), it became clear that the FDA is not yet prepared to accept therapeutic links between diet and disease. In fact, the FDA has taken steps to reclassify nutritional supplements as new drugs until their efficacy and safety can be scientifically proven. This position would seem to indicate that the functional food industry in the United States will be severely limited in its development.

In reality the FDA has taken measures to promote scientific investigation into the therapeutic potential of many botanical and herbal extracts and derivatives. The primary program is being conducted through the National Cancer Institute (NCI) and is focused on identifying, isolating, and testing the active therapeutic components of many common vegetables, fruits, and grains. During 1991, the first full year of the research program, the NCI evaluated six different food products in a series of independent studies. The studies focused on licorice, garlic, flax, soybeans, umbelliferous vegetables, and citrus fruits. The initial results are still under review, but there is indication that the NCI studies may partially validate some of the therapeutic claims for these products (Caragay 1992).

In early 1993 the National Institutes of Health announced the formation of a Bionutrition Initiative focused on the connection between nutritional status and human health and disease. According to the NIH, the Bionutrition Initiative will allow researchers an opportunity to focus on the relationship of nutrients and living organisms. The research will unite traditional nutritional studies with the latest advances in molecular, cellular, and genetic technologies. The formation of the Bionutrition Initiative further enhances the commitment of the United States government to developing an understanding of the causal relationship between nutrition and disease. In the long term, this may be a watershed movement in the development of functional food products in the United States.

The initial foray of the U.S. food industry into functional food products was through inclusion of enhanced levels of dietary fiber. The first products introduced were based on oat bran and were linked to reports that consumption of dietary fiber could potentially reduce the incidence of certain types of cancer. These products met with enormous market success and proliferated widely. The next wave of functional food products were based on other types of bran, particularly rice bran. They met with similar success as the oat bran-based products.

The bran products were followed by products based on ω-fatty acids. The popularity of the fish oil-based ω-fatty acids was based on studies that linked the compounds to reduced incidence of heart disease in Alaskan Eskimo tribes. Shortly after the studies received national attention, products began to appear that were claimed to contain the fatty acids, and references, in subtextual form, were made to the study conclusions. Currently, there is a great deal of interest within the food processing and food ingredient industries in the potential utilization of ω-fatty acids. Several major ingredient producers are examining ω-fatty acids as a means to enter or expand their presence in the functional food market.

Besides the ω-fatty acids and bran-based ingredients, antioxidant vitamins, such as β-carotene, represent the greatest growth potential in the U.S. functional food industry. Increasing consumer awareness of and media attention to the physiological effects of β-carotene are driving a renaissance in interest in vitamins and other free radical scavenging compounds.

The rapid market development of the bran-based products and the ω-fatty acids in the United States raises an interesting issue. What should be the role of the popular media in the advancement of therapeutic benefits of these and other products? In the United States, the popular media has a substantial influence on the buying decisions of the consumer, and it serves as one of the leading sources of information for the average citizen. When scientific studies highlight the therapeutic benefits of bran, dietary fiber, and the ω-fatty acids, the popular media is quick to publicize these results. A similar effect is now being seen with β-carotene and the antioxidants (Goldberg 1992).

The vast amount of media attention on the therapeutic properties of these ingredients was one of the motivating factors behind the passage of the NLEA and the establishment of the current FDA position. In some cases, both the FDA and independent state governments moved against companies and their products when questionable or drug claims were being made for the product or when unapproved ingredients were used in a functional food product. A case in point was the action taken by the State of Texas and the FDA against Kellogg's HeartWise breakfast cereal (Goldberg 1992; Lake 1990). The problem arose when Kellogg incorporated psyllium as a source of dietary fiber in the cereal formulation. The FDA has not granted GRAS (Generally Recognized As Safe) status to psyllium for use as a food ingredient because of the potential for allergic reaction and the laxative effect. Therefore, Kellogg was using an ingredient that was not approved for human consumption in the United States. The state attorney general of Texas seized warehoused supplies of HeartWise, and Kellogg subsequently removed the product from the market. The major concern of the regulatory officials, beyond the health claims made, was that by not informing the consumer of the laxative effects and potential for allergic reaction, Kellogg was misleading the consumer. Similar actions have been taken against other products that incorporate unapproved ingredients or make unsubstantiated claims.

The future of the functional food industry in the United States will lie in the development of narrowly focused products, specifically products for specific market segments or disorders. Particular emphasis will be placed on the development of products for an aging population. Other areas of functional food product development will focus on:

- Infant formula that more closely mimics breast milk
- Prescriptive food products for immunocompromised patients
- Food products that offer enhanced physiological benefits in inhibiting tumor formation.

All of these product areas reflect the growing interest in the United States in a life-long commitment to physical health through exercise and nutrition. Products developed specifically to capitalize on this interest can be expected to continue to increase in prevalence and popularity.

ROLE OF BIOTECHNOLOGY

The attitudes towards and the influence of biotechnology are also important factors in the development of the functional food business in Europe and the United States. Within the last decade, the importance of biotechnological developments in products, processes, and ingredients has affected every aspect of the traditional food and, in turn, the functional food industry (for review see Goldberg and Williams 1991). Developments and improvements in genetic engineering, fermentation, enzyme design and modeling, and other biotechnological methodologies have greatly increased the ability of the functional food industry to develop new generations of products for ever-changing consumer demands.

The importance of biotechnology in the development of products and processes is offset by public and regulatory opinion on the safety and application of this technology in the food industry. Traditionally, biotechnology and, in particular, genetic engineering have been viewed with a great deal of skepticism and concern over long-term health effects, environmental safety, and the possibility of uncontrollable events within the product or process. A leading causal factor in the current negative environment were the 27 deaths and over 1,500 cases of eosinophilia-myalgia syndrome which were linked to the use of Showa Denko's L-tryptophan nutritional supplement. This tragic situation and the massive amount of media coverage highlighted the dangers of biotechnology to many consumers and public advocates. Even though Showa Denko has since shown that the toxic dimer within the L-tryptophan molecule was produced through a change in ion exchange resin, initial reports to the public linked the problem to a biotechnological process. The concerns and outcries of consumers and of advocate organizations, such as Jeremy Rifkin's Pure Food Campaign, have now carried over to other biotechnological developments including bovine somatotropin (bsT) and Calgene Fresh's Flavr Savr tomato. Rifkin and other consumer advocacy groups are now calling on the general public to boycott any company that utilizes biotechnologically derived ingredients in its products.

To date, the attention paid to biotechnology in food applications has focused on the end-products and not the process itself. Compared to the controversies surrounding L-tryptophan, bsT, and the Flavr Savr tomato, the amount of attention paid to biotechnologically derived or enhanced process technologies has been minimal. The importance of these processes in the food industry should not be overlooked and can be expected to continue to increase as further technological advances are made. There are currently several food ingredients on the market that are produced through biotechnological procedures, including Kelcogel produced by the Kelco Division of Merck & Co., Inc. Further advances in process development will increase the number of ingredients being developed through biotechnology and their importance to the functional food industry.

A major stumbling block for the biotechnology industry and, therefore, for the functional food industry has been government or regulatory opinion or acceptance of biotechnology. For example, in recent years a law was introduced in the British Parliament that would have required all food products containing ingredients derived through genetic engineering to be clearly identified as such in their packaging. The ramifications of this law were obvious; foods that contained such ingredients would be clearly differentiated and isolated from "natural" products and therefore would have a much more difficult time in gaining consumer approval and acceptance. Fortunately, in the view of the biotechnology industry, this law was not passed. Another example lies in the German Gene Law which, while promoting basic and applied research in biotechnology and genetic engineering, places severe limitations on the types of genetic transfers used. Further limitations are placed on testing of genetically engineered plants and organisms. The impact of the German Gene Law on the biotechnology industry has not been as great or as widespread as was believed at the time of passage. Even now, there continues to be a strong level of support for limiting the application of biotechnology in European food products.

The United States, on the other hand, has recently taken steps to promote the application of biotechnology in all aspects of the food industry. In late 1992 the U.S. government and the FDA announced that biotechnology products would not be singled out from other products in the approval process as long as the changes made did not differ significantly from "natural" products. This decision was met with enthusiasm by the biotechnology industry and extreme outrage by public advocates. In recent months, the FDA has stepped back from the mandate set by the Bush administration to a more conservative stance of potentially requiring pre-market approval for all biotechnology products. It is possible that food products developed through biotechnology could be required to undergo an approval process similar to that of ethical products. Even with this apparent reversal of position, FDA leaders still publicly support the development and application of biotechnology in food processes, ingredients, and products.

One point of concern that has arisen in the United States lies in the proposed labeling of products containing dairy products produced by cows treated with bovine somatotropin. The proposed guidelines would require that all such products carry a labeling statement indicating that the dairy product was derived from bsT-treated cows. The biotechnology industry has moved to block the proposed guidelines on the basis that they would unfairly segregate biotechnology products from "natural" products and would led to unfounded consumer concerns over the safety of biotechnologically derived products. Public advocates support the proposed guidelines based on the reasoning that it will allow consumers to make personal choices in regard to purchasing biotechnologically derived products.

These guidelines clearly demonstrate that a pervasive negative attitude regarding the safety of biotechnology in the food industry exists in the United States as well as in Europe. The final impact of these attitudes on the functional food industry remains to be seen. In the short term, companies may continue to focus on the development of functional food products on natural ingredients such as antioxidant vitamins and ω-fatty acids. In the long term, as biotechnology becomes more of an accepted concept in the minds of consumers and regulatory officials, the future of biotechnology-based functional foods may be very bright.

SUMMARY

In summary, the major point of difference in development of the functional food industry in Europe versus the United States lies in the expectations, demands, and perceptions of the consumer. As the global marketplace continues to blur the lines between cultural differences and practices, a unique opportunity for the functional food product will unfold. We are already in an age where products are crossing international boundaries, experiencing widespread success and acceptance. As regulatory agencies continue to develop new understanding and acceptance of the role functional food products can play in offering physiological and even therapeutic benefits, the era of the global functional food product will be upon us.

References

Caragay, A.B. 1992. Cancer preventive foods and ingredients. *Food Technology*. April, pp. 65–68.

Goldberg, I., and Williams, R., eds. 1991. *Biotechnology and Food Ingredients*. pp. 1–577, New York: Van Nostrand Reinhold.

Goldberg, J.P. 1992. Nutrition and health communication: The message and the media over half a century. *Nutrit. Rev.* 50:71–77.

Jandacek, R.J. 1991. Developing a fat substitute. *Chemtech*. July, 398–402.

Lake, C. 1990. Up in the air over functional foods. *Food Fipp*. June.

Mentzer-Morrison, S. 1990. The market for fat substitutes. *National Food Rev*. April–June, pp. 24–30.

Chapter 20

The Potential Role of Functional Foods in Medicine and Public Health

Kathie L. Wrick

INTRODUCTION

The metabolic role that some nutrients and other nonnutrient constituents of foods may have in potentially preventing the premature onset of chronic diseases, or in preventing a disease or disorder from ever occurring at all, has captured the attention and imaginations of nutrition scientists, medical practitioners, the food and pharmaceutical industries, and the general public. The linkage of diet to many of the leading causes of death in the United States, including heart disease, cancers, stroke, and diabetes has received national attention at the highest levels of government and in prestigious scientific societies (U.S. Department of Health and Human Services 1988; National Research Council 1989). The definition of functional foods has evolved into one of foods for disease prevention. Food may be more important in medicine and public health than has previously been believed because of its chemical constituents that play a role in disease prevention or treatment.

Scientific evidence that elucidates the cellular and metabolic interactions between food constituents and biomarkers for disease has become increasingly compelling over the last decade. Our understanding of the role that nutrients, other less-studied constituents of foods, and even some drug compounds play at the cellular and subcellular level continues to grow. These substances may share similarities in biological behavior, because they either allow selected biochemical and cellular processes to occur or they prevent others from happening at all. However, the differences between food nutrients, other bioactive compounds present in foods, and drugs are obviously important. Nutrients such as carbohydrates and fats fuel all body cells. Proteins provide the amino acid building blocks for enzyme formation and tissue growth and repair. Without enough vitamins and minerals, critical metabolic chemistry would be impaired or stopped, as would the cellular and growth processes in which these chemical reactions play a part. Nutrients must be continually replenished in our diets. Drug compounds are also bioactive in that they also participate in metabolism, but they do so in different ways, by different mechanisms. They are deliberately introduced into the system to alter a metabolic reaction or other biological process that has gone awry. Once a drug has done its job, the body detoxifies and disposes of it. It need only be replenished until the condition it was intended to alleviate is corrected. Often, continued intake of a drug is typically of no value and in some circumstances may be harmful. Many nonnutrient constituents of foodstuffs, especially from various plant foods such as fruits, vegetables, grains, and some herbs and spices, are of interest because they also show bioactivity and participate in some aspects of human metabolism. Their presence in foods that have been consumed for centuries suggests that they are safe for consumption and do not have side effects, even though the bioactivity demonstrated is often more like that of drug compounds than nutrients. The coming decades will elaborate the metabolic functions of these naturally occurring compounds, the amounts required to be truly effective in delaying or preventing an undesirable biological effect such as disease onset or aging, the population segments in which these food compounds are really effective, and the presence of any undesirable side effects when consumed at levels required for efficacy.

THE INTEREST IN THE FUNCTIONAL FOODS CONCEPT

The scientific findings on functional foods that have captured the attention of so many have emerged at a time that also shows unprecedented public interest in matters of health and well-being, and a public fascination with diet in North America, Europe, and Japan. In addition, these findings come at a time when health care costs have escalated so dramatically that efforts to control expenditures have changed the face of medical care in the United States and some other countries to one of shorter hospital stays, more outpatient treatment, and a more financially accountable medical system. Attention has been drawn to the high cost of treating chronic diseases such as heart disease, hypertension, and cancer, which might reasonably be reduced in severity, delayed in onset, or even prevented in many people by changing one's behavior earlier in life. For example, nearly all public health dietary recommendations for disease prevention include the avoidance of tobacco products; regular exercise; maintenance of body weight within a desirable range for height, frame size, and age; and consumption of a diet lower in fat, especially saturated fat, and calories, and higher in fiber from fruits, vegetables, and cereal grains than has been traditionally consumed in Western society.

In the United States, the downward pressure on medical costs and a growing public disenchantment with the health care system has fueled the interest for some segments of the public to make their own decisions about some aspects of their medical care. A few are turning to alternative, nontraditional forms of medical therapy when they feel that their medical (and sometimes emotional) needs are inadequately met within the confines of the current medical system. For some, consumer products and services that address current health problems and disease prevention are becoming an accepted alternative to time-consuming visits to a health professional.

These social, environmental, public health, and economic forces are jointly operating as independent drivers to make the concept of functional foods very attractive. How nice it would be if the supermarket shelves would be lined with products that not only met consumer demands for good taste, but also helped ameliorate or even prevent chronic disease processes with regular use! But this is not so simple. There are scientific and public health issues that must be addressed, and there are also regulatory barriers to overcome.

SCIENTIFIC AND PUBLIC HEALTH ISSUES

Among the scientific and public health issues that surround the concept of functional foods is their effectiveness at truly improving the public health. Unlike scurvy, beriberi, or pellagra, the chronic diseases of concern today are not deficiency diseases that will be eliminated by adding specific nutrients or other chemical constituents to the diet. It is well known that there is a strong genetic component in assessing one's degree of risk for disease onset. However, the state of our scientific knowledge does not yet permit our medical system to accurately and inexpensively predict with much certainty the amount of time following dietary (and/or other lifestyle) change by which disease onset will be delayed in a given individual who is at some level of risk. We do not know how to predict whether an individual's genetic predisposition to a certain disease is so strong that it cannot be overcome by any dietary or lifestyle change, or whether his risk is so low that he will live his life free of a given disease regardless of his lifestyle.

Nonetheless, the hypotheses created by epidemiological findings, though expensive and difficult to test in clinical trials, suggest that dietary and other lifestyle changes should reduce the occurrence of chronic diseases in our population, and the public appears ready to accept a cause-and-effect linkage. It remains to be seen whether or not such a linkage will motivate long-term dietary change in enough of the high risk population to show an improvement in public health statistics over time. Marketplace dynamics will not ensure that those consumers who are most interested in changing their dietary pattern for health improvement are also the ones at high disease risk. The potential impact of medically sound functional foods in reducing chronic disease incidence will be greatest if they are used thoughtfully by those at high risk.

A SPECTRUM OF FUNCTIONAL FOODS THAT MEET MEDICAL AND PUBLIC HEALTH NEEDS

From a medical and public health standpoint, products providing a medically measurable health benefit should be developed for those segments of the population with a real medical need. This implies products that are highly focused for targeted, well-defined market

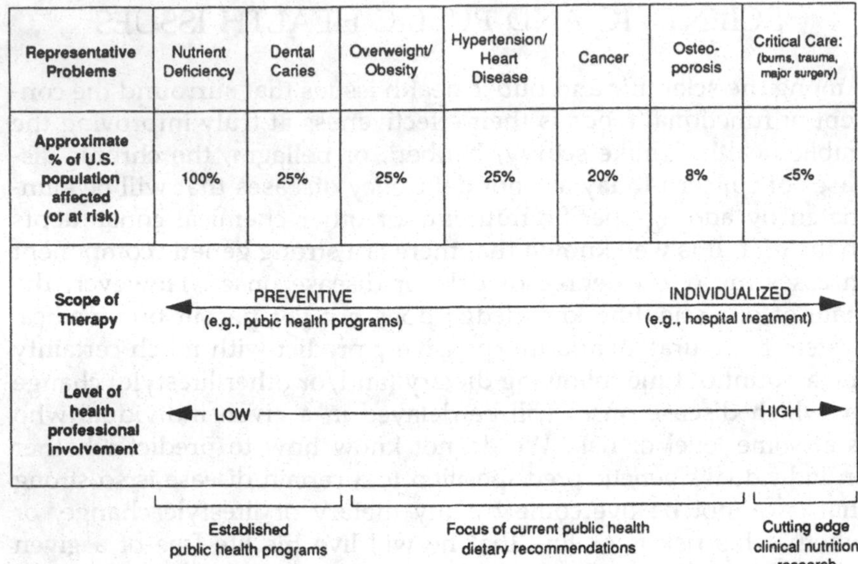

FIGURE 20-1. Spectrum of Medical Problems for Which Functional Foods Can Be Developed

segments, consisting of people who will see or feel a measurable benefit by themselves or with the help of a health professional, rather than for more broad-based markets defined by more traditional demographic consumer segmentations.

Fig. 20-1 shows a conceptual framework of the spectrum of medical and public health problems for which functional food products have been (or might be) developed and marketed. These disorders range from nutrient deficiency in the population at large to severe, life-threatening disease conditions such as burns, trauma, or other conditions of severe metabolic stress where nutrition support is needed. The current approach to addressing the prevention of broad scale nutrient deficiency has been the time-tested fortification programs of adding B vitamins and iron to cereal and flour products, or adding iodine to salt. These fortification programs have been in place for many years and have helped eliminate the public health problem of the deficiency of these nutrients in the United States. The entire population is exposed to these nutrients in the food supply, and it is likely that the population would be at risk for deficiency if these programs were discontinued. Another program that has demonstrated public health success is that of fluoridated water supplies in the reduction of the incidence and severity of dental caries.

At the other end of the spectrum, recent research in clinical nutrition and medicine has led to the development of enteral and parenteral products specifically tailored to help manage a specific medical disorder in the hospitalized patient. Manufacturers of these products apply the most recent research findings in their formulation and development so that the products help speed patient recovery, reduce the incidence of medical complications, and/or shorten the hospital length of stay. Among the means that can be used to foster these improved patient outcomes are the addition of tissue-specific nutrients such as glutamine, an amino acid that appears to be essential for the functional integrity of the intestinal tissue and is required in greater amounts during periods of metabolic stress from surgery or trauma. Another example is the use of the omega (ω)-3 fatty acids and selected nucleotides such as ribonucleic acid, or the pyrimidine base uracil, that appear to prevent or reverse the decline in immune function in patients that can occur during nutritional support of patients under metabolic stress. The use of high concentrations of branched-chain amino acids in formulations designed for patients with certain types of liver disease and for some critical care patients is another example. Clinical data demonstrating the effectiveness of formulations containing these "functional ingredients" in improving and shortening the recovery process is critical to acceptance of the approach to treatment, and therefore the product itself, by the health professionals who purchase, prescribe, or recommend and administer these products. The pharmaceutical companies that manufacture enteral or parenteral products routinely provide clinical data that supports any claims made for these products' health benefits.

THE CHALLENGE OF MARKETING FUNCTIONAL FOODS FOR PREVENTION OF CHRONIC DISEASE

In between these two extremes are the chronic diseases, such as obesity, osteoporosis, heart disease, hypertension, and cancer that affect large portions of the U.S. population, but for which not everyone is at risk. Some of these disorders are costly to the population as a whole, not only in terms of actual health care dollars spent, but also in terms of the burden on patients, family members, or other caretakers in terms of emotional stress and time, which is not as easily quantified. In each case, government agencies and/or prestigious scientific societies have formulated dietary recommendations for disease treatment, management, and/or delay in premature onset or prevention. The state of scientific knowledge does not yet permit our medical system to accurately and inexpensively screen and quantify everyone for their genetic risk of developing some of these diseases. As a result these same agencies and scientific societies have offered policy recommendations about diet to the public at large, and the media has helped spread these health messages to a receptive public that is interested in health, self-care, and fitness in general. These consumers have shown interest in products that claim to help delay or prevent diseases such as heart disease, cancer, and osteoporosis, regardless of whether they are at high risk for them or not. At Arthur D. Little, we have called this segment of consumers the "Worried Well." They are the health-concerned consumers who are technically not at risk or are at low risk for a particular disorder, but nonetheless are very vulnerable to public health and marketing messages that appeal to their interest in health or to their fears of contracting a certain disease later in life. The Worried Well can be both a blessing and a curse to food and pharmaceutical product marketers, and herein lies part of the challenge that faces these companies: Can the smaller market segment who is truly at risk and most likely to benefit from a well-designed functional food be as attractive and as profitable as the much larger segment of Worried Well consumers who will only benefit marginally, if at all? Those truly at risk will sustain the business of a product that meets their health needs in a measurable way. The Worried Well often drift from one health trend to another in their product purchase decisions.

MEDICAL AND PUBLIC HEALTH CONSIDERATIONS IN MARKETING FUNCTIONAL FOOD PRODUCTS

For Whom Should Functional Foods Be Developed?

From the standpoint of sound medical and public health policy, as well as good business policy, it makes sense to develop functional foods that address the real medical or nutritional needs of a defined population segment or patient group in a measurable way. As a company investigates the target audience selected, information will emerge that reveals the overall demographic characteristics of the group, and the most likely product forms that will be best received by this group (e.g., a product in the form of a pill or capsule, an enteral beverage to be administered by feeding tube, a breakfast cereal or a frozen dinner, etc.).

Does the Product Provide a Measurable Health Benefit or the Hope of Future Good Health?

Another major consideration for the technical development and marketing of functional foods is the benefit the product will deliver. One of the biggest challenges facing the long-term viability of functional foods in the marketplace is the delivery of a medically measurable benefit. Product categories that address only perceived nutritional and medical needs are likely to have a shorter life cycle or experience wide swings in sales volume relative to those products for which consumers and their health professionals can measure a difference. For example, consumers will abandon a pain reliever that does not make their headache go away for a brand that does. Use of products that claim to lower serum cholesterol or otherwise modify blood lipids to a more favorable profile can be evaluated for their effectiveness by taking a blood measurement. On the other hand, calcium supplements proliferated in the late 1980s in the face of extensive media coverage regarding calcium's potential role in preventing osteoporosis. However, the only benefit that could be offered at the time was the hope for disease prevention later in life. The intangibility of this promise, together with press coverage about products with questionable digestibility as measured by standard pharmaceutical tests, contributed to the downswing in the market for this product category.

Are the Needs of Those Who Influence Product Purchases Understood and Addressed?

It is also important to understand the role of those who influence product purchases, whether those influencers are family and friends, broadcast and printed media, or health professionals such as physicians, dietitians, and pharmacists.

Traditionally products purchased for a health benefit have been primarily pharmaceutical products. They were purchased because a health professional endorsed, suggested, recommended, or prescribed a product or product category. Now that food products and ingredients are under closer scrutiny for their role in disease prevention or management, consumers are making selections based on their own knowledge, which may or may not include sound medical advice from a health professional. The extent to which disease prevention will be an acceptable reason for use of a product for consumers will hinge strongly on the support for the product concept offered by the health professional community. The task, then, for a food or pharmaceutical company that seeks to produce functional food products is twofold. First, the company must correctly identify the target audience and develop a product that satisfies end user needs for taste, convenience, and appearance as well as the promised health benefit. Second, the requirements for efficacy or other measure of medical suitability expected by the health professional community must be addressed.

Careful identification of the target population that will benefit from the functional food product as well as an understanding of the needs and concerns of the health professionals who counsel or treat them can be critical to the marketing success of a functional food and to the ultimate availability of the product to those who will benefit the most from using it. When marketing and advertising programs target large segments of the population who in turn embrace a product that offers them no real medical benefit, the patience of the health professional community can be severely tested to the point where a product loses their support.

LESSONS LEARNED FROM OAT BRAN

A recent example of a consumer product that became embroiled in these issues is oat bran (Wrick 1991). Oat bran hot cereal had been marketed through health food channels with a message indicating that it might help lower serum cholesterol if consumed in the context of a low-fat diet. Sales levels were unremarkable until 1988 when an article in the April 1988 Journal of the American Medical Association (Kinosian and Eisenberg 1988) concluded that oat bran could be a cost-effective alternative for some individuals with high serum cholesterol levels relative to several cholesterol-lowering drugs. The article's conclusion was covered by the broadcast news media and all the major newspapers, and product sales soared. Oat milling companies were caught completely off guard. Within a year the demand for oat bran cereal and for oat bran as an ingredient in breads, bakery products, and other foods outstripped the capacity of the entire U.S. oat milling industry, and product had to be imported from Europe. During this time, oat bran was advertised to consumers concerned about their cholesterol levels as a means to help lower it. As publicity about product success grew, and as other food products utilized oat bran and tried to capitalize on the implied health claims for it, skepticism in the medical community also grew. Sound evidence was available, however, that oat bran could be effective in lowering cholesterol, provided that an individual with a high cholesterol level consumed it daily, in sufficient amounts, as part of a low-fat diet. The behavior of oat bran, as with other soluble fibers, is known to be effective in lowering a high cholesterol level by as much as 10–20%. Very little can be done to further lower a cholesterol level already in the normal range. In 1990, an article in the New England Journal of Medicine (Swain et al. 1990) reported that oat bran was not any more effective than other complex carbohydrates in lowering serum cholesterol in subjects with normal cholesterol levels. The broadcast and newspaper media discussed the report without mentioning that the study subjects all had cholesterol levels in the normal range at the start. Oat bran sales dropped as much as 30% according to some analyst reports (Ghez 1990). The history of oat bran hot cereal sales trends is shown in Fig. 20-2.

It is important to note that the rapid surge in oat bran sales following the Kinosian and Eisenberg (1988) article did not give the industry any time to evaluate a strategy that was so different from the way in which food products are typically marketed. The problem at that point was meeting consumer demand and correctly projecting future demand for the product. In this case, demand was created by the media, not product positioning. A closer look at the oat bran phenomenon highlights the challenge marketers face for cholesterol-lowering products, as well as other functional food products. Some useful approaches to identifying the best end user group for a product with a health benefit emerge from this example.

With regard to serum cholesterol, the adult population can be divided into two main groups: those who are concerned about serum cholesterol and those who are not (Fig. 20-3) (Wrick 1991). Those who are unconcerned will not change their behavior or product purchases because they don't care to, whether they know their cholesterol level is high or not. The cholesterol-concerned subgroup can

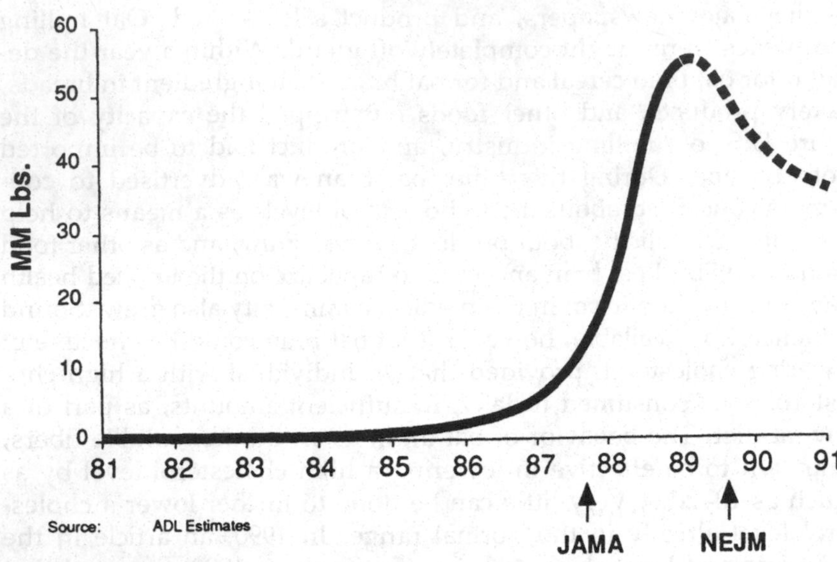

FIGURE 20-2. Estimated Volume (lb) of Oat Bran Used in U.S. Hot Cereals, All Brands

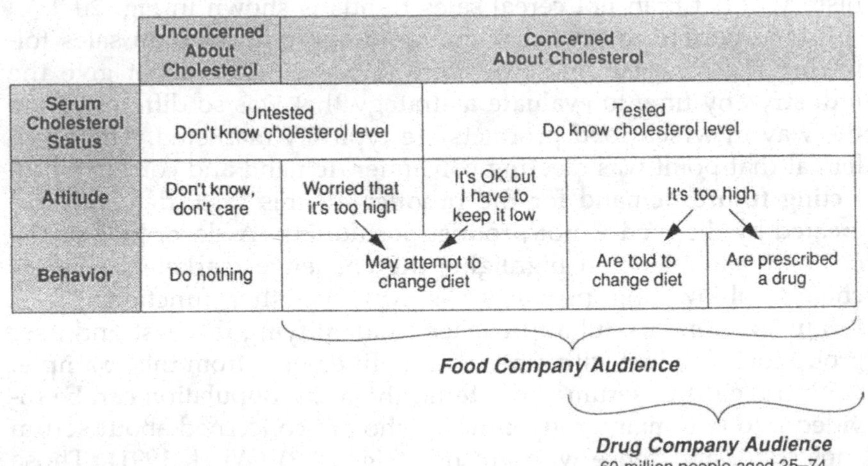

FIGURE 20-3. A Marketing Challenge for Cholesterol-Lowering Products is Selecting the Right Audience for the Product

be further divided into two segments—those who know, and those who don't know their cholesterol level. In the cholesterol-concerned subgroup, those who don't know their level will worry that it might be too high, and those who know that their level is within the normal range will worry about it getting higher. These two subsegments will attempt to change their diet even if doing so offers them minimal, if any, medical benefit. They do it for psychological insurance that they are doing what's right for them. In the right-hand portion of Fig. 20-3 is the subsegment of the cholesterol-concerned who have had their cholesterol tested and have been told by a health professional that it's too high. These individuals are told to change their diet and/or are prescribed a cholesterol-lowering drug because their risk for heart disease is high.

It is our theory that the cereal industry did not target the best end user audience for their oat bran products. They tried to reach the broad, cholesterol-concerned consumer audience, when only a small subsegment (those with a diagnosed high cholesterol level) could actually benefit. In addition, the medical community who counsels and treats those with high cholesterol received minimal, if any, attention from the industry about any demonstrated benefits of the oat bran products making cholesterol-lowering claims. Oat bran made many customers who lowered their cholesterol quite happy—however, the marketing message aimed at a more diverse audience that would see no benefit angered a segment of those health professionals. Had the industry targeted the audience of people with diagnosed high cholesterol levels and the health professionals who work with them, we believe the product abuses of oat bran would have been more limited and the of Swain and colleagues (1990) article would have had less impact, if it had been published at all. Whether or not oat bran sales would have been as great if the smaller audience typically reached by pharmaceutical companies had been targeted, will remain an unanswered question. However, one might speculate the market potential in that segment. If only 5% of the 60 million Americans aged 25–74 who have high cholesterol (Sempos 1991) ate oat bran as directed (2 ounces daily), over 136 million pounds of the hot cereal would be consumed annually. That is 2.5 times higher than the peak 1989 sales value shown in Fig. 20-2.

REGULATORY ISSUES

Historically, the U.S. Food and Drug Administration (FDA) did not permit any reference in food product labeling or print advertising that associated the intended use of the product with the diagnosis, cure, mitigation, or treatment of disease, because these properties were part of the legal definition of a drug. The 1973 Nutrition Labeling regulations expressly stated that a food shall be deemed misbranded if its labeling represents, suggests, or implies that the food, because of the presence or absence of certain dietary properties, is adequate or effective in the prevention, cure, mitigation, or treatment of any disease or symptom (currently 21 CFR 101.9 (i)(1)). The decade of the 1980s saw a period of lax enforcement of existing health claim labeling regulations, and it was during this period that oat bran sales rose and fell in the marketplace. The 1990 Nutrition Labeling and Education Act (NLEA) passed by the U.S. Congress in November of that year served as the legislative impetus for food labeling reform. The new law modified the Food, Drug and Cosmetic Act by adding section 403(r), which specifies in part that a food is misbranded if it bears a claim that expressly or by implication characterizes the relationship of certain nutrients to a disease or health-related condition unless the claim meets the requirements of a regulation authorizing its use (section 403(r)(1)(B)). The same section of the law directed the FDA to issue regulations authorizing health claims for nutrients in conventional foods and in dietary supplements in appropriate circumstances. In January 1993 a massive and comprehensive set of final regulations for revised labeling of foods was published that included a public comment period for selected portions, for the general requirements for health claims for foods (21 CFR 101.14).

The ruling permits health claims to be made for food products, under specified conditions, that are designed and developed for use in conjunction with a disease or health-related condition for which the U.S. population, or a subgroup of it, is at risk. There are four diet-disease linkages for which the agency has reviewed the publicly available scientific literature and concluded that health claims are warranted.

These are calcium and osteoporosis; diets low in saturated fat and cholesterol and/or high in soluble fiber from grain products, fruits and vegetables and coronary heart disease; sodium and hypertension; and diets low in fat and/or high in fiber-containing grain products, fruits and vegetables (which contain Vitamins A and C) and certain forms of cancer. A health claim relating folic acid intake in women of child-bearing age and neural tube defects in their offspring is still under consideration by the FDA at the time of this writing.

The FDA will consider petitions for health claims for substances that will be consumed as a conventional food or a dietary supplement provided they meet specified criteria. These include (but are not limited to):

- Claim substantiation by means of clinical and nonclinical data
- A statement explaining how the food component will help the consumer to attain a total dietary pattern or goal associated with the provided health benefit
- An explanation of how the component is associated with a disease or health-related condition for which the U.S. population or a definable subsegment of it is at risk
- Documentation that the food component, ingredient, or additive will maintain its original functional effect in the finished food (such as a contribution to taste, aroma, nutritive value, or other technical effect)
- Proof that use levels to justify a claim have been demonstrated as safe for consumption.

In addition, the FDA will authorize a claim by regulation only when the totality of publicly available scientific evidence indicates that, in the FDA's judgment, there is sufficient scientific agreement among experts qualified to evaluate such claims.

The FDA's approach to allowing health claims is conservative and makes sense from the standpoint of the FDA's legislative mandate to protect the public health and pocketbooks of the American people. It makes little sense from a regulatory or a marketing perspective to permit health claims to potentially proliferate throughout the food supply for diet-disease relationships that only affect a small population segment, or are unlikely to deliver a real health benefit that consumers or their health professionals can measure. Unfortunately, however, the regulations do nothing to encourage the investigation of isolated food constituents that have no other functional role in food than to provide a disease-preventative health benefit. It will still be required to file a petition for new drug status for these substances and to demonstrate their safety and efficacy for their intended use. The cost, time, and skills involved in bringing a drug product to market will deter nearly all food companies from pursuing this course with new ingredients, extracts, or isolates from foodstuffs or other agricultural commodities that contain chemical constituents whose addition to the diet is associated with a health benefit. Although pharmaceutical companies have the resources and know-how to bring a drug product to market, most of these firms do not compete in the food product marketplace, where profitability is generally much lower than for pharmaceuticals. A possible and sensible but highly speculative avenue for functional food products that contain nutrients or other bioactive food components is in the newly established regulatory category of "medical foods." These products are distinguished from other foods by the fact that they are intended for either oral or tube feeding in the management of a specific medical disorder, disease, or condition for which there are distinctive nutritional requirements, and the product is used under medical supervision. Medical food products in the marketplace today are primarily enteral nutrition formulas, discussed in detail elsewhere in this text. The markets for these products are being expanded to more and smaller-sized patient populations with diseases that can benefit from specialized nutrient formulations when administered under hospitalized or supervised home care conditions. Will the medical foods of the future shift their direction and focus on chronic diseases that affect larger segments of the population? If so, this category may emerge as one for functional foods by virtue of the functional ingredients used to deliver a health benefit.

References

DeFelice, S. 1991. The Foundation for Innovative Medicine, N.Y.

Ghez, N. 1990. *The Quaker Oats Company, Investment Research*. New York: Goldman Sachs, Inc.

Kinosian, B.P., and Eisenberg, J.M. 1988. Cutting into cholesterol; cost-effective alternatives for treating hypercholesterolemia. *JAMA* 259 (15):2249.

National Research Council, Committee on Diet and Health, Food and Nutrition Board, Commission on Life Sciences 1989. *Diet and Health: Implications for Reducing Chronic Disease Risk*. Washington, D.C.: National Academy Press.

Sempos, C. 1991. The prevalence of high blood cholesterol levels among adults in the United States. *JAMA* 261 (1):45.

Swain, J.F.; Rouse, I.L.; Curley, C.B.; and Sacks, F.M. 1990. Comparison of the effects of oat bran and low-fiber wheat on serum lipo-protein levels and blood pressure. *N. Engl. J. Med.* 322:147–52.

U.S. Department of Health and Human Services. 1988. *The Surgeon General's Report on Nutrition and Health*, Public Health Service, DHHS (PHS) Publication No. 88-50210.

Wrick, K.L. 1988. *The Food/Drug Interface*. Spectrum. Burlington, MA: Decision Resources, Inc.

———. 1991. The business challenge of positioning products at the food/drug interface. Conference Proceedings, Food Ingredients Europe. Paris: Expoconsult Publishers. Maarsen, The Netherlands.

Chapter 21

The Role of Marketing Communication in the Introduction of Functional Foods to the Consumer

Margaret P. Woods

MARKETING AND CONSUMER BUYING BEHAVIOR

A marketing orientation approach is concerned with the satisfaction of a consumer need. In product terms this need correlates with the benefit to be derived from that product, which, in marketing terms, can be considered to exist at different levels of association. The core product identifies the fundamental benefit to be derived, and when the product under consideration is food or drink the benefit is the satisfaction of a basic physiological need, that of hunger or thirst. This functional aspect is translated by manufacturers into the tangible physical product which, with other associated factors such as quality, design, image, packaging, and presentation, when presented to the consumer, offers a collection of stimuli aimed at eliciting a favorable response. The marketing objective must be that purchasers are attracted to that particular manufacturer's product in preference to others that are available. The marketing strategy is built around developing the appropriate marketing mix to achieve the desired consumer response.

Marketing is a pervasive social activity. The successful instrumentation of the marketing process requires an understanding of the behavior of consumers when they are involved in the purchase of products and services.

Engel, Blackwell, and Kollatt (1978) define consumer behavior as "Those acts of individuals directly involved in obtaining, using and disposing of economic goods and services including the decision processes that precede and determine these acts." An analysis of consumer behavior can provide an insight into current trends and help identify particular market segments and the marketing requirements necessary to reach those segments.

Several researchers have investigated the relationship between the psychological and physical activities involved in the decision-making processes of purchasing behavior. A number of models of buying behavior have been developed, including those by Howard and Shetch (1969), Kotler (1980), and Engel, Blackwell, and Kollat (1978). All provide useful theories concerning consumers and their buying behavior. It is the job of the marketer to apply this knowledge to the practical marketing situation.

Once the initial decision to purchase has been made (and this decision itself is influenced by any number of factors), there follows a further sequence of decisions related to the particular product or service under consideration. Factors that may influence this sequence of decisions include individual determinants such as attitude, motivation, perception, learning, knowledge, values, and beliefs, along with environmental aspects that are capable of influencing the individual's psychological condition and therefore the pattern of behavior. Those environmental aspects that could be considered to have the greatest influence include the current external physical, economic, and social environment combined with past experiences that may lead to specific reactions at a particular time and that may also influence future expectations, for example, with regard to health, income, and life style prospects.

FOOD PURCHASE

Whether people do their food shopping daily, weekly, or monthly, investigation of the contents of the shopping trolley will show that different individuals purchase different food products. Although price and availability may be a limiting factor in food and drink purchase, people in the same income bracket with similar life styles do not necessarily choose to eat or drink the same products. Some products are purchased by large numbers of people quite regularly and become brand leaders, but each individual's choice of purchase is exactly that—individual.

Why do people buy a particular food or beverage product? Application of the buying behavior models may be relevant up to a point, but in the purchase of food and drinks products, what is it that influences the purchase decision? The importance of the consumer to the food manufacturer and retailer is paramount. The success of any food or beverage product in the marketplace depends on its acceptance and repeated purchase by the consumer. Acceptable price, attractive packaging, and a strong promotion campaign may persuade consumers to purchase initially and possibly for a second time, but after that other factors inherent in the product itself and associated with consumer behavior come into play.

The final criteria by which food is judged and accepted by consumers relate to the sensory properties. How does it look, how does it smell, how does it taste? People use their senses first to determine whether the product is edible, and second to determine whether it pleases them. The first act is a judgment; the second is a reaction. The more favorable the reaction, then the more likely is the product to be a success.

According to Logue (1986), food choice, preference, and liking are not necessarily equivalent, but refer to overlapping behavior patterns. Responses are likely to be based on a combination of physiological and psychological stimuli elicited within particular external influences. It is difficult to accept that food or drink will be consumed it if is not liked, but situations often exist where, for instance, there is restricted availability and choice, and one product may be chosen in preference to another, but neither is really liked.

Translating this to the marketing of food and beverage products has implications for manufacturers and retailers alike. The shift in influence from manufacturer to retailer due to the purchasing power of the large retail groups and chains means that the retailers can "call the tune with regard to their requirements and product specifications." The retailers are therefore in a position of strength and their response to consumer demands can be passed back to the manufacturer.

Consumers, too, are in a position of strength. They have the power to select and choose, and although the majority of shoppers will make repeat purchases of their favorite food and beverage brands, there is always an opportunity to buy on impulse. There is no great risk involved in a food purchase—if the products are not liked for whatever reason then they will not be bought again. Similarly, if one usually purchased product is unavailable it is generally an easy matter to obtain a substitute. If this substitute is deemed to be satisfactory, or as good as or better than the original usual purchase, then a transfer of brand loyalty may occur, and the substitute may become the regular choice.

In general terms, purchase behavior of food and drink is not necessarily rational because choice is influenced by forces other than the need to purchase something to satisfy hunger or thirst. Also the timing of the majority of food purchases may influence the purchase decision in that foods are invariably bought in advance and therefore in anticipation of hunger or thirst. When snacks are purchased, they are often chosen for almost immediate consumption.

THE COMMUNICATION PROCESS

Any process of communication involves the original source of the message to be communicated, the message itself, the channels required and available for transmitting the message, the recipient, and the final interpreter of the message (who may or may not be the same person as the recipient).

For successful message composition, it is essential that what is intended is what is ultimately perceived. This requires a knowledge of consumer behavior; those involved in the marketing communication process must understand the potential receiver or target market. The aim of the message in the marketing communication process is:

1. To inform—for example, in the case of the introduction or launch of a new product
2. To remind—for example, where consumers who used to buy a product have ceased purchase and sales have fallen perhaps because of the emergence of a strong competitor
3. To persuade—nonbuyers to buy and lapsed buyers to buy again.

In all three situations an attempt is being made to influence consumer response with a view to that product being purchased.

Buttle (1986) identified three major types of communication problems for marketers: (1) cognitive, relating to lack of awareness and understanding, which in food and beverage products could relate to nutritional and health benefits; (2) affective, relating to attitudes concerning the product, which in food and beverage products could relate to a lack of interest or general desire; and (3) action or behavioral response, which for food and beverage products could influence the purchase decision resulting in nonpurchase or the purchase of an alternative. The communication media used may involve advertising, TV, newspapers, and magazines as well as other channels to reinforce the message.

Consumers are bombarded constantly with messages linking food habits to nutrition and health and attempting to emphasize the benefits of healthy eating in terms of a healthy family, improved self-esteem, and attractive life style. Thus, the marketing communication process involves the development and transmission of a message targeting the consumer with an identified product that confers benefits. The message must create awareness and interest, generating a desire for the product and leading to its ultimate purchase. Dyer (1982) suggests that advertising manipulates people into buying a way of life as well as goods, and this was reinforced by Williamson (1986) who drew an association between product features and values and human values and meanings. Concerning food products, Drew (1986) identified the "uniquely close relationship to the food we eat . . . we assimilate it; it becomes part of us."

The relationship between food and health is well established, particularly since the publication of U.K. government reports such as NACNE (Health Education Council 1983) and COMA (Department of Health and Social Security 1984). This has resulted in the manufacture of healthier products, giving consumers the opportunity to adopt a healthier life style. However, the messages regarding diet and health to which consumers are exposed may be confusing because of continuing debate concerning what is and is not healthy, because consumers incorrectly interpret the messages, or because a message is incorrect or confusing due to a lack of understanding by those who composed it in the first place. Often the legislative requirements of labeling, intended to give the consumer more information, may compound the confusion.

LEGISLATION OF FOOD ADVERTISING AND LABELING

In the United Kingdom, food legislation is controlled by the 1990 Food Safety Act where "advertisement" includes "any notice, circular, label, wrapper, invoice or other document and any public announcement made orally or by any means of producing or transmitting light or sound." For the purpose of food labeling as specified by The Food Labelling Regulations of 1984, SI 1305, "advertisement" has the same meaning as in the Act, but it does not include any form of labeling.

In addition to the controls laid down by The Food Safety Act and The Food Labelling Regulations, there are other constraints on advertisements. These are encountered in codes of practice adopted voluntarily by the advertising industry. The British Code of Advertising Practice is a voluntary code administered by The Advertising Standards Authority and financed by the advertising industry, but it does not cover television and radio advertising, which are administered by the Independent Broadcasting Authority. The Code applies to newspaper, magazine, and poster advertisements, direct mail, and cinemas. It does not apply to food labels or where an advertising claim is repeated on a label. The basic requirement of the Code is that all advertisements should be "legal, decent, honest and truthful." Guidelines are provided for health claims, vitamin and mineral claims, and slimming advertisements.

By implementing European Community (EC) Directives 79/112 EC, the U.K. Food Labelling Regulations 1984, SI 1305 apply to most prepackaged foods, deal with claims and misleading descriptions, and provide an extension to the Food Safety Act and the Trade Descriptions Act. To date the legislation covers:

1. Prohibited claims
2. General provisions relating to claims
3. Claims relating to foods for particular nutritional use
4. Claims relating to food for babies or young children
5. Diabetic claims
6. Slimming claims
7. Medicinal claims
8. Protein claims
9. Vitamin claims

10. Mineral claims
11. Polyunsaturated fatty acid claims
12. Cholesterol claims
13. Energy claims
14. Claims that depend on other food.

Restrictions in the use of certain words and descriptions in the labeling or advertising of food are also given. One specifically prohibited claim is advertising or labeling a food as having tonic properties.

The basic principle of control for claims is that they should be justified and supported by substantiating information. The regulations prohibit an express or implied claim in the labeling or advertising of food that it is capable of preventing or curing human disease unless a human medical license has been granted under the 1968 Medicines Act. Licences are granted only to products used wholly or mainly for medicinal purposes. Therefore medicinal claims for everyday normal foods are outside the law as it stands at present.

In the United Kingdom, the Food Advisory Committee report in its Review of Food Labelling and Advertising 1990, recommends additional controls over general claims and gives specific recommendations for health claims. In its acceptance of the majority of recommendations, the U.K. government decided to allow health endorsement schemes to be permitted with the proviso that such schemes be kept under review. This decision was reached in the belief that well-controlled health endorsement schemes could be beneficial to consumers in directing them towards a healthier diet.

Keeping within the law, the marketing communicator must offer the consumer the opportunity of association through eating and drinking with a healthy, desirable lifestyle couched in the labeling language of factual information. How does the consumer become knowledgeable when the product labels are full of technical information, and when two products issue the same health message, why is one chosen in preference to the other? But consumers are making their choices. We live in a more health-oriented society where manufacturers, retailers, and consumers are all on the healthy bandwagon.

The consumer awareness levels of the need for reduced fat, salt, and sugar and increased fiber in the diet have increased considerably over the past few years as demonstrated by the increasing availability of low- and reduced-fat products, diet food and drink products, high-fiber products, and reduced salt products. The terms "cholesterol," "saturates," and "polyunsaturates" are commonplace, and even if they do not understand how or why, consumers are at least aware of the messages associated with them. This does not mean that all consumers respond to the messages in a positive way, although the U.K. low-fat foods market has demonstrated a 44% growth between 1988 and 1990 (Industry and Market Reviews 1992). The additives issue has resulted in a response by manufacturers to consumer concern. We now have a wide range of food products with "no preservatives," "no added color," "no artificial additives," etc., because consumers have equated additives as "bad for you" in health terms. The intensity of the antiadditive approach has diminished somewhat (Consumer Focus 1991) but the word "natural" has a very positive impact on some consumers who perceive additives as chemicals and therefore suspect, and natural as healthy and therefore desirable. This is despite the fact that every food or drink product is chemical and we ourselves function as a result of a series of chemical reactions. Similarly, natural does not necessarily mean harmless or good, as there are innumerable naturally occurring allergens and toxins. In their report, the Food Advisory Committee has issued voluntary guidelines aimed at controlling the use of the word natural and similar words such as real, genuine, pure, etc., in the labeling or advertising of food in addition to the recommendations for the control of nutrition, health, and medicinal claims. EC Directive 90/496 relates to the nutrition labeling of foodstuffs; it makes nutrition labeling compulsory when a nutrition claim is made. In other instances, labeling in a prescribed format will be voluntary.

Marketing communicators have devised statements around established mathematical processes understood by all and have used these in food advertising. Subtraction or removal of fat, sugar, salt, and additives is desirable, as is addition of fiber, vitamins, and minerals. Substitution may be desirable or undesirable, depending on point of view. Substitution of sucrose by intense sweetener means a reduction in energy intake, but an increase in additive intake. Similarly, the requirement for greater use of additives in the production of low- or reduced-fat products could be viewed in the same way.

TARGETING OF FOOD WITH A HEALTH MESSAGE

The target market is of prime consideration in the marketing communication process. Segmentation is the process of dividing a large heterogeneous market into homogeneous segments, each with specific requirements. The market segments must be identifiable, substantial, and responsive and marketers must target each segment with the specific marketing mix aimed at that segment. The fortification of breakfast cereals with vitamins and minerals accompanied by the "fiber" message has resulted in positive consumer response. Even though vitamin and mineral fortification should not be necessary when a well-balanced diet is eaten, the general philosophy that "it can't be bad for you (and is more likely to be good for you)" is the one most likely to prevail. Targeting here is undifferentiated, i.e., it is targeted at everyone. We have come a long way since the days when breakfast cereal meant either porridge or cornflakes. Breakfast cereals and other foods such as crisps, soft drinks, and yogurts are now fortified. In the United Kingdom flour and margarine are fortified by law.

As far as targeting consumers with a health message is concerned, the responsiveness of the market segments is likely to be based on the particular experiences of that segment. Communication about product benefits for that segment must focus on the most relevant benefits to be derived. Those who are most likely to respond to products that contribute to a low fat/cholesterol diet fall into three categories:

1. Those who have experienced coronary illness either personally or by close association
2. Those who are at risk and who are concerned that they are at risk
3. Those who have an interest in healthy eating and preventive eating in a general sense.

Similarly, although many people may be aware of the desirability of reducing their salt intake, those who are most likely to respond to this in their purchasing behavior by purchasing reduced salt or no salt foods are those who suffer from hypertension or who have been advised to reduce their sodium intake. Awareness is highlighted when people experience for themselves the problems associated with particular clinical conditions. Because of this highlighted awareness they are likely to be more responsive to the marketing communication which is targeted at them specifically.

HEALTH CLAIMS AND FUNCTIONAL FOODS

In one sense all foods are functional. Even the most nonhealthy foods serve some positive, useful function even if it be a hedonistic total indulgence. As an integral part of our day-to-day living, foods and beverages form the basis for social functions and cultural influences. A study of food habits provide us with a fascinating insight into how people live.

In the specific terms of functional foods covered by this book, the origins of a range of foods providing some specific benefit in named diseases and clinical conditions may be found in *Culpeper's Herbal*, which provides "a comprehensive description of nearly all herbs with their medicinal properties." We see that asparagus, French beans, lettuce, cucumber, hops, licuorice, oats, onions, and rhubarb all have "Government and Virtues" that are beneficial to certain conditions. The association between scurvy and limes and anemia and raw liver is well documented. We now understand the scientific reasons behind these associations, and this must be the forerunner of today's functional foods.

Concern has been expressed regarding what specific claims might be made for functional foods. The main problem involves the question: Are functional foods really normal everyday foods, are they medicines, or are they foods for particular nutritional uses? Each of these categories has different regulatory requirements aimed primarily at protecting the consumer. If functional foods are deemed to be normal everyday foods (which was the original Japanese philosophy for these foods), then they must be treated as such. Any claim that the food is capable of preventing, treating, or curing human disease moves it out of the everyday food category and must be accompanied by a product licence issued under the provision of the Medicines Act.

Council Directive 189/398/EEC, (European Economic Community) concerning foods for particular nutritional uses, states that some specific claims may be permitted if they have been approved under the Directive. It sets out nine categories of foods for which more detailed directives will be drawn up covering aspects of composition and labeling requirements. One of the categories included is "foods intended to meet the expenditure of intense muscular effort, especially for sportsmen." Under legislation, vitamin and mineral supplements are considered as being for normal consumption and are not singled out. However, specific medicinal claims are unlikely to be given approval. A working draft for a directive on the use of claims relating to foodstuffs has been prepared in Brussels and a formal proposal will be sent to the Council of Ministers. No date has been agreed on for its finalization, but it specifically stipulates that no claim can be made about a foodstuff that it has the ability to prevent, treat, and/or cure a human disease. Although it prohibits references to links between food components such as normal ingredients and nutrients and certain clinical conditions, some claims may be allowed, particularly where there is valid scientific evidence internationally accepted to substantiate the claim. Enforcement of the directive, it is suggested, would be undertaken by an administrative body that can be approached by all who have a legitimate interest in monitoring food claims.

In the United States the FDA (Food and Drug Administration) does not approve health claims. However, in 1990 as a response to increasing consumer interest, it was agreed that the relationship between

- fat and heart disease
- fat and cancer
- fiber and heart disease
- fiber and cancer
- calcium and osteoporosis
- sodium and high blood pressure

would be examined with a view to drawing up guidelines for acceptable messages. Approval would be given where there is significant scientific agreement that can be presented in such a way as to give the consumer the whole truth. Those relationships where there was considered to be enough scientific evidence include:

- fat and heart disease
- fat and cancer
- sodium and high blood pressure
- calcium and osteoporosis.

Additional evidence is being sought on fiber and heart disease and fiber and cancer.

The majority of activity in the West is concentrated on health claims relating to prevention or recovery from disease; however, the Japanese approach to functional foods has wider implications. The 12 health-enhancing ingredients listed by the Japanese as having functional properties will be consumed for a variety of purposes including helping the body defense mechanisms and regulating body rhythms, as well as in the prevention and treatment of disease. (see chapter 18)

There is little doubt that although there may be dissent concerning permissible claims, the concept of functional foods is acceptable. We live in a more health-oriented society, and consumers in the main are paying more attention to what they eat and drink. Food manufacturers in the United Kingdom have responded with a plethora of new products, particularly since the publication of the NACNE and COMA reports.

SENSORY QUALITIES AND FOOD ACCEPTANCE

The sensory properties of foods play an important part in guaranteeing success and repeat purchase. It is unlikely that a healthy image backed by strong promotion can override deficiencies in aroma, flavor, and texture for any length of time. In promotional terms the ideal situation is one where the product benefits are a combination of healthy ingredients with functional properties and pleasing eating quality. There is no reason why something that has health advantages should not also be a pleasant experience.

Food manufacturers who aim to introduce functional foods to the consumer must remember that in the product development program attention must be given to the factors that govern food acceptance as well as to the active ingredients. A product should not only be packaged and presented in a way that will attract customers, it should also be attractive enough to stimulate the appetite after its removal from the package. Once sampled, the judgment of the product is made and the flavor assessed, and this assessment will influence future buying behavior.

Flavour is defined as the sensation realized when a food or beverage is placed in the mouth. It is a combination of taste, aroma, and kinaesthetic sensations associated with the product in the mouth—these are known as feeling factors. The taste sensation is perceived by the taste buds on the tongue, which respond to the primary tastes sweet, salt, sour, and bitter. When we taste, we taste in solution, and therefore the salivary glands play an important part in the mechanism. Chewing stimulates saliva production as do thought, sight, and smell—hence the well-known expression "It makes my mouth water." Aroma, which is the sensation produced by the volatile components in the food or beverage and perceived by the nose, is a vital component of flavor. The volatile constituents from the food impinge on the olfactory epithelium and stimulate recognition. The third component of flavor is the feeling factors, which include the kinaesthetic effects such as the bite and burn of pepper, the cooling of menthol, the pungency of alcohol, and the astringency of grapefruit. Thus, when a food is eaten, its chemical constituents register almost simultaneously as tastes, odors, and feelings; the combination of these produces the flavor.

The odor and appearance of food and drink products can attract or repel consumers and each is of prime importance to the food manufacturer. Just as a change in appearance can be an indication of a change in quality, so can a change in odor, e.g. development of a vinegary odor in wine, rancidity of fat, etc. Also important in influencing acceptability is the texture of products. For some foods the texture is the main influence on the assessment of quality. We expect meat to be tender, crisps to be crisp, ice cream to be smooth, etc. If the texture attributes are not perceived to be as they should, then no matter what quality the flavor, the product will not be considered acceptable.

These factors apply in exactly the same way to the development of both conventional and functional foods. A study of brand leaders undertaken as far back as 1965 by Arthur D. Little Incorporated, using the Flavor Profile Method of Analysis, found that those food products that had the most success in the marketplace demonstrated several flavor characteristics in common:

1. Early impact of appropriate flavor
2. Pleasant mouth sensation
3. Full body of highly blended flavor
4. Rapid disappearance of flavor on swallowing
5. Absence of unpleasant aftertaste.

Bearing these criteria in mind, it should be possible for a food manufacturer to develop a product that is acceptable in sensory terms as well as in health benefit terms.

PRODUCTS AND HEALTH: A PROFILE

Of the functional ingredients highlighted by the Japanese, some are already present in foods while others are being investigated. According to Potter (1990), the dairy industry in its quest for more beneficial products is particularly interested in the concept of functional foods, and Salji (1992) has described acidopholus milk products as foods with a third dimension with accumulating scientific evidence in support of their specific properties. Casein phosphopeptide (CPP) is claimed to increase absorption of calcium, iron, and other minerals to prevent loss of calcium from the bone. Lactitol, a sugar alcohol, is "tooth friendly," has lower energy value than other carbohydrates, and does not affect insulin or blood sugar levels. Butter with acidopholus and bifidus bacteria, yogurt with bifidus bacteria, and milk with added magnesium and vitamins have been produced by French companies. Cheese with added fiber has also been manufactured.

Despite the fact that legislative constraints prevent health and medicinal claims, there are ways of getting the message across. Some product names evoke an association of health, and where the name itself does not suggest a link, on-pack statements with subtle implications can serve the same purpose. Product descriptions such as "light," "low", and "diet," and product names such as Shape, Lean, Cuisine, and Weight Watchers are well understood and accepted.

Common Sense, a Kellogg's breakfast cereal based on oat bran that is said to be beneficial in cholesterol reduction, has been targeted specifically at the "cholesterol aware" market. There is no specific claim, but the slogan "The heart of a good breakfast" leaves us in no doubt as to its positioning in the marketplace. Similarly, Quaker Oats has included the statement "May reduce cholesterol when taken as part of a low fat diet" on its outer package.

This is exactly the same approach as that taken by the manufacturers of low- or reduced-calorie products. It is no longer allowed to call these "slimming products," but the implication is allowed that the product can help reduce weight when taken as part of a calorie-controlled diet. The growing brigade of slimmers is helping to boost the sales of both the manufacturers of diet products and the intense sweetener industries with an increased consumption of the wide range of diet products available. They are also deriving some perceived benefits as a result of this. As these products are targeted at the calorie conscious, with obesity being one of the diseases of affluence, diet products could be considered to be functional foods, although they may not fall within the Japanese definition stating that the functional ingredients are to be of natural origin.

Soft drinks manufacturers have found an interesting market niche in the development of sports and health drinks. In the United Kingdom this is exemplified by the repositioning of Beecham's Lucozade away from its dull and convalescent image to a thirst-quenching, energy-giving drink associated with vitality. The launch of Lucozade Sport, an isotonic drink formulated to provide a replacement for nutrients lost during strenuous activity, was supported by a promotion campaign using well-known and respected sports personalities accompanied by a complete revamp of the product image and package.

Introduction of a flavored Lucozade still drink in a soft pouch with a replaceable cap has contributed to the product's success, and the product package has an explanatory paragraph:

> "STILL LUCOZADE SPORT is a great tasting isotonic orange drink. It has been specially developed for use before, during and after exercise. Lucozade Sport has a unique isotonic formula which means that it is in balance with your body's own fluid. This special isotonic drink helps fight dehydration and fatigue by getting essential fluid into your body fast and carbohydrate energy to your working muscles—even before you feel thirsty or tired."

Lucozade Light, with "new citrus taste," with "new lower calories," with "calcium and rich in Vitamin C," has been launched as a 250-ml screw cap bottle. The only specific claim in very small print on the label is "Can help slimming or weight control only as part of a calorie controlled diet."

Introduction of other drink products such as Aqualibra by Callitheke (a herbal fruit drink), Dexters, Isostar, and Purdeys have been successful in what was virtually a nonexistent market 10 years ago. By the end of 1991 the sports drink sector had grown by 40% and the herbal fruit drink sector by 20% (Euromonitor 1992).

Although the new health-conscious consumer may have emerged, there are some products that, although considered to be more healthy, are not always chosen when possibly a less healthy equivalent is available. The U.K. bread market is an excellent example of this. Although the higher fiber content of brown bread may be desirable, there is still a preference for white bread. This is a perfectly healthy product, as white flour is still fortified with iron, niacin, calcium, and nicotinic acid, but the fiber content is less in white than in brown bread. The fact is that U.K. consumers on the whole actually prefer white bread. The introduction of Mighty White followed by other soft grain products with added fiber has gone some way to alleviate this. On a visit to the local supermarket it is noticeable that the white bread invariably sells first. Because bread is a substantial part of many family diets, it is understandable that what is preferred is purchased. Other breads and bread products with perhaps a more healthy image are purchased extra to the main choice.

CONSUMER PERSUASION

The benefits, psychological and physiological, accruing from eating and drinking must be clearly understood. A well-thought-out marketing mix of product, promotion, distribution, and price must be aimed at specific target markets. The sensory qualities must be such that a food imparts a pleasing sensation and feeling of well-being based on the eating or drinking sensation, combined with the knowledge that it is doing you good.

Consumers have confidence in the manufacturers of brand leaders. A well-known, respected name goes a long way in the acceptance of a new product. Consumers are more likely to make an early decision to try something new if they are "familiar" with other products from the same manufacturer, even if the new product is a totally different kind of experience.

A positive approach should be adopted in the promotion campaign, and far-reaching claims should be avoided at all costs. One question to be considered with regard to functional foods is: How are they to be eaten? A normal daily diet contains a wide range of foods consumed both as meals and snacks. Consumers could eat a complete meal composed totally of functional foods, they could select various items to combine with other products in the composition of the meal, or they could concentrate on functional snacks. This suggests a marketing opportunity for the provision of promotional leaflets, recipe suggestions, dietary advice, and demonstrations and promotions as already exist for a whole range of products.

On the basis of what exists already, it is likely that consumers will view functional foods as merely an extension to the healthy eating message. It is important that consumers are not misled or encouraged to consume something which may not be appropriate to a particular individual. The philosophy in Japan is that these foods may be consumed by a large percentage of the population in many cases as a preventive measure. For this to be successfully achieved requires an audience receptive to claims that can be substantiated. The Japanese are, however, already conditioned as a result of previous experience, current trends, more healthy life styles, and food and health promotions.

In competitive terms, success will go to those who deliver the most effective message in conjunction with a beneficial product that is a pleasing experience advertised in an ethical manner, or as according to The Advertising Standards Authority, in "an advertisement which is legal, decent, honest and truthful."

References

Buttle, F. 1986. Hotel and Food Service Marketing. Holt, Rinehart and Winston. London.

Consumer Focus No. 7. 1991. Additives and Attitudes—A UK Perspective. Leatherhead Food R.A.

Department of Health and Social Security. 1984. *Report of the Committee on Medical Aspects of Food Policy: Diet and Cardiovascular Disease*. London: Her Majesty's Stationary Office (usually H.M.S.O.).

Drew, K. 1986. The Role of Additives in Food Distribution. *Food Marketing* 1(1), pp. 3–24.

Dyer, G. 1982. Advertising as Communication. 2d ed. Methuen. London.

Engel, J.F.; Blackwell, R.D.; and Kollatt, D.T. 1978. Consumer Behaviour. 3rd ed. Dryden. Winsdale, Illinois.

Euromonitor 1992. Health and Sports Drinks. pp. 67–81.

Health Education Council 1983. *A discussion paper on proposals for nutritional guidelines for health education in Britain (NACNE)*. London: HEC.

Howard, J.A., and Sheth, J.N. 1969. *The Theory of Buyer Behaviour*. Wiley. New York.

Industry and Market Reviews No. 6. 1992. The UK Low Fat Foods Report. Leatherhead Food R.A.

Kotler, P. 1980. *Marketing Management*. Prentice Hall. New Jersey.

Logue, A. 1986. The Psychology of Eating and Drinking. Freeman. New York.

Potter, D. 1990. Functional Foods—A Major Opportunity for the Dairy Industry? *Dairy Industries International* 55(6), pp. 16–17.

Salji, J. 1992. Acidopholus milk products: Foods with a third dimension. *Food Science and Technology Today* 6(3).

Williamson, J. 1986. *Decoding Advertisements*. London: Marion Boyers.

Chapter 22

The Food Industry's Role in Functional Foods

Adolph S. Clausi

THE FOOD INDUSTRY'S ROLE IN FUNCTIONAL FOODS

As the food industry prepares itself for the challenges of the '90s and beyond into the twenty-first century, one hears much about a new class of foods that purport to represent an important opportunity for the food industry. These are the "functional foods." They are also referred to as "designer foods," or "nutraceuticals," "better-for-you" foods, or sometimes with the very unpretentious name of "super foods."

All foods serve a functional role by definition. They nourish the body, they provide energy, in many instances they prevent the onslaught of nutritionally based diseases, and very importantly, they also satisfy the psyche by the very nature of their aesthetic and hedonistic qualities.

If this is so, then what's new? The answer resides in the current fashion of taking a narrow definition of a functional food as meaning not only what we generally expect of foods, but much more specifically, a food that can *positively impact* an individual's health, physical performance, or state of mind. The active words here are "positively impact." We are talking about foods that contain ingredients that serve therapeutic roles or roles formerly held by therapeutic medicines.

In the past we have had therapeutic additives available for inclusion in foods. Certainly the essential vitamins would be considered such when added to foods to prevent the onslaught of certain deficiency diseases. If we take the current Food and Drug Administration's (FDA) nutritional labeling guidelines as a reference, we would also consider foods that have bioavailable calcium added for the prevention of osteoporosis, Foods with reduced sodium for mitigating effects on hypertension, and reduced-fat foods for mitigating effects on cardiovascular disease and/or cancer. With the possible exception of essential vitamin restoration and fortification or calcium fortification, these are primarily "eliminators of negatives" rather than positive effects. This has been the pattern in the past with respect to healthful food choices provided by the food industry.

Beginning in the 1970s, the direction of the food industry, according to the Institute of Food Technologists, was aimed toward reduction of fat, cholesterol, and sodium in the food. The availability of new technologies and ingredients have fostered this continually with a focus on the elimination or moderation of a negative health factor in foods (Anon. 1992).

Today one can find growing numbers of research studies that are supporting convincing arguments for ingredients showing a number of positive health effects. Examples are the positive effects of the antioxidant properties of vitamins A, C, and E in *preventing* certain cancers and in certain instances *reversing* premalignant conditions, or the cholesterol-lowering effects of certain fibers to *prevent* arterial plaque formation in humans, or the positive effects of folic acid in *preventing* certain pregnancy problems, or the beneficial impact of omega (ω)-3 fatty acids in the *prevention* of coronary heart diseases.

As more extreme examples, newer research studies are indicating that the slowing of biological aging, the use of foods to alter moods and nutrient control of the immune function could be possible. In addition, there is a whole range of other "positive effects" being experimentally shown for amino acids, oligosaccharides, fatty acids, alcohols, glycosides, etc.

The question arises as to the approach that the food industry will take with these opportunities. One thing is for certain, it will be a cautious approach because making a health claim for a foodstuff is not something that can or should be done lightly. Much is at stake in such a move, both from the manufacturer's viewpoint and from the traditional role of foods in the marketplace. The consumer's food habits change slowly. There is a built-in caution and skepticism on the part of most consumers with respect to anything new in the food supply, especially if some health benefit is suggested that cannot be felt immediately. Therefore, the first issue the food industry will face is whether or not to make a claim, even if they know that the material is safe in all respects and can contribute a positive impact on a person's well-being. Will the consumer understand and accept the health claim?.

A second question would be: Will this food be a replacement of current foods or will it be in addition to current foods as a second choice or as a line extension of the current product line? Introducing any new product, or line extension, is a costly operation and could cannibalize the parent business without any assurance of success. On the other hand, replacing an existing product with a more "functional" one is risking the whole ball game. Neither of these moves can be taken lightly.

The third question is: Will the food require FDA's clearance or not? This is a critical issue because the clearance process can be uncertain, extremely time consuming, and costly.

Thus, the food industry will have to deal with some tough questions with respect to "functional foods" and ingredients. Will they be "user friendly" and easily recognized like past ingredients such as vitamin-fortified foods, or will they be totally different and "strange" classes of foods and additives?

What will be the role of the food industry in the actual development of such functional foods? Clearly, the first consideration in any functional food development will be the role of the particular food in the diet. This should govern the kind of functionality it should deliver. There are certain things that we expect of "middle-of-the-plate" items that we would not expect from desserts or from a refreshment beverage. It is hard to conceive of a refreshment beverage, for instance, providing a functional role because by definition the refreshment role of the beverage would conflict with its believability and/or its appropriateness as a "functional food" in the consumer's mind.

Therefore, the industry first has to define the role of each food in the total diet and via that definition then define the "functionalities" that are appropriate to consider. The food industry has always been very conservative and has tended to follow rather than lead into new areas of food products. People's food habits change slowly, and merely saying that something is "better for you" will not necessarily ensure its acceptance. Paramount in the food industry's approach to the marketplace for these types of products will be their acceptance by the consumer.

On the other hand, there is a moral responsibility that the food industry must satisfy, in addition to its main purpose of providing products that the consumer will accept at a profit. That moral role is to deliver the "best" product that the industry can provide. If that includes a "functionality" from a health standpoint that was not possible before, then the industry has the moral obligation to explore every avenue for delivering that functionality to the consumer.

Yet the profit role of the industry will always result in a very careful cost-benefit equation being drawn before any new ingredient shown to have health functionality is introduced into the marketplace. This profit role includes not only a dollars and cents calculation, but also a determination of risk of image and/or risk of franchise if the customer does not accept the new product and as a result develops a negative attitude toward the product category. Thus, when it comes to developing new "functional foods," the food industry will be extremely cautious, will tend to follow rather than lead, and will very carefully consider the role of the particular food (plus its new functional ingredient) in the total diet and its appropriateness as a delivery vehicle. Finally, there could be a conflict between the moral role and the profit role, or risk of loss of image in any such new food adventure.

In the meantime, the food industry will carefully identify business opportunities in and around the periphery, such as the production of products with no or reduced fats, and the restoration and/or fortification of foods with micronutrients already approved and/or Generally Recognized As Safe (GRAS), etc. They will also "push" with aggressive marketing, on a health basis, those product categories that fit the health picture emerging, i.e., cruciferous vegetables, juices, seafoods, complex carbohydrates, fruit, etc. However, in my opinion, the food industry will not engage in or even sponsor a great deal of new basic nutritional research on functional foods for many years to come—if ever.

What then are the issues that need to be under control or resolved before we can expect a new wave of "functional foods" to become generally available in the marketplace? First of all, there need to be criteria and guidelines for what will constitute sufficient technological evidence to support health claims in foods. At the present time there are no such guidelines. It is easier to identify what technological evidence is needed to support a pharmaceutical drug than what is needed to support a nutraceutical "functional food." We do not know whether the same rules that are used for pharmaceuticals will be applied to functional foods, i.e., new drug application, clinical trials, etc. We might have to assume a more acquiescent posture for functional foods as, for instance, is suggested by the Foundation for Innovation in Medicine's publication *The nutraceutical initiative. A proposal for economic and regulatory reform* (Anon. 1991). In this report it is stated that "economic and regulatory reform for nutraceuticals is of vital concern for the delivery of important new advances in personal health care in this country, but is also essential for the U.S. food and drug industries to maintain their global competitiveness and leadership positions." This latter point is made because Japan has already established a system for approving nutraceuticals and major European countries now have systems that grant exclusivity for innovative medical and health claims based on proprietary research.

This report also states that the "economic and regulatory provision of the Nutrition Labeling and Education Act (NLEA) signed by President Bush on November 8, 1990 must now be broadened to include a more appropriate mechanism for establishing new and true claims based on innovative research" and that "the system must be designed to encourage rather than discourage investment in research, development and availability of important nutraceutical advances." (See Chapter 19)

The Foundation for Innovation in Medicine, an independent, nonprofit, and nonadversarial organization, proposes that a Nutraceutical Research and Development Act (NRDA) be passed by Congress to provide an economic and regulatory foundation for a vigorous research-based nutraceutical industry, accurate and sufficient product information to both physicians and consumers, and an assurance of consumer safety.

The NRDA would ensure seven years of marketing exclusivity for the sponsor of an innovative nutraceutical to provide economic incentive for investment in research and development. It would create a Nutraceutical Commission within the FDA that would work across government lines with the U.S. Department of Agriculture (USDA), Federal Trade Commission (FTC), etc., and deal with safety issues, labeling, and product claims and establish guidelines for submission and approval.

It would also establish a nutraceutical clinical research grant program within the National Institute of Health (NIH) to facilitate the broader clinical testing of potential nutraceuticals for medical and health benefits. This is an ambitious beginning. It is particularly relevant that this is being started by a medically oriented foundation, not a food industry group. However, we should be aware of a movement by food industry factors to establish a multi-disciplinary and multi-industry activity to deal with nutraceuticals. This movement proceeds along the lines of the Food Safety Council (Anon. 1992), an earlier group that established some definitive benchmarks and guidelines for the food safety issues that have proven useful in modern food safety decision making.

These will not be easy issues to resolve. I would like to give one example, based on my own experience, of what can happen given the unclear ground rules as they exist today. I am referring to the development of a "functional" beverage product (an orange-flavored breakfast drink mix) with which I was associated back in the 1970s. The basic research work was done at a very highly regarded and reputable university, the same one that had developed toothpaste containing stannis fluoride. It involved the addition of certain phosphate salts to a high acid, low pH beverage mix. These salts were found to mitigate dental erosion and showed the potential of actually preventing the overall development of new dental cavities even over a placebo water base.

This development occurred long before any societal discussion of "functional foods;" thus, the *only* avenue open was through the dental drug section of the FDA. An Investigational New Drug (IND) application was filed, which was followed by a large series of clinical trials of double-blind studies in a variety of environments and with a variety of populations. When positive results were observed and confirmed, a New Drug Application (NDA) was filed. This entire process took three years with an expenditure of over one million dollars.

The conclusion of the story is that the new drug application was not allowed for a variety of reasons; most of which could have been predicted up front. The main argument against approving the application was that the subject foods (beverages in this instance) would not be used in the same controlled environments as the clinical studies, therefore it would be erroneous to extrapolate those results into a real life situation. A logical conclusion perhaps, but illogical from the standpoint of food. How would one ever conduct a study that would satisfy that question?. A food, by definition is widely consumed under a great variety of conditions and circumstances.

The above example is cited because this could happen today. Nothing has changed from a regulatory standpoint other than a greater awareness of the problem. The conclusion of that particular beverage story was that the manufacturer stopped the pursuit of that product as a "functional beverage," but continued to add those generally recognized as safe phosphates anyway for moral reasons *without making any claim*. It is not conceivable that the same manufacturer would again invest the money that the project expended on another such venture today, given the ground rules as they exist and the added cost of such a development today.

The second issue is concerned with the kind of claims and the labeling requirements that would be required for such foods. When the 90-day period for public comment to the new NLEA labeling requirements ended in February of 1992, the FDA found itself with more than 30,000 written comments, many critical of the plan. The debate continues.

Moreover, as stated earlier, health claims can only be made for specific areas involving calcium, sodium, fat and cancer, and fat and cardiovascular diseases. The Council for Responsible Nutrition (Anon. 1992a) believes that these are too restrictive and that the FDA should permit many more health claims for those ingredients where there is substantial scientific evidence as long as the claims are qualified in a manner that accurately communicates the degree of scientific support and there is no safety risk.

As stated earlier, the Foundation for Innovation in Medicine (Anon. 1991) believes that "now with the arrival of the 'Nutraceutical revolution' we must establish new policies, based on the lessons learned from our three decades of experience with drugs." Their approach basically is to open the door to more claims as long as there is no safety risk.

This all leads to the third important issue that needs to be resolved. Are we really talking about a new class of "white coat" products; i.e., will these products be defined as pharmaceuticals? In the past, there was a great hesitancy on the part of the food industry to overtly make nutritional claims for their products because history had indicated that nutritional claims would lead the consumer to believe that the product would not taste as good as it should because of the nutritional fortification. Today the promotion of health claims for foods has become widespread. In the absence of an appropriate system of review and approval of scientific evidence, the net effect has been to promote by advertising rather than invest in new scientific research. According to the *New York Times* (Weiss 1992), 40% of all new food products introduced during the first half of 1989 included general or specific health claims, and a third of the annual expenditure for food advertising has presented some form of health message. Most of these health claims are elimination of negatives rather than positive health benefits. The NLEA essentially legalizes current promotional practices for a few selected instances, but does nothing to encourage new research on "functional foods."

Does this all spell the impending demise of the "functional foods" revolution? No, not really. The fact is that the food industry took a very long time to get to this new awareness, and many functional foods and/or near functional foods were made for hospital environments and distributed through health food stores long before the mass market food industry was aware of the "health" opportunity.

As we delve deeper into the health and "functionality" dimensions of foods, we may have to create a new class of foods, perhaps even to be distributed through new outlets in a manner more similar to that of over-the-counter drugs than that of mass marketed food products. These may be the forerunners of any longer term mass market in functional foods.

We already see signs of this. ω-3 fatty acids are very much in demand for the fortification of infant formulas since they more closely simulate mother's breast milk. ω-3 fatty acid-fortified infant formulas will be on the market soon, but these are products that are recommended by pediatricians and generally sold through over-the-counter drug distribution points with some share moving through supermarkets.

If the preceding discussion has sounded negative, perhaps it is because we are facing formidable issues yet to be resolved before the food industry will take a leading role in the development of functional foods and in the mass marketing of those foods. But having said all that, I also firmly believe that what the consumer in the food industry wants, the consumer gets sooner or later. Every one of these issues about functional foods—the need for research and technological supporting evidence, what claimology is permitted by the government, or how these products get distributed to the consumer—would all be resolved very quickly and easily if there was an overwhelming demand by the consumer for such products in the marketplace. Much of the industry caution, government restrictions, and technological issues would fall by the wayside because when the consumer speaks, these sectors respond.

The debate is joined and the media is picking it up. Perhaps the best we can hope for is that the consuming public will experience a broader awareness and understanding of what we mean by "functional foods" to the point of their demanding that more of these foods be permitted in the marketplace. At that point, the role of the food industry will be clear. It will provide those foods to the consumer.

References

Anon. 1982. *A proposed food safety evaluation process*. Food Safety Council, June.

Anon. 1991. *The nutraceutical initiative. A proposal for economic and regulatory reform.* The Foundation for Innovation in Medicine. pp. 1–40. December.

Anon. 1992. *Medical foods. A scientific status summary*. Institute of Food Technologists. April.

Anon. 1992a. General comments on the use of health claims. Council for Responsible Nutrition. June.

Weiss, R. 1992. Storm rages over nutrition labels for foods. *The New York Times*, March 25.

Part V

Consumer's Viewpoint

Chapter 23

Consumers' Views on Functional Foods

Kristen McNutt

Ask a U.S. consumer in 1993 "*What do you think of functional foods?*" The universal response is, "*What in the world are you talking about?*"

Perhaps, therein lies the most important lesson from this chapter for readers who work in the scientific community and in the business world. Consumers do indeed have views, which are discussed below, about foods which might provide health benefits beyond the traditional functions of vitamins, minerals, and macronutrients. Your neighbors and mine will ultimately determine whether there is a profitable market for foods promoted for "functional" benefits. Voters who pay taxes will also influence whether government agencies and university faculty have funding to research such functions.

But consumers are not numbers or percentages. They are people. The way in which we talk with them and listen to *them* will shape the future of "functional foods."

WHAT ARE WE TALKING ABOUT?

Nutritionists often blame consumer confusion about diet and health or about food safety on the "layman's" limited understanding of science. Far too often, a truer analysis reveals that consumers are confused because the "experts" have limited communications skills. If we hope to communicate successfully with consumers about functional foods, scientists, marketers, and health professionals have some homework to do.

The Japanese translation of "functional foods" may be a household word in Osaka. The French equivalent may be meaningful in Paris. But the phrase "functional foods" probably will not mean much to consumers in the United States, unless we make some changes. The consumer semantics problem simply reflects the confusion that still exists among scientists and product developers regarding the meaning of functional foods. The abstract for this book broadly defined functional foods as "any food that can positively impact an individual's health, physical performance, or state of mind, in addition to its nutritive values." That's a reasonable draft definition, at this stage of our science. But it doesn't say much to many people I know.

Another indication of the need to sharpen our own thinking is that although our working phrase uses the word "foods," most professionals involved in this arena are talking to each other in entirely different contexts. For example, the first section of this book is organized around *physiological* functions and the second deals with *components* of food. Foods are foods; consumers can understand that. But cancer-prevention and isoprenoids are *not foods*, functional or otherwise.

The confusion is exacerbated because some scientists about "functional foods," and others use the term "nutraceuticals," without explaining the difference; still another phrase is "pharmafoods." And the National Cancer Institute's project name, "Designer Foods," which is catching the media's attention, only further adds to the consumer dilemma. None of these words really captures what we're trying to say.

Meanwhile, other changes in the food supply seem to be moving so fast that consumers reasonably feel uncomfortable and out of control. They hear about biotechnology and then something about hormones being added to the milk they give their children. Then there is food irradiation; consumers understand what it means to "nuke" their food in a microwave, but now what they eat is being "zapped," without their being sure of when, much less understanding why. Proponents of functional foods may think that none of this really relates to functional foods. But those who want to understand *consumer views of food* must step outside of the mental boundaries restricting *their own view of food*.

Consumer assessment of functional foods can only be understood if one is sensitive to the others things happening to the food supply, a whirling world for most people.

The take-home message is that scientists must first clarify functional foods terminology for their own purposes. They must do so in the context of other changes in the food marketplace. Only then, can we select the proper words to communicate this concept to non-scientist consumers.

LOOK BACK, THEN LOOK AHEAD

Lest readers become discouraged by the consumer communications challenge, a glimpse into history provides perspective. Relatively speaking, our situation isn't that bad.

Most nutrition scientists realize that, shortly after the turn of the century, our knowledge of vitamins was as garbled as is our current understanding of everything we are now putting under the umbrella term, functional foods. "Vital amines" eventually became vitamins. Fat soluble A and water soluble B turned out to be a dozen-plus unique constituents, each distributed in various categories of food and each with different metabolic and/or physiologic roles. But it took a while—until 1942, in fact—for the last structure of a vitamin, B_{12}, to be determined. Patience, along with more scientific information and a little language creativity, will, in time, resolve our terminology task.

My crystal ball says that a whole new nomenclature is coming. It will come as part of the evolution of our understanding of what we now broadly call functional foods. We're progressing on that road by having separated fiber into two types—soluble and insoluble. Much more teasing apart of fiber-component/function subdivisions will occur in the years ahead. We're beginning to realize that saturated versus unsaturated doesn't tell the whole story when it comes to the health risk of fatty acids. Similarly, we are learning that the HDL and LDL divisions are an oversimplification of cholesterol. Someday, consumers who shop for groceries, as well as students of biomedical science (or whatever it is called at that time), will marvel that in the early 1990s we didn't have the food/function words they then use so casually.

CATEGORIES OF CONCEPTS

For the time being, the food-health linkages that seem to place a food under the functional foods umbrella might best be thought of as a spectrum of categories, rather than a single set of relationships. As I understand functional foods terminology, foods containing components that prevent vitamin and mineral deficiency diseases, as historically defined (for example, scurvy and anemia), fall *outside of our definition*.

At the far left, within the spectrum, are foods with components that are linked primarily to heart disease and cancer. These relationships have been better established during the last two decades. For example, foods containing fat, cholesterol, and/or fiber are functional foods because of their effects on these diseases or conditions. ***Consumers are well aware of these relationships.***

The components of food falling *in the middle of the spectrum* are those that contain a substance that now is proposed as having a newly discovered "function," often somehow related to the immune system. The components that are most frequently cited and are now receiving more scientific attention are antioxidants and phytochemicals. The foods per se are usually fruits and vegetables, and to a lesser degree, legumes. ***This body of science, in fairly general terminology, is reaching the public in varying degrees.***

At the far right of the functional foods spectrum are those foods containing components with a possible health benefit that is only beginning to be explored by scientists. The components of these foods have names that are still relatively unknown to the general public. Glycosides, phenolic acids, and isoflavones are not, at the time of this writing, words consumers generally know. If fact, if asked, ***most people would probably guess that these long names, these chemical-sounding words, are things that should be avoided as a health risk, rather than eaten because of potential health benefits.***

MEDIA MESSAGES

Controversy over the proposed Food and Drug Administration regulations regarding disease-prevention claims for branded food products and dietary supplements has stimulated a significant number of electronic and print media reports on some functional foods. The media in the United States currently has a much greater influence than do health professionals on consumer perceptions of health in general, and nutrition in particular. But advertising is only one means by which the media is telling consumers about what professionals call functional foods. Editorial content, which is Constitutionally beyond the reach of regulatory agencies, is spreading the word on magazine and newspaper pages, which face food company advertisements. Sometimes articles help substantiate direct or indirect claims for "functional" attributes of food; sometimes editors question whether science supports such messages.

The *Consumer Magazines Digest*, a monthly publication of Consumer Choices, Inc., summarizes for food marketer and health professional subscribers what consumers are reading in the popular press about food and health. Analysis of approximately 250 articles per month from 50 U.S. and Canadian magazines reveals that many journalists are reporting to their readers and viewers recommendations from various health policy documents. During the first two quarters of 1992, the emphasis in health-related magazine articles shifted away from dietary cholesterol as a heart disease risk and toward the importance of "reducing fat, especially saturated fat," as recommended by most nutrition experts. Some articles are now recommending that consumers count grams of fat, rather than calories.

Mood foods are showing up in consumer magazines, probably more because of the rhyme of the term than because of the science to support the concept. One reason for the popularity of this potential functional food topic is the linkage to premenstrual syndrome; anything about PMS is always of major interest to female readers. Another women-only, functional foods topic that is very popular with the media is the protective effect of yogurt on vaginal yeast infections.

Another functional foods-related topic of consumer interest, for better or for worse, is that of the "smart drugs." Bars selling cocktails of amino acids, vitamins, and other ingredients *touted by scientists* as improving mental powers, are becoming more popular with consumers, especially in California. Trends such as this are very appealing to the world of journalism. Therefore, this science story is reaching millions of consumers as news.

Research related to the ability of certain foods to extend life is also gaining momentum in the press. This message is something anyone would like to believe; therefore, it makes a good story for the press. Another significant observation that partly explains why consumers are learning about the longevity "function of food" is that research from the USDA Center on Aging at Tufts University is broadly covered by the U.S. press. In 1992, the Tufts public relations office did a much better job of regularly sending press releases to the media than did publicists employed by most academic institutions. Thus, it is not surprising that the anti-aging research by this group of scientists is reported more widely to consumers than the findings of other researchers working on functional foods.

A final media trend merits strategic consideration by health educators and advertisers. The wave of interest in functional foods is carrying with it renewed interest in alternative therapies, which are generally opposed rather than endorsed by the traditional medical establishment. In the context of articles about "healing foods," consumers today are reading more than they did a decade ago about holistic medicine, herbology, toxin-cleansing fasting, African medicine, and yin foods versus yang foods. Journalists are painting a broader picture of functional foods than some scientists would do, if the scientists could control the press.

EVOLVING COMMUNICATIONS TECHNIQUES

Newer techniques and technologies are evolving for communicating with consumers. In recent years, public information projects of professional associations such as the American Dietetic Association have created joint venture efforts to spread the word about the effects of diet on heart health and cancer. The editorial content of Special Advertising sections in magazines are now authored by professional association staff, paid consultants, or member volunteers. Publishers sell advertising space on facing pages to food companies, frequently those marketing products based on their left-of-the-spectrum functional properties.

Public service announcements, developed by professional organizations and funded by the food industry, are also becoming more popular as a means of communicating the benefits of functional foods. In 1992, Dole Foods Company launched such a project in cooperation with the American Academy of Pediatrics. McDonald's has announced a similar PSA partnership with the Society for Nutrition Education.

Television infomercials are yet another technique for spreading the word about functional foods. The most recent and far-reaching examples in the United States are programs that demonstrate food-juicer machines for home kitchens. These 15- or 30-minute programs, which appear to viewers to be entertainment rather than commercials, include the television hosts' advice about the "functional" benefits of juice from fruits and vegetables.

WHAT DRIVES MEDIA COMMUNICATIONS

The situations described above suggest some principles about what will, and will not, be told to consumers about functional foods in the future.

- *What People Would Like to Believe* is what journalists like to print.
- *Words with a Ring* are appealing to the media. Mood foods, for example, is a memorable phrase.
- *Women Feel Slighted by Biomedical Scientists*. Many are angry that most research is done only on men, although sometimes they acknowledge legitimate research design reasons for this situation. The resultant reality is that any scientific knowledge that responds to women wanting to know female-specific, health information will be well received.
- *Counter-Consensus is Appealing*. Everyone, to a certain degree, likes to champion the underdog. Therefore, hypotheses that challenge conventional wisdom are welcomed by many people. Journalists know this; health professionals should.
- *KISSing Works Well*. Keep-It-Simple-Stupid is probably the hardest consumer communications lesson for most scientists to learn. If advocates of functional foods want to sell their ideas to consumers, they must learn to drop their scientific jargon and tell their story, as they would to their next door neighbor.

WHAT CONSUMERS NOW KNOW

Some of the left-of-the-spectrum diet-health relationships under our umbrella term are indeed already well understood by consumers. Market research conducted in 1991 by MRCA Information Services shows that most U.S. consumers have accepted the notion that diet does affect health. For example, 82% believe that they can avoid future health problems by eating healthy, whereas only 24% believe that they can't have any control over their future health. Approximately one-third of respondents selected "don't know" as their answers to a nutrition quiz in the MRCA study, but, in general, more people gave correct, rather than incorrect, answers to questions about the relationship of fat, cholesterol, and fiber to heart disease and cancer. They generally scored even better on questions about food/disease relationships that fall outside our current umbrella definition of functional foods.

Consumer acceptance of the life-extension benefits of food, other than reduction of cancer and heart disease risk, are more difficult to assess. Approximately three-fourths of the MRCA respondents choose healthy foods because those people think such foods increase the chance of living longer; about the same percentage do this because healthy foods make them feel better about themselves. Fewer respondents, approximately half of them, select healthy foods because they make them feel healthier. Feeling more productive and alert, a newer potential "function" being attributed by scientists to some foods, was a reason for less than half of respondents' choosing healthier foods.

BARRIERS TO ACCEPTANCE

Scientists who want their future to include continued funding for functional foods research and product developers whose mouths water at the thought of potential profit margins may think that functional foods are their ticket to job security in the twenty-first century. Perhaps this is true, perhaps not.

Value Added?

A basic requirement for the success of any food product is that product's ability to fill a real consumer need, or a perceived need that might have been created by health professionals or advertisers. In 1992, the consumer benefit question, as it relates to functional foods, is far from decided. R&D to develop functional foods and basic research to document marketing claims for functional foods are expensive. Profitability, therefore, will probably require premium pricing. Unless the world economy takes a dramatic upward turn, most consumers are not going to pay value-added prices for functional foods, unless they believe that the value is really there for them.

The potential profitability of functional foods might well suffer because of consumer experience with alternative sweeteners and fat substitutes. If these products fail to help consumers lose weight or to actually reduce their risk of diseases people now know about, it is unlikely that those same consumers will believe marketing claims for products that supposedly have even more esoteric health benefits.

Credibility?

Most consumers belief that what they eat affects their health. That's not the credibility issue. Where the problem lies is that consumers don't believe that the government is protecting their interest, when it comes to the food supply. Furthermore, they don't believe marketing claims from food companies. The pervasiveness of consumers not believing government or the food industry is demonstrated by a *Saturday Evening Post* cartoon, showing a couple in the supermarket, looking at a box of cereal. The man in the cartoon says to the woman, *"It's labeled high fat, high sodium and low fiber. I think we should reward their honesty and buy it."*

But the lack-of-confidence dilemma is hardly limited to government and the food industry; it reaches academic and health voluntary circles, as well. Only 17% of consumers in the MRCA nutrition market research disagreed or strongly agreed with the statement, "I wonder if the experts really know what good nutrition is." By contrast, 50% of respondents agreed or strongly agreed with this assessment of nutrition experts' wisdom.

ETHICAL QUESTIONS

Individuals, regardless of who signs their paycheck, incorporate certain personal values into career-related decisions. Some people have decided not to care about other people. Some people just never have developed any sensitivity to others. Some care and are sensitive, but think such questions only apply to other people. Others, however, in both industry and academic, do feel that ethical behavior is an important consideration in their professional, as well as personal, life.

The concept of functional foods raises ethical issues.

On the positive side, one could argue that functional foods offer significant potential benefits to public health and, therefore, there is an ethical obligation to make this information known to the public. If that assumption is true, other questions follow:

- To what degree does the potential public health benefit justify withholding from consumers limitations in the scientific documentation for the benefits of functional foods?
- If there are potential dangers at normal, or at higher, levels of intake, should the responsible person make this information known?
- Does the person who is lucky enough to have a greater scientific understanding of functional foods have an ethical obligation to anticipate areas of consumer confusion and help people avoid making marketplace mistakes?
- If yes, is that obligation different when clarification might decrease, not increase, consumer acceptance of the benefits of functional foods?

The reader who is on this ethical wavelength can come up with many more such questions that should be answered.

COMFORT VERSUS CREDIBILITY

One problem raised by our evolving knowledge of functional foods presents more of a dilemma to the scientific community, than to food marketer readers. Some scientists have for years aggressively attacked theories (and sometimes people who champion such theories) that conflict with conventional biomedical wisdom. Today, in the context of functional foods research, many theories previously considered to be misinformation or blatant quackery are accepted as legitimate hypotheses, some with sufficient preliminary evidence to justify major research funding.

In the mid-1970s, for example, the supposed health benefits of garlic was one of several topics addressed in a Nutrition Misinformation and Quackery Special Supplement to

Nutrition Reviews

Today, garlic is one of the foods (along with cabbage, soybeans, ginger, and umbelliferae) being studied by the National Cancer In-based copy, to help responsible journalists correctly report to the public the real-world implications of the research findings.

- Scientists can learn how to explain to consumers that our understanding of the effects of food on health is an *evolving body of knowledge*. This will give biomedical researchers a way of comfortably saying to consumers, "We were not lying to you yesterday just because that message proves false today."
- Scientists and health educators, including journalists, can get *quantitative* in their messages to consumers. How much of which functional foods are beneficial? How much might be harmful?
- Industry and academicians can support actions which strengthen consumer *confidence* in the powers that should protect their interests. From renewed confidence in government will flow renewed credibility for the scientific and business communities.

Scientists and food marketers will, in the future, be able to offer consumers more and more foods which might positively impact on individual health, physical performance or state of mind, in addition to their nutritive values. Time will tell whether our neighbors will believe that the benefits of such foods outweigh their costs and whether they and other consumers accept our offer.

ACKNOWLEDGEMENTS

Dr. McNutt acknowledges with appreciation the help of one of her consultant clients, Dr. Carol Smejda, Vice President and Director, Nutritional Marketing Information Services, MRCA Information Services. Dr. Smejda graciously provided the proprietary consumer nutrition market research data discussed in this chapter.

boxed copy, to help responsible journalists correctly report to the public the real-world implications of the research findings.

- Scientists can learn how to explain to consumers that our understanding of the effects of food on health is an ***evolving body of knowledge***. This will give biomedical researchers a way of comfortably saying to consumers, "We were not lying to you yesterday just because that message proves false today."
- Scientists and health educators, including journalists, can ***get quantitative*** in their messages to consumers. How much of which functional foods are beneficial? How much might be harmful?
- Industry and academicians can support actions which ***strengthen consumer confidence*** in the powers that should protect their interests. From renewed confidence in government will flow renewed credibility for the scientific and business communities.

Scientists and food marketers will, in the future, be able to offer consumers more and more foods which might "positively impact on individual health, physical performance or state of mind, in addition to their nutritive values." Time will tell whether our neighbors will believe that the benefits of such foods outweigh their costs, and whether they and other consumers accept our offer.

ACKNOWLEDGEMENTS

Dr. McNutt acknowledges with appreciation the help of one of her consultant clients, Dr. Carol Smeja, Vice President and Director, Nutritional Marketing Information Services, MRCA Information Services. Dr. Smeja generously provided the proprietary consumer nutrition market research data discussed in this chapter.

Part VI

Future Prospects

Chapter 24

Future Prospects for Functional Foods

Jeffrey S. Bland and
Darrell G. Medcalf

A NEW AGE OF NUTRITION

A new definition of nutrition has begun to take shape within the past decade. This definition goes beyond the traditional role of foods and nutrition in health promotion and disease prevention. It was only a few decades ago that Atwater completed his classic studies to determine the calorie content of foods and differentiate the energy content of the macronutrients protein, carbohydrate, and fat (Davidson et al. 1975). Organic chemistry gave rise to the new view that food is made up of a complex array of organic molecules, all somehow contributing to the composition, nutritional value, and hedonistic qualities of our diet. At the start of the twentieth century, organic chemists isolated trace constituents of food and found that they contained small molecules that had amino groups that could prevent and treat disease. These substances were called life-giving amines, or vitamins. Following the discovery of vitamins A through K came the view that nutrients could be used to treat many diseases beyond the deficiency diseases of scurvy, beriberi, pellagra, xerophthalmia, and rickets. From the 1930s through the 1970s, however, research indicated that for individuals consuming the standard Western diet it would be very hard to produce a clinical deficiency with regard to these various nutrients. The determination that nutrients could not be used to treat diseases other than the traditional nutrient-deficiency diseases resulted in significant loss of interest in nutrition in medical school education. As a consequence, in the 1970s no U.S. medical school had a required nutrition course in its curriculum.

In the 1980s, however, new questions were raised concerning the role of nutrition and foods in health. Rather than looking at nutrients as drug-like substances used to treat specific deficiency diseases, the research changed to different types of epidemiologic and animal studies. These studies demonstrated that inappropriate distribution of both micronutrients and macronutrients could result in increased incidence of various chronic degenerative diseases, including coronary heart disease, certain forms of cancer, and hypertensive disorders, and that it could possibly contribute to adult-onset diabetes through excessive calorie consumption and obesity. This remarkable reappraisal of the role of nutrition in promoting health set the stage for new research strategies that could evaluate the effect of food and nutrition upon health and disease throughout the aging process.

Human function is now recognized to be the result of the operation of literally thousands of biochemical processes working in synchrony. Each individual has differences in his or her biochemistry as determined by his/her own genetic structure. This means that what is required for optimal function for one individual may be very different from that which is required for another. The enlightened health practitioner of the future will be involved with monitoring and prescribing interventions for patients to maintain optimal function of their specific biochemical systems over a much longer period of their life span than is currently possible for most individuals. This can be likened to matching the individual's "health span" to his or her "life span." The term "functional medicine" has been coined to describe this approach to health enhancement and disease prevention. Functional medicine uses individually tailored diet and nutrition, life style, and exercise recommendations, based on unique functional testing protocols to maintain optimal physiological function and thus reduce or remove the risk of chronic diseases associated with aging. Functional medicine is focused upon reducing

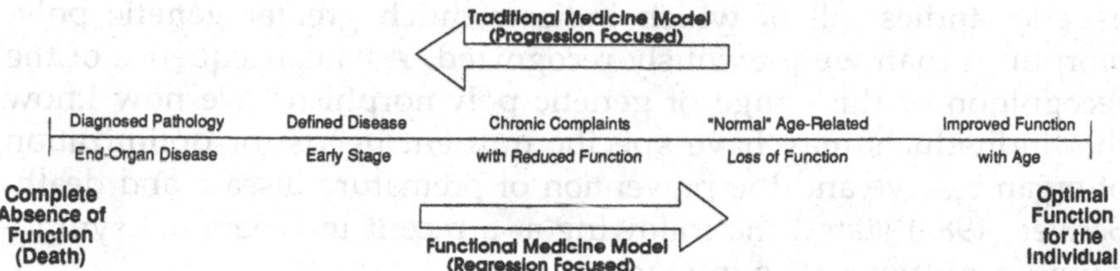

FIGURE 24-1. A Diagram of the Traditional Medicine Model Versus the Functional Medicine Model.

premature disease while improving vitality and physiological performance through the various life stages. The contrast of the traditional medical model, which focuses upon differential diagnosis and treatment of disease, versus the functional medical model, which focuses upon assessing and improving physiological function, can be seen in Fig. 24-1.

Nutrition, foods, and diet in the functional medicine model are employed to help improve physiological function of the individual by matching unique biochemical needs with nutrient intake. In this way the functional medicine model utilizes food and nutrition to help promote disease regression. An example of this is represented by the work of Ornish (1990) who demonstrated in intervention trials that regression of existing coronary artery disease occurred in a group of heart disease patients who followed a very low-fat diet while participating in stress reduction and exercise programs.

FUNCTIONAL MEDICINE AND GENETIC POLYMORPHISM

Ample data are available to support the concept that specific nutrients can be used to reduce an individual's risk for various chronic diseases. However, understanding how nutrition impacts the individual's biological aging process requires a better understanding of genetic variability within human populations. Recent advances in genetics have pinpointed significant variability in biochemical and immunological characteristics in individuals, involving enzymes, proteins, blood groups, and other defensive physiological systems. Human variability has been demonstrated by linkage, other family studies, somatic cells, genetic hybridization studies, and molecular genetic studies, all of which indicate much greater genetic polymorphism than we previously recognized. As a consequence of the recognition of this range of genetic polymorphism, we now know that individuals may have specific nutrient needs for optimization of organ reserve and the prevention of premature disease and death. Scriver (1988) stated the following at a recent international symposium on nutrient-gene interactions:

1. There is no average individual. The risk of having a nutrient-related disease is individualized.
2. Changes in nutritional environment will affect the expression of various genotypes. This expression is seen as the physical characteristics and functioning of an individual.
3. Understanding of single gene variance is useful to measure determinants of disease, both genetic and environmental, and how they relate to nutrient-involved multifactorial diseases.

In their recent book, *Genome,* Bishop and Waldholz (1990) comment, "Unmasking the identity of genes that render people susceptible to a host of chronic and crippling diseases is an important matter. These aberrant genes do not, in and of themselves, cause disease. By and large, their impact on an individual's health is minimal until the person is plunged into a harmful environment. The person who inherits the genes that interfere with normal processing of cholesterol, for example, may be little affected unless he or she habitually consumes a high-fat diet. Because such a person is genetically unable to handle an overload of fat efficiently, the arteries begin clogging up with cholesterol-laden deposits. The ultimate result is a heart attack, perhaps a fatal one early in life. Similarly, a person who inherits a genetic susceptibility to cancer may never develop a malignancy, but if that person smokes cigarettes, eats or fails to eat certain foods, or is exposed to carcinogenic chemicals, the chances of developing cancer are several times higher than the cancer risk faced by the genetically nonsusceptible person . . .. It is possible and would seem desirable to know early in life, perhaps at birth, if an individual is genetically susceptible to one or more of the degenerative diseases. Avoidance of risky environments could be taught early in childhood and thus would become a natural part of the individual's lifestyle and diet without the discomfort and difficulties of giving up cherished habits."

As we learn more about the human genome, many molecular variants will be determined for which individualized disease risk can be established. Many of these individualized risks will be found to relate to nutritional variables, so that by employing the correct approach to optimize the nutrient-gene interaction a more favorable phenotype with reduced risk for chronic disease and improved functional capability throughout the various stages of life will be achieved.

As an example, it has recently been reported that nearly 10% of the population may carry a genetic predisposition for coronary heart disease as a consequence of the poor metabolism of the amino acid homocysteine. This cholesterol-independent risk factor for heart disease can be modified in those who carry this genetic risk by increasing the dietary intake of vitamins B_6, folic acid, and B_{12} to levels in excess of the U.S. Recommended Dietary Allowances (RDA) (Stampfer et al. 1992). This enhanced intake of these nutrients in genetically susceptible individuals promotes the proper conversion in the body of homocysteine to the noncardiotoxic amino acid methionine. The levels of the nutrients required to promote this desirable result in genetically susceptible individuals are not so high as to cause risk of toxicity in "normal" individuals. It can be imagined that nutrient-tailored foods could be produced with augmented levels of these nutrients to meet the specific needs of these susceptible individuals without risk to unaffected populations.

FUNCTIONAL MEDICINE AND DYING HEALTHY

Too often we fail to come to grips with the fact that as long as we are alive, we are aging. For most people, getting older is not the problem! The problem is living with the result of aging as manifested in poor health, lower vitality, and reduced mental capacity, among other things. These issues usually manifest themselves as "age-related diseases" in our "mature" years, beyond age 50. In reality, however, this process is ongoing, and diseases of the elderly are truly diseases associated with a decline in physiological function that occurs in most people with increasing chronological age.

The recognition that individuals are genetically unique and that a diet that promotes optimal function for one person may not be optimal for another has resulted in a re-examination of the objectives of nutritional science. In the past, those objectives were to determine the level of nutrient intake that would meet the needs of "practically all healthy people." This language, taken from the Recommended Dietary Allowances of the U.S. National Academy of Sciences, has been the standard nutritional research focus in the United States over the past 30 years (National Academy of Sciences 1989). Nutritional adequacy has traditionally meant protection against nutritional deficiency disorders. In examining the relationship between health and nutrition in various countries, however, scientists found that there were certain epidemiological associations between diet patterns and improved health. Investigators like Keys and colleagues were among the first to examine diet-disease relationships (Keys, Kimura, and Larsen 1958). Their investigations resulted in increasing diet-health messages being promulgated by agencies within the federal government. For nearly a decade, various groups have advised Americans to limit their fat intake to no more than 30% of total calories, and that recommendation has been a major objective of the National Cholesterol Education Program (The Surgeon General's Report 1988) and the more general Dietary Guidelines for Americans (1990). These public health nutrition policy statements were designed to improve general health, but they have not dealt with the specifics of individualized nutrition, which go beyond the question of "adequacy."

Only within the past two decades have we had to deal with the fact that people will now live for many years after their retirement, as a consequence of increasing life expectancy. The increasing number of older-aged individuals has also placed an increasing demand upon the disease-care delivery system. Gerontology research is now coming of age, however, and the message is that we should not expect our mean average life expectancy to increase significantly over the next 20 years. What we can accomplish is an extension of the "health span" and a reduction in premature disease and disability. As Callahan (1990) points out in *What Kind of Life?*, our health care system will certainly change in the coming generation in order to deal with the individual questions associated with aging. The next two decades will challenge medicine and the health-care delivery system to find better ways to reduce the incidence of premature disease and improve health and vitality of an aging population. Gori and Richter (1978) explained that the only cost-efficient way to reduce health-care costs is to institute prevention so as to avoid premature diseases of the aged. Olshansky, Carnes, and Cassel (1990) have pointed out that, "Further gains in life expectancy will occur, but these gains will be modest. However, it is not clear whether a longer life implies better health. In fact, we may be trading off a longer life for a prolonged period of frailty and dependency . . ..

Current research efforts by the medical community are now focused on prolonging life rather than preserving and improving the quality of life. An obvious conclusion, therefore, is that the time has come for a shift toward ameliorating the nonfatal diseases of aging."

In order to reduce the nonfatal diseases of aging, scientists must learn more about human performance throughout the aging process. They can gain this knowledge by studying optimum individual metabolic function and its relationship to diet. Stevens and Blattler (1986) believe that for virtually every substance there is an optimum level, determined by an individualized parabolic dose-response relationship. The particular shape and slope of this dose-response parabola is related to the genetic uniqueness of the individual.

Under ideal circumstances, the objective of medical science would be to evaluate the shape of this multidimensional array of nutrients that the individual needs to consume in his or her diet in order to promote improved function and organ reserve. In the past the methodology for exploring this complex interaction matrix was not available. However, with the advent of molecular biological techniques for probing genetic uniqueness, it is very likely that these techniques will become available in the near future.

Schneider and his colleagues (1986) have asked whether the micronutrient levels established by the RDA adequately consider the interactions between nutrition and aging. His group has suggested that expanded considerations should be taken into account in formulating the RDA for older populations. Those considerations can be summarized as follows:

1. The RDA for the older-aged group should be aimed at the maintenance of optimal physiological function and the prevention of age-dependent diseases and disorders.
2. The effects of nutritional factors on physiological function throughout the life span and on disease development in later life should be considered in formulating RDAs for all age groups.
3. The optimum serum or tissue levels of nutrients, as well as the balance of all nutrients, should be determined by the criteria described in 1 and 2 above.
4. The substantial heterogeneity (genetic polymorphism) of the older population should be examined.

This theme is mirrored in a discussion by Holliday (1984), who suggests that understanding the mechanism of aging is a key problem in biomedical research related to virtually every discipline in science. He suggests that research on the pathological features and treatment of age-related diseases is often fruitless if only the origin of these diseases is studied without attempting to understand the underlying mechanism of aging itself and how it relates to nutrition and environment. He also suggests that the ultimate aim of such research would not be to stretch out the life span by retarding the aging process, but specifically to prevent the onset of innumerable debilitating diseases associated with middle to old age.

It is now recognized that many diseases of the aged have specific nutritional links. Not only do we see the major causes of death—coronary heart disease, stroke, cancer, and maturity-onset diabetes—related to nutrition, but specific age-related conditions such as cataracts, chronic kidney disease, gastrointestinal disorders, immune function, motoneuron diseases, and possibly even Alzheimer's disease may have a nutrition link. Recent research from the U.S. Department of Agriculture has indicated that nutrition may also play a very important role in neurophysiological performance in the aged, and therefore cognitive function and maintenance of central nervous system activity may be very closely tied to optimum nutritional states throughout the life stages (Tucker et al. 1990).

It has also been recently suggested (Cutler 1991) that antioxidant status, with specific reference to vitamin C, beta(β)-carotene, and vitamin E, may have a relationship to biological aging. Vitamin E, for instance, is known to reduce the amount of protein glycation that occurs with aging. Protein glycation has been suggested as one of the major determinants of biological aging, and therefore its prevention by improved antioxidant status may have significant impact upon many disorders associated with increased age. Cutler (1991) pointed out that oxygen radicals are implicated in accelerated aging, and the antioxidant status of an individual could be important in determining the frequency of age-dependent diseases and general health maintenance. From the cross-cultural studies of Gey, Brubacher, and Stahelin (1987) we know that those individuals who have the highest plasma levels of antioxidant vitamins have the lowest incidence of ischemic heart disease and cancer. It is recognized that vitamin C and vitamin E inhibit the oxidative modification of low-density lipoprotein cholesterol and therefore may help prevent the atherosclerotic process. Over the past 10 years, many other extended cardiovascular disease risk factors have been identified that demonstrate the principle of individual nutritional needs. These include the recognition that a greater number of people than was previously recognized have the heterozygous form of homocystinuria.

As a result, their risk of heart disease is increased. That risk can be reduced by consuming levels of pyridoxine (vitamin B_6), folate, and vitamin B_{12} in excess of the U.S. RDA (Boers et al. 1985). By evaluating blood or urine homocysteine status, the need for this tailored nutrition can be identified. Conversely, there are people for whom traditional risk factors do not apply, and a dietary restriction or modification is unwarranted. An example is a recent report describing an 88-year-old man who ate 25 eggs per day and had a normal blood cholesterol level (Kern 1991).

How do we determine what particular health risk groups require for maintenance of optimal health? How do we differentiate those individuals with specific needs for improved function from those who do not have these needs? What are the more prevalent health risk genotypes for which specific nutritional modifications will decrease premature disease?

Referring to these questions as a whole, specific nutritional relationships have been identified for many disorders associated with premature aging. These relationships are clearly related to individualized risk factors and genetic sensitivities. It is the emerging view of many researchers that genetic polymorphism may set in motion a unique relationship between the individual and his or her environment and diet. Throughout years of interaction this relationship results in the expression of reduced functional states, which we see as premature disease and debilitation. The future focus of research and development in this area would be to find better methods of evaluating genetic uniqueness and more tailored nutritional programs to minimize individualized health risk by effectively modifying the individual's nutritional environment. By accomplishing this goal, a new form of nutrition related to individualized intervention and management would be developed, and its result would be maintenance of organ reserve and reduced risk of chronic disease.

FUNCTIONAL FOODS—A RESPONSE TO FUNCTIONAL MEDICINE

The preceding sections of this chapter have attempted briefly to set out the premise that nutrient tailoring to meet the needs of individual genotypes can be used to design diets that would have the potential to reduce the risk of chronic disease and loss of physiological function associated with aging. These diets would be individualized based on tests designed to measure unique aspects of the individual's functional biochemical systems. Functional foods to improve physiological function, therefore, would be those food products designed and marketed based on the presence of specific ingredients deemed relevant to improved biochemical functionality.

If functional foods are to play an effective role in improving human functionality, at least three critical factors must be met: (1) the public must be convinced of the relevance of the information relating diet to functional improvement, (2) biochemical assessments that are easy to administer and interpret must be made available, and (3) biochemical assessment and acceptable products must be made available in the marketplace from which individual diets can be constructed.

Education of the Public

It is apparent from published surveys that consumer interest in nutrition continues to grow. Data also suggest that nutrition labeling of food products and the presence of health claims on labels have significantly heightened consumer awareness and interest in products with perceived specific health benefits. The recent success of products making low-fat, fat-free, lower-in-saturated-fat, and low- or no-cholesterol claims certainly supports this observation. The publicity and educational effort that will accompany the new U.S. food labeling regulations scheduled to go into effect in 1994 can only add to this awareness. In 1992, reports in *Time* and other consumer publications concerning the benefits of specific nutrients related to health demonstrated that the consuming public can be made aware of the benefits and that consumers have a strong interest in gathering and acting upon information related to their health.

Another force acting to increase consumer interest in food-related health issues is the increasingly negative press related to the drug industry. The increasing awareness of the total cost of medical care and the role of drug costs in this inflationary spiral, coupled with increasing publicity related to drug side effects and efficacy, have made consumers skeptical of traditional drug-based medical therapies. This is borne out by the startling statistic that only about 60% of all prescriptions written by doctors are filled and only about 50% are actually consumed according to directions. People are looking for acceptable alternatives and will be responsive to information that gives them another way to approach the health issue.

Biochemical System Functional Tests

The development of inexpensive, readily available, and (presumably) easily interpretable tests on which to base individualized action is critical to establishing consumer interest in functional foods. Does anyone doubt that the market for lower-fat and cholesterol-free products was significantly enhanced by the fact that a number of Americans were suddenly aware of their blood cholesterol numbers? It was as if we all were given a gift! For many individuals, knowledge of this issue (high risk of cardiovascular diseases associated with elevated cholesterol) was a call to action based not on some abstract "It might happen to me" premise, but on the presumption of real "It's going to happen to me unless I do something" knowledge. Faced with this information, many consumers did take action. The conclusion is unmistakable: personalize the information and people will be motivated to change. The driver for this "personalization" of the message was the development of the finger-stick blood cholesterol measurement technique.

The challenge is to develop other "functional tests" for all of the critical systems that meet the criteria outlined above. These tests must be sensitive enough to track meaningful changes from an individual's norm so that motivation and the opportunity for action to alter the diet are provided. Sensitivity is also critical to monitoring improvements based on any dietary changes an individual makes. Equally critical to appropriate sensitivity is the availability and/or ease of completing the test. Ideally, the tests would be noninvasive and the results interpretable by an untrained individual. It is clear from the cholesterol experience that there is room for some compromise in terms of these criteria. But, clearly, the easier it is to administer and interpret, the better chance a test has of leading to proactive change. Although few such tests exist today, the basis for them is already in the scientific literature. Recently, HealthComm, Inc., in Gig Harbor, Washington, has been instrumental in developing a series of tests to measure liver detoxication function (Bland and Bralley 1993). Although still not "consumer friendly" enough to be used in regard to functional foods, these tests and others we are developing are likely to play an important role in promoting functional medicine. As functional medicine progresses, the basis for the development of new value-added food products will also be promoted.

Availability of Specific Functional Foods

If the concept of functional medicine expands, as it seems almost certain it will, and the known factors in the marketplace (such as the new labeling regulations) continue on course, there will be an expanding need to provide food products containing sufficient quantities of specific nutrients to meet a broad range of dietary variables. These products will fit the definition of functional foods in every sense. In order to meet the needs of a wide range of consumers, the product range and the nutrient variables within these products will have to be quite broad. At the extreme, the number of product variables is almost infinite and certainly unmanageable in a profitable way. Consolidation or focus will be required, and the viability of products will have to be determined by good market research. However, there seems little doubt as to the validity of the benefits such products can bring if they are used widely and wisely by an informed public. We believe the science of functional medicine and functional foods will continue to grow, thereby increasing consumer interest and the demand for new nutrient-tailored food products. As the average age of individuals in developed countries continues to increase, the interest in health enhancement will grow as well.

Will food science and agriculture be able to respond to this need by providing the technology and information needed? There seems little doubt that a technological community that has delivered the green revolution, low-saturated-fat oil, high-quality fat-free products, and pulsed-light sterilization, to name but a fraction of the recent advances, will deliver the needed technology and in a form acceptable to a consuming public that has learned to be careful of technologies that carry hidden burdens on the environment or energy resources. The development of biotechnology concepts may see some of their application in the production of nutrient-tailored or nutrient-augmented foods or ingredients.

LIMITS TO THE FUTURE OF FUNCTIONAL FOODS

It seems clear that the limits to the future for functional foods are not the basic science, the food technology, or consumer awareness and interest. The limiting factor would appear to be the ability to inform consumers of the benefits attainable by using products formulated to provide specific health benefits, i.e., the regulations related to what ingredients can be legally incorporated and what can be said about the products in advertising, promotion, and labeling copy. At present, these areas are severely limited by regulations in all the developed countries. The widely heralded new functional food regulations that recently went into effect in Japan have been much more restrictive to date than was anticipated. Health claims under the new food labeling regulations in the United States strictly control what can be said and allow health claims in only a few areas (fat and cancer, fat and heart disease, sodium and hypertension, calcium and osteoporosis, and possibly claims for the benefits of fiber). It would be very easy to conclude that regulations will be so severely limiting that the ability to mass market functional foods is doomed. However, we believe such conclusions are short-sighted and fail to take into account the tremendous force that a combination of sound science and strong consumer demand can evoke. Regulations usually lag behind science because, by nature, they must be relatively conservative to protect the public interest. The issue is that, in most situations, the regulatory status quo becomes central dogma and change, even scientifically well-documented change, becomes almost impossible.

Recently, in the United States, an organization called the Foundation for Innovation in Medicine published a "white paper" proposing major reforms in the regulatory framework for dealing with "Nutraceuticals," a term coined by this foundation (1991). As stated in the white paper, a nutraceutical can be defined as any substance that may be considered a food or part of a food and provides medical or health benefits, including prevention and treatment of disease. Such products may range from isolated nutrients, dietary supplements, and diets, to genetically engineered "designer foods," herbal products, and processed foods such as cereals, soups, and beverages. Although this definition may exceed the scope of functional foods, all functional foods would seem to be included in this broad definition. The reforms suggested in this document are broad and would greatly enhance both the economic incentives for companies to develop nutraceutical programs and the regulatory framework for bringing products containing nutraceuticals to the market. Regardless of the status of this particular proposal, its widespread distribution and the discussions it is generating are sure to increase the probability that major reforms in regulations in the area of functional foods will become a reality.

SUMMARY

Based on the depth of discussions in the preceding chapters of this book, together with the framework rationale for these discussions using the concept of functional medicine that we have proposed, the future prospects for functional foods to play a major role in improving the health and vitality of the world's population seem very bright indeed. The freedom to control our own destiny in terms of health and vitality as we age is within our grasp. While the path ahead is filled with unmet research needs, economic barriers, and regulatory hurdles, the compelling need to achieve an improved "health span" corresponding to our increased life span will ensure success. Because of the recognition that the cost of disease treatment is increasing at a too-rapid rate and the population is getting older, new health promotion strategies focused upon functional medicine must be implemented. By the time we enter into the new millennium in the year 2000, functional foods will be commonplace in most marketplaces in the developed world. We believe this picture of the future is inevitable, and it is a future that benefits us all.

References

Bishop, J.E., and Waldholz, M. 1990. *Genome*, pp. 17–18. New York: Simon and Schuster.

Bland, J. and Bralley, J.A. 1993. Nutritional upregulation of hepatic detoxication enzymes. *Journal of Applied Nutrition* January (publication pending).

Boers, G.H.J.; Smals, A.G.H.; Trijbels, F.J.M.; Fowler, B.; Bakkeren, J.A.J.M.; Schoonderwaldt, H.C.; Kleijer, W.J.; and Kloppenborg, P.W.C. 1985. Heterozygosity for homocystinuria in premature peripheral and cerebral occlusive arterial disease. *N. Engl. J. Med.* 313:709–15.

Callahan, D. 1990. *What Kind of Life?* p. 135. New York: Simon and Schuster.

Cutler, R.G. 1991. Antioxidants and aging. *American Journal of Clinical Nutrition*. 53:3735–95.

Davidson, S.; Passmore, R.; Brock, J.F.; and Truswell, A.S. 1975. *Human Nutrition and Dietetics*. p. 16. New York: Churchill Livingstone.

Dietary Guidelines for Americans. 1990. Washington, D.C.: U.S. Government Printing Office.

Foundation for Innovation in Medicine. *The Nutraceutical Initiative: A Proposal for Economic and Regulatory Reform*. 1991. New York: Foundation for Innovation in Medicine.

Gey, K.F.; Brubacher, G.B.; and Stahelin, H.B. 1987. Plasma levels of antioxidant vitamins in relation to ischemic heart disease and cancer. *American Journal of Clinical Nutrition*. 45:1368–77.

Gori, G., and Richter, B.S. 1978. Macroeconomics of disease prevention in the United States. *Science* 200:1124–9.

Holliday, R. 1984. The aging process is a key problem in biomedical research. *Lancet*. Dec. 15; 1386.

Kern, F. 1991. Normal plasma cholesterol in an 88-year-old man who eats 25 eggs a day. *N. Engl. J. Med.* 324:896–9.

Keys, A.; Kimura, N.; and Larsen, N. 1958. Lessons from serum cholesterol studies in Japan, Hawaii and Los Angeles. *Annals of Internal Medicine* 48:83–94.

National Academy of Sciences. *U.S. Recommended Dietary Allowances*. 1989. Washington, D.C.: National Academy Press.

Olshansky, S.J.; Carnes, B.A.; and Cassel, C. 1990. In search of Methuselah: Estimating the upper limits to human longevity. *Science* 250:634–9.

Ornish, D. 1990. *Dr. Dean Ornish's Program for Reversing Heart Disease*. New York: Random House.

Schneider, E.L.; Vining, E.A.; Hadley, E.C.; and Farnham, S.A. 1986. Recommended dietary allowances and the health of the elderly. *N. Engl. J. Med.* 314:157–60.

Scriver, C.R. 1988. Nutrient-gene interactions: The gene is not the disease and vice versa. *American Journal of Clinical Nutrition* 48:1505–9.

Stampfer, M.J.; Malinow, M.R.; Willett, W.C.; Newcomer, L.M.; Upson, B.; Ullmann, D.; Tishler, P.V.; and Hennekens, C.H. 1992. A prospective study of plasma homocyst(e)ine and risk of myocardial infarction in U.S. physicians. *Journal of the American Medical Association*. 268:887–91.

Stevens, H.A., and Blattler, D.P. 1984. Human performance as a function of optimum metabolic concentrations in serum. In *Aging—Its Chemistry: Proceedings of the Third Conference in Clinical Chemistry*, Arnold O. Beckman, ed., pp. 390–93. Washington, D.C.: American Association for Clinical Chemistry Press.

The Surgeon General's Report on Nutrition and Health. 1988. pp. 93–94. Washington, D.C.: U.S. Government Printing Office.

Tucker, D.M.; Penland, J.G.; Sandstead, H.H.; Milne, D.B.; Heck, D.G.; and Klevay, L.M. 1990. Nutrition status and brain function in aging. *American Journal of Clinical Nutrition* 52:93–102.

Gori, G. and Richter, B.S. 1978. Macroeconomics of disease prevention in the United States. *Science* 200:1124–[illegible]

Hollands, R. 1986. The [illegible] process in [illegible] biomedical research. *Lancet* [illegible]1386.

[illegible] 1991. [illegible] cholesterol [illegible] in an 88-year-old man who eats 25 eggs a day. *N. Engl. J. Med.* 324:896–[illegible]

[illegible] 19[illegible]. Dose-response [illegible] of [illegible] and [illegible]. *American Journal of [illegible]* [illegible]

[illegible] of Sciences. [illegible]. *Recommended Dietary Allowances*, 10th ed. Washington, D.C.: National Academy Press.

[illegible] 1994. [illegible] Estimating [illegible] limits [illegible] 2306–[illegible]

[illegible] 1990. [illegible] *The [illegible] for [illegible] Healthy Heart Program*. New York: Random House.

[illegible] Virtue, [illegible] Hadley, [illegible] and [illegible] 1985. Recommended dietary allowances and the health of the elderly. *N. Engl. J. Med.* 314:[illegible]

[illegible] The [illegible] nutrient [illegible] the disease and [illegible] *American Journal of Clinical Nutrition* 48:[illegible]

[illegible] 1992. A prospective study of [illegible] risk of [illegible] in U.S. physicians. *Journal of the American Medical Association* [illegible]

[illegible] H. [illegible] and [illegible], D.P. 1986. Human [illegible] as a function of [illegible] concentrations [illegible] *Proceedings of the Third [illegible]*, [illegible] ed., pp. [illegible]–93. Washington, D.C.: [illegible] Chemistry Press.

[illegible] *Report on Nutrition and Health*. 1988. pp. [illegible]. Washington, D.C.: U.S. Government Printing Office.

[illegible] 1986. [illegible] status and [illegible] function in aging. *American Journal of [illegible]* [illegible]–102.

Index